Published for
**OXFORD INTERNATIONAL
AQA EXAMINATIONS**

International A Level
FURTHER
MATHEMATICS
with Statistics

T0177768

Brian Gaulter

Mark Gaulter

Brian Jefferson

John Rayneau

OXFORD
UNIVERSITY PRESS

Great Clarendon Street, Oxford, OX2 6DP, United Kingdom

Oxford University Press is a department of the University of Oxford. It furthers the University's objective of excellence in research, scholarship, and education by publishing worldwide. Oxford is a registered trade mark of Oxford University Press in the UK and in certain other countries

British Library Cataloguing in Publication Data
Data available

978-0-19-837599-9

9 10

Paper used in the production of this book is a natural, recyclable product made from wood grown in sustainable forests. The manufacturing process conforms to the environmental regulations of the country of origin.

Printed and bound by CPI Group (UK) Ltd, Croydon, CR0 4YY

Acknowledgements
The publishers would like to thank the following for permissions to use their photographs:

Cover: Colin Anderson/Getty Images.

Header: Shutterstock.

Although we have made every effort to trace and contact all copyright holders before publication this has not been possible in all cases. If notified, the publisher will rectify any errors or omissions at the earliest opportunity.

Links to third party websites are provided by Oxford in good faith and for information only. Oxford disclaims any responsibility for the materials contained in any third party website referenced in this work.

AQA material is reproduced by permission of AQA.

Contents

1 Loci, Graphs and Algebra

Introduction

Polynomial functions always form a continuous curve with no breaks. However, when you divide one polynomial by another, the graph of the new function can have breaks in it and is said to be **discontinuous**. An example of such graphs is a **conic section**, which is the curve formed when a plane intersects a right circular cone. Some of the curves you meet the most in the real world are examples of conic sections, such as the ellipse that describes Earth's orbit around the Sun and the parabola that models the path of a football.

Recap

You will need to remember how to...
► Solve construction problems involving loci.
► Solve equations, including quadratics.
► Sketch basic graphs such as $y = x^2$.
► Find the distance between two points in Cartesian coordinates.
► Solve simple inequalities such as $4x + 7 > 3(x - 4)$ and $x^2 - 7x + 10 \geq 0$.
► Transform graphs using stretches, reflections and translations.

Objectives

By the end of this chapter, you should know how to:
► Sketch graphs of rational functions.
► Find equations of asymptotes to graphs.
► Solve inequalities involving rational functions.
► Describe and sketch various conic sections.
► Find the points of intersection between conic sections and coordinate axes and various straight lines.
► Find the Cartesian equation of simple loci that are described verbally.

1.1 Loci

In the context of graphs, a **locus** (plural **loci**) is a set of points that follow a given rule. Therefore, a locus can be represented by an equation.

For example, a locus is given as the points that are a distance of four units from the point (2, 3). This locus forms a circle with centre (2, 3). You know from previous studies that the equation of a circle is given in the form $(x - a)^2 + (y - b)^2 = r^2$, where r is the radius of the circle with centre (a, b). Therefore, the locus described above can be given as the Cartesian equation $(x - 2)^2 + (y - 3)^2 = 16$.

To find the Cartesian equation of a given locus, consider a general point on the curve (x, y) and use what you know about loci to help you formulate an appropriate equation.

Example 1

Find the Cartesian equation of the locus of points that are equidistant from the point (−1, 4) and the line $x = 2$.

Question

(x, y) is a general point that obeys the rule.

Distance of (x, y) from line $x = 2$ is $|x - 2|$.

Distance from (x, y) to $(-1, 4)$ is

$\sqrt{(x+1)^2 + (y-4)^2}$

$(x+1)^2 + (y-4)^2 = (x-2)^2$

$x^2 + 2x + 1 + y^2 - 8y + 16 = x^2 - 4x + 4$

Equation of the locus is $(y-4)^2 = -6x + 3$.

> **Note**
> The two distances are equal.

> **Note**
> You will discover later in this chapter that this is a **conic equation** of a parabola.

Exercise 1

1 Find the Cartesian equation of the locus of points which are equidistant from the point $(3, -2)$ and the line $y = 5$.

2 Find the Cartesian equation of the locus of points which are a distance of four units from the point $(2, -3)$.

3 Find the Cartesian equation of the locus of points which are equidistant from the point $(-5, 3)$ and the line $x = 2$.

4 Find the Cartesian equation of the locus of points which are a distance of $4\sqrt{2}$ units from the point $(4, -4)$.

1.2 Rational functions

During your A-level Mathematics studies you will have learned how to sketch curves. In this chapter, you will learn how to sketch curves for functions that are more complicated than trigonometric or polynomial ones.

The graph of a rational function will always have a horizontal asymptote provided that the degree of x in the denominator is the same as or larger than the degree of x in the numerator. In this chapter, you will only deal with cases where there is a horizontal asymptote.

Sketching rational functions with a linear denominator

In order to sketch rational functions equations of the form $y = \frac{ax+b}{cx+d}$, you need to start by finding the asymptotes.

You should remember that in the context of sketching a curve, an **asymptote** is a line that becomes a tangent to the curve as x or y tends to infinity. (Vertical asymptotes are the lines where the graph is undefined.)

For example, take the curve of $y = \frac{4x - 8}{x + 3}$.

In order for $y \to \pm\infty$, the denominator of this function must tend to zero.

That is, as $x + 3 \to 0$, $x \to -3$. Hence, $x = -3$ is an asymptote.

To find the asymptote as $x \to \pm \infty$, you express the function as

$y = \dfrac{4 - \dfrac{8}{x}}{1 + \dfrac{3}{x}}$

As $x \to \pm\infty$, $\frac{3}{x} \to 0$ and $\frac{8}{x} \to 0$.

Therefore, $y \to \dfrac{4}{1} = 4$.

> **Tip**
> The asymptote occurs where the graph is undefined, so to find the vertical asymptote you need to equate the denominator of the rational function to zero.

> **Tip**
> Divide the top and bottom by x.

Hence, $y = 4$ is also an asymptote.

Notice that as $x \to \pm\infty$, the largest terms in the numerator and the denominator are $4x$ and x respectively, and so $y \approx 4x \div x = 4$.

$x = -3$ is a **vertical asymptote**, as it is parallel to the y-axis.

$y = 4$ is a **horizontal asymptote**, as it is parallel to the x-axis.

To be able to sketch $y = \frac{4x-8}{x+3}$, you also need to find where it crosses the x- and y-axes.

When $x = 0$: $y = -\frac{8}{3}$ When $y = 0$: $4x - 8 = 0 \implies x = 2$

In summary, you proceed as follows:

1. Draw the asymptotes using dashed lines.

2. Mark the points where the curve crosses the axes; as the numerator and the denominator of the function each contain only a linear term in x, the curve cannot cross either asymptote.

3. Considering the curve for $x > -3$, you can see that y tends to $-\infty$ as x approaches -3 from values of x greater than -3. Hence, the curve tends to $+\infty$ as x approaches -3 from values of x less than -3.

4. You can now complete the curve of $y = \frac{4x-8}{x+3}$.

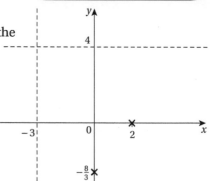

Tip

When the degrees of the numerator and denominator are the same, you can find the horizontal asymptote by dividing the leading coefficient of the numerator by the leading coefficient of the denominator.

To sketch the curve of a rational function in the form $y = \frac{ax+b}{cx+d}$:

1. Find the vertical asymptote by equating the denominator to zero; find the horizontal asymptote by dividing the numerator and denominator by x,

   that is, expressing the function as $y = \frac{a + \frac{b}{x}}{c + \frac{d}{x}}$ and then considering what happens when x tends to infinity.

2. Substitute $x = 0$ and $y = 0$ into the function to find where the curve crosses the axes.

3. If necessary, consider the curve for the x-values, to see what happens to the y-values as x tends to $\pm\infty$.

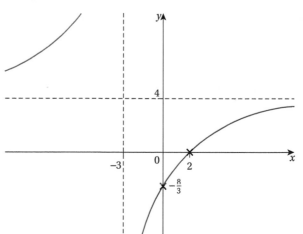

Example 2

Question

Sketch the graph of $y = \dfrac{2x-6}{x-5}$

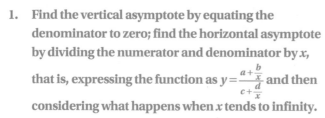

Answer

Horizontal asymptote:
as $x \to \pm\infty$, $y \to \frac{2}{1}$, $y = 2$

Vertical asymptote:
as $y \to \pm\infty$, $x - 5 \to 0$, $x = 5$

When $x = 0$: $y = \dfrac{-6}{-5} = \dfrac{6}{5}$

Note

First, find the asymptotes.

Note

Next, find where the curve crosses the axes.

When $y = 0$: $2x - 6 = 0$ $\Rightarrow$ $x = 3$

Note that the asymptotes here are parallel to the coordinate axes.

Note

Now copy and complete the sketch of $y = \frac{2x-6}{x-5}$ by also considering what will happen to the y-values as the x-values change.

Exercise 2

In questions **1**, **2** and **3**, state the equations of the asymptotes for the curve.

1 $y = \dfrac{x-1}{2x+6}$

2 $y = \dfrac{5}{x+2}$

3 $y = \dfrac{2x+7}{x-3}$

4 Show that $y = 2$ is an asymptote for the curve $y = \dfrac{2x+3}{x+4}$.

5 Show that $y = -4$ is an asymptote for the curve $y = \dfrac{8-4x}{x+3}$.

6 Sketch $y = \dfrac{6x-3}{x+4}$.

7 Sketch $y = \dfrac{3x-6}{x-1}$.

8 Sketch $y = \dfrac{2x+8}{3x-5}$, stating the equations of the asymptotes of the curve.

Sketching rational functions with a quadratic denominator

When sketching rational functions with equations of the form $y = \frac{ax^2+bx+c}{dx^2+ex+f}$, where both the numerator and the denominator are quadratics, you proceed as before by finding the asymptotes, where the graphs cross the axes, and consider the shape of the curve. However, with these functions, the shape of the curve requires consideration of stationary points and the number of asymptotes depends on the number of roots of the quadratic denominator.

▶ Two different roots will result in two vertical asymptotes.
▶ One repeated real root will result in one vertical asymptote.
▶ No roots indicates that there are no vertical asymptotes.

Regardless of the number of asymptotes, there may be values of y that are not realised by any value of x. Therefore, you need to find the **range** of the rational function in order to determine the set of possible values of y and in turn find the maximum and minimum points (the stationary points) of the curve.

Example 3

Find the range of possible values of y when $y = \dfrac{3x-4}{x^2+3x-4}$.

Note

To find the range of values of y, you need to find the values for which $y = \frac{3x-4}{x^2+3x-4}$ has real solutions for x.

Note

For x to be real, we know that $b^2 - 4ac \geq 0$.

Cross-multiplying,

$yx^2 + 3yx - 4y = 3x - 4$

$\Rightarrow yx^2 + (3y-3)x + 4 - 4y = 0$

Therefore,

$(3y-3)^2 - 4y(4-4y) \geq 0$

(continued)

Loci, Graphs and Algebra ⑤

(continued)

$\Rightarrow \quad (y-1)(9y-9+16y) \le 0$

$\Rightarrow \quad (y-1)(25y-9) \le 0$

$\dfrac{9}{25} \le y \le 1$

The range of possible values of y is $\dfrac{9}{25} \le y \le 1$.

To find the coordinates of the maximum and minimum points, substitute the minimum and maximum values from the range of y into the original equations. The solution of $1 = \dfrac{3x-4}{x^2+3x-4}$ is $x = 0$, and so one turning point is $(0, 1)$. The other turning point is $\left(\dfrac{8}{3}, \dfrac{9}{25}\right)$.

To discover which is a maximum and which is a minimum, you need to understand the shape of the curve, which means that you need to consider its asymptotes.

Curves with two vertical asymptotes

Some functions will have more than one vertical asymptote, for example the curve $y = \dfrac{(x-3)(2x-5)}{(x+1)(x+2)}$.

> When the denominator is a quadratic expression that can be factorised:
> ▶ There are always two vertical asymptotes if there are two distinct factors, and
> ▶ The curve will normally cross the horizontal asymptote.

> **Tip**
>
> The two vertical asymptotes could coincide, as in Example 5.

Hence, in addition to finding the asymptotes and the points where the curve crosses the axes, you need to establish where the curve crosses the horizontal asymptote.

Hence, there are four stages to sketching the curve with equation $y = \dfrac{(x-3)(2x-5)}{(x+1)(x+2)}$:

1. To find the horizontal asymptote of $y = \dfrac{(x-3)(2x-5)}{(x+1)(x+2)}$, express the equation as
   $$y = \frac{\left(1 - \frac{3}{x}\right)\left(2 - \frac{5}{x}\right)}{\left(1 + \frac{1}{x}\right)\left(1 + \frac{2}{x}\right)}.$$
   As $x \to \pm\infty$, $\dfrac{1}{x} \to 0$ and $y \to 2$. Therefore, the horizontal asymptote is $y = 2$.

2. To find the vertical asymptotes, equate the denominator to zero, which gives $(x+1)(x+2) = 0$. Hence, the vertical asymptotes are $x = -1$ and $x = -2$.

3. To find where the curve cuts the axes, you have:
   When $x = 0$: $y = \dfrac{15}{2}$ When $y = 0$: $x = 3$ and $x = \dfrac{5}{2}$

4. To find where the curve crosses the horizontal asymptote, $y = 2$, you have
   $$2 = \frac{(x-3)(2x-5)}{(x+1)(x+2)}$$
   $$2(x^2 + 3x + 2) = 2x^2 - 11x + 15$$
   $$\Rightarrow \qquad x = \frac{11}{17}$$

5. To sketch the curve, you need to insert all four points, as well as the three asymptotes.

It is important to note that:

▶ The curve can cross an axis or an asymptote only at the points found.
▶ If one branch of the curve goes to +∞, the next branch must return from −∞. The exception to this is when the two vertical asymptotes coincide as the result of a squared factor in the denominator. See Example 5.

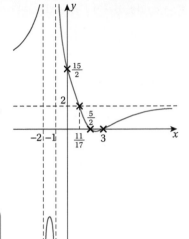

Example 4

Question

Sketch $y = \dfrac{(x+1)(x-4)}{(x-2)(x-5)}$.

Note

Equate the denominator to zero.

Answer

Horizontal asymptote: $y = 1$

Vertical asymptotes: $x = 2$ and $x = 5$

Curve crosses axes at $x = 0$, $y = -\dfrac{4}{10}$, and at $y = 0$, $x = -1$ and 4.

Curve crosses horizontal asymptote when $y = 1$,

$$1 = \frac{(x+1)(x-4)}{(x-2)(x-5)}$$

$$x^2 - 7x + 10 = x^2 - 3x - 4$$

$$\Rightarrow \quad x = \frac{7}{2}$$

Note

Express the equation as

$$y = \frac{\left(1 - \frac{1}{x}\right)\left(1 - \frac{4}{x}\right)}{\left(1 - \frac{2}{x}\right)\left(1 - \frac{5}{x}\right)}$$

or divide the leading coefficient of the numerator by the leading coefficient of the denominator, as both are of the same degree.

Example 5

Question

Sketch the curve $y = \dfrac{(x-1)(3x+2)}{(x+1)^2}$.

Answer

Horizontal asymptote: $y = 3$.

Vertical asymptotes: $x = -1$ (twice).

Curve crosses the axes at $x = 0$, $y = -2$, and at $y = 0$, $x = 1$ and $-\dfrac{2}{3}$.

Curve crosses the horizontal asymptote when $y = 3$,

$$3 = \frac{3x^2 - x - 2}{x^2 + 2x + 1}$$

$$3(x^2 + 2x + 1) = 3x^2 - x - 2$$

$$\Rightarrow \quad x = -\frac{5}{7}$$

Note

Equate the denominator to zero.

Note

Find $x = 0$ and $y = 0$.

Note

Use methods as before.

Note

Since $x = -1$ is a repeat asymptote, and the curve tends to +∞ as x approaches the value of −1 from the right (that is, x tends to −1 from above), it also tends to +∞ as x approaches the value of −1 from the left (that is, from below).

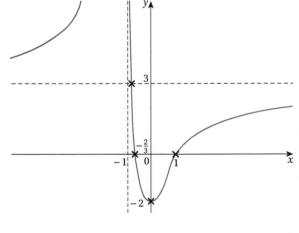

Curves with no vertical asymptotes

Not all curves with equations of the form $y = \frac{ax^2 + bx + c}{px^2 + qx + r}$ have vertical asymptotes. If the roots of $px^2 + qx + r = 0$ are not real, the curve will not have a vertical asymptote.

Example 6

Given that $y = \dfrac{2x^2 + 5x + 3}{4x^2 + 5x + 3}$

a   sketch the curve

b   find the range of possible values for y

c   find the maximum and minimum points.

a   Horizontal asymptote: as $x \to \infty$, $y \to \dfrac{1}{2}$.

Vertical asymptote: $4x^2 + 5x + 3 = 0$, which gives $x = \dfrac{-5 \pm \sqrt{-23}}{8}$.

The roots are not real. Therefore, the curve does not have a vertical asymptote.

> **Note**
>
> Express the function as $y = \dfrac{2 + \frac{3}{x} + \frac{3}{x^2}}{4 + \frac{5}{x} + \frac{3}{x^2}}$ to find the horizontal asymptote, or divide the leading coefficient of the numerator by that of the denominator.

The curve cuts the axes when $y = 0$,

$2x^2 + 5x + 3 = 0$

$(2x + 3)(x + 1) = 0$

$\Rightarrow \qquad x = -1$ and $-\dfrac{3}{2}$

and when $x = 0$, $y = 1$.

> **Note**
>
> Equate the denominator to zero.

The curve crosses the horizontal asymptote $y = \dfrac{1}{2}$ when

$\dfrac{1}{2} = \dfrac{2x^2 + 5x + 3}{4x^2 + 5x + 3}$

$4x^2 + 5x + 3 = 4x^2 + 10x + 6 \quad \Rightarrow \quad x = -\dfrac{3}{5}$

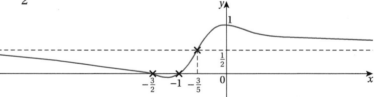

b   Cross-multiplying $y = \dfrac{2x^2 + 5x + 3}{4x^2 + 5x + 3}$:

$4yx^2 + 5yx + 3y = 2x^2 + 5x + 3$

$\Rightarrow (4y - 2)x^2 + (5y - 5)x + 3y - 3 = 0$

From the quadratic formula, $b^2 - 4ac \geq 0$ for the roots of x to be real. Therefore,

$(5y - 5)^2 - 4(4y - 2)(3y - 3) \geq 0$

$\Rightarrow 23y^2 - 22y - 1 \leq 0$

$\Rightarrow (23y + 1)(y - 1) \leq 0$

$\Rightarrow -\dfrac{1}{23} \leq y \leq 1$

The range of possible values of y is $-\dfrac{1}{23} \leq y \leq 1$.

> **Note**
>
> To find the coordinates of the maximum and minimum points, substitute these values of y into the original equations.

c   Hence, the maximum value of y is 1, and the minimum value is $-\dfrac{1}{23}$.

If $1 = \dfrac{2x^2 + 5x + 3}{4x^2 + 5x + 3}$, then $4x^2 + 5x + 3 = 2x^2 + 5x + 3$.

Therefore $2x^2 = 0$, and $x = 0$, so the maximum is $(0, 1)$.

If $-\dfrac{1}{23} = \dfrac{2x^2 + 5x + 3}{4x^2 + 5x + 3}$, then $-(4x^2 + 5x + 3) = 23(2x^2 + 5x + 3)$.

> **Note**
>
> To find the range of values of y, you need to find the values for which $y = \dfrac{2x^2 + 5x + 3}{4x^2 + 5x + 3}$ has real roots for x.

(continued)

(continued)

Therefore $-4x^2 - 5x - 3 = 46x^2 + 115x + 69$

$0 = 50x^2 + 120x + 72$

$0 = 25x^2 + 60x + 36 = (5x+6)^2$

Then $x = -\dfrac{6}{5}$, and so the minimum point is $\left(-\dfrac{6}{5}, -\dfrac{1}{23}\right)$.

Solving inequalities involving rational functions

Sketching the graph of a rational function can help you to solve an inequality involving that function.

When solving an inequality involving rational functions, you can add, subtract, multiply and divide positive numbers as if there were an equals symbol. However, to multiply or divide by a negative number, you must change the sign of the inequality.

For example, $3 > 2$ but $-3 < -2$ and $-2x > 4 \Rightarrow x < -2$.

Hence, an inequality such as $\dfrac{ax+b}{cx+d} > 2$ cannot be solved simply by multiplying both sides of the inequality by $cx + d$, since you do not know whether $cx + d$ is positive, giving $ax + b > 2(cx+d)$, or negative, giving $ax + b < 2(cx+d)$.

> To solve inequalities such as $\dfrac{ax+b}{cx+d} > k$, you can use either of these two methods:
>
> 1. **Algebraic method: multiply both sides of the inequality by $(cx+d)^2$, which must be non-negative.**
>
> 2. **Graphical method: sketch $y = \dfrac{ax+b}{cx+d}$, then solve $\dfrac{ax+b}{cx+d} = k$ and then, by comparing the two results, write the solution to the inequality.**

Either method can be used to solve any problem, unless you are asked specifically to use a particular method.

Example 7

Solve the inequality $\dfrac{5x-9}{x+3} > 2$ using an algebraic method.

Multiplying by $(x+3)^2$,

$\dfrac{5x-9}{x+3}(x+3)^2 > 2(x+3)^2$

$\Rightarrow (5x-9)(x+3) > 2(x+3)^2$

$\Rightarrow (5x-9)(x+3) - 2(x+3)^2 > 0$

$(x+3)[5x-9-2(x+3)] > 0$

$\Rightarrow (x+3)(3x-15) > 0$

$\Rightarrow (x+3)(x-5) > 0$

$\Rightarrow x > 5 \quad \text{or} \quad x < -3$

> **Note**
>
> $(x+3)$ is a factor, so factorise the term to simplify the inequality.

> **Note**
>
> Consider the signs of $x+3$ and of $x-5$ to deduce when their product is positive.

Example 8

Solve the inequality $\dfrac{(x+1)(x+4)}{(x-1)(x-2)} < 2$ using a graphical method.

Given the curve of $y = \dfrac{(x+1)(x+4)}{(x-1)(x-2)}$,

the horizontal asymptote is $y = 1$, and

the vertical asymptotes are $x = 1$ and $x = 2$.

The curve crosses the axes when $y = 0$, $x = -1$ and -4, and when $x = 0$, $y = 2$.

The curve crosses the horizontal asymptote

when $y = 1$,

$$\dfrac{(x+1)(x+4)}{(x-1)(x-2)} = 1$$

$\Rightarrow \quad x^2 + 5x + 4 = x^2 - 3x + 2$

$\Rightarrow \quad 8x = -2$

$\qquad x = -\dfrac{1}{4}$

when $y = 2$,

$$\dfrac{(x+1)(x+4)}{(x-1)(x-2)} = 2$$

$\Rightarrow \quad x^2 + 5x + 4 = 2(x^2 - 3x + 2)$

$\Rightarrow \quad 0 = x^2 - 11x$

$\Rightarrow \quad x = 0$ and 11

Therefore, we have

$$\dfrac{(x+1)(x+4)}{(x-1)(x-2)} < 2$$

when $x > 11$, $1 < x < 2$, $x < 0$.

> **Note**
>
> Use either of the methods seen earlier in the chapter.

> **Note**
>
> Equate the denominator to zero.

> **Note**
>
> Use $x = 0$ and $y = 0$.

> **Note**
>
> Use the graph.

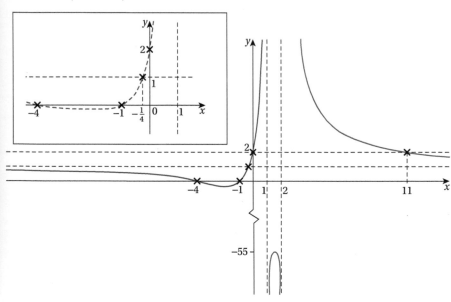

Exercise 3

In questions **1**, **2**, **3** and **4** write the equations of the asymptotes for the graph of each function.

1 $y = \dfrac{x+1}{(x+4)(x-1)}$

2 $y = \dfrac{5}{(x-3)(x+2)}$

3 $y = \dfrac{(x+8)(x-2)}{(x-3)^2}$

4 $y = \dfrac{(2x-3)(4x+5)}{(x-1)(x+7)}$

Sketch the graph of each of these functions.

5 $y = \dfrac{(x-3)(x-1)}{(x+2)(x-2)}$

6 $y = \dfrac{(2x-1)(x+4)}{(x-1)(x-2)}$

7 $y = \dfrac{(x+4)(x-5)}{(x-2)(x-3)}$
    8 $y = \dfrac{(x+1)(2x+5)}{(x+2)(x-5)}$

9 Find the range of values of

  a $y = \dfrac{4x^2 - x - 3}{2x^2 - x - 3}$
    b $y = \dfrac{x^2 + x - 1}{x^2 + x - 3}$

10 State the maximum and minimum points of the curves in question **10**.

In questions **11** to **13**, solve each of the inequalities for x.

11 **a** $\dfrac{x+3}{x+2} < 2$
    b $\dfrac{x+5}{x-3} > 1$

12 **a** $\dfrac{(x-1)(x-2)}{(x+1)(x+2)} > 1$
    b $\dfrac{(x+2)(x-5)}{(x-3)(x-2)} > 1$

  c $\dfrac{(x-1)(x-4)}{(x+1)(x-5)} > 2$

13 **a** $\dfrac{x^2 + x - 3}{x^2 + x - 2} > 1$
    b $\dfrac{2x^2 + x - 5}{2x^2 + x - 3} < 1$

1.3 Conic sections

If you take a solid, right circular cone and cut a plane section through it in any direction, you obtain a curve that is a member of the class of curves known as **conics** or **conic sections**.

The shape of the curve obtained in this way is determined by the direction in which you make the cut. In other words, it is dependent on the inclination, θ, of the plane section to the axis.

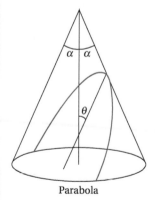 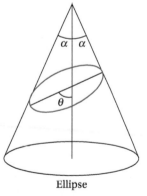 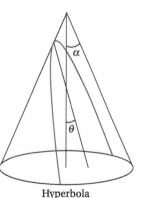

Parabola   Ellipse   Hyperbola

With the cone standing on a horizontal plane, if you cut:

▶ In a direction parallel to the slant height of the cone, whereby $\theta = \alpha$, you obtain a parabola

▶ In a direction for which $\alpha < \theta < \dfrac{\pi}{2}$, you obtain an ellipse

▶ In a direction, not through the vertex, for which $\theta < \alpha$, you obtain a hyperbola

▶ Horizontally through the cone $\left(\text{that is, } \theta = \dfrac{\pi}{2}\right)$, you obtain a circle.

In this chapter, you will see how to sketch each of these types of curve (except the circle), and how to apply transformations. The transformations of more complicated functions follow the same rules linear transformations do.

If $y = f(x)$, then

▶ $y = f(x) + a$ results in a positive translation in the y-direction
▶ $y = f(x) - a$ results in a negative translation in the y-direction
▶ $y = f(x + a)$ results in a negative translation in the x-direction
▶ $y = f(x - a)$ results in a positive translation in the x-direction
▶ $y = af(x)$ results in a stretch parallel to the y-axis, with scale factor a
▶ $y = f(ax)$ is a stretch parallel to the x-axis, with scale factor $\dfrac{1}{a}$
▶ $y = -f(x)$ is a reflection in the x-axis
▶ $y = f(-x)$ is a reflection in the y-axis.

Parabola

The standard equation of a conic parabola is $y^2 = 4ax$.

The graph shows $y^2 = 4x$.

Increasing the value of a causes the parabola to stretch in the x-direction.

Note that parabolas do not have asymptotes.

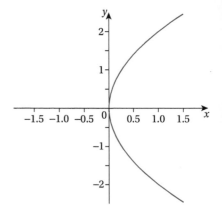

Example 9

Sketch $(y - 3)^2 = 4x$.

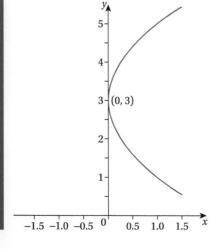

> **Note**
>
> Start with the graph $y^2 = 4x$, then translate +3 units in the y-direction.

Ellipse

The standard equation of a conic ellipse is $\dfrac{x^2}{a^2} + \dfrac{y^2}{b^2} = 1$.

Ellipses do not have asymptotes.

The graph is of $\dfrac{x^2}{4} + \dfrac{y^2}{9} = 1$.

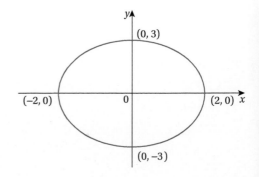

Example 10

Sketch the ellipse $16x^2 - 64x + 64 + 9y^2 = 144$.

$16(x-2)^2 + 9y^2 = 144$

$\dfrac{(x-2)^2}{9} + \dfrac{y^2}{16} = 1$

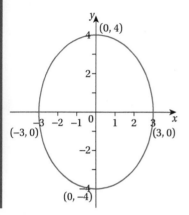
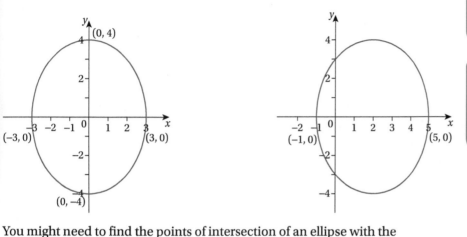

Start by sketching the ellipse $\frac{x^2}{9} + \frac{y^2}{16} = 1$. Since this ellipse is centered at $(0, 0)$, the easiest way to sketch it is to find the x- and y-intercepts, and to draw the ellipse through those points. The intercepts with the axes are at $(\pm 3, 0)$ and $(0, \pm 4)$.

Then translate the sketch by $+2$ units in the x-direction to get the final sketch.

You might need to find the points of intersection of an ellipse with the coordinate axes or other straight lines.

Example 11

The number of intersection points of the ellipse $x^2 + \frac{y^2}{16} = 16$ with the straight line $y + 3x = k$ varies according to the value of k. Calculate the value(s) of k that causes the line to be tangent to the ellipse.

Substituting for y,

$x^2 + \dfrac{(k-3x)^2}{16} = 16$

$16x^2 + k^2 - 6kx + 9x^2 = 256$

$25x^2 - 6kx + k^2 - 256 = 0$

A quadratic equation has equal roots when $b^2 - 4ac = 0$.

This means

$(6k)^2 - 4(25)(k^2 - 256) = 0$

$25\,600 - 64k^2 = 0$

$400 - k^2 = 0$

Therefore the values of k that cause the line to be a tangent of the ellipse are ± 20.

You can use your knowledge of the number of solutions to a quadratic to find that the line intersects the ellipse at two distinct points when $-20 < k < 20$.

Depending on the value of k, this quadratic could have zero, one or two solutions. Zero solutions would indicate that the line and the ellipse do not intersect; two distinct solutions would indicate that the line and the ellipse intersect twice. You need to find the value(s) of k that causes the line and the ellipse to intersect exactly once, so that the line is tangent to the ellipse.

Hyperbola

The standard equation of a hyperbola is $\dfrac{x^2}{a^2} - \dfrac{y^2}{b^2} = 1$.

As x and y become large, you have

$$\frac{x^2}{a^2} \to \frac{y^2}{b^2} \quad \Rightarrow \quad y \to \pm\frac{bx}{a}$$

Therefore, the asymptotes of a hyperbola

are $y = \pm\dfrac{b}{a}x$.

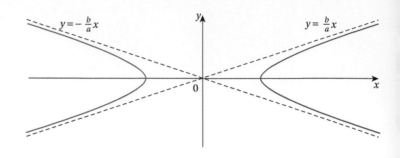

Example 12

Sketch $\dfrac{y^2}{9} - \dfrac{x^2}{25} = 1$.

The equation $\dfrac{x^2}{9} - \dfrac{y^2}{25} = 1$ is in standard form.

Its intercepts with the x-axis are $(\pm 3, 0)$.

Its asymptotes are $y = \pm\dfrac{5}{3}x$.

Its intercepts with the y-axis are $(\pm 5, 0)$.

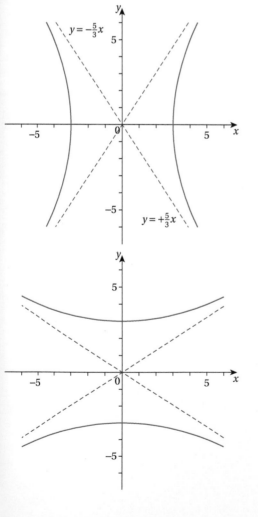

Note

Start by sketching

$\dfrac{x^2}{9} - \dfrac{y^2}{25} = 1$.

Note

By reflecting in the line $y = x$, this transforms into the required graph.

Note

It is also possible to sketch the graph using first principles, by finding the intercepts and asymptotes of the original equation.

Example 13

a Describe the geometrical transformation by which the hyperbola
$x^2 - 9y^2 = 1$ can be obtained from the hyperbola $x^2 - y^2 = 1$.

b Explain the geometrical transformation by which the hyperbola
$x^2 - y^2 - 6y + 8 = 0$ can be obtained from the hyperbola $x^2 - y^2 = 1$.

Note

Convert the hyperbola's equation into the standard form using
$(y+3)^2 = y^2 + 6y + 9$

a $x^2 - 9y^2 = 1$ can be rewritten as $x^2 - (3y)^2 = 1$ which means that this is
obtained by a stretch parallel to the y-axis, scale factor $\frac{1}{3}$.

b $x^2 - (y+3)^2 - 9 + 8 = 0$

$x^2 - (y+3)^2 = 1$

Thus the geometrical transformation is a translation
of -3 units in the positive y-direction.

Rectangular hyperbola

The standard equation of a conic
rectangular hyperbola is $xy = c^2$.

A rectangular hyperbola is a hyperbola with
asymptotes $y = \pm x$ that has been rotated by $45°$.

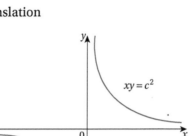
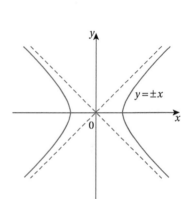

Example 14

Sketch $(x+2)y = 4$.

Note

First, sketch $xy = 4$.

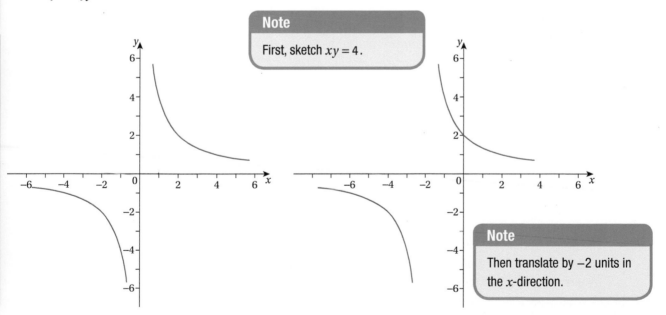

Note

Then translate by -2 units in
the x-direction.

Exercise 4

1 Sketch the parabola $y^2 = 16x$.

2 Sketch the parabola $(y-5)^2 = 4(x-2)$.

3 State the asymptotes of the curve $xy = 25$ and sketch the curve.

4 State the asymptotes of the curve $(x-3)(y+2) = 12$ and sketch the curve.

In questions **5** to **7**, sketch the curve and clearly mark the points where the curve crosses the coordinate axes.

5 $\dfrac{x^2}{36} + \dfrac{y^2}{25} = 1$

6 $\dfrac{(x-4)^2}{25} + \dfrac{(y-3)^2}{16} = 1$

7 $\dfrac{(x-2)^2}{25} - \dfrac{(y+6)^2}{16} = 1$

8 A straight line through $(1, 0)$ with gradient m intersects the hyperbola $\dfrac{x^2}{9} - \dfrac{y^2}{25} = 1$ at point P. Show that the x-coordinate of point P satisfies the equation $(25 - 9m^2)x^2 + 18m^2x - (9m^2 + 225) = 0$.

9 Write the asymptotes of $(x-5)(y+3) = 6$, and sketch the curve.

10 An ellipse has the equation $\dfrac{x^2}{4} + \dfrac{y^2}{25} = 1$.

    a Sketch the ellipse.

    b Given that the line $y = x + k$ intersects the ellipse at two distinct points, show that $-\sqrt{29} < k < \sqrt{29}$.

    c The ellipse is translated by the vector $\begin{pmatrix} a \\ b \end{pmatrix}$ to form another ellipse whose equation is

$$25x^2 + 4y^2 + 50x - 24y = c$$

      Find the values of a, b and c.

Summary

- A locus is a set of points that obey a certain rule, and a locus can be expressed graphically, verbally or in the form of an equation.

- Asymptotes show the 'end behaviour' of a graph as x or $y \to \pm\infty$.

- To sketch the graphs of rational, parabola, ellipse and hyperbola equations you might need to find the:
  - Asymptotes (if applicable; parabolas and ellipses do not have asymptotes)
  - Intercepts with the axes, if any
  - Coordinates of any maxima or minima (if applicable).

- You can solve a rational inequality by:
  - Using algebra to multiply both sides of the inequality by $(cx + d)^2$
  - Sketching $y = \dfrac{ax+b}{cx+d}$, then solving $\dfrac{ax+b}{cx+d} = k$ and comparing the two results to find the solution.

- Conic sections are a family of curves with standard equations, and include the:
  - Parabola, $y^2 = 4ax$
  - Ellipse, $\dfrac{x^2}{a^2} + \dfrac{y^2}{b^2} = 1$
  - Hyperbola, $\dfrac{x^2}{a^2} - \dfrac{y^2}{b^2} = 1$
  - Rectangular hyperbola, $xy = c^2$.

▶ The transformations of more complicated curves follow the same rules linear transformations do.

If $y = f(x)$, then

- $y = f(x) + a$ results in a positive translation in the y-direction
- $y = f(x) - a$ results in a negative translation in the y-direction
- $y = f(x + a)$ results in a negative translation in the x-direction
- $y = f(x - a)$ results in a positive translation in the x-direction
- $y = af(x)$ results in a stretch parallel to the y-axis, with scale factor a
- $y = f(ax)$ is a stretch parallel to the x-axis, with scale factor $\frac{1}{a}$
- $y = -f(x)$ is a reflection in the y-axis
- $y = f(-x)$ is a reflection in the x-axis.

Review exercises

1 Find the Cartesian equation of the locus of points which are equidistant from the point $(5, -1)$ and the y-axis.

2 Sketch the graph of $y = \dfrac{2x^2 + 3x - 5}{x^2 - x - 2}$.

3 Sketch the graph of $y = \dfrac{3x^2 + 4x + 4}{x^2 - 2x - 3}$.

4 Solve $\dfrac{2x - 1}{x + 3} > 3$.

5 Solve $\dfrac{x^2 - x - 2}{x^2 + 3x + 2} > 1$.

6 Sketch $\dfrac{(x - 3)^2}{36} - \dfrac{(y + 2)^2}{25} = 1$.

7 The curve $x^2 + \dfrac{y^2}{9} = 1$ is translated by k units in the positive y-direction.

   a Show that the equation of the curve after this translation is
$x^2 + \dfrac{(y - k)^2}{9} = 1$.

   b Show that if the line $x + y = 3$ intersects the translated curve, the y-coordinate of the points of intersection satisfies the equation
$10y^2 - (54 + 2k)y + k^2 + 72 = 0$.

Practice examination questions

1 a i Write down the equations of the two asymptotes of the curve
$y = \dfrac{1}{x - 3}$.   (2 marks)

   ii Sketch the curve $y = \dfrac{1}{x - 3}$, showing the coordinates of any points of intersection with the coordinate axes.   (2 marks)

   iii On the same axes, again showing the coordinates of any points of intersection with the coordinate axes, sketch the line $y = 2x - 5$.   (1 mark)

   b i Solve the equation $\dfrac{1}{x - 3} = 2x - 5$.   (3 marks)

   ii Find the solution of the inequality $\dfrac{1}{x - 3} < 2x - 5$.   (2 marks)

AQA MFP1 June 2010

2 A parabola P has equation $y^2 = x - 2$.

   a **i** Sketch the parabola P.   (2 marks)

       ii On your sketch, draw two tangents to P which pass through the point $(-2, 0)$.   (2 marks)

   b **i** Show that, if the line $y = m(x + 2)$ intersects P, then the x-coordinates of the points of intersection must satisfy
$$m^2 x^2 + (4m^2 - 1)x + (4m^2 + x) = 0.$$ (3 marks)

       ii Show that, if this equation has equal roots, then $16m^2 = 1$.   (3 marks)

      iii Hence find the coordinates of the points at which the tangents to P from the point $(-2, 0)$ touch the parabola P.   (3 marks)

AQA MFP1 June 2010

3 The diagram shows the hyperbola

$\dfrac{x^2}{a^2} - \dfrac{y^2}{b^2} = 1$ and its asymptotes.

The constants a and b are positive integers.

The point A on the hyperbola has coordinates $(2, 0)$.

The equations of the asymptotes are $y = 2x$ and $y = -2x$.

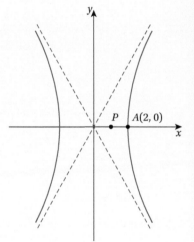

   a Show that $a = 2$ and $b = 4$.   (4 marks)

   b The point P has coordinates $(1, 0)$. A straight line passes through P and has gradient m. Show that, if this line intersects the hyperbola, the x-coordinates of the points of intersection satisfy the equation
$$(m^2 - 4)x^2 - 2m^2 x + (m^2 + 16) = 0$$ (4 marks)

   c Show that this equation has equal roots if $3m^2 = 16$.   (3 marks)

   d There are two tangents to the hyperbola which pass through P. Find the coordinates of the points at which these tangents touch the hyperbola.

      (No credit will be given for solutions based on differentiation.)   (5 marks)

AQA MFP1 January 2010

4 A curve has equation
$$y = \frac{x^2 - 2x + 1}{x^2 - 2x - 3}$$

   a Find the equations of the three asymptotes to the curve.   (3 marks)

   b **i** Show that if the line $y = k$ intersects the curve, then
$$(k - 1)x^2 - 2(k - 1)x - (3k + 1) = 0$$ (1 mark)

       ii Given that the equation $(k - 1)x^2 - 2(k - 1)x - (3k + 1) = 0$ has real roots, show that $k^2 - k \geq 0$.   (3 marks)

    iii Hence show that the curve has only one stationary point
      and find its coordinates. (No credit will be given for solutions based
      on differentiation.)   (4 marks)

  c Sketch the curve and its asymptotes.   (3 marks)

AQA MFP1 June 2013

5 An ellipse is shown on the right.

The ellipse intersects the x-axis at the points A and B. The equation
of the ellipse is $\dfrac{(x-4)^2}{4} + y^2 = 1$.

  a Find the coordinates of A and B.   (2 marks)

  b The line $y = mx \ (m > 0)$ is a tangent to the ellipse, with point of contact
     P.

    i Show that the x-coordinate of P satisfies the equation

      $(1 + 4m^2)x^2 - 8x + 12 = 0$ (3 marks)

    ii Hence find the exact value of m.   (4 marks)

    iii Find the coordinates of P.   (4 marks)

AQA MFP1 January 2013

6 The curve C has equation $y = \dfrac{x}{(x+1)(x-2)}$.

The line L has equation $y = -\dfrac{1}{2}$.

  a Write down the equations of the asymptotes of C.   (3 marks)

  b The line L intercepts C at two points. Find the x-coordinates of these two
     points.   (2 marks)

  c Sketch C and L on the same axes.
     (You are given that the curve C has no stationary points).   (3 marks)

  d Solve the inequality $\dfrac{x}{(x+1)(x-2)} \le -\dfrac{1}{2}$.   (3 marks)

AQA MFP1 June 2012

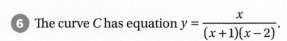

2 Complex Numbers

Introduction

You have always been told that finding the square root of a negative number is not possible since the square of any real number is always positive. Whilst it is the case that there is no **real number** that is $\sqrt{-1}$, there is an **imaginary number** with this value: i. A **complex number** is a number consisting of an imaginary part and a real part. You can use the idea that $i = \sqrt{-1}$ to perform calculations in the real world. For example, in mechanics you can use the theory of complex numbers to explain the motion of springs that have resistance.

Recap

You will need to remember...

▶ How to solve a quadratic equation using the quadratic formula.

▶ How to simplify expressions containing surds, such as $\dfrac{2 + \sqrt{3}}{5 - \sqrt{2}}$.

▶ That the displacement of a point with position vector **a** to a point with position vector **b** is given by the vector **b** − **a**.

Objectives

By the end of this chapter, you should know how to:

▶ Find complex roots of quadratic equations.

▶ Find the sum, difference, product and quotient of two complex numbers.

▶ Compare real and imaginary parts of two given complex numbers.

▶ Write a complex number either in the Cartesian form, $x + iy$, or in polar coordinate form, that is, in terms of its modulus and argument.

▶ Represent a complex number by a point on an Argand diagram, and draw simple loci in the complex plane.

2.1 What is a complex number?

A complex number consists of an imaginary part and a real part, for example $3 + 5i$.

> A complex number is a number of the form $a + ib$, where a and b are real numbers and $i^2 = -1$.

If $a = 0$, the number is said to be **wholly imaginary**. If $b = 0$, the number is **real**. If a complex number is 0, both a and b are 0.

In all your previous mathematics work, you have assumed that it is not possible to have a square root of a negative number. Therefore, when you considered the solution of quadratic equations in the form $ax^2 + bx + c = 0$, you noted that the equation has no real roots when $b^2 - 4ac$ is less than zero.

However, if you consider the imaginary number i, such an equation actually has two **complex roots**.

Example 1

Find the complex roots of the quadratic equation $x^2 + 2x + 3 = 0$.

$$x = \frac{-2 \pm \sqrt{4-12}}{2}$$

$$= \frac{-2 \pm \sqrt{-8}}{2}$$

$$= \frac{-2 \pm \sqrt{8}\sqrt{-1}}{2}$$

$$= \frac{-2 \pm 2\sqrt{2}\sqrt{-1}}{2}$$

$$= -1 \pm \sqrt{2}\sqrt{-1}$$

$$-1 \pm \sqrt{2}i$$

or $-1 + \sqrt{2}i$ and $-1 - \sqrt{2}i$

> **Note**
> Use the quadratic formula.

> **Note**
> Simplify $\sqrt{-8}$ using i as one of the factors.

> **Note**
> Simplify $\sqrt{8}$.

> **Note**
> Denote the value of $\sqrt{-1}$ as i.

Of course, i does not actually exist, but you can perform calculations using the simple rule that $i^2 = -1$.

So, to find the complex roots (in the form $a + ib$) of quadratic equations where there are no real roots ($b^2 - 4ac$ is strictly less than zero), you factorise the negative root by $\sqrt{-1}$ and then denote $\sqrt{-1}$ as i.

> **Tip**
> j is very occasionally used in place of i.

It is common to use $x + iy$ to represent an unknown complex number, and in turn, z is used to represent $x + iy$. So, when the unknown in an equation is a complex number, you denote it by z. In a similar way, you can use w to represent a second unknown complex number, where $w = u + iv$.

The complex conjugate

The complex number $x - iy$ is called the **complex conjugate** of $x + iy$, and is denoted by z^* or $\bar{z}$. For example, $2 - 3i$ is the complex conjugate of $2 + 3i$, and the complex conjugate of $-8 - 9i$ is $-8 + 9i$.

> If $z = x + iy$ is a root of a quadratic equation which has real coefficients, then $z^* = x - iy$ is also a root of the equation, where z^* is the conjugate of z.

You saw this in Example 1, where the solutions of the quadratic $x^2 + 2x + 3 = 0$ were found to be $-1 + \sqrt{2}i$ and $-1 - \sqrt{2}i$.

Example 2

Solve $z^2 - 4z + 40 = 0$.

> **Note**
> Use the quadratic formula.

$$z = \frac{4 \pm \sqrt{16-160}}{2} = \frac{4 \pm 12i}{2}$$

Roots are $2 \pm 6i$.

Exercise 1

1 Simplify each of these.

    a i^3 **b** i^4 **c** i^6

2 Express each of these complex numbers in the form $a + ib$.

    a $3 + 2\sqrt{-1}$ **b** $6 - 3\sqrt{-1}$

    c $-4 + \sqrt{-9}$ **d** $\sqrt{-100} - \sqrt{-64}$

3 Write the complex conjugate of z when z is

    a $3 + 4i$ **b** $2 - 6i$ **c** $-4 - 3i$

4 Find the solution of each of these equations.

    a $x^2 + 4x + 7 = 0$ **b** $x^2 + 2x + 6 = 0$

5 Solve each of these equations.

    a $z^2 + 2z + 4 = 0$ **b** $z^2 - 3z + 6 = 0$ **c** $2z^2 + z + 1 = 0$

2.2 Calculating with complex numbers

When you work with complex numbers you can use the same algebraic methods that you would use with 'ordinary' numbers. So, you cannot combine the real number part of a complex number with the i-term. For example, $2 + 3i$ cannot be simplified.

For two complex numbers to be equal, *both* the real parts must be equal and the imaginary parts must be equal. This is a **necessary condition** for the equality of two complex numbers.

Hence, if $a + ib = c + id$, then $a = c$ and $b = d$. For example, if $2 + 3i = x + iy$, then $x = 2$ and $y = 3$.

Addition and subtraction

When adding two complex numbers, add the real terms and *separately* add the i-terms. For example,

$$(3 + 7i) + (4 - 6i) = (3 + 4) + (7i - 6i)$$
$$= 7 + i$$

Generally, for complex numbers:
- ▶ $(x + iy) + (u + iv) = (x + u) + i(y + v)$ for addition
- ▶ $(x + iy) - (u + iv) = (x - u) + i(y - v)$ for subtraction.

Example 3

Question

Subtract $8 - 4i$ from $7 + 2i$.

Answer

$7 + 2i - (8 - 4i) = 7 - 8 + (2i + 4i)$
$= -1 + 6i$

Sometimes you will need to compare the real and imaginary parts of complex numbers in order to find the required solution.

Example 4

Question

Find x and y if $x + 2i + 2(3 - 5iy) = 8 - 13i$.

Answer

$x + 6 = 8$
$\Rightarrow \quad x = 2$
$2 - 10y = -13$
$\Rightarrow \quad 15 = 10y$
$\Rightarrow \quad y = 1\frac{1}{2}$

> **Note**
> Equate the real terms.

> **Note**
> Equate imaginary terms.

Multiplication

For multiplication of complex numbers, apply the general algebraic method for multiplication.

Example 5

Question

Simplify $(2 + 3i)(4 - 5i) = 2(4 - 5i) + 3i(4 - 5i)$.

Answer

$(2 + 3i)(4 - 5i) = 2(4 - 5i) + 3i(4 - 5i)$
$= 8 - 10i + 12i - 15i^2$

Since $i^2 = -1$,

$8 - 10i + 12i - 15 \times -1 = 8 - 10i + 12i + 15$
$= 23 + 2i$

Generally, for the multiplication of complex numbers,
$(a + ib)(c + id) = ac - bd + i(ad + bc)$ since $i^2 = -1$.

> **Note**
> Expand.

> **Note**
> Simplify.

> **Tip**
> It can be simpler to multiply out the numbers every time than to memorise this formula.

Division

To be able to divide by a complex number, you have to change it to a real number. Take, for example, the fraction $\frac{2+3i}{4+5i}$.

Recall from your A-level Mathematics course that you can simplify an expression such as $\frac{1}{1+\sqrt{3}}$ by multiplying the numerator and the denominator by $1 - \sqrt{3}$. Similarly, to simplify an expression such as $\frac{2+3i}{4+5i}$ you multiply its numerator and its denominator by $4 - 5i$, which is the **complex conjugate** of the denominator. Thus,

$$\frac{2+3i}{4+5i} = \frac{(2+3i)(4-5i)}{(4+5i)(4-5i)}$$

$$= \frac{8+12i-10i-15i^2}{4^2-(5i)^2}$$

$$= \frac{23+2i}{16+25}$$

$$= \frac{23}{41} + \frac{2}{41}i$$

> **Note**
>
> $-(5i)^2 = -(-25) = 25$

To divide by a complex number, write the calculation as a fraction, and multiply the top and bottom of the fraction by the complex conjugate of the divisor.

Example 6

Question

Simplify $\frac{3+i}{7-3i}$.

Answer

$$\frac{3+i}{7-3i} = \frac{(3+i)(7+3i)}{(7-3i)(7+3i)}$$

$$= \frac{21+7i+9i+3i^2}{7^2-(3i)^2}$$

$$= \frac{21+16i-3}{49+9}$$

$$= \frac{18}{58} + \frac{16i}{58}$$

$$= \frac{9}{29} + \frac{8}{29}i \quad \text{or} \quad \frac{1}{29}(9+8i)$$

> **Note**
>
> $-(3i)^2 = -(-9) = +9$

> **Note**
>
> Multiply the numerator and denominator of the fraction by the complex conjugate of $7 - 3i$, which is $7 + 3i$.

Equations involving complex conjugates

Equations involving complex conjugates can be solved by writing $z = a + bi$, and equating real and imaginary parts.

Example 7

Question

Solve $2z + z^* = 1 - i$.

Answer

If $z = a + ib$,

then $2a + 2bi + a - bi = 1 - i$.

Equating real and imaginary parts: $3a = 1$, $b = -1$.

Therefore $z = \dfrac{1}{3} - i$.

> **Note**
>
> Write z in the general form and substitute in z and z^* in the equation given.

Exercise 2

1 Simplify each of these.

   a $(8 + 4i) + (2 - 6i)$ **b** $(-7 + 3i) + (8 - 4i)$ **c** $2 - 4i + 3(-1 + 2i)$

2 Evaluate each of these expressions.

   a $(3 + i)(2 + 3i)$ **b** $(4 - 2i)(5 + 3i)$ **c** $i(2 - 3i)(i + 4)$

3 Express each of these fractions in the form $a + ib$, where $a, b \in \mathbb{R}$.

   a $\dfrac{2 + 3i}{4 - i}$ **b** $\dfrac{4 + 3i}{5 + i}$

4 Solve each of these equations in x and y.

   a $x + iy = 4 - 2i$ **b** $x + iy + 3 - 2i = 4(-2 + 5i)$

   c $x + iy = \dfrac{7 + i}{2 - i}$

5 If $z = 2 + 3i$, find the value of $z + \dfrac{1}{z}$.

6 Find z when

   a $2z + z^* = 9 + 6i$ **b** $3z - 5z^* = 7 + 2i$

   c $7z - 3z^* = 4 + 6i$ **d** $3z - 6z^* = 4 + 8i$

2.3 Argand diagram

The **Argand diagram** is a graphical representation of complex numbers. It is a way to represent complex numbers on a graph that looks like a real, two-dimensional graph.

> In the Argand diagram, the complex number $a + ib$ is represented by the point (a, b).

Real numbers are represented on the x-axis and imaginary numbers, on the y-axis. Therefore, the general complex number $(x + iy)$ is represented by the point (x, y).

> In an Argand diagram, the position of the complex conjugate z^* can always be obtained by reflecting the point representing z in the real axis.

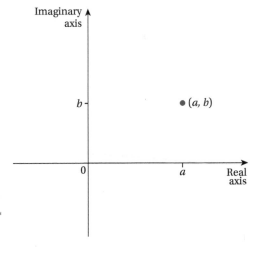

Example 8

Question

Represent the complex number $2 + 3i$ on an Argand diagram. Show its complex conjugate.

$2 + 3i$ is represented by the point $P(2, 3)$.

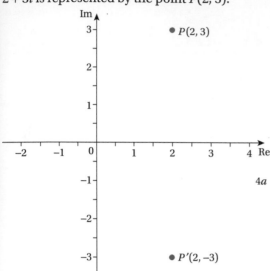

4a

> **Note**
>
> $2 + 3i$ is represented by the point $P(2, 3)$.

> **Note**
>
> The complex conjugate is $2 - 3i$, which is represented by the point $P'(2, -3)$.

Polar form of complex numbers

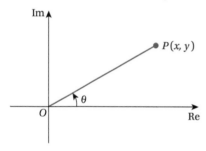

The form of the complex number you have seen so far, $z = x + iy$, is known as the **Cartesian form**. This can be easily plotted on an Argand diagram to find its modulus and argument.

The position of point $P(x, y)$ on the Argand diagram can be given in terms of OP, the distance of P from the origin (the **modulus**), and θ, the angle in the anticlockwise sense that OP makes with the positive real axis (the **argument**).

The length of the line OP is the modulus of z and is denoted by $r = |z|$; it is always taken to be positive.

The angle θ (normally in radians) is the argument of z, denoted by arg z. The **principal value** of θ is taken to be between $-\pi$ and π.

Connection between the Cartesian form and the polar form

From the diagram,

$r = |z| = \sqrt{x^2 + y^2}$

$x = r \cos \theta$ and $y = r \sin \theta$

So,

$z \equiv x + iy = r \cos \theta + ir \sin \theta$

So, the Cartesian form can now be written using the polar form as $z = r(\cos \theta + i \sin \theta)$.

Also, θ can be found using $\tan \theta = \dfrac{y}{x}$.

Take care when either x or y is negative. (See part **b** in Example 9.)

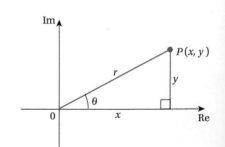

Example 9

Find the modulus and argument of each of these complex numbers.

a $2+2\sqrt{3}i$ b $-1-i$

Note

Sketch a diagram to help.

a

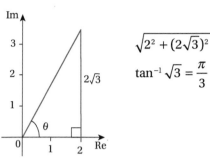

$\sqrt{2^2+(2\sqrt{3})^2}=4$

$\tan^{-1}\sqrt{3}=\dfrac{\pi}{3}$

Note

Remember that the angle is measured anticlockwise from the positive real axis.

b

The modulus of $-1-i$ is $\sqrt{1^2+1^2}=\sqrt{2}$.

The angle ϕ is $\dfrac{\pi}{4}$. Therefore, the argument

is $-\pi+\dfrac{\pi}{4}=-\dfrac{3\pi}{4}$.

Note

If the angle in Example 9 is measured anticlockwise from the positive real axis, its value is $\dfrac{5\pi}{4}$, but this is not between π and $-\pi$. Thus, you take the clockwise angle, which is $-\dfrac{3\pi}{4}$. The minus sign denotes that the angle is measured in the clockwise sense.

Example 10

Express the complex number z given by modulus 4 and argument $\dfrac{\pi}{3}$ in the form $a+ib$, where a and b are real numbers.

Using $x=r\cos\theta$, $x=4\cos\dfrac{\pi}{3}=2$.

Using $y=r\sin\theta$, $y=4\sin\dfrac{\pi}{3}=2\sqrt{3}$.

Hence $z=2+2i\sqrt{3}$.

Exercise 3

1 Represent each of these on an Argand diagram.

   a $2+2i$ b $-3+3i$ c $-2+2\sqrt{3}i$

   d $-1-i$ e $4i$ f $5+12i$

   g -4 h $6+\sqrt{13}i$

2 Find the modulus and the argument of each of the complex numbers in question **1**.

3 Given that $z = 3 + 4i$,

   a calculate

      i z^2 **ii** z^3

   b find

      i $|z|$ **ii** $|z^2|$ **iii** $|z^3|$

   c evaluate

      i $\arg z$ **ii** $\arg z^2$ **iii** $\arg z^3$

4 Express the complex number z in its $a + ib$ form when

   a $|z| = 2$ and $\arg z = \dfrac{\pi}{3}$

   b $|z| = 4$ and $\arg z = \dfrac{\pi}{4}$

5 Given that $z = \dfrac{7 + 2i}{3 + 4i}$, find the modulus and argument of z.

2.4 Loci in the complex plane

You know from your previous work on vector geometry that the vector $\mathbf{a} - \mathbf{b}$ connects the point with position vector $\mathbf{b}$ to the point with position vector $\mathbf{a}$.

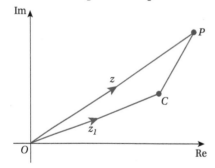

Similarly, in the complex plane, $z - z_1$ joins the point z_1 to the point z.

From the diagram, $\overrightarrow{OC} = z_1$ and $\overrightarrow{OP} = z$.

Therefore,

$\overrightarrow{CP} = \overrightarrow{CO} + \overrightarrow{OP} = -z_1 + z = z - z_1$

Using this fact, you can identify a number of loci that you need to learn and recognise.

$|z - z_1| = r$

$|z - z_1|$ is the modulus or length of $z - z_1$, that is, the length of the line joining z_1 to a variable point z.

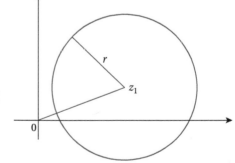

Therefore, $|z - z_1| = r$ is the locus of a point z, moving so that the length of the line joining a fixed point z_1 to z is always r. Hence, the locus of z is a circle, with centre z_1 and radius r.

$|z| = r$

In the special case in which $z_1 = 0$, the centre of the circle becomes the origin, and its radius is r.

Example 11

State and sketch the locus of $|z - 2 - 3i| = 3$.

The locus is $|z - (2 + 3i)| = 3$, which is a circle of centre $(2, 3)$ and radius 3.

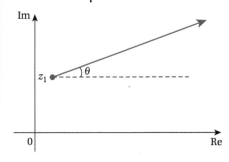

Note

Convert $|z - 2 - 3i|$ to $|z - (2 + 3i)|$ since the form above is always written as $|z - z_1|$. When sketching this locus, show clearly that the circle touches the x-axis and cuts the y-axis twice.

arg $(z - z_1) = \theta$

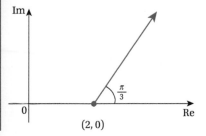

The point z satisfies the equation of this locus when the line joining z_1 to z has argument θ.

So, the locus is the **half-line**, starting at z_1, inclined at θ to the real axis. (It is called a half-line because you only want the part of the line that starts at z_1.)

Example 12

State and sketch the locus of $\arg(z - 2) = \dfrac{\pi}{3}$.

This locus is the half-line starting at $(2, 0)$, inclined at an angle of $\dfrac{\pi}{3}$ to the real axis.

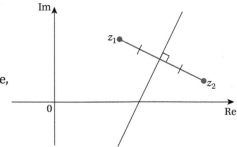

$|z - z_1| = |z - z_2|$

The line joining z to z_1 is equal in length to the line joining z to z_2. Therefore, the locus of z is the **perpendicular bisector** of the line joining z_1 to z_2.

Example 13

State the locus of $|z - 3 + i| = |z - 2i|$.

$|z - (3 - i)| = |z - 2i|$

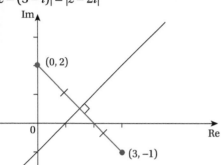

The locus is the perpendicular bisector of the line joining $3 - i$ to $+2i$.

Example 14

On the same diagram, show the locus of z when

a $|z - 4| = 4$

b $\arg z = \dfrac{\pi}{4}$

c Find the point that satisfies both loci.

a, b

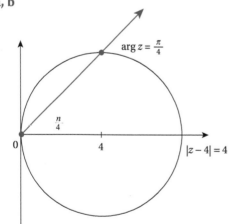

c The point that satisfies both loci is $(4, 4)$ or $(4 + 4i)$.

Note

Usually, it is possible to find a common point on two separate loci by using simple geometry and common sense. In this example, the point $(4, 4)$ can readily be seen to be on both loci. To calculate a common point may involve complicated algebra.

Example 15

Find the locus of $\dfrac{\pi}{4} < \arg(z - 2) < \dfrac{\pi}{3}$.

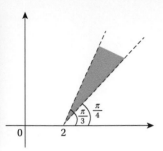

Note

Draw the two separate loci
$\frac{\pi}{4} = \arg(z - 2)$ and $\frac{\pi}{3} = \arg(z - 2)$,
then make sure that you select the correct sector.

Exercise 4

1 Sketch the locus of z when

  a $|z| = 5$ **b** $|z| = 3$

  c $|z - 2| = 3$ **d** $|z - 2i| = 4$

2 Sketch the locus of z when

  a $\arg z = \dfrac{\pi}{3}$ **b** $\arg z = -\dfrac{3\pi}{4}$

  c $\arg(z + 2) = \dfrac{\pi}{2}$

3 Sketch the locus of z when

  a $|z - 6| = |z + 3|$ **b** $|z - i| = |z - 2i|$

  c $|z + 2i| = |z - 2|$ **d** $|z| = 2$ and $\arg z = \dfrac{5\pi}{6}$

  e $|z| = 6$ and $\arg z = \dfrac{7\pi}{6}$

Summary

► For quadratic equations where there are no real roots ($b^2 - 4ac$ strictly less than zero), you find the complex roots in the form $x + iy$.

► If $z = x + iy$ is a root of a quadratic equation that has **real coefficients**, then $z^* = x + iy$ is also a root of the equation, where z^* is the conjugate of z.

► Generally, for complex numbers in Cartesian form:

  • $(x + iy) + (u + iv) = (x + u) + i(y + v)$ for addition

  • $(x + iy) - (u + iv) = (x - u) + i(y - v)$ for subtraction

  • $(a + ib)(c + id) = ac - bd + i(ad + bc)$ for multiplication

  • to divide by a complex number, write the division as a fraction, and multiply the top and bottom of the fraction by the complex conjugate of the divisor.

► Comparing the real and imaginary parts of two given complex numbers can help you to solve problems, including equations that involve complex conjugates.

► In an Argand diagram, real numbers are represented on the x-axis and imaginary numbers, on the y-axis, such that the complex number $a + ib$ is represented by the point (a, b).

- The Cartesian form of a complex number, $z = x + iy$, can be easily plotted on an Argand diagram to find its modulus and argument. The modulus is $|z|$ and the argument is θ, the angle in the anticlockwise sense from the positive real axis. When the modulus and argument are given, the complex number is said to be in polar form.
- The Cartesian form of a complex number can also be written in terms of the polar form (that is, in terms of its modulus and argument) $z = r(\cos\theta + i\sin\theta)$. The angle θ can be found using $\tan\theta = \frac{y}{x}$.
- Generally, for complex numbers in polar form, to find the:
  - Product of two complex numbers, multiply their moduli and add their arguments
  - Quotient of two complex numbers, divide their moduli and subtract their arguments.
- Some loci in the complex plane that you need to recognise are:
  - $|z| = r$
  - $|z - z_1| = r$
  - $\arg(z - z_1) = \theta$
  - $|z - z_1| = |z - z_2|$

Review exercises

1 Simplify each of these.

   a $(8 + 3i) - (7 + 2i)$ **b** $(4 - 2i)(5 + 3i)$ **c** $i(2 - 3i)(i + 4)$

2 Express each of these fractions in the form $a + ib$, where $a, b \in$.

   a $\dfrac{8 - i}{2 + 3i}$ **b** $\dfrac{2 + 5i}{-3 + 2i}$

3 If $z = 3 + i$, find the value of $z + \dfrac{1}{z}$.

4 Find the solution of each of these equations.

   a $2x^2 + 6x + 9 = 0$ **b** $x^2 - 5x + 25 = 0$

5 Find z when

   a $|z + 2 + 2i| = 2\sqrt{2}$ **b** $|z + 3 - \sqrt{3}i| = 2\sqrt{3}$

   c $2|z - i| = 3$

6 Sketch the locus of z when

   a $\arg z = \dfrac{\pi}{3}$ **b** $\arg z = -\dfrac{3\pi}{4}$

   c $\arg(z + 2) = \dfrac{\pi}{2}$

7 Sketch the locus of z when

   a $\left|\dfrac{z - 4i}{z + 4}\right| = 1$ **b** $|z| = 1$ and $\arg z = -\dfrac{\pi}{2}$

   c $|z| = 4$ and $\arg z = \dfrac{3\pi}{4}$

8 Given that $z = \dfrac{2 - 5i}{3 + 4i}$, find the modulus and argument of z.

Practice examination questions

1 Find the complex number z such that $5iz + 3z^* + 16 = 8i$.

Give your answer in the form $a + bi$, where a and b are real. (6 marks)

AQA MFP1 June 2014

2 a Solve the equation $w^2 + 6w + 34 = 0$, giving your answers in the form $p + qi$, where p and q are integers. (3 marks)

b It is given that $z = i(1 + i)(2 + i)$.

  i Express z in the form $a + bi$, where a and b are integers. (3 marks)

  ii Find integers m and n such that $z + mz^* = ni$. (3 marks)

AQA MFP1 January 2013

3 It is given that $z = x + iy$, where x and y are real numbers.

a Find, in terms of x and y, the real and imaginary parts of $i(z + 7) + 3(z^* - i)$. (3 marks)

b Hence find the complex number z such that $i(z + 7) + 3(z^* - i) = 0$. (3 marks)

AQA MFP1 June 2012

4 It is given that $z = x + iy$, where x and y are real.

a Find, in terms of x and y, the real and imaginary parts of $(z - i)(z^* - i)$. (3 marks)

b Given that $(z - i)(z^* - i) = 24 - 8i$ find the two possible values of z. (4 marks)

AQA MFP1 June 2011

5 Two loci, L_1 and L_2, in an Argand diagram are given by

$L_1 : |z + 6 - 5i| = 4\sqrt{2}$

$L_2 : \arg(z + i) = \dfrac{3\pi}{4}$

The point P represents the complex number $-2 + i$.

a Verify that the point P is a point of intersection of L_1 and L_2. (2 marks)

b Sketch L_1 and L_2 on one Argand diagram. (6 marks)

c The point Q is also a point of intersection of L_1 and L_2. Find the complex number that is represented by Q. (2 marks)

AQA MFP2 January 2013

6 a Draw on an Argand diagram:

  i the locus of points for which $|z - 2 - 3i| = 2$ (3 marks)

  ii the locus of points for which $|z + 2 - i| = |z - 2|$ (3 marks)

b Indicate on your diagram the points satisfying both $|z - 2 - 3i| = 2$ and $|z + 2 - i| \le |z - 2|$. (1 mark)

AQA MFP2 June 2012

3 Roots and Coefficients of a Quadratic Equation

Introduction

Quadratic equations are used to model events in the real world. For example, if you assume that the number of items sold decreases linearly with the price of each item, then the total revenue is a quadratic function. The height of a tennis ball thrown into the air can be modelled using a quadratic function of time. Indeed, quadratic equations are an important part of mathematics.

Objectives

By the end of this chapter, you should know:

► How to use the relationship between the roots and the coefficients of a quadratic equation.

► How to create an equation that has roots related to the roots of a given equation; for example, if an equation has roots α and β, you should be able to find an equation with roots α^2 and β^2.

Recap

You will need to remember...

► How to solve quadratic equations using the formula or by factorisation.

► How to expand expressions such as $(\alpha + \beta)^2$ and $(\alpha + \beta)^3$ using algebraic identities.

► Algebraic identities such as:

- Cube of a binomial $(a \pm b)^3 = a^3 \pm b^3 \pm 3a^2b + 3ab^2$

- Square of a binomial $(a \pm b)^2 = a^2 \pm 2ab + b^2$

- Difference of squares $(a - b)(a + b) = a^2 - b^2$

- Sum of cubes $(a + b)(a^2 - ab + b^2) = a^3 + b^3$.

· ·

3.1 Roots of a quadratic equation

You know that the roots of a polynomial equation in x are the values of x that make that polynomial equation equal to zero.

If α and β are the roots of a quadratic equation $f(x) = ax^2 + bx + c = 0$, then it follows that the equation must be of the form $f(x) = k(x - \alpha)(x - \beta)$ for some constant k.

Equating these two forms of a quadratic equation makes it possible to establish a relationship between the roots of a quadratic and the coefficients of that quadratic.

$$k(x - \alpha)(x - \beta) \equiv ax^2 + bx + c$$
$$\Rightarrow \quad k(x^2 - [\alpha + \beta]x + \alpha\beta) \equiv ax^2 + bx + c$$

Equating the coefficients of x^2 gives $k = a$.

Equating the coefficients of x gives $-k(\alpha + \beta) = b$.

Equating the constants gives $k\alpha\beta = c$.

Therefore,

$$\alpha + \beta = -\frac{b}{a} \quad \text{and} \quad \alpha\beta = \frac{c}{a}$$

Given a quadratic equation $ax^2 + bx + c = 0$, the sum of the roots is $-\frac{b}{a}$ and the product of the roots is $\frac{c}{a}$.

From this, it follows that you can also write the quadratic equation as
$x^2 - (\text{sum of roots})x + (\text{product of roots}) = 0$.

The relationship between the roots and the coefficients of a quadratic can be used to help you find the sum and product of the roots if given the equation, or to find the equation if given the sum and product of the roots.

Example 1

For the equation $3x^2 - 7x + 11 = 0$, find

a the sum of the roots

b the product of the roots.

a Since $\alpha + \beta = -\frac{b}{a}$, the sum of the roots is $\alpha + \beta = -\frac{-7}{3} = +\frac{7}{3}$.

b Since $\alpha\beta = \frac{c}{a}$, the product of the roots is $\alpha\beta = \frac{11}{3}$.

Example 2

Find a quadratic equation whose roots have a sum of $\frac{1}{2}$ and a product of $-\frac{5}{2}$.

$$x^2 - \frac{1}{2}x - \frac{5}{2} = 0$$

or $2x^2 - x - 5 = 0$

Note

Use $x^2 - (\text{sum of roots})x + (\text{product of roots}) = 0$.

Example 3

The equation $x^2 + 9x + 5 = 0$ has roots α and β.

a Write the value of $\alpha + \beta$ and the value of $\alpha\beta$.

b Show that $2(\alpha + \beta) = -18$.

a $\alpha + \beta = -\frac{b}{a} = -\frac{9}{1} = -9$; $\alpha\beta = \frac{c}{a} = \frac{5}{1} = 5$

b $2(\alpha + \beta) = 2 \times -9 = -18$

For some questions, you will first need to rearrange the equation into the standard form $ax^2 + bx + c = 0$.

Example 4

Question

For the equation $x + 5 = \frac{7}{x}$, show that the sum and the product of the roots are -5 and -7 respectively.

Answer

$x + 5 = \dfrac{7}{x}$

$x^2 + 5x - 7 = 0$

Sum of roots $= \alpha + \beta = -\dfrac{b}{a} = -\dfrac{5}{1} = -5$

Product of roots $= \alpha\beta = \dfrac{c}{a} = \dfrac{-7}{1} = -7$

Note

First write the equation in the standard form by multiplying by x and rearranging.

Exercise 1

1 Find the sum and the product of the roots of each of these equations.

   a $x^2 + 3x - 7 = 0$ **b** $x^2 - 11x + 5 = 0$

   c $x^2 + 5x - 4 = 0$ **d** $3x^2 + 11x + 2 = 0$

2 Write a quadratic equation whose roots have the sum and the product given below.

   a   Sum 7; product 15   **b**   sum -3; product 5

   c   sum -2; product -4 **d**   sum -5; product -11.

3 Write the sum and the product of the roots of each of these equations.

   a $x + 2 = \dfrac{5}{x}$ **b** $2x^2 = 7 - 4x$

   c $x - 7 = \dfrac{-4}{x}$ **d** $8 = -3x^2 + 2x - 7$

4 Show that the sum and product of the equation $-x^2 + 4x = 7$ are 4 and 7 respectively.

3.2 Finding an equation with roots that are a function of existing roots

If an equation has roots α and β, you can use the sum and product of the roots to find equations with roots that are functions of α and β. First you need to express the new sum and product of the roots in terms of $\alpha + \beta$ and $\alpha\beta$ by manipulating algebraic identities. Then you substitute in the values of $\alpha + \beta$ and $\alpha\beta$ to find the new sum and product, and hence the new equation.

Example 5

Question

The equation $4x^2 + 7x - 5 = 0$ has roots α and β.
Find the equation whose roots are α^2 and β^2.

Answer

Because $4x^2 - 7x - 5 = 0$, $\alpha + \beta = -\dfrac{7}{4}$ and $\alpha\beta = -\dfrac{5}{4}$.

Sum of the new roots: $\alpha^2 + \beta^2 = (\alpha + \beta)^2 - 2\alpha\beta$

$$\alpha^2 + \beta^2 = \left(-\frac{7}{4}\right)^2 - 2 \times -\frac{5}{4} = \frac{89}{16}$$

Product of the new roots: $\alpha^2\beta^2 = (\alpha\beta)^2$

$$(\alpha\beta)^2 = \left(-\frac{5}{4}\right)^2 = \frac{25}{16}$$

Therefore, the new equation is $x^2 - \dfrac{89}{16}x + \dfrac{25}{16} = 0$

or $16x^2 - 89x + 25 = 0$.

> **Note**
> Substitute the values of $\alpha + \beta$ and $\alpha\beta$ in the right-hand side.

> **Note**
> First find the values of the sum and product of the roots of the given equation.

> **Note**
> Because $(\alpha + \beta)^2 = \alpha^2 + 2\alpha\beta + \beta^2$.

> **Note**
> Substitute in the value for $\alpha\beta$.

Example 6

Question

The equation $3x^2 + 8x - 4 = 0$ has roots α and β.
Find a quadratic equation with integer coefficients whose roots are α^3 and β^3.

Answer

$\alpha + \beta = -\dfrac{8}{3}$ and $\alpha\beta = -\dfrac{4}{3}$

Sum of new roots is

$$\alpha^3 + \beta^3 = (\alpha + \beta)^3 - 3\alpha^2\beta - 3\alpha\beta^2$$
$$= (\alpha + \beta)^3 - 3\alpha\beta(\alpha + \beta)$$
$$= \left(-\frac{8}{3}\right)^3 - 3 \times -\frac{4}{3} \times -\frac{8}{3} = -\frac{800}{27}$$

Product of the new roots is $\alpha^3\beta^3 = (\alpha\beta)^3 = \left(-\dfrac{4}{3}\right)^3 = -\dfrac{64}{27}$.

Therefore, the new equation is

$$x^2 - \left(-\frac{800}{27}\right)x - \frac{64}{27} = 0$$
$$x^2 + \frac{800}{27}x - \frac{64}{27} = 0$$

Alternatively, you could give $27x^2 + 800x - 64 = 0$.

> **Note**
> Use the identity $(\alpha + \beta)^3 = \alpha^3 + \beta^3 + 3\alpha^2\beta + 3\alpha\beta^2$.

> **Note**
> Factorise so the expression is in terms of $\alpha + \beta$ and $\alpha\beta$.

Example 7

Question

The equation $3x^2 + 8x - 4 = 0$ has roots α and β. Find the equation whose roots are $\dfrac{1}{\alpha}$ and $\dfrac{1}{\beta}$.

Roots and Coefficients of a Quadratic Equation 37

$3x^2 + 8x - 4 = 0$ has roots α and β.

Substituting $y = \dfrac{1}{x}$, we find that $\dfrac{3}{y^2} + \dfrac{8}{y} - 4 = 0$

has roots $\dfrac{1}{\alpha}$ and $\dfrac{1}{\beta}$.

Therefore $3 + 8y - 4y^2 = 0$ has roots $\dfrac{1}{\alpha}$ and $\dfrac{1}{\beta}$.

Simplifying, $4y^2 - 8x - 3 = 0$ has the required roots.

Alternatively, $4x^2 - 8x - 3 = 0$.

Example 8

The equation $3x^2 + 8x - 4 = 0$ has roots α and β. Find the quadratic equation with integer coefficients whose roots are $\alpha + \dfrac{2}{\beta}$ and $\beta + \dfrac{2}{\alpha}$.

$\alpha + \beta = -\dfrac{8}{3}$ and $\alpha\beta = -\dfrac{4}{3}$

Sum of the new roots: $\alpha + \dfrac{2}{\beta} + \beta + \dfrac{2}{\alpha} = (\alpha + \beta) + \dfrac{2}{\alpha} + \dfrac{2}{\beta}$

$$= (\alpha + \beta) + 2\dfrac{\beta + \alpha}{\alpha\beta}$$

$$= -\dfrac{8}{3} + 2 \times \dfrac{-\dfrac{8}{3}}{-\dfrac{4}{3}} = \dfrac{4}{3}$$

Product of the new roots: $\left(\alpha + \dfrac{2}{\beta}\right)\left(\beta + \dfrac{2}{\alpha}\right) = \alpha\beta + 4 + 4\left(\dfrac{1}{\alpha\beta}\right) = -\dfrac{1}{3}$.

Therefore, the new equation is $x^2 - \dfrac{4}{3}x - \dfrac{1}{3} = 0$.

Alternatively, $3x^2 - 4x - 1 = 0$.

Exercise 2

1 The equation $4x^2 + 7x - 5 = 0$ has roots α and β. Find an equation whose roots are α^2 and β^2.

2 The equation $5x^2 + 7x - 12 = 0$ has roots α and β.

    a   Write the value of $\alpha + \beta$ and the value of $\alpha\beta$.

    b   Show that $\alpha^2 + \beta^2 = \dfrac{169}{25}$.

3 The equation $3x^2 + 5x - 6 = 0$ has roots α and β. Find an equation whose roots are $\dfrac{1}{\alpha}$ and $\dfrac{1}{\beta}$.

4 If α and β are the roots of the equation $x^2 - 5x + 3 = 0$, find the values of

    a $\alpha + \beta$ **b** $\alpha^2 + \beta^2$ **c** $\alpha^3 + \beta^3$

5 The quadratic equation $x^2 - 4x + 9 = 0$ has roots α and β.

    a   Write the value of $\alpha + \beta$ and the value of $\alpha\beta$.

    b   Find a quadratic equation which has roots $\dfrac{1}{\alpha^2} + \alpha^2$ and $\dfrac{1}{\beta^2} + \beta^2$.

6 The quadratic equation $4x^2 - 9x + 8 = 0$ has roots α and β.

    a   Write the value of $\alpha + \beta$ and the value of $\alpha\beta$.

    b   Find a quadratic equation, with integer coefficients, which has roots $\alpha + \beta^2$ and $\beta + \beta^2$.

Summary

▶ For a quadratic equation of the form $ax^2 + bx + c = 0$ with two roots α and β, the relationship between the roots and the coefficients of the equation is as follows:

- The sum of the roots is $\alpha + \beta = -\dfrac{b}{a}$.

- The product of the roots is $\alpha\beta = \dfrac{c}{a}$.

▶ The relationship between the roots and the coefficients can be used to find the sum and product of the roots if given the equation, or to find the equation if given the sum and product of the roots.

▶ If an equation has roots α and β, you can use the sum and product of the roots to find equations with roots that are functions of α and β by expressing the new sum and product of the roots in terms of $\alpha + \beta$ and $\alpha\beta$ and substituting in their values.

Review exercises

1 Write the sum and the product of the roots of each of these equations.

    a $x + 3 = \dfrac{1}{x}$ **b** $3x^2 = 12 - 2x$

2 The equation $2x^2 + 7x + 8 = 0$ has roots α and β.

    a   Write the value of $\alpha + \beta$ and the value of $\alpha\beta$.

    b   Show that $\alpha^2 + \beta^2 = \dfrac{17}{4}$.

3 The equation $3x^2 + 12x - 7 = 0$ has roots α and β. Find a quadratic equation whose roots are $\alpha + \dfrac{2}{\beta}$ and $\beta + \dfrac{2}{\alpha}$.

4 The quadratic equation $6x^2 - 8x + 3 = 0$ has roots α and β.

    a   Write the value of $\alpha + \beta$ and the value of $\alpha\beta$.

    b   Show that $\alpha^3 + \beta^3 = \dfrac{10}{27}$.

    c   Find a quadratic equation with integer coefficients which has roots $\alpha^2 + \beta$ and $\alpha + \beta^2$.

Practice examination questions

1 The equation $2x^2 + 3x - 6 = 0$ has roots α and β.

  a Write down the value of $\alpha + \beta$ and the value of $\alpha\beta$.   (2 marks)

  b Hence show that $\alpha^3 + \beta^3 = -\dfrac{135}{8}$.   (3 marks)

  c Find a quadratic equation, with integer coefficients, whose roots are $\alpha + \dfrac{\alpha}{\beta^2}$ and $\beta + \dfrac{\beta}{\alpha^2}$.   (6 marks)

AQA MFP1 June 2013

2 The quadratic equation $5x^2 - 7x + 1 = 0$ has roots α and β.

  a Write down the value of $\alpha + \beta$ and the value of $\alpha\beta$.   (2 marks)

  b Show that $\dfrac{\alpha}{\beta} + \dfrac{\beta}{\alpha} = \dfrac{39}{5}$.   (3 marks)

  c Find a quadratic equation, with integer coefficients, which has roots $\alpha + \dfrac{1}{\alpha}$ and $\beta + \dfrac{1}{\beta}$.   (5 marks)

AQA MFP1 June 2012

3 The quadratic equation $2x^2 + 7x + 8 = 0$ has roots α and β.

  a Write down the value of $\alpha + \beta$ and the value of $\alpha\beta$.   (2 marks)

  b Show that $\alpha^2 + \beta^2 = \dfrac{17}{4}$.   (2 marks)

  c Find a quadratic equation, with integer coefficients, which has roots $\dfrac{1}{\alpha^2}$ and $\dfrac{1}{\beta^2}$.   (5 marks)

AQA MFP1 January 2012

4 The equation $4x^2 + 6x + 3 = 0$ has roots α and β.

  a Write down the value of $\alpha + \beta$ and the value of $\alpha\beta$.   (2 marks)

  b Show that $\alpha^2 + \beta^2 = \dfrac{3}{4}$.   (2 marks)

  c Find an equation, with integer coefficients, which has roots $3\alpha - \beta$ and $3\beta - \alpha$.   (5 marks)

AQA MFP1 June 2011

5 The quadratic equation $x^2 - 6x + 18 = 0$ has roots α and β.

  a Write down the value of $\alpha + \beta$ and the value of $\alpha\beta$.   (2 marks)

  b Find a quadratic equation, with integer coefficients, which has roots α^2 and β^2.   (4 marks)

  c Hence find the values of α^2 and β^2.   (1 mark)

AQA MFP1 January 2011

> **Tip**
>
> For part **c**, you will need to draw on your knowledge of complex numbers and complex conjugates; see Chapter 2 *Complex Numbers* for a reminder if you need to.

4 Series

Introduction

It is likely that you will recognise series from your earliest school days, particularly triangular numbers: 1, 3, 6, 10, 15.... The nth triangular number is the sum $1 + 2 + 3 + 4 + \ldots + n$. In this chapter, you will generalise the concept of summing series, for example to find the sum of the first n squares or cubes. Summing squares is a useful tool when finding lines of best fit.

Objectives

By the end of this chapter, you should know how to:

▶ Use formulae to determine the sum of the squares and cubes of the natural numbers.

▶ Sum mixed sums such as $\Sigma r^2(r+2)$.

▶ Add finite sums using the method of differences.

▶ Add various infinite series using the method of differences.

Recap

You will need to remember how to ...

▶ Read and use sigma notation, for example $\displaystyle\sum_{r=0}^{n} 1 = n + 1$.

▶ Use simple formulae for summation notation, such as:

- $\displaystyle\sum_{i=i_0}^{n} ca_i = c \sum_{i=i_0}^{n} a_i$, where c is any number

- $\displaystyle\sum_{i=i_0}^{n} (a_i \pm b_i) = \sum_{i=i_0}^{n} a_i \pm \sum_{i=i_0}^{n} b_i$.

▶ Determine the impact of limits, for example, as $n \to \infty$, you should be able to say what happens to $\frac{1}{n}$.

4.1 Summation formulae

You know from your A-level Mathematics studies that a series is the sum of the terms in a sequence. You will have used formulae to help you find the sum of arithmetic and geometric progressions.

There will be times, however, when you need to find the sum of a series that is not as familiar to you. In these cases, you will need to use a different set of summation formulae.

Important summation formulae

$$\sum_{r=1}^{n} r = \frac{1}{2} n(n+1)$$

$$\sum_{r=1}^{n} r^2 = \frac{1}{6} n(n+1)(2n+1)$$

$$\sum_{r=1}^{n} r^3 = \frac{1}{4} n^2 (n+1)^2$$

$$\sum_{r=1}^{n} r^3 = \left(\sum_{r=1}^{n} r \right)^2$$

Formulae booklet

The first three are given in the *Formulae and Statistical Tables* booklet.

A useful summation formula that you should already know is

$$\sum_{r=1}^{n} 1 = 1 + 1 + 1 + \cdots + 1 = n \text{ (total of } n \text{ terms of 1)}.$$

You can use these formulae to find the sums of many series. Split the given term into parts (by breaking up the summation across the sum or difference) and factor out any constants before applying the relevant summation formula. When the upper limit is given as n, you will find the sum as a polynomial expression.

Example 1

Question

Find the sum of $\displaystyle\sum_{r=1}^{n} (4r^2 + 1)$.

> **Note**
> Split the term into its parts using
> $$\sum_{i=l_0}^{n} (a_i + b_i) = \sum_{i=l_0}^{n} a_i + \sum_{i=l_0}^{n} b_i$$

> **Note**
> $\displaystyle\sum_{r=1}^{n} 1 = n$ (since you are summing n copies of 1)

Answer

$$\sum_{r=1}^{n} (4r^2 + 1) = 4\sum_{r=1}^{n} r^2 + \sum_{r=1}^{n} 1$$

Therefore

$$\sum_{r=1}^{n} (4r^2 + 1) = 4 \times \frac{1}{6} n(n+1)(2n+1) + n$$

$$= \frac{2}{3} n(n+1)(2n+1) + n$$

$$= \frac{1}{3} [2n(n+1)(2n+1) + 3n]$$

$$\Rightarrow \sum_{r=1}^{n} (4r^2 + 1) = \frac{1}{3} n[2(n+1)(2n+1) + 3]$$

Therefore,

$$\sum_{r=1}^{n} (4r^2 + 1) = \frac{1}{3} n(4n^2 + 6n + 5)$$

> **Note**
> Use the summation formula
> $$\sum_{r=1}^{n} r^2 = \frac{1}{6} n(n+1)(2n+1)$$

Example 2

Question

Find the sum of $\displaystyle\sum_{r=1}^{n} (2r^3 + 3r^2 + 1)$.

Answer

$$\sum_{r=1}^{n} (2r^3 + 3r^2 + 1) = \sum_{r=1}^{n} 2r^3 + \sum_{r=1}^{n} 3r^2 + \sum_{r=1}^{n} 1$$

$$= 2\sum_{r=1}^{n} r^3 + 3\sum_{r=1}^{n} r^2 + \sum_{r=1}^{n} 1$$

> **Note**
> Split the term into its parts and factor out the constants.

Therefore

$$\sum_{r=1}^{n} (2r^3 + 3r^2 + 1) = 2 \times \frac{1}{4} n^2(n+1)^2 + 3 \times \frac{1}{6} n(n+1)(2n+1) + n$$

$$= \frac{n}{2} [n(n+1)^2 + (n+1)(2n+1) + 2]$$

> **Note**
> Simplify.

Factorising,

$$\sum_{r=1}^{n} (2r^3 + 3r^2 + 1) = \frac{n}{2} (n^3 + 4n^2 + 4n + 3)$$

Example 3

Find the value of $\displaystyle\sum_{r=n+1}^{2n} (4r^3 - 3)$.

$$\sum_{r=n+1}^{2n} (4n^3 - 3) = \sum_{r=1}^{2n} (4r^3 - 3) - \sum_{r=1}^{n} (4r^3 - 3)$$

$$\sum_{r=n+1}^{2n} (4r^3 - 3) = 4\sum_{1}^{2n} r^3 - 3\sum_{1}^{2n} 1 - \left(4\sum_{1}^{n} r^3 - 3\sum_{1}^{n} 1 \right)$$

$$= 4 \times \frac{1}{4}(2n)^2(2n+1)^2 - 3 \times 2n - \left[4 \times \frac{1}{4} n^2 (n+1)^2 - 3n \right]$$

$$= 4n^2(2n+1)^2 - 6n - n^2(n+1)^2 + 3n$$

$$= 4n^2(4n^2 + 4n + 1) - n^2(n^2 + 2n + 1) - 3n$$

$$= n^2(15n^2 + 14n + 3) - 3n$$

Therefore,

$$\sum_{r=n+1}^{2n} (4r^3 - 3) = 15n^4 + 14n^3 + 3n^2 - 3n$$

> **Note**
>
> Split the given term into its parts.

> **Note**
>
> Each term must be split into the sum from 1 to $2n$; then subtract the sum from 1 to n.

When you are given the value of the end term, you will need to find the value of the sum rather than a polynomial expression.

Example 4

Find $\displaystyle\sum_{r=1}^{8} (r^2 + 2)$.

$$\sum_{r=1}^{8} (r^2 + 2) = \sum_{1}^{8} r^2 + \sum_{1}^{8} 2$$

Since $\displaystyle\sum_{r=1}^{n} 1 = n, \sum_{r=1}^{n} 2 = 2n,$

$$\sum_{r=1}^{n} (r^2 + 2) = \frac{1}{6} n(n+1)(2n+1) + 2n$$

Since $n = 8$,

$$\sum_{r=1}^{8} (r^2 + 2) = \frac{1}{6} \times 8 \times 9 \times 17 + 16 = 220$$

Therefore $\displaystyle\sum_{r=1}^{8} (r^2 + 2) = 220$.

> **Note**
>
> Split the given term.

> **Note**
>
> Use
> $$\sum_{r=1}^{n} r^2 = \frac{1}{6} n(n+1)(2n+1)$$

Exercise 1

1 Find $\displaystyle\sum_{r=1}^{n} (2r^2 + 2r)$.

2 Find $\displaystyle\sum_{r=1}^{n} (2r^3 + r)$.

3 Find $\displaystyle\sum_{r=1}^{n} (r+1)(r-2)$.

4 Find $\displaystyle\sum_{r=1}^{15} (3r-1)(2r+3)$.

4.2 Method of differences

For some series, it is possible to use the **method of differences** to find the sum of a finite series because most of the terms cancel out. You will usually be given a relationship, which you may be asked to prove, and then each term of the series you are trying to sum is split into two terms which will cancel out.

In Example 5 you will see how to use a given identity $r \equiv \frac{1}{2}[r(r+1)-(r-1)r]$ to find the sum $\displaystyle\sum_{r=1}^{n} r$.

There is no expectation that you will 'guess' the initial identity. Rather, insert $r = 1$ into the given identity, then insert $r = 2$ into the given identity, then insert $r = 3$ and so on until you can see clearly which pairs of terms cancel with each other.

Then you have to ensure that you notice which terms at the end of the series do not cancel with the earlier terms.

Example 5

Use the identity $r \equiv \dfrac{1}{2}[r(r+1)-(r-1)r]$ to find the sum $\displaystyle\sum_{r=1}^{n} r$.

Making the given substitution, we obtain

$$\sum_{r=1}^{n} r = \sum_{r=1}^{n} \frac{1}{2}[r(r+1)-(r-1)r]$$

$$= \frac{1}{2}(1\times 2 - 0\times 1)+\frac{1}{2}(2\times 3 - 1\times 2)+\frac{1}{2}(3\times 4 - 2\times 3)+\cdots$$

$$+ \frac{1}{2}[(n-1)n-(n-2)(n-1)]+\frac{1}{2}[n(n+1)-(n-1)n]$$

$$= \frac{1}{2}(1\times 2)-\frac{1}{2}(0\times 1)+\frac{1}{2}(2\times 3)-\frac{1}{2}(1\times 2)+\frac{1}{2}(3\times 4)-\frac{1}{2}(2\times 3)+\cdots$$

$$+ \frac{1}{2}(n-1)n-\frac{1}{2}(n-2)(n-1)+\frac{1}{2}n(n+1)-\frac{1}{2}(n-1)n$$

Therefore

$$\sum_{r=1}^{n} r = \frac{1}{2}[-0\times 1 + n(n+1)]$$

$$= \frac{1}{2}n(n+1)$$

> **Note**
>
> We notice that almost all the terms cancel one another out.

Finding the sum of infinite series using differences

An **infinite series** is a sum such as $\displaystyle\sum_{r=1}^{\infty}\left(\frac{1}{r}-\frac{1}{r+1}\right)$, which looks like the sum of infinitely many terms. Of course, you are not expected to add infinitely many terms. Instead, you are expected to find the limit as $n \to \infty$ of the **partial sums** $\displaystyle\sum_{r=1}^{n}\left(\frac{1}{r}-\frac{1}{r+1}\right)$, provided that this limit exists.

If the limit of partial sums exists, then we can say that the series **converges** and can give the value of the sum of the infinite series. If it does not, we say that the series **does not converge**.

By definition, $\displaystyle\sum_{r=1}^{\infty} f(r) = \lim_{n\to\infty}\sum_{r=1}^{n} f(r)$, provided this limit exists.

Example 6

Question

Determine whether the sum $\displaystyle\sum_{r=1}^{\infty}\left(\frac{1}{r}-\frac{1}{r+1}\right)$ exists. If it does, give its value.

Answer

$$\sum_{r=1}^{n}\left(\frac{1}{r}-\frac{1}{r+1}\right)=\left(1-\frac{1}{2}\right)+\left(\frac{1}{2}-\frac{1}{3}\right)+\left(\frac{1}{3}-\frac{1}{4}\right)+\cdots+\left(\frac{1}{n}-\frac{1}{n+1}\right)$$

$$=1-\frac{1}{n+1}$$

Therefore, the infinite series converges and

$$\sum_{r=1}^{\infty}\left(\frac{1}{r}-\frac{1}{r+1}\right)=\lim_{n\to\infty}\sum_{r=1}^{n}\left(\frac{1}{r}-\frac{1}{r+1}\right)=\lim_{n\to\infty}\left(1-\frac{1}{n+1}\right)=1$$

> **Note**
>
> As before, do not simplify each term so that you can tell which terms will cancel.

Exercise 2

1. Show that $r(r+1)(r+2)-(r-1)r(r+1)=3r(r+1)$.

   Hence find the sum of the series $\displaystyle\sum_{r=1}^{n}r(r+1)$.

2. Show that
   $$r(r+1)(r+2)(r+3)-(r-1)r(r+1)(r+2)=4r(r+1)(r+2)$$

   Hence find the sum of the series $\displaystyle\sum_{r=1}^{n}r(r+1)(r+2)$.

3. If $S=x+2x^3+3x^5+\cdots+nx^{2n-1}$ find $S-x^2S$. Use this result to find the sum $\displaystyle\sum_{r=1}^{n}rx^{2r-1}$.

4. Find the value of A for which $(2r+1)^2-(2r-1)^2=Ar$.

   Hence find $\displaystyle\sum_{r=1}^{n}r$.

5. Find $\displaystyle\sum_{r=1}^{\infty}\left(\frac{2}{r}-\frac{2}{r+2}\right)$.

Summary

▶ You can find the sum of unfamiliar series using a set of summation formulae:

- $\displaystyle\sum_{r=1}^{n}r=\frac{1}{2}n(n+1)$

- $\displaystyle\sum_{r=1}^{n}r^2=\frac{1}{6}n(n+1)(2n+1)$

- $\displaystyle\sum_{r=1}^{n}r^3=\frac{1}{4}n^2(n+1)^2$

- $\displaystyle\sum_{r=1}^{n}r^3=\left(\sum_{r=1}^{n}r\right)^2$

▶ Split the given term into parts (by breaking up the summation across the sum or difference) and factor out any constants before applying the relevant summation formula.

▶ These formulae can be adjusted to change the start and end points of the values of r.

- You can also find the sum of a series using the method of differences.
- In some cases, it is possible to extend summations to infinite sums, because the partial sums converge.

Review exercises

1 Find $\displaystyle\sum_{r=1}^{10} (7r+3)^2$.

2 Find $\displaystyle\sum_{r=1}^{n} (2r-1)(r-5)$.

3 Show that $\displaystyle\sum_{1}^{n} \left[(r+1)!-r!\right] = r \times r!$, and hence find $\displaystyle\sum_{1}^{n} r \times r!$

Practice examination questions

1 Use the formulae for $\displaystyle\sum_{r=1}^{n} r^3$ and $\displaystyle\sum_{r=1}^{n} r^2$ to find the value of

$$\sum_{r=3}^{60} r^2(r-6)$$
      (4 marks)

AQA MFP1 June 2014

2 a Use the formulae for $\displaystyle\sum_{r=1}^{n} r^3$ and $\displaystyle\sum_{r=1}^{n} r^2$ to show that

$$\sum_{r=1}^{n} r^2(4r-3) = kn(n+1)(2n^2-1) \text{ where } k \text{ is a constant.}$$
      (5 marks)

  b Hence evaluate $\displaystyle\sum_{r=20}^{40} r^2(4r-3)$.
      (2 marks)

AQA MFP1 January 2012

3 a Show that $\displaystyle\sum_{r=1}^{n} r^3 + \sum_{r=1}^{n} r$ can be expressed in the form

    $kn(n+1)(an^2+bn+c)$ where k is a rational number and a, b and c are integers.
      (4 marks)

  b Show that there is exactly one positive integer n for which

$$\sum_{r=1}^{n} r^3 + \sum_{r=1}^{n} r = 8\sum_{r=1}^{n} r^2$$
      (5 marks)

AQA MFP1 January 2010

4 a Given that $f(r) = r^2(2r^2-1)$, show that $f(r)-f(r-1) = (2r-1)^3$.
      (3 marks)

  b Use the method of differences to show that

$$\sum_{r=n+1}^{2n} (2r-1)^3 = 3n^2(10n^2-1)$$
      (4 marks)

AQA MFP2 June 2013

5 a Given that $f(r) = \frac{1}{4}r^2(r+1)^2$, show that $f(r)-f(r-1) = r^3$.
      (3 marks)

  b Use the method of differences to show that

$$\sum_{r=n}^{2n} r^3 = \frac{3}{4}n^2(n+1)(5n+1)$$
      (5 marks)

AQA MFP2 January 2009

6 a Given that $f(r) = (r-1)r^2$, show that $f(r+1)-f(r) = r(3r+1)$.
      (3 marks)

  b Use the method of differences to find the value of

$$\sum_{r=50}^{99} r(3r+1)$$
      (4 marks)

AQA MFP2 June 2007

5 Trigonometry

Introduction

Trigonometric functions such as $\sin x$ are periodic. The equation $\sin x = \frac{1}{2}$ has one solution in a calculator, but as sin is periodic it actually must have infinitely many solutions. In this chapter, you will learn how to find all the solutions of standard trigonometric equations. These methods can be applied to many situations in which an oscillation occurs.

Recap

You will need to remember how to...

▶ Understand angles given in radians or degrees.
▶ Use your calculator to solve equations such as $\sin x = 0.4$.

Objectives

By the end of this chapter, you should know how to:

▶ Find general solutions to trigonometric equations such as $\sin x = a$, where a is a given number between -1 and 1.
▶ Find the general solution of similar equations that are given in terms of cosine or tangent.

▶ The general solution of $\cos \theta = x$ is given by:
  • $\theta = 360n° \pm \cos^{-1} x$ for any integer n, if θ and $\cos^{-1} x$ are measured in degrees
  • $\theta = 2n\pi \pm \cos^{-1} x$ for any integer n, if θ and $\cos^{-1} x$ are measured in radians
▶ The general solution of $\sin \theta = x$ is given by:
  • $\theta = 360n° + \sin^{-1} x$ and $\theta = 360n° + 180° - \sin^{-1} x$ for any integer n, if θ and $\sin^{-1} x$ are measured in degrees
  • $\theta = 2n\pi + \sin^{-1} x$ and $\theta = 2n\pi + \pi - \sin^{-1} x$ for any integer n, if θ and $\sin^{-1} x$ are measured in radians
▶ The general solution of $\tan \theta = x$ is given by:
  • $\theta = 180n° + \tan^{-1} x$ for any integer n, if θ and $\tan^{-1} x$ are measured in degrees
  • $\theta = n\pi + \tan^{-1} x$ for any integer n, if θ and $\tan^{-1} x$ are measured in radians.

In general calculators are used to find the values of the trigonometric ratios; however in some cases you should be able to recall certain ratios. These values are given in the table.

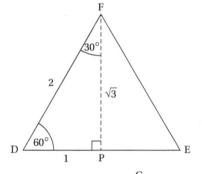

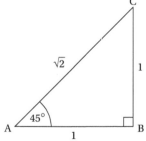

Angle (°)	Radians	Sin θ	Cos θ	Tan θ
0	0	0	1	0
30	$\dfrac{\pi}{6}$	$\dfrac{1}{2}$	$\dfrac{\sqrt{3}}{2}$	$\dfrac{1}{\sqrt{3}}$
45	$\dfrac{\pi}{4}$	$\dfrac{1}{\sqrt{2}}$	$\dfrac{1}{\sqrt{2}}$	1
60	$\dfrac{\pi}{3}$	$\dfrac{\sqrt{3}}{2}$	$\dfrac{1}{2}$	$\sqrt{3}$
90	$\dfrac{\pi}{2}$	1	0	not defined

5.1 General solutions of trigonometric equations

You already know that it is possible to solve the equation $\cos\theta = \frac{1}{2}$ in a number of ways, including using a calculator to find the solution $60°$ and then using the graphs of $y = \cos\theta$ and $y = \frac{1}{2}$ to obtain the other solutions.

When you are asked to find solutions, it is normally within a given range. When there are several solutions to find, this method can be very time-consuming and as a result tends to lead to errors.

Trigonometric functions are **periodic** (cosine and sine repeat every $360°$ and tangent repeats every $180°$). Since there is at least one solution of $\cos\theta = \frac{1}{2}$, there will in fact be infinitely many solutions of that equation.

A way to find more than one solution of trigonometric equations, or indeed to find an *infinite* number, is to use the **general solution** for that trigonometric equation. The general solution of a given trigonometric equation is one that applies to all the values of θ that satisfy the equation.

General solutions for cosine curves

When $\cos\theta = \frac{1}{2}$, you can use the graph of $y = \cos\theta$ to see that the solutions for θ are

..., $-300°$, $-60°$, $60°$, $300°$, $420°$, $660°$, $780°$, $1020°$, $1140°$, ...

which can be written as

..., $-360° + 60°$, $-60°$, $60°$, $360° - 60°$, $360° + 60°$, $720° - 60°$, $720° + 60°$, $1080° - 60°$, $1080° + 60°$, ...

Notice that for each of the solutions, $\theta = 360n° \pm 60°$ for any integer n.

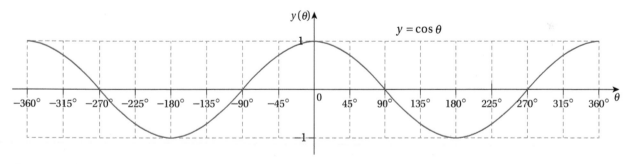

To summarise,

> The general solution of $\cos\theta = \cos\alpha$ is given by
>
> ▸ $\theta = 360n° \pm \alpha$ for any integer n, where θ and α are measured in degrees
> ▸ $\theta = 2n\pi \pm \alpha$ for any integer n, where θ and α are measured in radians.

You can use the general solution to solve any equation involving $\cos\theta = \cos\alpha$. You start by removing the $\cos\alpha$ term by finding one solution using your calculator, then substituting this value into the general solution in order to find all the other solutions.

Example 1

Find the values of θ from $0°$ to $720°$ for which $\cos\theta = \dfrac{1}{\sqrt{2}}$.

$\alpha = \cos^{-1}\left(\dfrac{1}{\sqrt{2}}\right) = 45°$

$\theta = 360n° \pm 45°$

When $n = 0$, $\theta = 45°$.

When $n = 1$, $\theta = 315°$ or $405°$.

When $n = 2$, $\theta = 675°$.

There are four solutions: $\theta = 45°, 315°, 405°$ and $675°$.

Note

Use your calculator to find the first solution ($45°$).
Substitute $\alpha = 45°$ into the general solution,
$\theta = 360n° \pm \alpha$.

Note

Use different integer values of n until you have found all the solutions in the required range.

Example 2

Find the values of θ in radians from 0 to 4π for which $\cos\theta = 0$.

$\alpha = \cos^{-1} 0 = \dfrac{\pi}{2}$

$\theta = 2\pi n° \pm \dfrac{\pi}{2}$

When $n = 0$, $\theta = \dfrac{\pi}{2}$.

When $n = 1$, $\theta = \dfrac{3\pi}{2}$ or $\dfrac{5\pi}{2}$.

When $n = 2$, $\theta = \dfrac{7\pi}{2}$.

There are four solutions: $\theta = \dfrac{\pi}{2}, \dfrac{3\pi}{2}, \dfrac{5\pi}{2}$ and $\dfrac{7\pi}{2}$.

Note

Use your calculator to find the first solution (α).
Substitute $\alpha = \dfrac{\pi}{2}$ into the general solution.

Example 3

Find the values of θ from $0°$ to $360°$ for which $\cos 5\theta = \dfrac{\sqrt{3}}{2}$.

$\alpha = \cos^{-1}\left(\dfrac{\sqrt{3}}{2}\right) = 30°$

So, with $\alpha = 30°$,

$\qquad 5\theta = 360n° \pm 30°$

$\Rightarrow \quad \theta = 72n° \pm 6°$

Therefore,

When $n = 0$, $\theta = 6°$ $\qquad\qquad$ When $n = 3$, $\theta = 210°$ or $222°$

Note

In this case, the general solution is an equation in 5θ.

Note

You can always check these values on a graphing calculator, after having selected the correct range or view window.

(continued)

(continued)

When $n = 1$, $\theta = 66°$ or $78°$ ⠀⠀⠀When $n = 4$, $\theta = 282°$ or $294°$

When $n = 2$, $\theta = 138°$ or $150°$ ⠀When $n = 5$, $\theta = 354°$

There are 10 solutions: $\theta = 6°$, $66°$, $78°$, $138°$, $150°$, $210°$, $222°$, $282°$, $294°$ and $354°$.

General solutions for sine curves

When $\sin \theta = \frac{1}{2}$, you will find from the graph of $y = \sin \theta$ that the solutions for θ are

$\ldots, -330°, -210°, 30°, 150°, 390°, 510°, 750°, \ldots,$

which can be written as

$\ldots, -360° + 30°, -180° - 30°, 30°, 180° - 30°, 360° + 30°, 540° - 30°, 720° + 30°, \ldots$

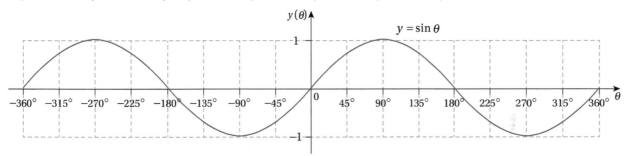

To summarise,

> **The general solution of $\sin \theta = \sin \alpha$ is given by**
>
> ► $\theta = 180n° + (-1)^n \alpha$ for any integer n, where θ and α are measured in degrees
>
> ► $\theta = n\pi + (-1)^n \alpha$ for any integer n, where θ and α are measured in radians.

You can apply this general solution in the same way you did for $\cos \theta = \cos \alpha$.

Example 4

Find the values of θ between $0°$ and $720°$ for which $\sin \theta = \dfrac{\sqrt{3}}{2}$.

$\alpha = \sin^{-1}\left(\dfrac{\sqrt{3}}{2}\right) = 60°$

$\theta = 180n° + (-1)^n 60°$

Therefore,

When $n = 0$, $\theta = 60°$

When $n = 1$, $\theta = 180° - 60° = 120°$

When $n = 2$, $\theta = 360° + 60° = 420°$

When $n = 3$, $\theta = 540° - 60° = 480°$

When $n = 4$, $\theta = 720° + 60° = 780°$

Therefore, there are four solutions: $\theta = 60°$, $120°$, $420°$ and $480°$.

Note

Find the first solution.

Note

Use the general solution, $\theta = 180n° + (-1)^n \alpha$.

Note

$\theta = 780°$ is out of the required range.

Example 5

Find the values of θ between $0°$ and $360°$ for which $\sin 3\theta = \dfrac{1}{\sqrt{2}}$.

$\alpha = \sin^{-1}\left(\dfrac{1}{\sqrt{2}}\right) = 45°$

With $\alpha = 45°$,

$3\theta = 180n° + (-1)^n 45°$

$\Rightarrow \quad \theta = 60n° + (-1)^n 15°$

Therefore,

When $n = 0$, $\theta = 15°$ When $n = 3$, $\theta = 165°$

When $n = 1$, $\theta = 45°$ When $n = 4$, $\theta = 255°$

When $n = 2$, $\theta = 135°$ When $n = 5$, $\theta = 285°$

There are six solutions: $\theta = 15°$, $45°$, $135°$, $165°$, $255°$ and $285°$.

> **Note**
>
> The general solution is an equation in 3θ.

> **Note**
>
> Divide both sides by 3 to make a general solution in terms of θ.

General solutions for tangent curves

When $\tan \theta = 1$, you can find from the graph of $y = \tan \theta$ that the solutions for θ are

$$..., -135°, 45°, 225°, 405°, ...$$

You can see that $\theta = 180n° + 45°$ for any integer n.

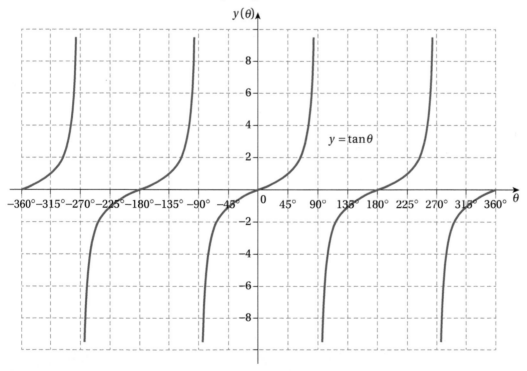

The general solution of $\tan \theta = \tan \alpha$ is given by

▶ $\theta = 180n° + \alpha$ for any integer n, where θ and α are measured in degrees

▶ $\theta = n\pi + \alpha$ for any integer n, where θ and α are measured in radians.

The general solution can be used as before.

Example 6

Find the values between 0 and 2π for which $\tan 4x = -\sqrt{3}$.

$\tan^{-1}(-\sqrt{3}) = -\dfrac{\pi}{3}$

The general solution has the form $\theta = n\pi + \alpha$, which in this case gives

$\quad 4x = \pi n - \dfrac{\pi}{3}$

$\Rightarrow \quad \theta = \dfrac{\pi}{4}n - \dfrac{\pi}{12}$

Therefore, the solutions are $\dfrac{\pi}{6}, \dfrac{5\pi}{12}, \dfrac{2\pi}{3}, \dfrac{11\pi}{12}, \dfrac{7\pi}{6}, \dfrac{17\pi}{12}, \dfrac{5\pi}{3}$ and $\dfrac{23\pi}{12}$.

Exercise 1

For Questions **1** to **4**

a find the general solution in **i** radians **ii** degrees

b find the solutions, in degrees, that lie within the interval 0° to 360°.

① $\sin\theta = \dfrac{1}{\sqrt{2}}$

② $\cos\theta = -\dfrac{1}{2}$

③ $\sin 2\theta = \dfrac{1}{2}$

④ $\tan 3\theta = 1$

⑤ Find the general solution in radians of the equation $\tan\theta = \dfrac{\sqrt{3}}{3}$.

5.2 Solving equations involving more complicated terms

Example 7

Find the general solution for x, in radians, of $\sin\left(3x + \dfrac{\pi}{3}\right) = 1$.

$\alpha = \sin^{-1}(1) = \dfrac{\pi}{2}$

General solution of $\sin x$

$\quad x = n\pi + (-1)^n\alpha$

$\quad 3x + \dfrac{\pi}{3} = n\pi + (-1)^n\dfrac{\pi}{2}$

Hence $3x = n\pi - \dfrac{\pi}{3} + (-1)^n\dfrac{\pi}{2}$

$\quad x = n\dfrac{\pi}{3} - \dfrac{\pi}{9} + (-1)^n\dfrac{\pi}{6}$

Note

Substitute the value of α into the general solution and create an equation in $\left(3x + \dfrac{\pi}{3}\right)$.

Note

You could then use this general solution for different values of n to find a range of solutions as before.

Trigonometric powers such as $\sin^2 x$ and trigonometric identities

Given an equation involving $\sin^2 x$ you should first try to find the value of the underlying trigonometrical function. So, in the case of $\sin^2 x$ you would first find the value of $\sin x$. It might be necessary to use trigonometrical identities such as $\sin^2 x + \cos^2 x = 1$.

Example 8

Question

Find the general solution for θ, in radians, of $4\cos^2 2\theta + 5\sin 2\theta = 5$.

Answer

$4\cos^2 2\theta + 5\sin 2\theta = 5$

Therefore $4(1 - \sin^2 2\theta) + 5\sin 2\theta = 5$

$$4 - 4\sin^2 2\theta + 5\sin 2\theta = 5$$

$$4\sin^2 2\theta - 5\sin 2\theta + 1 = 0$$

$$(\sin 2\theta - 1)(4\sin 2\theta - 1) = 0$$

$$\sin 2\theta = 1 \text{ or } \frac{1}{4}$$

$$\sin^{-1} 1 = \frac{\pi}{2} \text{ and } \sin^{-1}\left(\frac{1}{4}\right) = 0.253\ldots$$

General solution with equation in terms of 2θ: $2\theta = n\pi + (-1)^n \frac{\pi}{2}$
or $2\theta = n\pi + (-1)^n 0.253$.

Therefore, general solution with equation in terms of θ: $\theta = \frac{n\pi}{2} + (-1)^n \frac{\pi}{4}$
or $\theta = \frac{n\pi}{2} + (-1)^n 0.126$.

> **Note**
>
> Factorise to get the equation in terms of $\sin 2\theta$.

> **Note**
>
> Use the identity $\cos^2 \theta = 1 - \sin^2 \theta$.

> **Note**
>
> Equate each factor to zero.

Exercise 2

1. Find the general solution, in radians, of $\sin\left(2x + \frac{\pi}{4}\right) = 1$.

2. Find the general solution, in degrees, of $\cos(x + 30°) = -\frac{1}{\sqrt{2}}$.

3. a Find the general solution, in radians, of $\cos\left(3x - \frac{\pi}{3}\right) = \frac{1}{2}$.

    b Use your general solution to find the solutions in the range $0 \le x \le 4\pi$.

4. Find the general solution in degrees of $\cos^2 \theta + \sin \theta + 1 = 0$.

5. Find the general solution of each equation in   i radians   ii degrees.

    a $\tan\left(\frac{\pi}{3} - 2x\right) = 1$ b $\cos(3x - 1) = -0.2$

    c $\sin^2 4\theta = \frac{1}{2}$ d $\sin^2 3\theta + \cos 3\theta + 1 = 0$

    e $\cos 2\theta = \cos\theta - 1$

6. Find the general solution of $\sin\left(2x + \frac{\pi}{3}\right) = \cos\left(2x + \frac{\pi}{3}\right)$.

Summary

▶ Trigonometric functions are periodic, meaning that there are infinitely many solutions to many trigonometric equations; a general solution is one that applies to all the values of θ that satisfy the equation.

▶ The general solution of $\cos\theta = \cos\alpha$ is given by:
  - $\theta = 360n° \pm \alpha$ for any integer n, where θ and α are measured in degrees
  - $\theta = 2n\pi \pm \alpha$ for any integer n, where θ and α are measured in radians.
  So, if $\theta = x$ is a solution of $\cos x = y$, then the other solutions are of the form $360n \pm \theta$ or $2n\pi \pm \theta$.

▶ The general solution of $\sin\theta = \sin\alpha$ is given by:
  - $\theta = 180n° + (-1)^n\alpha$ for any integer n, where θ and α are measured in degrees
  - $\theta = n\pi + (-1)^n\alpha$ for any integer n, where θ and α are measured in radians.
  So, if $\theta = x$ is a solution of $\sin x = y$, then the other solutions are of the form $360n + \theta$ or $360n + 180 - \theta$, $2n\pi + \theta$ or $2n\pi \pm \theta$.

▶ The general solution of $\tan\theta = \tan\alpha$ is given by:
  - $\theta = 180n° + \alpha$ for any integer n, where θ and α are measured in degrees
  - $\theta = n\pi + \alpha$ for any integer n where θ and α are measured in radians.
  So, if $\theta = x$ is a solution of $\tan x = y$, then the other solutions are of the form $180n + \theta$ or $n\pi + \theta$.

▶ To find a general solution for a given trigonometric equation, first you find one solution, and then you substitute this value into the appropriate general solution to find the rest of the solutions within the required range.

▶ When the trigonometric equations are more complicated, first express the equation in terms of $\sin\theta$, $\cos\theta$ or $\tan\theta$, as appropriate, then use the corresponding general solution and manipulate as required to get an equation in terms of θ.

▶ The general solution of $\cos\theta = x$ is given by:
  - $\theta = 360n° \pm \cos^{-1}x$ for any integer n, if θ and $\cos^{-1}x$ are measured in degrees
  - $\theta = 2n\pi \pm \cos^{-1}x$ for any integer n, if θ and $\cos^{-1}x$ are measured in radians.

▶ The general solution of $\sin\theta = x$ is given by:
  - $\theta = 360n° + \sin^{-1}x$ and $\theta = 360n° + 180° - \sin^{-1}x$ for any integer n, if θ and $\sin^{-1}x$ are measured in degrees
  - $\theta = 2n\pi + \sin^{-1}x$ and $\theta = 2n\pi + \pi - \sin^{-1}x$ for any integer n, if θ and $\sin^{-1}x$ are measured in radians.

▶ The general solution of $\tan\theta = x$ is given by:
  - $\theta = 180n° + \tan^{-1}x$ for any integer n, if θ and $\tan^{-1}x$ are measured in degrees
  - $\theta = n\pi + \tan^{-1}x$ for any integer n, if θ and $\tan^{-1}x$ are measured in radians.

Review exercises

1 Find the general solution, in degrees, for the equation $\sin(3x + 20) = \sin 50$

2 Find the general solution of the equation $\sin\left(4x - \frac{\pi}{6}\right) = \frac{1}{2}$ giving your answer in terms of π.

3 Find the general solution of the equation $\cos 4\left(x - \frac{\pi}{6}\right) = \frac{1}{2}$ giving your answer in terms of π.

4 Find, in radians, the general solution of the equation $\cos\left(\frac{x}{3} + \frac{\pi}{6}\right) = \frac{\sqrt{3}}{2}$ giving your answer in terms of π.

5 Find the general solution of the equation $\tan\left(3x - \frac{\pi}{4}\right) = 1$ giving your answer in terms of π.

Practice examination questions

1 **a** Find the general solution of the equation $\sin\left(2x + \frac{\pi}{4}\right) = \frac{\sqrt{3}}{2}$ giving your answer in terms of π. (6 marks)

   b Use your general solution to find the exact value of the greatest solution of this equation which is less than 6π. (2 marks)

   AQA MFP1 January 2013

2 Find the general solution of the equation $\sin\left(4x + \frac{\pi}{4}\right) = 1$. (4 marks)

   AQA MFP1 January 2010

3 **a** Find the general solution of the equation $\cos\left(\frac{5}{4}x - \frac{\pi}{3}\right) = \frac{\sqrt{2}}{2}$ giving your answer for x in terms of π. (5 marks)

   b Use your general solution to find the **sum** of all the solutions of the equation $\cos\left(\frac{5}{4}x - \frac{\pi}{3}\right) = \frac{\sqrt{2}}{2}$ that lie in the interval $0 \leq x \leq 20\pi$. Give your answer in the form $k\pi$, stating the exact value of k. (4 marks)

   AQA MFP1 June 2014

4 Find the general solution of the equation $\sin\left(4x - \frac{2\pi}{3}\right) = -\frac{1}{2}$ giving your answer in terms of π. (6 marks)

   AQA MFP1 January 2011

5 Find the general solution of each of the following equations:

   a $\tan\left(\frac{x}{2} - \frac{\pi}{4}\right) = \frac{1}{\sqrt{3}}$ (4 marks)

   b $\tan^2\left(\frac{x}{2} - \frac{\pi}{4}\right) = \frac{1}{3}$ (3 marks)

   AQA MFP1 January 2012

6 Calculus

Introduction

Calculus can help to explain rates of change such as speed. If you are driving at 140 km/h, you will not realistically be driving 140 km each hour. The reading taken is not your average speed over an hour, nor over a minute, and not even over one second. Instead, calculus helps to explain that your *instantaneous* speed is the limit of smaller and smaller measurements of your average speed.

Recap

You will need to remember how to . . .

▶ Find the derivative of a function, f(x), and understand that this is the gradient of the tangent to the graph of $y = f(x)$ at a point, and interpret it as a rate of change.
▶ Differentiate polynomials.
▶ Differentiate functions in the form x^n, and related sums and differences, where n is a rational number.
▶ Understand the gradient of a tangent as a limit.
▶ Apply differentiation to gradients, tangents and normal lines, maxima, minima and stationary points, and increasing and decreasing functions.
▶ Use the notation of, and calculations of, limits.
▶ Integrate polynomials and x^n, where n is a rational number not equal to −1, and related sums and differences.
▶ Evaluate definite integrals and interpret them as the area under a curve.

Objectives

By the end of this chapter, you should know how to:

▶ Calculate the gradient of a tangent to a curve at a point, using first principles.
▶ Use a small change in x to estimate the corresponding change in y.
▶ Connect rates of change in order to find a desired rate of change.
▶ Solve some integrals with infinite domains.

• •

6.1 Gradient of a tangent to a curve

You should know from previous studies that the **derivative** of a function at a point P is the gradient (or slope) of the tangent of the graph at that point.

If P is a point on the curve $y = f(x)$, where f(x) is a simple polynomial, then it is possible to estimate the gradient of the tangent of the curve at point P using **first principles**. Choose a point on the curve, Q, near to point P. The gradient of the line segment PQ is an approximation of the gradient of the tangent to the curve at P.

To measure the gradient of the segment PQ, you divide the change in the y-coordinate by the change in the x-coordinate. In differentiation from first

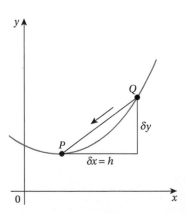

principles in previous studies, you used δy to denote the small change in the y-coordinate, and δx to denote the small change in the x-coordinate, so the gradient of $PQ = \frac{\delta y}{\delta x}$.

The difference in x-coordinate from P to Q is often represented by h. In the limit, as h approaches 0, in other words, as Q slides closer and closer to P, the gradient of PQ approaches the gradient of the tangent at P.

> If P is a point on the curve $y = f(x)$, where $f(x)$ is a simple polynomial, and Q is a point on the curve near to P, the gradient of the curve at P is gradient
> $PQ = \lim\limits_{h \to 0} \frac{\delta y}{h}$, where h is the difference in the x-coordinates from P to Q.

It is possible to use this fact to differentiate without using the standard formula $\frac{d}{dx}(x^n) = nx^{n-1}$.

Example 1 concerns a specific point on a curve, and Example 2 shows how the method can be used for a general point on a curve.

Example 1

A curve C has equation $y = x^3$.

a   Find the equation of the gradient of the line PQ, where P is the point $(2, 8)$, and Q is the point on C with x-coordinate $2 + h$.

b   Hence, find the gradient of the tangent to the curve C at the point P.

a Q is the point $(2 + h, (2+h)^3)$.

The gradient of PQ is

$$\frac{\delta y}{\delta x} = \frac{(2+h)^3 - 8}{h} = \frac{(8 + 12h + 6h^2 + h^3) - 8}{h} = \frac{12h + 6h^2 + h^3}{h} = 12 + 6h + h^2$$

Note

h represents a small increase in the x-coordinate of P, where P is the point $(2, 8)$.

b   As Q approaches P, h approaches 0, and so the gradient of the tangent at P is

$$\lim_{h \to 0} \frac{\delta y}{\delta x} = \lim_{h \to 0}(12 + 6h + h^2) = 12$$

Example 2

Find the gradient of $y = x^2 - 3x$ at the point (x_0, y_0) using differentiation from first principles.

Let P be the point $(x_0, x_0^2 - 3x_0)$ and Q be the point $(x_0 + h, (x_0 + h)^2 - 3(x_0 + h))$.

The gradient of PQ is

$$\frac{\delta y}{\delta x} = \frac{\left[(x_0 + h)^2 - 3(x_0 + h)\right] - (x_0^2 - 3x_0)}{h}$$

$$= \frac{(x_0^2 + 2x_0 h + h^2 - 3x_0 - 3h) - (x_0^2 - 3x_0)}{h}$$

$$= \frac{2x_0 h + h^2 - 3h}{h} = 2x_0 + h - 3$$

$$\frac{dy}{dx} = \lim_{h \to 0}\frac{\delta y}{\delta x} = \lim_{h \to 0}(2x_0 + h - 3) = 2x_0 - 3$$

Exercise 1

1 A curve C has equation $y = x^2 - 4x + 7$.

The points A and B on the curve have x-coordinates 2 and $2 + h$, respectively.

a In terms of h, find the gradient of the line AB.

b Hence find the gradient of the curve at A.

2 A curve C has the equation $f(x) = 2x^3 + 3x^2 - 2x - 4$.
Express $f(x + h) - f(x)$ in terms of h, and hence find the gradient of the curve at the point $x = 3$.

3 A curve has equation $y = x^2 + 3$.

a Find the gradient of the line passing through the point $(4, 19)$ and the point on the curve for which $x = 4 + h$. Give your answer in terms of h.

b Show how your answer can be used to find the gradient of the curve at the point $(4, 19)$. State the value of this gradient.

4 A curve has equation $y = x^3 - 27x$.

The point P on the curve has coordinates $(3, -54)$. The point Q on the curve has x-coordinate $3 + h$.

a Show that the gradient of the line PQ is $(h^2 + 9h)$.

b Explain how the result of part **a** can be used to show that P is a stationary point on the curve.

6.2 Rates of change

You know from previous studies that the gradient of a line, or the derivative of a function, is a **rate of change**. For example, $\frac{dy}{dx}$ is the rate of change of y with respect to x.

How a small change in one variable affects another

You can use the value of $\frac{dy}{dx}$ to estimate the change in y that occurs after a small change in x. To do this, you are assuming that the tangent line approximates the true value of y.

> If δx represents a small change in x, and δy represents the corresponding change in y, then $\frac{dy}{dx} \approx \frac{\delta y}{\delta x}$.

It follows, for example, that if you know the value of δx, you can estimate the value of δy using $\delta y \approx \frac{dy}{dx} \delta x$.

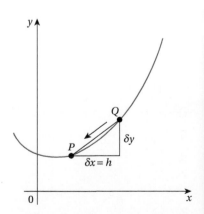

Example 3

Question

Find an approximate value of $(100\,002)^{\frac{1}{5}}$.

Answer

Let $y = x^{\frac{1}{5}}$.

$\dfrac{dy}{dx} = \dfrac{x^{-\frac{4}{5}}}{5}$

$\dfrac{\delta y}{\delta x} \approx \dfrac{dy}{dx}$

Therefore, $\delta y \approx \dfrac{dy}{dx}\,\delta x$

$\delta y \approx \dfrac{x^{-\frac{4}{5}}}{5}\,\delta x$

Therefore $\delta y \approx \dfrac{1}{5} \times \dfrac{1}{10\,000} \times 2 = 0.00004$.

Hence $(100\,002)^{\frac{1}{5}} \approx 10.00004$.

You can also use small changes in x to estimate possible errors in y caused by small errors in the value of x. (You will see this in question **8** of Exercise 2.)

> **Note**
>
> You know that $100\,000^{\frac{1}{5}} = 10$, so you will look at the value of $x^{\frac{1}{5}}$ near $x = 100\,000$.

> **Note**
>
> Now use that $x = 100\,000$ and $\delta x = 2$.

> **Note**
>
> Add δy to the value of y that corresponds to x.

Rates of change of connected variables

Connecting rates of change is a method of calculating a rate of change by connecting it to two known rates of change using the chain rule.

For instance, if you fill a cylindrical glass of water from a tap, then the rate at which the height of the water changes depends on how narrow the glass is. You can use calculus to calculate the rate of change of the height of the water (height per minute), provided you know the volume of water released by the tap each minute (the rate of water released) and the radius of the cylindrical glass. You use the relationship between the volume of the cylinder and its height to create a third rate of change, which you can do because you know the radius of the glass. Using each connected rate of change and the chain rule allows you to find the rate of change of the height of the water in the glass.

> If y is a function of x, then you know from the chain rule that $\dfrac{dy}{dx} = \dfrac{dy}{dt} \times \dfrac{dt}{dx}$.
> The variables y and x relate to the problem you are trying to solve.

Connecting rates of change allows you to solve many problems that involve rates of change:

▶ First, identify the connected variables that are changing.
▶ Identify the relationship between the variables that are changing, and use this to create a third rate of change.
▶ Substitute these into the chain rule to find the desired rate of change.

Example 4

A perfectly spherical weather balloon is inflated at a rate of 0.1 m^3 per minute.
How fast is the radius growing when the balloon has radius

a 0.2 m **b** 1 m

Let $V =$ volume of balloon in cubic metres, after t minutes.

Let $r =$ radius of balloon in metres.

> **Note**
>
> Identify the relationship between the changing variables. Here, the changing variables are the volume and radius of the spherical balloon, so use the relationship between the volume and the radius of a sphere, that is, the formula for calculating the volume of a sphere given its radius.

$$V = \frac{4}{3}\pi r^3$$

> **Note**
>
> Your aim is to use the chain rule in the form $\frac{dV}{dt} = \frac{dV}{dr} \times \frac{dr}{dt}$. You want to find $\frac{dr}{dt}$ (rate of change of radius per minute), and you know $\frac{dV}{dt}$ (the rate of change of the volume per minute), so you need to use the relationship between the volume of the balloon and its radius to find $\frac{dV}{dr}$ (rate of change of volume given the change in radius).

> **Note**
>
> First you should identify the connected variables that are changing.

Differentiate the equation for volume of a sphere: $\frac{dV}{dr} = 4\pi r^2$.

$$\frac{dV}{dt} = 0.1 \text{ (given)}$$

$$\frac{dV}{dt} = \frac{dV}{dr} \times \frac{dr}{dt}$$

$$0.1 = 4\pi r^2 \frac{dr}{dt} \qquad [1]$$

a $r = 0.2$

$$0.1 = 4\pi(0.04)\frac{dr}{dt}$$

$$\frac{dr}{dt} = \frac{5}{8\pi}$$

Therefore, the radius is increasing at $\frac{5}{8\pi}$ m per minute.

b $r = 1$

$$0.1 = 4\pi \frac{dr}{dt}$$

$$\frac{dr}{dt} = \frac{1}{40\pi}$$

The radius is increasing at $\frac{1}{40\pi}$ m per minute.

> **Note**
>
> Substitute in the rates you have. Now you have an equation that can be used for the given values of r.

> **Note**
>
> Since the volume of a sphere is proportional to the cube of its radius, for a given increase in volume you would expect the radius of a sphere to increase more slowly when the radius of the sphere is larger.

These ideas can be applied to more complicated examples involving trigonometry. Provided that there is an underlying relationship between two variables, you may be able to exploit it.

Exercise 2

1. By considering the derivative of the graph $y = x^{\frac{3}{2}}$ when $x = 4$, estimate the value of $4.01^{\frac{3}{2}}$.

2. By considering the derivative of the graph $y = \sqrt{x}$ when $x = 9$, estimate the value of $\sqrt{8.94}$.

3. If $q = (2r + 3)^3$ find $\dfrac{dq}{dt}$ when $r = 1$ given that $\dfrac{dr}{dt} = 3$.

4. Find an approximate value of $(81.1)^{\frac{1}{4}}$.

5. The area of a circle is increasing at 3 cm²/s. How quickly is the radius increasing when its radius is 8 cm?

6. The volume of a spherical balloon is increasing at a rate of 2 cm³/s. Find the rate at which the surface area of the spherical balloon is increasing when the radius is 2 cm.

7. Each side of a square is increasing at a rate of 0.4 cm/s. Find the rate at which the area of the square is increasing when each side is of length 3 cm.

8. The radius of a sphere is measured to be 3 cm and hence its volume is calculated as 36π cm³. The measurement of the radius is correct to the nearest 0.1 cm. Find the maximum error in the calculated volume.

6.3 Improper integrals

You already know that integration can be used to find the area under a curve using the general expression $\int_b^a f(x)\,dx$, where a and b are the real number boundaries of the interval, and $f(x)$ is known as the **integrand**. In this case, both the function, $f(x)$, and the interval have finite boundaries. However, there are cases where either the function or the interval do not have finite boundaries. In these cases, you have an improper integral.

An improper integral is one that has either
- A limit of integration of $\pm\infty$, or
- An integrand (function) that is infinite at either of its limits of integration, or between these limits.

It is often possible to evaluate improper integrals, though sometimes you will discover that the integral cannot be found.

When the limit of integration is $\pm\infty$

In this example of an improper integral, the integrand, $f(x)$, has no finite boundaries since there is a limit of integration of $\pm\infty$, where ∞ represents infinity.

However, you can treat it as if there is an upper/lower boundary by replacing $\pm\infty$ with n, for example. This allows you to find the limit of the integral as $n \to \pm\infty$. When this limit is **finite**, the integral *can* be found. When this limit is **not finite**, the integral *cannot* be found.

Example 5

Determine $\displaystyle\int_{1}^{\infty} \frac{1}{x^2}\, dx$.

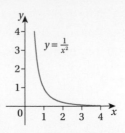

$\displaystyle\int_{1}^{\infty} \frac{1}{x^2}\, dx = \lim_{n\to\infty}\int_{1}^{n} \frac{1}{x^2}\, dx.$

$\displaystyle = \lim_{n\to\infty}\left[-\frac{1}{x} \right]_{1}^{n} = \lim_{n\to\infty}\left(-\frac{1}{n} + 1 \right)$

As $n \to \infty,\ \dfrac{1}{n} \to 0$, which gives $\displaystyle\lim_{n\to\infty}\left(-\frac{1}{n} + 1 \right) = 1.$

$\displaystyle\int_{1}^{\infty} \frac{1}{x^2}\, dx = 1$

> **Note**
>
> The upper limit is ∞, so replace it with n and find the limit of the integral as $n \to \infty$.

> **Note**
>
> This shows that the area under the curve $y = \dfrac{1}{x^2}$ is finite even though the boundary is infinitely long.

Example 6

Determine $\displaystyle\int_{1}^{\infty} x^{-\frac{1}{4}}\, dx$.

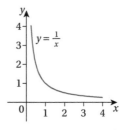

$\displaystyle\int_{1}^{\infty} x^{-\frac{1}{4}}\, dx = \lim_{n\to\infty}\int_{1}^{n} x^{-\frac{1}{4}}\, dx$

$\displaystyle = \lim_{n\to\infty}\left[-4x^{\frac{3}{4}} \right]_{1}^{n}$

$\displaystyle = \lim_{n\to\infty}\left(-4n^{\frac{3}{4}} + 4 \right)$

This is not finite since $\displaystyle\lim_{n\to\infty}\left(-4n^{\frac{3}{4}} \right) = \infty$.

Therefore, the integral does not exist.

> **Note**
>
> The upper limit is ∞, so replace it with n.

> **Note**
>
> This shows that the area under $y = \dfrac{1}{x}$ is not finite although the curve looks very similar to $y = \dfrac{1}{x^2}$, which does have a finite area.

When the integrand is infinite

In this example of an improper integral, the integrand is infinite at one of its limits of integration, or between these limits, because there is a value for which the function becomes infinite.

If you replace one of the limits of integration with p, for example, then you can find the limit of the integral as p tends to the value of the limit it has replaced. You can see from Example 8 that the area under a curve can be finite (bounded), even if one of its boundaries is infinite.

Example 7

Question
Answer

Determine $\displaystyle\int_0^1 \frac{1}{\sqrt{x}}\,dx$.

$$\int_0^1 \frac{1}{\sqrt{x}}\,dx = \lim_{p\to 0}\int_p^1 \frac{1}{\sqrt{x}}\,dx$$

$$= \lim_{p\to 0}\left[2x^{\frac{1}{2}}\right]_p^1 = \lim_{p\to 0}(2-2\sqrt{p})$$

Since $\displaystyle\lim_{p\to 0} 2\sqrt{p}=0$, $\displaystyle\lim_{p\to 0}(2-2\sqrt{p})=2$.

Therefore, $\displaystyle\int_0^1 \frac{1}{\sqrt{x}}\,dx = 2$.

> **Note**
>
> This is an improper integral since the integrand, $\frac{1}{\sqrt{x}}$, is infinite when $x=0$. So, you need to replace the *lower limit* with p.

Exercise 3

Find the value, where it exists, of each of these.

1 $\displaystyle\int_0^1 \frac{1}{x^{\frac{1}{3}}}\,dx$

2 $\displaystyle\int_0^1 \frac{1}{x^{\frac{3}{2}}}\,dx$

3 $\displaystyle\int_0^\infty \frac{1}{x^{\frac{1}{2}}}\,dx$

4 $\displaystyle\int_0^\infty \frac{1}{x^{\frac{4}{3}}}\,dx$

Summary

▶ If P is a point on the curve $y=f(x)$, where $f(x)$ is a simple polynomial, and Q is a point on the curve near P, the gradient of the curve at P is gradient $PQ = \lim_{h\to 0}\frac{\delta y}{h}$, where h is the difference in x-coordinate from P to Q. You can use this to find the gradient of the tangent to a point on a curve using first principles.

▶ If δx represents a small change in x, and δy represents the corresponding change in y, then $\frac{dy}{dx}\approx\frac{\delta y}{\delta x}$.

▶ If y is an integrand of x, then you know that $\frac{dy}{dt}=\frac{dy}{dx}\times\frac{dx}{dt}$. You can use this to connect rates of change in order to find a desired rate of change.

▶ An improper integral is one that has either
  • A limit of integration of $\pm\infty$, or
  • An integrand (function) that is infinite at one of its limits of integration, or between these limits.

▶ When the limit of integration is $\pm\infty$, you can find the limit of the integral as $n\to\pm\infty$ by replacing ∞ with n; when this limit is **finite**, the integral *can* be found. When this limit is **not finite**, the integral *cannot* be found.

▶ It is possible to solve some integrals with infinite domains; indeed, you can sometimes find the total area bounded by a curve, even though the length of that curve is infinite.

Review exercises

1 Curve C has equation $y = 2x^2 - x + 1$.

The points A and B on the curve have x-coordinates 3 and $3 + h$ respectively.

  a In terms of h, find the gradient of the line AB.

  b Hence find the gradient of the curve at A.

2 The curve C has the equation $f(x) = x^2 - 4$. Express $f(x + h) - f(x)$ in terms of h, and hence find the gradient of the curve at the point $x = 3$.

3 A sphere is being inflated at a rate of 4 cm³/s. At the instant when the sphere has radius 3 cm, find the rate of increase of the surface area of the sphere.

4 Sand is poured on the ground at a rate of 20 cm³/s. It forms a circular cone. The volume of the cone is $\frac{1}{4}\pi r^3$ when its height is r cm. Find the rate at which the height of the cone is increasing when its height is 8 cm.

5 Either calculate the integral given, or state why it does not exist.

$$\int_1^\infty \frac{1}{x^{\frac{5}{2}}}\,dx$$

Practice examination questions

1 A curve C has equation $y = x(x + 3)$.

  a Find the gradient of the line passing through the point $(-5, 10)$ and the point on C with x-coordinate $-5 + h$. Give your answer in its simplest form.   (3 marks)

  b Show how the answer to part **a** can be used to find the gradient of the curve C at the point $(-5, 10)$. State the value of this gradient.   (2 marks)

AQA MFP1 June 2014

2 **a** Expand $(5 + h)^3$.   (1 mark)

  b A curve has equation $y = x^3 - x^2$.

    i Find the gradient of the line passing through the point $(5, 100)$ and the point on the curve for which $x = 5 + h$. Give your answer in the form $p + qh + rh^2$ where p, q and r are integers.   (4 marks)

    ii Show how the answer to part **b i** can be used to find the gradient of the curve at the point $(5, 100)$. State the value of this gradient.

(2 marks)

AQA MFP1 June 2011

3 A curve has equation $y = x^3 - 12x$.

The point A on the curve has coordinates $(2, -16)$.

The point B on the curve has x-coordinate $2 + h$.

  a Show that the gradient of the line AB is $6h + h^2$.   (4 marks)

  b Explain how the result of part **a** can be used to show that A is a stationary point on the curve.   (2 marks)

AQA MFP1 June 2010

④ Show that the improper integral $\int_{25}^{\infty} \frac{1}{x\sqrt{x}}\,\mathrm{d}x$ has a finite value and find that value. (4 marks)

AQA MFP1 January 2013

⑤ Show that only one of the following improper integrals has a finite value, and find that value:

a $\int_{8}^{\infty} x^{-\frac{2}{3}}\,\mathrm{d}x$;

b $\int_{8}^{\infty} x^{-\frac{4}{3}}\,\mathrm{d}x$. (5 marks)

AQA MFP1 January 2012

⑥ a Explain why $\int_{0}^{\frac{1}{16}} x^{-\frac{1}{2}}\,\mathrm{d}x$ is an improper integral. (1 mark)

b For each of the following improper integrals, find the value of the integral **or** explain briefly why it does not have a value:

i $\int_{0}^{\frac{1}{16}} x^{-\frac{1}{2}}\,\mathrm{d}x$; (3 marks)

ii $\int_{0}^{\frac{1}{16}} x^{-\frac{5}{4}}\,\mathrm{d}x$. (3 marks)

AQA MFP1 January 2010

⑦ An oil slick is formed on a lake by an underwater leak of a light grade of refined oil. The oil is leaking at the rate of 2000 cm³ per minute. Assume that the oil forms a circular slick on top of the water. The height of the slick is 3×10^{-6} m.

a Write down an equation that expresses the volume of the slick V, measured in cubic metres, in terms of the radius r of the slick, measured in metres. (1 mark)

b Find the rate at which the radius of the slick is increasing when its radius is 1 m. (5 marks)

Introduction

Matrices store mathematical information in a concise form. For example, they can store the coordinates of the vertices of a shape, and any linear transformation in two or three dimensions can be presented using a matrix. In this chapter, the focus is on 2×2 matrices but the ideas presented could be used for many different applications, such as programming a video game to calculate the reflection of an object in water.

Objectives

By the end of this chapter, you should know how to:

▶ Use the matrix algebra of 2×2 matrices and some 3×3 matrices.

▶ Use matrices to represent linear transformations of two dimensions ($\mathbf{R}_2$).

▶ Decide how the determinant of a 2×2 matrix impacts the scale factor of a transformation.

▶ Explain the difference between lines of invariant points and invariant lines.

▶ Recognise a variety of matrices that represent linear transformations, including specifically shears parallel to the x- or y-axis.

Recap

You will need to remember how to...

▶ Work with vectors in two dimensions, including the different notation, algebraic operations, position vectors and scalar products.

▶ Know that $\mathbf{i}$ is the unit vector $\begin{pmatrix} 1 \\ 0 \end{pmatrix}$ and $\mathbf{j}$ is the unit vector $\begin{pmatrix} 0 \\ 1 \end{pmatrix}$.

▶ Recognise and carry out transformations such as rotations, reflections, stretches and shears.

7.1 Introduction to matrices

A **matrix** stores mathematical information in a concise way. The information is written down in a rectangular array of rows and columns of terms, called **elements** or **entries**, each of which has its own precise position in the array.

For example, $\begin{pmatrix} 4 \\ 8 \\ 7 \end{pmatrix}$ is a matrix, but its meaning depends on the context.

This matrix could represent a vector, meaning $4\mathbf{i} + 8\mathbf{j} + 7\mathbf{k}$. In football, it could represent the number of goals scored by three different clubs. In a shop, it could represent the number of packets of three different items bought.

The order of a matrix

The order of a matrix describes its size and shape. For example, the matrix $\begin{pmatrix} 6 & -2 & 7 \\ 4 & 3 & -5 \end{pmatrix}$ has order 2×3, since its elements are arranged in two rows and three columns.

When stating the order of a matrix, you must *always give the number of rows first*, followed by the number of columns.

$\begin{pmatrix} 4 \\ 8 \\ 7 \end{pmatrix}$ is a **column matrix** and has order 3×1, since its elements are arranged

in three rows and only one column.

The matrix $(4 \ 8 \ 7)$ has order 1×3 and is a **row matrix**.

When the number of rows and the number of columns are equal, the matrix is called a **square matrix**.

Note that $(4, 8, 7)$ with the numbers separated by commas is a point; $(4 \ 8 \ 7)$ with no commas is a matrix.

Addition and subtraction of matrices

Only when two matrices are of the *same order* can you add them or subtract them.

To add two matrices of the same order, you proceed as follows, element by element:

$$\begin{pmatrix} a & b & c \\ d & e & f \\ g & h & i \end{pmatrix} + \begin{pmatrix} p & q & r \\ s & t & u \\ v & w & x \end{pmatrix} = \begin{pmatrix} a+p & b+q & c+r \\ d+s & e+t & f+u \\ g+v & h+w & i+x \end{pmatrix}$$

You subtract two matrices of the same order in a similar way.

You *cannot* evaluate $\begin{pmatrix} a \\ b \end{pmatrix} + \begin{pmatrix} c & d \\ e & f \end{pmatrix}$ because the matrices are *not of the same order*.

Multiplication of matrices by a scalar

To multiply a matrix by, for example, k, you multiply *every* element of the matrix by k. Hence, you have

$$k \begin{pmatrix} a & b & c \\ d & e & f \\ g & h & i \end{pmatrix} = \begin{pmatrix} ka & kb & kc \\ kd & ke & kf \\ kg & kh & ki \end{pmatrix}$$

Example 1

Find $3\mathbf{A} + 2\mathbf{B}$ when $\mathbf{A} = \begin{pmatrix} 4 & 7 & -1 \\ 8 & 1 & 5 \end{pmatrix}$ and $\mathbf{B} = \begin{pmatrix} 3 & 2 & 4 \\ -1 & -3 & 2 \end{pmatrix}$.

$$3\mathbf{A} + 2\mathbf{B} = 3 \begin{pmatrix} 4 & 7 & -1 \\ 4 & 1 & 5 \end{pmatrix} + 2 \begin{pmatrix} 3 & 2 & 4 \\ -1 & -3 & 2 \end{pmatrix}$$

$$3\mathbf{A} + 2\mathbf{B} = \begin{pmatrix} 12 & 21 & -3 \\ 24 & 3 & 15 \end{pmatrix} + \begin{pmatrix} 6 & 4 & 8 \\ -2 & -6 & 4 \end{pmatrix}$$

$$3\mathbf{A} + 2\mathbf{B} = \begin{pmatrix} 18 & 25 & 5 \\ 22 & -3 & 19 \end{pmatrix}$$

> **Note**
>
> You simply multiply every element of each of the matrices by the relevant scalar.

Multiplying matrices

Not all pairs of matrices can be multiplied together. To allow multiplication, the orders of the two matrices concerned must conform to this rule:

The number of columns in the first matrix must be the same as the number of rows in the second matrix.

For example, if the first matrix has order 3×3, the second must have order $3 \times n$, as in the case of **A** and **B** below, which you will multiply together:

$$A = \begin{pmatrix} 2 & 3 & 1 \\ 0 & -2 & 3 \\ 0 & 2 & 3 \end{pmatrix}, \quad B = \begin{pmatrix} 1 & 2 & 0 \\ 1 & -2 & 1 \\ 0 & 2 & 1 \end{pmatrix}$$

To multiply **A** by **B**, you start by taking the first row of matrix **A**, (2 3 1), and the first column of matrix **B**, $\begin{pmatrix} 1 \\ 1 \\ 0 \end{pmatrix}$.

You then multiply the first element of the row by the first element of the column, the second element of the row by the second element of the column, and the third element of the row by the last element of the column. You then add up these three products.

This gives the element in the top left-hand corner of the matrix **AB**, which is $2 \times 1 + 3 \times 1 + 1 \times 0 = 5$.

So, you have

$$AB = \begin{pmatrix} 5 & ? & ? \\ ? & ? & ? \\ ? & ? & ? \end{pmatrix}$$

Next, you take the second row of matrix **A**, (0 −2 3), and the first column of matrix **B**, $\begin{pmatrix} 1 \\ 0 \\ 0 \end{pmatrix}$.

Again, you multiply each element of the row by the corresponding element of the column and add up the products.

This gives the second element of the first column of matrix **AB**, which is $0 \times 1 - 2 \times 1 + 3 \times 0 = -2$.

So, now you have

$$AB = \begin{pmatrix} 5 & ? & ? \\ -2 & ? & ? \\ ? & ? & ? \end{pmatrix}$$

You repeat the procedure on the second and third columns of matrix **B**, eventually obtaining

$$AB = \begin{pmatrix} 5 & 0 & 4 \\ -2 & 10 & 1 \\ 2 & 2 & 5 \end{pmatrix}$$

It is important to note that the product of two matrices depends on which one comes first:

$$AB \neq BA$$

Therefore, you must ensure that you write the matrices in the correct sequence.

As you would expect from scalar multiplication, A^2 means $A \times A$ and A^n is defined in a similar way for any natural number n. A^n can exist only if A is a square matrix.

Determinant of a matrix

Only a square matrix has a determinant. In this chapter you look only at determinants of 2×2 matrices.

The 2×2 determinant of matrix $\begin{pmatrix} a & b \\ c & d \end{pmatrix}$ is written as $\begin{vmatrix} a & b \\ c & d \end{vmatrix}$ and is calculated as $ad - bc$.

For example,

$$\begin{vmatrix} 3 & 4 \\ 7 & 8 \end{vmatrix} = 3 \times 8 - 4 \times 7 = 24 - 28 = -4$$

Determinant of the product of two matrices

The determinant of the product AB is the same as the product of the determinant of A and that of B:

$$\det(AB) = \det A \times \det B$$

Exercise 1

1. The matrices A and B are given by

$$A = \begin{pmatrix} 4 & 2 \\ 1 & 5 \end{pmatrix}, B = \begin{pmatrix} 1 & -1 \\ 4 & 3 \end{pmatrix}$$

   The matrix $M = A + 3B$. Find M.

2. Evaluate PQ and QP, where

$$P = \begin{pmatrix} 6 & 4 \\ 2 & 3 \end{pmatrix} \quad \text{and} \quad Q = \begin{pmatrix} 1 & -2 \\ 2 & 3 \end{pmatrix}$$

   What do you conclude from your results, and why has it happened?

3. The matrices A and B are defined by

$$A = \begin{pmatrix} 0 & k \\ 1 & 1 \end{pmatrix}, B = \begin{pmatrix} -1 & 0 \\ 0 & -1 \end{pmatrix}$$

   a. Calculate $B^2 - A^2$, and find the value(s) of k for which $\det(B^2 - A^2) = 0$.

   b. Calculate $(B + A)(B - A)$.

4. Given that $X = \begin{pmatrix} 1 & 2 \\ 3 & 0 \end{pmatrix}$, find X^2. Show that $X^3 - 7X = nI$ for some integer n, and state the value of n.

5 The matrices **A** and **B** are given by

$$\mathbf{A} = \begin{pmatrix} 0 & 3 \\ 3 & 0 \end{pmatrix}, \mathbf{B} = \begin{pmatrix} 3 & 0 \\ 0 & -3 \end{pmatrix}$$

a Calculate the matrix **AB**.

b Find $\mathbf{A}^2$.

c Show that $(\mathbf{AB})^2 \neq \mathbf{A}^2\mathbf{B}^2$.

7.2 Transformations

Using matrices to represent a linear transformation

A **linear translation** T is a transformation that can be represented by a corresponding matrix **M**. Each linear transformation in two dimensions can be fully explained by describing to where the points $(1, 0)$ and $(0, 1)$ are transformed.

For example, $T(r, s) = rT(1, 0) + sT(0, 1)$.

To find the image of (x, y) under the transformation use the equation

$$\mathbf{M}\begin{pmatrix} x \\ y \end{pmatrix} = \begin{pmatrix} x_1 \\ y_1 \end{pmatrix} \text{ to obtain the image } (x_1, y_1).$$

In two dimensions, you would represent T by the matrix

$$\mathbf{M}\begin{pmatrix} a & b \\ c & d \end{pmatrix}$$

> **Tip**
>
> Recall that **i** is the unit vector $\begin{pmatrix} 1 \\ 0 \end{pmatrix}$ and **j** is the unit vector $\begin{pmatrix} 0 \\ 1 \end{pmatrix}$.

Hence, to find, under T, the image of the point with position vector **i**, you calculate

$$\begin{pmatrix} a & b \\ c & d \end{pmatrix}\begin{pmatrix} 1 \\ 0 \end{pmatrix} = \begin{pmatrix} a \\ c \end{pmatrix}$$

So, under T, the image of the point $(1, 0)$ is (a, c), which you can see is the first column of **M**.

To find which type of transformation is represented by a matrix, you find the images of the vectors $(1, 0)$ and $(0, 1)$. Common linear transformations are rotations about the origin, reflections in lines through the origin, stretches and shears.

The meaning of the determinant in a linear transformation

Given a linear transformation T, consider to where the unit square is mapped.

The area of the new parallelogram $OA'B'C'$ is given by the value of $|\det(\mathbf{M})|$.

In fact, you can convince yourself of this by starting with the area of the large rectangle and subtracting the area of three triangles to calculate the area of the shaded triangle $OA'C'$.

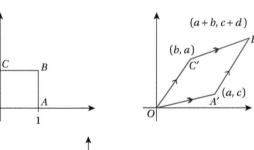

$$\text{Area } OA'C' = ad - \tfrac{1}{2}ac - \tfrac{1}{2}bd - \tfrac{1}{2}(a-b)(d-c) = ad - \tfrac{1}{2}ad - \tfrac{1}{2}bc$$

$$= \tfrac{1}{2}ad - \tfrac{1}{2}bc$$

Therefore the area of $OA'B'C' = ad - bc = \det(\mathbf{M})$.

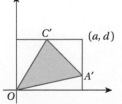

When powers of x and y are known

When the expected relationship between two variables is modelled by an equation in the form $y^2 = ax^3 + b$, where the exponents of the variables x and y are known but the coefficient of x^3 and the constant are not known, you need to estimate the values of a and b in order to suggest the equation that models the relationship.

You can estimate the values of a and b by reducing the relationship between x and y to a linear law. That is, you plot the straight line of y^2 against x^3 using given values for x and y. The gradient of the straight line gives you an estimate for a and the intercept on the y-axis gives you an estimate for b.

> When a relationship is modelled by an equation of the form $f(y) = ag(x) + b$ where $f(y)$ and $g(x)$ are known but a and b are not,
>
> ▶ Plot the straight line of $f(y)$ against $g(x)$.
> ▶ Gradient $= a$, intercept (on the vertical axis) $= b$.

When the relationship is modelled by an equation of the form $\frac{1}{x} + \frac{1}{y} = k$ it is important to remember that $\frac{1}{x}$ is also a power of x, and therefore can be treated in the same way. The same is true for $\sqrt{x}$.

Example 1

Question

In an experiment, the values of x and y are measured and recorded. The expected relationship between the two values is of the form $y^2 = ax^3 + b$.

Draw a suitable straight line graph, and estimate the values of a and b. Hence, suggest a model for the relationship between x and y.

x	1	1.5	2	2.5	3
y	1.05	3.22	5.39	7.70	10.25

Answer

Plot a graph of y^2 against x^3. Using the table, you will get these points:
(1, 1.1025); (3.375, 10.3684); (8, 29.0521); (15.625, 59.29); (27, 105.0625).

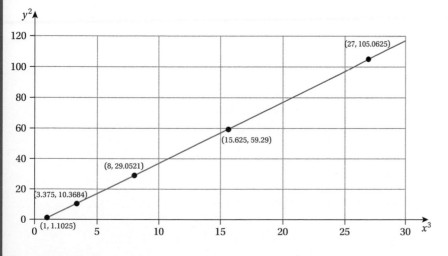

From the graph:

Gradient $= 4$, y^2-intercept $= -3$

This suggests the model $y^2 = 4x^3 - 3$.

> **Note**
>
> Gradient estimates a, intercept estimates b.

Example 2

You are told that the data in the table can be modelled by the relationship $\frac{1}{s} + \frac{b}{t} = c$. Estimate the values of b and c and hence suggest a model for the relationship between s and t.

t	0.5	1	1.5	2
s	−0.32	0.51	0.27	0.22

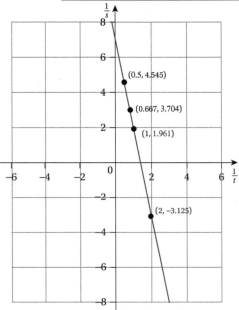

$\frac{1}{s} = -\frac{b}{t} + c$

$\frac{1}{s} = -b\left(\frac{1}{t}\right) + c$

Approximate values from the graph: gradient $= -5$ and intercept $= 7$.

So, the relationship between s and t appears to be $\frac{1}{s} + \frac{5}{t} = 7$.

Note

You are told the relationship between the variables s and t. Rearrange the equation to show the powers of s and t that should be related by a straight line, in other words, rearrange to get it in the form $f(s) = ag(t) + b$.

Note

Plot the graph of $\frac{1}{s}$ against $\frac{1}{t}$; using the data in the table you will get coordinates: (2, −3.125); (1, 1.961); (0.667, 3.704); (0.5, −4.545). Draw a straight line and estimate the gradient of the line and the intercept with the vertical axis.

Sometimes you have to be inventive in deciding which functions should be plotted. For instance, when the expected relationship between x and y is in the form $y^2 = ax^3 + \frac{b}{x^2}$, the equation appears to have three terms that are variables. To obtain an equation of the type $f(y) = ag(x) + b$, it is necessary to amend the equation so that it will contain only two terms that are variables. In this example, you will notice that if the equation is multiplied by x^2, the expected equation connecting x and y is $x^2y^2 = ax^5 + b$, which has only two variables. Hence, you would plot x^2y^2 against x^5.

Exercise 1

1. In an experiment, the heights of a number of trees, h m, and their ages in years, x, are recorded.

   The expected relationship between the two values is of the form $h = ax^2 + b$.

   a   Draw a suitable straight line graph, and estimate the values of a and b.

   b   Hence, suggest a model for the relationship between h and x.

x	2	3	4	5
h	2.4	2.9	3.6	4.5

2 The variables x and y are related by an equation of the form $y^2 = ax^2 + b$ where a and b are constants. The approximate values of x and y, shown in the table, have been found.

x	1	2	3	4
y	2.6	3.6	4.8	6.1

a Estimate the values of a and b.

b Find the value of x when $y = 5$.

3 The variables x and y are known to be related by an equation of the form $y^2 = ax^4 + bx$ where a and b are constants.

a The variables X and Y are defined by $X = x^3$ and $Y = \dfrac{y^2}{x}$. Show that $Y = aX + b$.

b The approximate values of x and y have been measured.

x	2.1	3.2	4.1	5.0
y	7.9	17.9	29.3	43.4

Estimate the values of a and b and hence suggest an equation to model the relationship between X and Y.

4 The variables x and y are known to be related by an equation of the form $y^2 = ax^3 + \dfrac{b}{x}$ where a and b are constants.

a Given that X and Y are defined by $X = x^4$ and $Y = y^2 x$, show that $Y = aX + b$.

b The approximate values of x and y have been measured.

x	2.2	3.1	4.0	6.0
y	4.8	8.1	11.8	21.8

Estimate the values of a and b and hence suggest an equation to model the relationship between X and Y.

8.2 When the power of x is unknown, or when x is in the exponent

When the expected relationship is of the form $y = bx^n$, where n is unknown, you will need to manipulate the equation more than you did in the previous examples. The only way to turn equations of this form into a linear relationship is to take **logarithms** of both sides, which will make n more manageable.

Similarly, if you are given an equation of the form $y = ab^x$, where x is the exponent, you can also use logarithms so that you can rewrite the equation without having x as an exponent.

When the relationship is modelled by an equation of the form $y = bx^n$, where n is unknown:

▶ Take the logarithms of both sides of the equation.
▶ Simplify the right-hand side using the law of logarithms to give
$\log y = \log b + n \log x$.

- ▶ Plot the straight line graph of $\log y$ against $\log x$.
- ▶ From the formula above, the gradient $= n$.
- ▶ From the formula above, the y-intercept is $\log b$.

For relationships modelled by an equation of the form $y = ab^x$, where x is the exponent:

- ▶ Take the logarithms of both sides of the equation.
- ▶ Simplify the right-hand side using the law of logarithms to give $\log y = \log a + x \log b$.
- ▶ Plot the straight line graph of $\log y$ against x.
- ▶ The gradient is $\log b$, so can be used to find b.
- ▶ The y-intercept is $\log a$, so can be used to calculate a.

Example 3

You are told that the values in the table are related by the formula $y = bx^n$.

x	1	1.2	1.4	1.6	1.8	2
y	5.98	7.90	9.90	12.16	14.51	16.98

Estimate b and n and hence suggest an equation to model the relationship between x and y.

Plot the graph of $\log y$ against $\log x$; using the table, you will get the points $(0, 0.7767)$; $(0.07918, 0.8976)$; $(0.1461, 0.9956)$; $(0.2041, 1.0849)$; $(0.25527, 1.16167)$; $(0.30103, 1.2299)$.

$\log y = \log (bx^n)$
$\log y = \log b + n \log x$

From the graph:

Gradient $= 1.5$ and y-intercept $= 0.78$

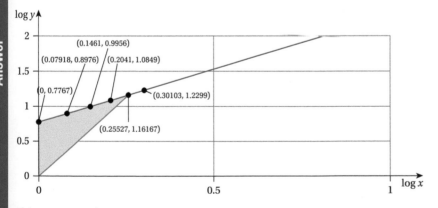

This suggests that $n = 1.5$.

$\log b = 0.78$, so $b = 10^{0.78} \cong 6$

Therefore, a suggested model for the relationship between x and y is $y = 6x^{1.5}$.

Note

Use the law of logarithms to make the right-hand side more manageable. First take logarithms of both sides of the given equation.

Note

Use $\log_a(xy) = \log_a x + \log_a y$ and $\log_a(x^k) = k \log_a x$.

Note

This gives a linear relationship between $\log y$ and $\log x$.

Note

In the linear relationship, n is the gradient. In the linear relationship, $\log b$ is the intercept on the y-axis. Substitute the values of b and n into the original equation.

Example 4

You are told that x and y are related by the formula $y = kb^x$. The measured values of x and y are given.

x	1.3	1.6	1.9	2.2	2.5
y	54.1	83.2	126.0	191.1	286.9

Estimate the values of k and b and hence suggest an equation to model the relationship between x and y.

$\log y = \log(kb^x)$

$\log y = \log k + x \log b$

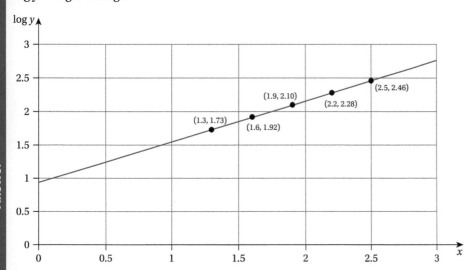

Note

Take logarithms of both sides of the equation given, and use the rules of logarithms to make the right-hand side more manageable.

Note

In the linear relationship, $\log b$ is the gradient and $\log k$ is the y-intercept. Substitute the values of k and b back into the original equation.

From the graph:

Gradient $= 0.60$ and y-intercept $= 0.95$

Therefore $\log b = 0.60$

$b = 10^{0.60} \cong 4$

Similarly, $k = 10^{0.95} \cong 9$.

The relationship between x and y can be modelled approximately by $y = 9 \times 4^x$.

Although you have used base 10 logarithms in the above work, there is no reason why you cannot take logarithms in any base of both sides, provided of course that you use the same logarithm for both sides of the equation!

Exercise 2

1 You are told that these values are related by the formula $y = bx^n$.

x	1	2	3	4	5
y	2	8	18	32	50

Estimate b and n and hence suggest an equation to model the relationship between x and y.

2 You are told that x and y are related by the formula $y = kb^x$. The measured values of x and y are given.

x	1	2	3	4	5
y	5.4	9.7	17.5	31.5	56.7

Estimate the values of k and b and hence suggest an equation to model the relationship between x and y.

3 The variables x and y are known to be related by an equation of the form $y = ab^x$ where a and b are constants.

a Given that $Y = \log_{10} y$, show that x and Y must satisfy an equation of the form $y = mx + c$.

b These measurements of x and y have been found.

x	1	2	3	4.3
y	14	28	63	74

Estimate the values of a and b and hence suggest an equation to model the relationship between x and y.

4 The variables x and y are known to be related by an equation of the form $y = ax^n$ where a and n are constants.

a Given that $Y = \log_{10} y$ and $X = \log_{10} x$, show that X and Y must satisfy an equation of the form $Y = mx + c$.

b These values of x and y have been measured.

x	1.5	2.5	3.5	4.5
y	17	78	215	460

Estimate the values of a and n and hence suggest an equation to model the relationship between x and y.

Summary

▶ If there are data values for two variables x and y, and an expected relationship between the two variables, you can find an equation to model the relationship by first reducing the expected relationship to a linear law, using substitutions and logarithms.

▶ To reduce an equation to a linear law, you need to ensure it contains only two terms that are variables.

▶ When a relationship is modelled by an equation in the form $f(y) = ag(x) + b$ where $f(y)$ and $g(x)$ are known but a and b are not,
  • Plot the straight line of $f(y)$ against $g(x)$
  • Gradient $= a$, intercept (on the y-axis) $= b$.

Substituting the values of a and b into the original equation will give you a model of the relationship between x and y.

▶ When the relationship is modelled by an equation of the form $y = bx^n$, where n is unknown:
  - Take the logarithms of both sides of the equation.
  - Simplify the right-hand side using the law of logarithms to give $\log y = \log b + n \log x$.
  - Plot the straight line graph of $\log y$ against $\log x$.
  - From the formula above, the gradient equals n.
  - From the formula above, the y-intercept is $\log b$.

▶ For relationships modelled by an equation of the form $y = ab^x$, where x is the exponent:
  - Take the logarithms of both sides of the equation.
  - Simplify the right-hand side using the law of logarithms to give $\log y = \log a + x \log b$.
  - Plot the straight line graph of $\log y$ against x.
  - The gradient is $\log b$, so can be used to find b.
  - The y-intercept is $\log a$, so can be used to calculate a.

▶ You will need to use measurements that you are given to estimate the explicit equation of a given linear law, or to estimate the value of one variable, given the other.

Review exercises

1 An advertising company records the monthly sales of a washing machine (y), x months after the start of an advertising campaign.

The expected relationship between the two values is of the form $y = ax^2 + b$.

a Draw a suitable straight line graph, and estimate the values of a and b.

x	1	2	3	4
y	136	440	940	1650

b Suggest a suitable equation to model the relationship between the monthly sales of the washing machine and the time, x months, after the start of the advertising campaign.

2 The variables x and y are related by an equation of the form $y^2 = ax^5 + bx^3$, where a and b are constants.

The approximate values of x and y have been found.

x	1	2	3	4
y	3.1	15.4	41.6	85.0

Estimate the values of a and b and hence suggest an equation to model the relationship between x and y. Draw a suitable straight line graph.

3 The variables y and x are related by an equation of the form $y = ax^n$, where a and n are constants.

Let $Y = \log_{10} y$ and $X = \log_{10} x$.

Show that there is a linear relationship between X and Y.

The approximate values of x and y have been found.

x	1	2	3	4
y	2	32	162	512

Estimate the values of a and n.

4 The variables x and y are known to be related by an equation of the form $y = ab^x$, where a and b are constants.

a Given that $Y = \log_{10} y$, show that x and Y must satisfy an equation of the form $Y = mx + c$.

b These approximate values of x and y have been found.

x	1	2	3	4
y	4.4	4.8	5.3	5.8

Estimate the values of a and b.

Practice examination questions

1 The variables x and y are related by an equation of the form $y = ax^2 + b$ where a and b are constants.

The following approximate values of x and y have been found.

x	2	4	6	8
y	6.0	10.5	18.0	28.2

a Complete the table below, showing values of X, where $X = x^2$.   (1 mark)

x	2	4	6	8
X				
y	6.0	10.5	18.0	28.2

b On the diagram below, draw a linear graph relating X and y.   (2 marks)

c Use your graph to find estimates, to two significant figures, for:

   i the value of x when $y = 15$;   (2 marks)

   ii the values of a and b.   (3 marks)

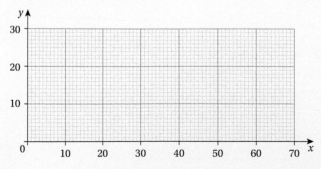

AQA MFP1 June 2010

2 The variables y and x are related by an equation of the form $y = ax^n$ where a and n are constants.

Let $Y = \log_{10} y$ and $X = \log_{10} x$.

a Show that there is a linear relationship between X and Y. **(3 marks)**

b The graph of Y against X is shown in the diagram.

Find the value of n and the value of a. **(4 marks)**

AQA MFP1 January 2013

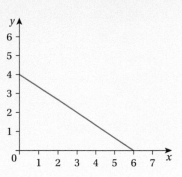

3 The variables x and y are known to be related by an equation of the form $y = ab^x$ where a and b are constants.

a Given that $Y = \log_{10} y$, show that x and y must satisfy an equation of the form $Y = mx + c$. **(3 marks)**

b The diagram shows the linear graph which has equation $Y = mx + c$.

Use this graph to calculate:

i an approximate value of y when $x = 2.3$, giving your answer to one decimal place;

ii an approximate value of x when $y = 80$, giving your answer to one decimal place.

(You are not required to find the values of m and c.) **(4 marks)**

AQA MFP1 June 2009

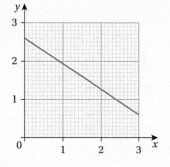

4 The variables x and y are related by an equation of the form
$$y = ax + \frac{b}{x+2}$$
where a and b are constants.

a The variables X and Y are defined by $X = x(x+2)$, $Y = y(x+2)$. Show that $Y = aX + b$. **(2 marks)**

b The following approximate values of x and y have been found:

x	1	2	3	4
y	0.40	1.43	2.40	3.35

i Complete the table below, showing values of X and Y. **(2 marks)**

ii Draw a linear graph relating X and Y. **(2 marks)**

iii Estimate the values of a and b. **(3 marks)**

x	1	2	3	4
y	0.40	1.43	2.40	3.35
x	3			
y	1.20			

AQA MFP1 June 2008

9 Numerical Methods

Introduction

You already know how to find the exact solution of any quadratic equation. There are also complicated methods to find the solutions of any cubic or quartic equations. However, it is possible to prove that there is no general formula to solve polynomial equations of degree 5 or higher. Many equations that model situations in mechanics involve trigonometric functions or differential equations, and are correspondingly even harder to solve. In this chapter, you will start to understand how computers can be programmed to solve these more complex equations to a given accuracy.

Recap

You will need to remember…

▶ How to find the solution of any quadratic equation with real coefficients, including those with complex roots.

▶ That if $f(a) = 0$ then $(x - a)$ is a factor of $f(x)$.

▶ That the maximum number of possible roots of an nth-degree polynomial is n.

▶ How to find the solutions of polynomial equations, and that occasionally you need to estimate the solutions to equations.

Objectives

By the end of this chapter, you should know how to:

▶ Determine if roots of $f(x)$ exist in the interval $a < x < b$, by considering the signs of $f(a)$ and $f(b)$, provided the function $f(x)$ is continuous.

▶ Find roots using interval bisection, linear interpolation and the Newton–Raphson method.

▶ Use Euler's step-by-step method to solve differential equations of the form $\frac{dy}{dx} = f(x, y)$.

You learned about the complex roots of quadratic equations in Chapter 2 *Complex Numbers*.

9.1 Solutions of polynomial equations

Most equations cannot be solved using algebraic procedures that give exact solutions, which means you have to turn to numerical methods to solve them. While several distinct numerical methods are available to use, they all have one property in common: if you repeatedly apply any of the methods to a problem, you will normally be able to obtain the solution to any desired degree of accuracy.

Initially, you need to determine an interval in which the root lies. Hence, generally:

> For a curve with equation $y = f(x)$ that is continuous between two points where $x = \alpha$ and $x = \beta$, if $f(\alpha)$ and $f(\beta)$ are of opposite signs then a solution of $f(x) = 0$ exists for at least one value of x in the interval $\alpha < x < \beta$.

So, it follows that so long as $y = f(x)$ is continuous and changes sign within an interval, then $f(x) = 0$ must have a root in that interval. However, if $y = f(x)$ is not continuous, nothing can be proved. If $y = f(x)$ is not continuous, it might be that $f(1)$ and $f(-1)$ are of opposite signs and $f(x) \neq 0$ for any value between -1 and 1, as shown in the diagram.

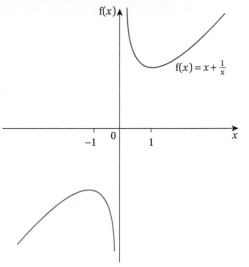

Example 1

Show that there is a root of $f(x) = x^3 + 5x - 9 = 0$ in the interval $1 < x < 2$.

$f(1) = 1 + 5 - 9 = -3$

$f(2) = 8 + 10 - 9 = 9$

A solution to $f(x) = 0$ exists in the interval $1 < x < 2$ since $y = f(x)$ is continuous for this interval and the signs of $f(1)$ and $f(2)$ are opposite, informing us that the x-axis has been crossed and therefore there is a value of x for which $f(x) = 0$.

If you were asked to find the value of the root more accurately, you would repeat the method in Example 1, finding $f(1.1)$, $f(1.2)$, $f(1.3)$, and so on. You would find that the value of $f(x)$ changes sign between 1.3 and 1.4 indicating that the root lies between these two values of x; you would then repeat to find $f(1.31)$, $f(1.32)$, and so on. You would repeat this process to find a smaller and smaller interval in which the root lies until you find the interval in which you can state the root to the required degree of accuracy.

The method of finding a more accurate root by using repeatedly smaller values of x is time-consuming. The procedures that are normally used to solve polynomial equations are **interval bisection**, **linear interpolation** and the **Newton–Raphson method**.

> **Note**
>
> You know that $y = f(x)$ is continuous in the interval $1 < x < 2$ because all polynomial functions are continuous. So, if the values of $f(1)$ and $f(2)$ are of opposite signs, then a solution of $f(x) = 0$ must lie in the interval, as per the key point above.

Interval bisection

As the name suggests, you use this method when you know that a root exists within a given interval.

> If you know that there is a root of $f(x) = 0$ between $x = \alpha$ and $x = \beta$, you try
>
> $x = \dfrac{(\alpha + \beta)}{2}$ and repeat as required. The sign of $f(x)$ determines which side
>
> of $\dfrac{(\alpha + \beta)}{2}$ the root lies on, that is, between $x = \alpha$ and the bisecting value,
>
> or between $x = \beta$ and the bisecting value.

The method is repeated until you obtain the root to the degree of accuracy required.

Example 2

The solution of $f(x) = x^3 + 5x - 9 = 0$ is within the interval $1 < x < 2$.

Find, by interval bisection, this solution, correct to two significant figures.

$f(1) = -3$; $f(2) = 9$; set $x = 1.5$

$f(1.5) = 1.875$

Therefore, the root lies between $x = 1$ and $x = 1.5$.

$f(1.25) = -0.796\ 875$

$f(1.25)$ and $f(1.5)$ are of opposite signs:
root between $x = 1.25$ and 1.5

$f(1.375) = 0.474\ 609\ 375$

$f(1.25)$ and $f(1.375)$ of opposite signs:
root between $x = 1.3125$ and 1.375

$f(1.3125) = -0.176\ 513\ 67$

$f(1.3125)$ and $f(1.375)$ of opposite signs:
root between $x = 1.3125$ and 1.375

$f(1.343\ 75) = 0.145\ 111$

$f(1.343\ 75)$ and $f(1.3125)$ of opposite signs:
root between $x = 1.3125$ and $1.343\ 75$

Therefore, $f(x) = 0$ when $x = 1.3$ (2sf).

Interval bisection is a very long and generally slow method. Therefore, it is worth considering other methods.

> **Note**
>
> Bisect the given interval using $x = \dfrac{(\alpha + \beta)}{2}$.

> **Note**
>
> Continue to bisect the interval in which you know the root lies, until you obtain the required accuracy.

> **Note**
>
> The *actual* value of the solution is 1.329 744 122 to 10 significant figures.

> **Note**
>
> There is a change in sign from f(1) to f(1.5).

> **Note**
>
> $x = \dfrac{(1 + 1.5)}{2} = 1.25$

> **Note**
>
> $x = \dfrac{(1.25 + 1.5)}{2} = 1.375$

> **Note**
>
> Only now are you able to state that the solution of $f(x) = 0$ is $x = 1.3$ to two significant figures since both α and β approximate to the same value at this degree of accuracy.

Linear interpolation

If given the same question as in Example 2, a more efficient method of progressing from $f(1) = -3$ and $f(2) = 9$ is to deduce that the root of $f(x) \equiv x^3 + 5x - 9 = 0$ is likely to be much nearer to 1 than to 2, since $|f(2)| > |f(1)|$.

This intuitive approach is formalised in **linear interpolation**, where the two points $(1, -3)$ and $(2, 9)$ are joined by a straight line and the x-value of the x-intercept of this line is calculated.

Using similar triangles, with the root at $x = 1 + p_1$, you have

$$\frac{p_1}{3} = \frac{1 - p_1}{9} \quad \Rightarrow \quad p_1 = \frac{1}{4}$$

Therefore, a better approximation of the root of $f(x) = 0$ is 1.25, $(1 + p_1)$, which gives $f(1.25) = -0.796\ 875$.

Hence, the root is between 1.25 and 2.

To find the root to two decimal places:

Using similar triangles again, you have

$$\frac{p_2}{0.796\ 875} = \frac{0.75 - p_2}{9}$$

$$\Rightarrow \quad 9.796\ 875 p_2 = 0.75 \times 0.796\ 875$$

$$\Rightarrow \quad p_2 = 0.061\ 004\ 784$$

Therefore, the second approximation of the root of $f(x) = 0$ is 1.311 004 784, $(1.25 + p_2)$, which gives $f(1.\ 311\ 004\ 784) = -0.191\ 708\ 181$

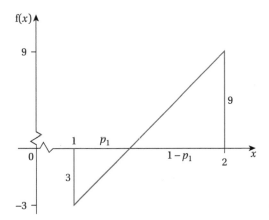

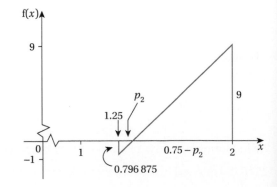

Hence, the root is between 1.311 004 784 and 2.

Repeating the procedure again, you obtain

$$\frac{p_3}{0.191\,708\,181} = \frac{0.688\,995\,216 - p_3}{9}$$

$\Rightarrow \quad 9.191\,708\,181 p_3 = 0.688\,995\,216 \times 0.191\,708\,181$

$\Rightarrow \quad p_3 = 0.014\,370\,127$

Therefore, the third approximation of the root is 1.325 374 912, which gives

$f(1.325\,374\,912) = -0.044\,947\,145$

Hence, the root is between 1.325 374 912 and 2.

Repeating the procedure yet again, you obtain

$$\frac{p_4}{0.044\,947\,145} = \frac{0.674\,625\,088 - p_4}{9}$$

$\Rightarrow \quad 9.044\,947\,145 p_4 = 0.674\,625\,088 \times 0.044\,947\,145$

$\Rightarrow \quad p_4 = 0.003\,352\,421\,099$

Therefore, the fourth approximation of the root is 1.328 727 333.

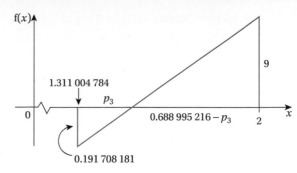

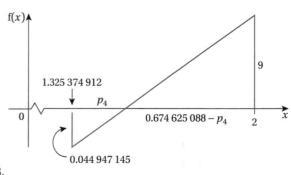

Both the fourth and third approximations are 1.33 correct to two decimal places. To check that this is the correct answer to two decimal places, you find $f(1.335)$:

$f(1.335) = 0.054\,27$ which has the opposite sign to $f(1.326)$.

Hence, the root is 1.33 correct to two decimal places. Notice that it is 1.3 (2sf), which corresponds to the answer found in Example 2.

Although linear interpolation is much quicker than interval bisection, it still does not take into account the shape of the graph of $f(x)$ between the starting points.

The procedure that does do this is the **Newton–Raphson method**.

Newton–Raphson method

If α is an approximate value for the root of $f(x) = 0$, then $\alpha - \frac{f(\alpha)}{f'(\alpha)}$ is generally a better approximation than α.

Consider the graph of $y = f(x)$. Draw the tangent at P, where $x = \alpha$, and let the tangent meet the x-axis at T.

You see that the x-value at T is closer than α is to the x-value at N, where the graph cuts the axis.

Using triangle PTQ, you have

Gradient of tangent $= \dfrac{PQ}{QT} \quad \Rightarrow \quad f'(\alpha) = \dfrac{f(\alpha)}{QT}$

$\Rightarrow \quad QT = \dfrac{f(\alpha)}{f'(\alpha)}$

The x-value of the point T is $\alpha - QT = \alpha - \frac{f(\alpha)}{f'(\alpha)}$ which is a better approximation of the root of $f(x) = 0$.

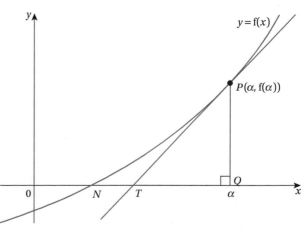

However, when the root of $f(x) = 0$ is not close to α, the method might fail. For example, in Figure A, the next x-value found is at T, which is further from the root than α is. In Figure B, $f'(\alpha) = 0$, which is unhelpful.

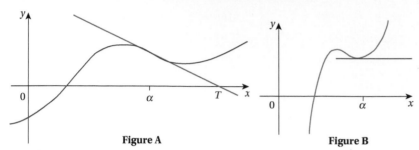

Figure A   **Figure B**

In its iterative form, the Newton–Raphson method gives $\alpha_{n+1} = \alpha_n - \dfrac{f(\alpha_n)}{f'(\alpha_n)}$.

Example 3

Question

Use the Newton–Raphson method to find a root of $f(x) = x^3 + 5x - 9 = 0$ to three significant figures. Start with an initial value of $x = 1$.

Answer

Let α be the required root.

$f(x) = x^3 + 5x - 9$, so $f'(x) = 3x^2 + 5$

$f(\alpha_1) = 1 + 5 - 9 = -3$

$f'(\alpha_1) = 3 + 5 = 8$

Using Newton–Raphson, $\alpha_{n+1} = \alpha_n - \dfrac{f(\alpha_n)}{f'(\alpha_n)}$.

$\alpha_2 = \alpha_1 - \dfrac{f(\alpha_1)}{f'(\alpha_1)} = 1 + \dfrac{3}{8} = 1.375$

Hence,

$\alpha_3 = 1.375 - \dfrac{f(1.375)}{f'(1.375)} = 1.375 - \dfrac{0.474\ 609\ 375}{10.671\ 875} = 1.330\ 5271$

$\alpha_4 = 1.330\ 5271 - \dfrac{f(1.330\ 5271)}{f'(1.530\ 5271)} = 1.330\ 5271 - \dfrac{0.008\ 070\ 770\ 27}{10.310\ 907} = 1.329\ 744\ 36$

So, the root of $f(x) \equiv x^3 + 5x - 9$ is $x = 1.33$ (3sf).

> **Note**
>
> Differentiate $f(x)$.

> **Note**
>
> Substitute in $\alpha_1 = 1$.

> **Note**
>
> α_3 and α_4 are now so close together that you can say that the root is 1.33 to three significant figures.

> **Note**
>
> The root is actually 1.329 744 to seven significant figures. Or, were you to repeat the procedure a few more times, you would find that the root is 1.329 744 122, correct to 10 significant figures.

Exercise 1

1. Show that a root of the equation $x^3 = 7 - 5x$ lies in the interval $1 < x < 2$.
   Use linear interpolation to find this root correct to two decimal places.

2. Show that a root of the equation $xe^{3x} = 12$ lies in the interval $0 < x < 1$.
   Use linear interpolation to find this root correct to two decimal places.

3. A root of the equation $x^3 - 4x = 5$ lies in the interval $2 < x < 3$.
   Use interval bisection to find this root correct to two decimal places.

4. Show that a root of the equation $3x^3 - 2^x = 0$ lies in the interval $0 < x < 1$.
   Use interval bisection to find this root correct to two decimal places.

5 Using the Newton–Raphson method, find the real root of $x^3 + 3x - 7 = 0$ correct to two decimal places.

6 Using the Newton–Raphson method, find the real solution to the square root of $x + x^{\frac{1}{5}} = 8$ correct to two decimal places.

9.2 Step-by-step solution of differential equations

A **differential equation** is one that relates an expression, f(x), to its derivatives. Most first-order differential equations cannot be solved exactly; they need a step-by-step approach that provides an approximation of the solution.

These approaches depend on drawing lines parallel to the y-axis a distance of h apart: h is called the **step length**.

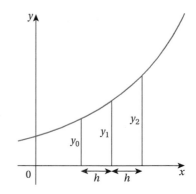

One such method is a **single-step approximation** known as **Euler's method**. This is the idea that when two points of a curve are very close together then the gradient of the curve between these points can be approximated by the gradient of the tangent at one of the points.

With reference to the figure, $P(x_0, y_0)$ is a point on the curve $y = f(x)$ and $Q(x_1, y_1)$ is another point on the curve close to P, where $x_1 - x_0 = h$ and h is small. You see that the gradient of the chord PQ is approximately the same as the gradient of the tangent at P. Hence, you have that the gradient of PQ is $\frac{y_1 - y_0}{h}$ which gives the gradient of the tangent at $P \left(\frac{dy}{dx} \right)_0 \approx \frac{y_1 - y_0}{h}$.

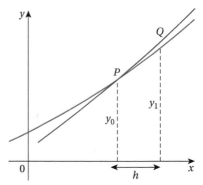

If $\left(\frac{dy}{dx} \right)_0 \approx \frac{y_1 - y_0}{h}$ then it follows that $y_1 \approx y_0 + h \left(\frac{dy}{dx} \right)_0$, which means that if you know the value of h, an initial value of (x, y) and the differential equation, you can calculate y_1.

To use Euler's method you need:
▶ An initial point (x_0, y_0)
▶ The step length, h
▶ The differential equation that relates y to x.

If you were given the above information and asked to find f(3) for example, you would use the initial point to find the point $\left((x + h), y + h \left(\frac{dy}{dx} \right) \right)$, where h is the step length. You then use this *new* point in the same way to find the next point and so on, until you reach a point with an x-coordinate of 3. This will then give you the corresponding value of y that is the solution. In this way, Euler's method effectively joins together small line segments such that they approximate the curve of the actual function and allow you to find an approximate solution for a given value of the function. The smaller the value of the step length, the more accurate the solution obtained.

> In function notation Euler's method is $y_{n+1} \approx y_n + h f(x_n, y_n)$ and $x_{n+1} = x_n + h$, where the values of x_0, y_0 and h are known.

You can use Euler's step-by-step method to solve differential equations of the form $\frac{dy}{dx} = f(x, y)$.

Example 4

Use a step length of 0.2 and Euler's method to estimate $y(0.6)$ for $\frac{dy}{dx} = \sqrt{x+y}$, given that $y = 3$ when $x = 0$.

$x_0 = 0$, $y_0 = 3$, $\frac{dy}{dx} = \sqrt{x+y}$ and $h = 0.2$

Start at $(0, 3)$. Then $x_1 = x_0 + h = 0.2$.

When $x = 0$, $\frac{dy}{dx} = \sqrt{0+3}$ and so $y_1 = y_0 + h\frac{dy}{dx} = 3 + 0.2\sqrt{3} = 3.346\,410\,162$.

Taking the next step, $x_2 = x_1 + h = 0.4$.

$$\frac{dy}{dx} \approx \sqrt{0.2 + 3.346\,410\,162} = 1.883\,191\,483$$

Therefore $y_2 = y_1 + 1.883\,191\,483\,h = 3.723\,048\,458$

Finally, at x_2, we know that $\frac{dy}{dx} \approx \sqrt{0.4 + 3.723\,048\,458} = 2.030\,529\,108$.

So $y_3 = y_2 + 2.030\,529\,108\,h = 4.129\,154\,280\ldots$

Example 5

Use a step length of 0.1 and Euler's method to find $y(0.3)$ for $\frac{dy}{dx} = \sqrt{x+y}$ given that $y = 2$ when $x = 0$.

Using $\left(\frac{dy}{dx}\right)_0 \approx \frac{y_1 - y_0}{h}$, $y_1 \approx y_0 + h\left(\frac{dy}{dx}\right)_0$.

Therefore,

y at new value when $x = 0.1 = y$ at original

value of $x + h \times \frac{dy}{dx}$ at original value of x.

Therefore

$y(0.1) \approx 2 + 0.1\sqrt{0+2}$

$\Rightarrow y(0.1) \approx 2.141\,42$

$y(0.2) \approx y(0.1) + h\left(\frac{dy}{dx}\right)_{x=0.1}$

$\Rightarrow y(0.2) \approx 2.141\,42 + 0.1\sqrt{0.1 + 2.141\,42}$

$\Rightarrow y(0.2) \approx 2.291\,133$

Note

Continue to $x = 0.3$,

$y(0.3) \approx y(0.2) + h\left(\frac{dy}{dx}\right)_{x=0.2}$

$\Rightarrow y(0.3) \approx 2.291\,133 + 0.1$ $\sqrt{0.2 + 2.291\,133}$

$\Rightarrow y(0.3) \approx 2.448\,96$ or 2.4490 (to 4dp)

Note

Now repeat this procedure with the values obtained for y and $\frac{dy}{dx}$ when $x = 0.1$ now being treated as the original values, and the new value for y being found for $x = 0.2$.

Exercise 2

In questions **1–4** use Euler's formula, $y_{r+1} \cong y_r + hf(x_r, y_r)$.

1 Find an approximation for $y(1.1)$ to four decimal places, when $h = 0.1$, $\frac{dy}{dx} = 3x^3 + 2y$ and $y(1) = 1$.

2 Given that $h = 0.2$, $\frac{dy}{dx} = x^2 + 3y^3$ and $y(1) = 1$, find an approximation for $y(1.2)$ to four decimal places.

3 Given that $h = 0.1$, $\frac{dy}{dx} = e^{3x} + e^y$ and $y(1) = 1$, find an approximation for $y(1.2)$ to three decimal places.

4 Find an approximation for $y(1.4)$ to three decimal places, when $h = 0.2$, $\frac{dy}{dx} = 3\cos x + y$ and $y(1) = 1$.

Summary

- For a function, $f(x)$, that is continuous between two points α and β, if $f(\alpha)$ and $f(\beta)$ are of opposite signs then a solution of $f(x) = 0$ exists for at least one value of x in the interval $\alpha < x < \beta$.

- You can find the roots of equations to any number of decimal places by carefully using interval bisection, linear interpolation or the Newton–Raphson method.

- You can use Euler's step-by-step method to solve differential equations of the form $\frac{dy}{dx} = f(x, y)$.

- These are methods that are normally used to solve polynomial equations to any degree of accuracy.

  - **Interval bisection**: if you know that there is a root of $f(x) = 0$ between $x = \alpha$ and $x = \beta$, you try $x = \frac{(\alpha+\beta)}{2}$ and repeat as required. The sign of $f(x)$ determines which side of $\frac{(\alpha+\beta)}{2}$ the root lies on, that is, between $x = \alpha$ and the bisecting value, or between $x = \beta$ and the bisecting value. The method is repeated until you obtain the root to the degree of accuracy required.

  - **Linear interpolation** is much quicker than interval bisection. In this method, the two points representing the limits of the interval are joined by a straight line and the x-value of the x-intercept of this line is calculated. Using similar triangles, approximations closer and closer to the root are obtained.

  - **Newton–Raphson method**: this method takes into account the shape of the graph of $f(x)$ between the values of α and β. It is based on the idea that if α is an approximate value for the root of $f(x) = 0$, then $\alpha - \frac{f(\alpha)}{f'(\alpha)}$ is generally a better approximation than α. In its iterative form, the Newton–Raphson method gives $\alpha_{n+1} = \alpha_n - \frac{f(\alpha_n)}{f'(\alpha_n)}$. Again, α_1 is used to get a more accurate approximation of the root, α_2, which is in turn used to get an even more accurate approximation, α_3, and so on.

▶ You can use Euler's step-by-step method to solve differential equations of the form $\frac{dy}{dx} = f(x, y)$. You use a starting value, the step length and the differential equation to find the required value of y by repeatedly applying the same process to previously obtained values of x and y. In function notation, Euler's method is $y_{n+1} \approx y_n + hf(x_n, y_n)$ and $x_{n+1} = x_n + h$, where the values of x_0, y_0 and h are known.

Review exercises

1 The function $y(x)$ satisfies the differential equation $\frac{dy}{dx} = f(x, y)$ where $f(x, y) = \sqrt{3x} + \sqrt{2y}$ and $y(2) = 3$.

Use the Euler formula $y_{r+1} = y_r + hf(x_r, y_r)$ with $h = 0.1$ to obtain an approximation to $y(2.1)$, giving your answer to four decimal places.

2 The function $y(x)$ satisfies the differential equation $\frac{dy}{dx} = x(2 - y^2)$ and $y(1) = 3$.

Use the Euler formula $y_{r+1} = y_r + hf(x_r, y_r)$ with $h = 0.2$ to obtain an approximation of $y(1.6)$, giving your answer to four decimal places.

3 The cubic equation $x^3 - 3x + 1 = 0$ has a root in the interval $0 < x < 1$. Use the Newton–Raphson method with starting value $x_0 = 0.5$ to find this root, giving your answer correct to six decimal places.

Practice examination questions

1 The function $y(x)$ satisfies the differential equation $\frac{dy}{dx} = f(x, y)$ where $f(x, y) = \sqrt{x^2 + y + 1}$ and $y(3) = 2$.

Use the Euler formula $y_{r+1} = y_r + hf(x_r, y_r)$ with $h = 0.1$, to obtain an approximation to $y(3.1)$, giving your answer to four decimal places. (3 marks)

AQA MFP3 June 2009

2 The equation $y = x^3 - x^2 + 4x - 900$ has exactly one real root, α.

Taking $x_1 = 10$ as the first approximation to α, use the Newton–Raphson method to find a second approximation, x_2, to α. Give your answer to four significant figures. (3 marks)

AQA MFP1 June 2013

3 The equation $y = x^3 + 2x^2 + x - 100\ 000 = 0$ has one real root. Taking $x_1 = 50$ as a first approximation to this root, use the Newton–Raphson method to find a second approximation, x_2, to the root. (3 marks)

AQA MFP1 January 2011

4 a Show that the equation $x^3 + x - 7 = 0$ has a root between 1.6 and 1.8. (3 marks)

b Use interval bisection **twice**, starting with the interval in part (a), to give this root to one decimal place. (4 marks)

AQA MFP1 June 2007

5 A curve passes through the point $(0,1)$ and satisfies the differential equation $\frac{dy}{dx} = \sqrt{1 + x^2}$.

Starting at the point $(0,1)$, use a step-by-step method with a step length of 0.2 to estimate the value of y at $x = 0.4$. Give your answer to five decimal places. (5 marks)

AQA MFP1 January 2009

6 The equation $24x^3 + 36x^2 + 18x - 5 = 0$ has one real root, α.

a Show that α lies in the interval $0.1 < x < 0.2$. (2 marks)

b Starting from the interval $0.1 < x < 0.2$, use interval bisection **twice** to obtain an interval of width 0.025 within which α must lie. (3 marks)

c Taking $x_1 = 0.2$ as a first approximation to α, use the Newton–Raphson method to find a second approximation, x_2, to α. Give your answer to four decimal places. (4 marks)

AQA MFP1 June 2012

7 A curve satisfies the differential equation $\frac{dy}{dx} = 2^x$.

Starting at the point $(1, 4)$ on the curve, use a step-by-step method with a step length of 0.01 to estimate the value of y at $x = 1.02$. Give your answer to six significant figures. (5 marks)

AQA MFP1 January 2008

10 Bayes' Theorem

Introduction

In your study of AS-level Mathematics, you learned about the basics of probability theory including conditional probability. In this chapter, these ideas will be developed to include a discussion of probability tree diagrams and the very important Bayes' Theorem, which can be applied in many areas of science including population genetics.

Objectives

By the end of this chapter, you should know how to:

▶ Solve problems relating to independent and dependent events using probability tree diagrams.

▶ Identify and solve problems relating to dependent events where Bayes' Theorem can be applied.

Recap

You will need to remember...

▶ The definitions of the union of two sets, $A \cup B$ (either A or B or both), and the intersection of two sets, $A \cap B$ (both A and B)

▶ If $A \cap B = \varnothing$, A and B are mutually exclusive; if $A \cup B = S$, where S is the sample space, A and B are exhaustive

▶ Kolmogorov's three axioms of probability; for an experiment, E, with associated events A and B defined on a sample space S,

1. $P(S) = 1$

2. $0 \leq P(A) \leq 1$ for $A \subseteq S$

3. $P(A \cup B) = P(A) + P(B)$; $A \cap B = \varnothing$

▶ For an experiment, E, with associated events A and B defined on a sample space S, the probability of both A and B occurring when E is performed once is given by

$P(A \cap B) = P(A) \times P(B \mid A)$.

10.1 Tree diagrams

You have met probability tree diagrams in your study of GCSE Mathematics and will now consider them again in a more rigorous way.

Tree diagrams for independent events

Tree diagrams are often used in probability questions when random experiments can be considered as processes consisting of multiple stages. Suppose that a fair coin is flipped three times and you wish to show all possible outcomes. The three-stage nature of this problem suggests a **tree diagram** approach.

At the first flip, the possible outcomes are heads (H) or tails (T). For each of these, the next flip can either be H or T, giving HH, HT, TH or TT. The result of the third flip follows similarly and the whole process can be shown diagrammatically, with probabilities for each stage, as shown in the figure.

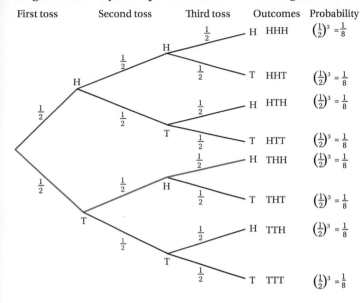

A **branch** is represented by a single line segment; it is shown here in red. A set of branches is called a **path** and indicates a possible outcome of the whole experiment; it is shown here in blue and demonstrates the possible path of THH, in that order.

For three **independent events**, A, B and C, the probability of all three occurring is given by the equation

$P(A \cap B \cap C) = P(A) \times P(B) \times P(C)$

Therefore, the probabilities of each of the final outcomes (HHH, HHT, ...) are all $\left(\frac{1}{2}\right)^3 = \frac{1}{8}$. These are shown at the ends of the paths.

Suppose now that you want to use your tree diagram to calculate the probability of a total of two heads being obtained. In general, for mutually exclusive events D and E, $P(D \cup E) = P(D) + P(E)$.

According to the tree diagram, exactly two heads can occur in three mutually exclusive ways: HHT, HTH or THH. Therefore,

$P(2 \text{ heads}) = P(\text{HHT or HTH or THH}) = \frac{1}{8} + \frac{1}{8} + \frac{1}{8} = \frac{3}{8}$

For independent events represented on a tree diagram, to find the probability of:

► A given final outcome you *multiply* the probabilities along the relevant path

► Any one of a set of final outcomes, you *add* the relevant final probabilities.

Tip

The probabilities of each outcome are shown.

Tip

Axiom 3 of Kolmogorov's three axioms.

Tip

Remember: multiply the probabilities along the paths, but add between them.

Tree diagrams for dependent events

A more interesting set of problems occurs when the events in the stages are not independent of each other. Although the theory is significantly different, the process of solving the problems is similar. Consider this example.

In a large drugs trial, patients are divided into two groups: group A who receive the drug and a control group, B, who receive a placebo instead. Success for any given patient is defined as five-year survival. The two groups are of the same size and analysis of the results shows that, for group A, the success rate is 71%. Group B achieved a success rate of 49%.

Tip

A placebo looks like the drug being tested but has no pharmaceutical effect.

Suppose a person is chosen at random from all of the patients in the trial, and you wanted to find the probability of these outcomes:

▶ The person receives the drug and survives five years
▶ The person does not receive the drug and survives five years
▶ The person survives five years (regardless of whether they did or did not take the drug).

This problem consists of two stages: assigning a person to one of two groups, followed by survival or non-survival during the five years. Therefore, you can use a tree diagram.

First, you must define the events: A is a patient assigned to group A; S is the five-year survival for a patient. A' and S' represent the complementary events.

A tree diagram for this situation is as follows.

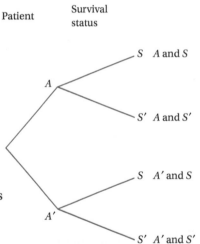

The diagram shows four possible outcomes for any given patient, represented by the four paths. For example, A and S is 'the chosen patient is assigned to group A *and* survives five years'.

It is important to recognise that, in each path, the second of each pair of branches is dependent or **conditional** on the first of the pair. For example, in the top path with outcome A and S, the first branch represents the (unconditional) event 'the patient is assigned to group A', and the second branch represents the *conditional* event 'the patient survives *given that the patient is assigned to group A*'.

The two survival rates 71% and 49% are therefore **conditional probabilities**. They are

$P(S\,|\,A) = 0.71$ and $P(S\,|\,A') = 0.49$

Since the only two possibilities for the next stage, given that A occurs, are S and S', you can also write that $P(S'\,|\,A') = 1 - 0.71 = 0.29$ from Axiom 1, the total probability theorem.

Similarly, $P(S'\,|\,A') = 1 - 0.49 = 0.51$.

Tip

You have met this notation in your AS-level Mathematics studies. $P(S\,|\,A)$ is read 'the probability that S occurs given that A occurred' or 'the probability that S occurs under the hypothesis that A occurred'.

Since the two groups are of the same size, $P(A) = P(A') = \frac{1}{2}$. These probabilities can be added to the above diagram.

You know from your study of A-level probability theory that, if A and B are two events defined on sample space S, then $P(A \cap B) = P(A) \times P(B\,|\,A)$.

Using this result for the top outcome in the diagram (group A *and* survival),

$P(A \cap S) = P(A) \times P(S\,|\,A) = 0.5 \times 0.71 = 0.355$

This approach gives each of the final probabilities in the diagram.

You are now able to find the required probabilities:

▶ The probability that the patient receives the drug and survives five years is given by $P(A \cap S) = 0.5 \times 0.71 = 0.355$.

▶ The probability that the patient does not receive the drug and survives five years is $P(A' \cap S) = 0.5 \times 0.49 = 0.245$.

▶ The probability of five-year survival can come about in one of two mutually exclusive ways: *either* 'group A and five-year survival' *or* 'group B and five-year survival'. This suggests $P(S) = 0.355 + 0.245 = 0.6$.

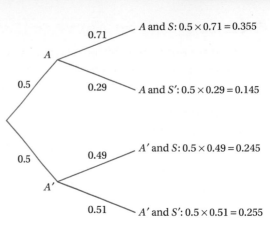

For dependent events, to find the probability of:

▶ **A given final outcome you *multiply* the probabilities along the relevant path**

▶ **Any one of a set of final outcomes, you *add* the relevant final probabilities.**

Note that this was also the case for sets of independent events. The important difference is, when you are creating the tree, to consider what probabilities should go along the branches and if they are conditional on the previous branch or not.

Sometimes the use of tree diagrams is not as obvious as in the above case. For example, the problem may not have a clearly defined set of stages. It is also possible in some cases for the tree diagram to contain paths of different lengths.

Example 1

A lift has a brake with a probability of failing of 0.001. If the brake fails, an emergency brake is activated. The probability of the emergency brake failing is 0.0004. On one-half of all occasions when you travel to the ground floor, you take the stairs; on the other occasions you take the lift. The stairs have a 99% safety record.

By drawing a fully labelled tree diagram, find the probability that on a randomly chosen journey to the ground floor, you arrive safely. Give your answer to 4 dp.

> **Note**
>
> Identify the different paths. The 'stairs' path has two branches, and the 'lift' path has two or three branches. The probability of each path is found as before by multiplication.

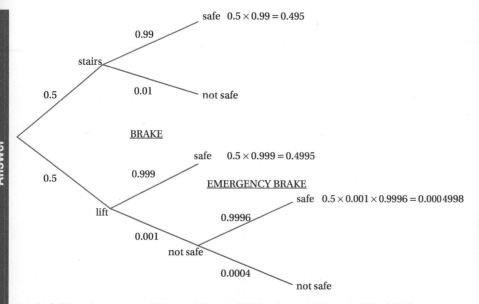

> **Note**
>
> Add the relevant final probabilities.

Probability of arriving safely: $0.495 + 0.4995 + 0.000\,4998 = 0.9950$ (4dp).

Exercise 1

1 There are two colours of counter, red (R) and black (B). Draw a tree diagram to show the possible **outcomes** if two counters are selected at random, with replacement.

2 You choose a card, *A* or *B*. If you choose card *A*, you then flip a fair coin. If you choose card *B*, you then flip a coin that has a head on both sides. Draw a tree diagram for this experiment showing all possible **outcomes**.

3 You have two fair dice. One dice is four-sided and numbered 1–4, and one dice is six-sided numbered 1–6. You choose a dice at random and roll it.

   a Draw a tree diagram to represent the possible outcomes and their probabilities.

   b Calculate the probability that you obtain a 3.

4 Two cards are chosen with replacement from a normal pack of cards. P is the event 'a picture card was chosen'.

   a Draw a tree diagram to show the possible outcomes.

   b What is the probability that one card is a picture card and the other is not?

5 Three fair coins are flipped and the outcomes noted. Draw a tree diagram to illustrate this and find the probability of obtaining exactly two heads.

6 A box contains eight red beads and six blue beads. Two beads are chosen at random without replacement and their colours noted. By drawing a fully labelled tree diagram, find the probability that the beads are of different colours.

7 A survey into school travel arrangements is carried out at school in a busy city.

35% of all students walk to school. Students who do not walk to school travel by bus, car or bicycle in these proportions: 45%, 35% and 20% respectively. The proportion of late arrivals for each type of travel is as follows: walk (7%), bus (12%), car (9%) and bicycle (4%).

   a Draw a fully labelled tree diagram to show this information.

   b A student is chosen at random from the whole school. What is the probability that the student

     i travels by bus and is not late

     ii is late

     iii travels by motorised transport and is late.

   Give your answers to 3 dp.

8 A test for a genetic disorder gives either a positive (disease present) or negative (no disease present) result. It correctly identifies 95% of cases where the disease is present. The percentage of 'false positives' (cases where no disease is present but the test result is positive) is 4%. All test results are either positive or negative.

By drawing a tree diagram, find the percentage of all cases where there is a positive result if the prevalence of the disease among people being tested is 21%.

> **Tip**
>
> A normal pack of playing cards contains 52 cards. There are four suits: hearts (red), diamonds (red), spades (black) and clubs (black). Each suit has 13 cards: an ace, nine number cards (numbered 2–10) and three picture cards: jack, queen and king.

10.2 Bayes' Theorem

The situations involving dependent events discussed so far have contained more than one 'stage' and you were finding probabilities of events in a later 'stage' that were *conditional* on events occurring in the previous stage.

Bayes' Theorem provides a way to calculate the probability of an event occurring in a *particular way*, given that it has occurred in the first place. Note that the use of 'given' here implies conditional probability.

For example, say you rolled a dice and scored an even number. Given that you rolled an even number, you could calculate the probability that the outcome 'even number' was specifically achieved by rolling a 2 (as opposed to rolling a 4 or a 6).

Recall that the probability of an event A occurring given that event B occurred is given by the formula

$$P(A \mid B) = \frac{P(B \cap A)}{P(B)}$$

Bayes' Theorem generalises this result by allowing B to occur through a set of possible outcomes, A_j, $j = 1, 2, ..., n$. As shown in the diagram, the events A_j are **mutually exclusive** and **exhaustive**.

So, the probability of any one of these events occurring, conditional on B, is

$$P(A_j \mid B) = \frac{P(B \cap A_j)}{P(B)} = \frac{P(A_j) \times P(B \mid A_j)}{P(B)}$$

However, from the diagram of the sample space, you have

$$P(B) = \sum_{i=1}^{n} P(B \cap A_i) = \sum_{i=1}^{n} P(A_i) \times P(B \mid A_i)$$

Together, these two equations give Bayes' Theorem.

> If a sample space consists of a set of n mutually exclusive and exhaustive events, A_j, $j = 1, 2, ..., n$, and event B is defined on this sample space, then Bayes' Theorem states that
> $$P(A_j \mid B) = \frac{P(A_j) \times P(B \mid A_j)}{\sum_{i=1}^{n} P(A_i) \times P(B \mid A_i)}$$

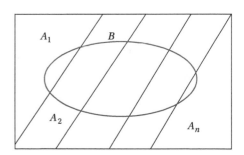

Example 2

Question

A, B, C and D are events associated with a random experiment and $A \cup B \cup C = S$. A, B and C are mutually exclusive and are exhaustive events. The probabilities of A and B are 0.4 and 0.3 respectively.

$P(D \mid A) = 0.2$, $P(D \mid B) = 0.1$ and $P(D \mid C) = 0.4$. Find

a $P(C)$ b $P(B \mid D)$

Answer

a $A \cup B \cup C = S$ and $A \cap B \cap C = \varnothing$

$P(S) = 1$; $P(A \cup B) = P(A) + P(B)$ since $A \cap B = \varnothing$

$P(C) = 1 - (P(A) + P(B)) = 1 - (0.4 + 0.3) = 0.3$

b $P(B \mid D) = \dfrac{P(B) \times P(D \mid B)}{P(A) \times P(D \mid A) + P(B) \times P(D \mid B) + P(C) \times P(D \mid C)}$

$= \dfrac{0.3 \times 0.1}{0.4 \times 0.2 + 0.3 \times 0.1 + 0.3 \times 0.4} = \dfrac{3}{23} = 0.11 \text{ (2dp)}$

It is also possible to apply Bayes' Theorem to real-life problems.

Note

Use Axioms 1 and 3 and rearrange.

Note

Use Bayes' Theorem.

Note

Substitute in the values of $P(D \mid A)$, $P(D \mid B)$ and $P(D \mid B)$, which are given in the question.

Example 3

In a simple model, $\frac{1}{100}$ males are colour-blind, and $\frac{1}{10\,000}$ females are colour-blind.

If a person chosen at random from the population is colour-blind, find the probability that this person is male. Assume equal numbers of males and females in the population.

Answer

C denotes that the person is colour-blind; M, that the person is male.

$$P(M\mid C) = \frac{P(M)P(C\mid M)}{P(M)P(C\mid M) + P(M')P(C\mid M')}$$

$$P(C\mid M) = \frac{1}{100} \text{ and } P(C\mid M') = \frac{1}{10\,000}$$

$$P(C\mid M) = \frac{\frac{1}{2}\times\frac{1}{100}}{\frac{1}{2}\times\frac{1}{100} + \frac{1}{2}\times\frac{1}{10\,000}} = \frac{100}{101}$$

> **Note**
>
> Define the events.

> **Note**
>
> The mutually exclusive and exhaustive events that constitute the sample space are M and M', so Bayes' Theorem can be used.

> **Note**
>
> It is very important for you to recognise that the probabilities given in the question are not absolute but conditional on gender. For example, $\frac{1}{100}$ is the probability of a person being colour-blind *given that they are male*.

Example 4

Two bags contain coloured balls. Bag 1 contains six black, two red and four white balls. Bag 2 contains three black, one red and two white balls. A ball is chosen at random from Bag 1 and then placed in Bag 2. A ball is then chosen at random from Bag 2. Find the probability that

a the first ball is white and the second is red

b the first ball is white given that the second is red.

> **Note**
>
> Identify what values were given in the question.

> **Note**
>
> Substitute the known values into Bayes' Theorem.

Answer

$W_i, R_i, B_i =$ the ith ball is white, red, black, respectively; $i = 1, 2$

a $P(W_1) = \frac{4}{12} = \frac{1}{3}; \; P(R_2\mid W_1) = \frac{1}{7}$

$$P(W_1 \cap R_2) = P(W_1) \times P(R_2\mid W_1)$$
$$= \frac{1}{3}\times\frac{1}{7} = \frac{1}{21}$$

> **Note**
>
> Define the events.

> **Note**
>
> If a white ball is chosen from Bag 1 and placed in Bag 2, there are now seven balls in Bag 2, one of which is red.

b $P(W_1\mid R_2) = \dfrac{P(W_1)P(R_2\mid W_1)}{P(B_1)P(R_2\mid B_1) + P(R_1)P(R_2\mid R_1) + P(W_1)P(R_2\mid W_1)}$

$$= \frac{\frac{1}{3}\times\frac{1}{7}}{\frac{1}{2}\times\frac{1}{7} + \frac{1}{6}\times\frac{2}{7} + \frac{1}{3}\times\frac{1}{7}}$$

$$= \frac{2}{7}$$

> **Note**
>
> The mutually exclusive and exhaustive events that constitute the sample space are W_1, R_1 and B_1; use Bayes' Theorem.

Exercise 2

1 Given that $P(P) = P(Q)$, $P(T|Q) = \frac{1}{4}$, $P(T \mid P) = \frac{2}{3}$ and P and Q are mutually exclusive, exhaustive events, use Bayes' Theorem to find $P(Q \mid T)$.

2 The sample space for an experiment is made up of two mutually exclusive events, A and B. An event R is a subset of the same sample space. If $P(A) = 0.3$, $P(R \mid A) = 0.2$ and $P(R \mid B) = 0.6$, find

  a $P(B)$ **b** $P(A \mid R)$

3 Two mutually exclusive and exhaustive events, B_1 and B_2, are defined on a sample space. An event A is also defined on the sample space. Given that $P(B_1) = 0.4$ and the conditional probabilities of A given B_1 and A given B_2 are respectively 0.3 and 0.5, find

  a $P(B_2)$ **b** $P(B_1 \mid A)$ **c** $P(B_2 \mid A)$ **d** $P(A)$

4 A bag contains the seeds of three different varieties of flower, P, Q and R, in proportions $1 : 2 : 7$. The germination rates for these varieties are 84%, 55% and 20%, respectively. One seed chosen at random is planted. Given that it germinates, what is the probability that it was of variety P?

5 You have three coins: two of them are fair and one has a head on both sides. You choose one of them at random and flip it.

  a Find the probability that a head is obtained.

  b Given that a head occurred, find the probability that a fair coin was tossed.

6 In a random experiment, an event, S, can come about in one of three exhaustive and mutually exclusive ways, P, Q and R. Given that the probabilities of P, Q and R are 0.1, 0.4 and 0.5 respectively and that $P(S \mid P) = 0.3$, $P(S \mid Q) = 0.4$ and $P(S \mid R) = 0.5$, show that $25P(Q \mid S) = 16P(R \mid S)$.

7 You choose a card from a set of nine cards numbered 1–9. If the card shows a number

  ▶ Less than 3, you flip a fair coin

  ▶ Between 3 and 6 (inclusive), you flip a coin for which the probability of a head is $\frac{2}{3}$

  ▶ Greater than 6, you flip a coin for which the probability of a head is $\frac{1}{3}$.

    a Find the probability that the coin shows a tail.

    b If it shows a head, calculate the probability that the coin most likely to produce a head was tossed.

8 A test is developed for a disease. When the disease is present, the probability of the test detecting it is 0.80. The probability of a false-positive result, where the test wrongly suggests the presence of the disease, is 0.15. The disease is present in 38% of those tested.

  a Given that a patient is tested positive, what is the probability that that patient does *not* have the disease?

  b What is the probability that the test gives the wrong result?

9 Three technicians, E, F and G, each assemble identical electrical components. The probabilities of each technician producing a defective component are 0.18, 0.15 and 0.03, respectively. The technicians produce these components at the same rate.

Two components are taken at random from the day's output and neither is found to be defective. Find the probability that both components were produced by technician G.

10 An urn contains four white and six black balls. A ball is drawn at random and its colour noted. The ball, and another ball of the same colour, are put back into the jar. A second ball is then drawn and its colour noted.

Given that the second ball was black, what is the probability that the first ball was white?

> **Note**
>
> This is an example of a Polya urn model: the probability of getting a ball on the second draw that is the same colour as the first ball is *increased* rather than staying constant or decreasing.

Summary

▶ You can use a tree diagram to represent the possible outcomes of experiments that have more than one stage, and where the events are independent or dependent. When the stages are:
  • Independent, the probabilities along a branch are not affected by the outcome of the preceding branch
  • Dependent, the probabilities along a given branch will be *conditional* on the outcome of the preceding branch.

▶ For *any* tree diagram, to find the probability of:
  • A given final outcome, multiply the probabilities along the relevant path
  • Any one of a set of final outcomes, add the relevant final probabilities.

▶ If a sample space consists of a set of n mutually exclusive and exhaustive events, B_j, $j = 1, 2, \ldots, n$, and event A is defined on this sample space, then **Bayes' Theorem** states that $P(A_j \mid B) = \dfrac{P(A_j) \times P(B \mid A_j)}{\sum_{i=1}^{n} P(A_i) \times P(B \mid A_i)}$.

Review exercises

1 A fair coin is flipped twice.
  a Draw a tree diagram to show the possible outcomes of this experiment.
  b Use your tree diagram to calculate the probability that the results of the two flips are different.

2 An experiment can result in one of two possible outcomes, E_1 and E_2, with equal probability. The experiment is performed three times. If the outcomes of the three experiments are independent of each other, find, by drawing a probability tree diagram, the probability that the outcome of the second experiment is different from those of the first and third.

3 Two urns contain red and green counters. Urn 1 contains eight red and four green; urn 2 contains six red and three green. A counter is chosen at random from urn 1 and its colour noted, and then it is placed in urn 2. A counter is then chosen at random from urn 2. By drawing a tree diagram, find

   a The probability that both counters are green

   b The probability that the first counter is red and the second, green

   c The probability that the counters are of different colours.

4 In a random experiment, an event, D, can come about in one of three exhaustive and mutually exclusive ways, A, B and C. Given that the probabilities of A, B and C are 0.3, 0.3 and 0.4 and that $P(D\,|\,A) = 0.3$, $P(D\,|\,B) = 0.4$ and $P(D\,|\,C) = 0.5$, find the probability $P(C\,|\,D)$.

5 Two mutually exclusive and exhaustive events, B and C, are defined on a sample space. Event A is also defined on the sample space. Given that $P(B) = 0.4$, $P(A\,|\,B) = 0.6$ and $P(A\,|\,C) = 0.5$, find

   a $P(C)$ **b** $P(B\,|\,A)$ **c** $P(C\,|\,A)$ **d** $P(A)$

Practice examination questions

1 On a rail route between two stations, A and B, 90% of trains leave A on time and 10% of trains leave A late. Of those trains that leave A on time, 15% arrive at B early, 75% arrive on time and 10% arrive late. Of those trains that leave A late, 35% arrive at B on time and 65% arrive late.

   a Represent this information by a fully-labelled tree diagram.   (3 marks)

   b Hence, or otherwise, calculate the probability that a train:

      i arrives at B early or on time;

      ii left A on time, given that it arrived at B on time;

      iii left A late, given that it was not late in arriving at B.   (7 marks)

   c Two trains arrive late at B. Assuming that their journey times are independent, calculate the probability that exactly one train left A on time.   (4 marks)

AQA MS03 June 2013

2 In a large drug trial, a certain disease is treated with either one of two drugs, D_1 or D_2, or a placebo, P, with equal numbers in each group. The cure rates for these treatments are 0.81, 0.52 and 0.58, respectively.

   a Find the probability that a patient chosen at random is cured.   (3 marks)

   Another patient is chosen at random.

   b Given that this patient is cured, what is the probability that she received the placebo?   (4 marks)

3 It is proposed to introduce, for all males at age 60, screening tests, A and B, for a certain disease. Test B is administered only when the result of Test A is inconclusive. It is known that 10% of 60-year-old men suffer from the disease. For those 60-year-old men suffering from the disease:

▶ Test A is known to give a positive result, indicating a presence of the disease, in 90% of cases, a negative result in 2% of cases and a requirement for the administration of Test B in 8% of cases;

▶ Test B is known to give a positive result in 98% of cases and a negative result in 2% of cases.

For those 60-year-old men not suffering from the disease:

▶ Test A is known to give a positive result in 1% of cases, a negative result in 80% of cases and a requirement for the administration of Test B in 19% of cases;

▶ Test B is known to give a positive result in 1% of cases and a negative result in 99% of cases.

a Draw a tree diagram to represent the above information. (4 marks)

b i Hence, or otherwise, determine the probability that:

    A a 60-year-old man, suffering from the disease, tests negative;

    B a 60-year-old man, not suffering from the disease, tests positive. (2 marks)

  ii A random sample of ten thousand 60-year-old men is given the screening tests. Calculate, to the nearest 10, the number who you would expect to be given an **incorrect** diagnosis. (2 marks)

c Determine the probability that:

  i a 60-year-old man suffers from the disease given that the tests provide a positive result;

  ii a 60-year-old man does not suffer from the disease given that the tests provide a negative result. (5 marks)

AQA MS03 June 2010

4 An investigation was carried out into the type of vehicle being driven when its driver was caught speeding. The investigation was restricted to drivers who were caught speeding when driving vehicles with at least four wheels.

An analysis of the results showed that 65% were driving cars (C), 20% were driving vans (V) and 15% were driving lorries (L).

Of those driving cars, 30% were caught by fixed speed cameras (F), 55% were caught by mobile speed cameras (M) and 15% were caught by average speed cameras (A).

Of those driving vans, 35% were caught by fixed speed cameras (F), 45% were caught by mobile speed cameras (M) and 20% were caught by average speed cameras (A).

Of those driving lorries, 10% were caught by fixed speed cameras (F), 65% were caught by mobile speed cameras (M) and 25% were caught by average speed cameras (A).

a Represent this information by a tree diagram on which are shown labels and percentages or probabilities. (3 marks)

b Hence, or otherwise, calculate the probability that a driver, selected at random from those caught speeding:

i was driving either a car or a lorry and was caught by a mobile speed camera;

ii was driving a lorry, given that the driver was caught by an average speed camera;

iii was not caught by a fixed speed camera, given that the driver was not driving a car. (8 marks)

c Three drivers were selected at random from those caught speeding by **fixed speed cameras**.

Calculate the probability that they were driving three different types of vehicle. (4 marks)

AQA MS03 June 2014

5 A manufacturer produces three models of washing machine: basic, standard and deluxe. An analysis of warranty records shows that 25% of faults are on basic machines, 60% are on standard machines and 15% are on deluxe machines.

For basic machines, 30% of faults reported during the warranty period are electrical, 50% are mechanical and 20% are water-related.

For standard machines, 40% of faults reported during the warranty period are electrical, 45% are mechanical and 15% are water-related.

For deluxe machines, 55% of faults reported during the warranty period are electrical, 35% are mechanical and 10% are water-related.

a Draw a tree diagram to represent the above information. (3 marks)

b Hence, or otherwise, determine the probability that a fault reported during the warranty period:

i is electrical (2 marks)

ii is on a deluxe machine, given that it is electrical. (2 marks)

c A random sample of 10 electrical faults reported during the warranty period is selected. Calculate the probability that exactly 4 of them are on deluxe machines. (3 marks)

AQA Teacher Support Materials MS03 2008

11 Discrete Uniform and Geometric Distributions

Introduction

Probability distributions are mathematical descriptions, or models, of random phenomena. The most important distributions model commonly occurring, real-life situations, for example the height of a randomly chosen person from a particular population or the number of random events in a given period of time. In this chapter, you will be introduced to two discrete probability distributions, the discrete uniform and geometric distributions.

Objectives

By the end of this chapter, you should know how to:

▶ Recognise situations that can be modelled by a discrete uniform distribution or a geometric distribution.

▶ Calculate probabilities, means and variances for these distributions.

▶ Solve problems related to these distributions.

Recap

You will need to remember…

▶ The formulae for the mean and variance of a random variable:

$$E[X] = \mu = \sum x_i p_i$$

$$\sigma^2 = \text{Var}[X] = \sum x_i^2 p_i - \mu^2.$$

▶ The summation of infinite geometric series, first term a, common ratio r: $S_\infty = \frac{a}{1-r}; |r| < 1$.

▶ The binomial expansion of expressions of the form $(a + x)^n$ for positive integers n: $(a + x)^n \equiv a^n + na^{n-1}x + \frac{n(n-1)a^{n-2}x^2}{2!} + \cdots$

▶ The definition of a Bernoulli trial: any random experiment that can be considered as having only two outcomes, often labelled 'success' and 'failure'; Bernoulli trials are often considered in sets of n that are all identical and independent in the context of a binomial random variable.

▶ Basic rules of probability; for an experiment E, with associated events A and B defined on a sample space S,

1. $P(S) = 1$

2. $0 \leq P(A) \leq 1$ for $A \subseteq S$

3. $P(A \cup B) = P(A) + P(B); A \cap B = \varnothing$.

11.1 Discrete uniform distribution

In any experiment modelled by a **discrete uniform distribution**, the random variable can take one of a finite number of discrete values, each with the same probability. These values are often, but not always, consecutive integers 1 to n.

Probability model and probability distribution

Consider the classic random experiment, E, where a fair dice is rolled once. Since the dice is fair, each of the six outcomes, integers 1–6, have an equal probability of occurring of $\frac{1}{6}$.

If X is a random variable for the score obtained when the dice is rolled once, then its probability distribution is given by $P(X = x) = \frac{1}{6}$; $x = 1, 2, \ldots, 6$.

This is a special case of a discrete uniform distribution; in general, the x-values do not have to be consecutive integers.

If a random variable, U, has a uniform distribution taking values u_i, $i = 1, 2, \ldots, m$, it follows that $P(U = u_i) = k$ where k is some constant. However, you know from Axiom 1 (the total probability theorem) that $P(S) = 1$.

Therefore, $mk = 1$ and so $k = \dfrac{1}{m}$.

> A discrete uniform random variable, U, taking values u_i, $i = 1, 2, \ldots, m$, has probability distribution given by $P(U = u_i) = \frac{1}{m}$; $i = 1, 2, \ldots, m$; $u_i \in \mathbb{R}$.

You can use this rule to calculate the probabilities of certain outcomes. If the random variable X takes m different values each with equal probability, then the probability of any one of them occurring when the experiment is performed is $\frac{1}{m}$.

Example 1

Two fair tetrahedral dice, each numbered 1–4, are rolled once. The random variables T_1 and T_2 are the scores on the two dice. Find the probability of obtaining

a a total of 5   b a product of 4.

$P(T_1 = i) = P(T_2 = i) = \dfrac{1}{4}$; $i = 1, 2, 3, 4$

a Outcomes for a total of 5:
(1, 4), (2, 3), (3, 2), (4, 1)
The probability of a total of 5 is $4 \times \dfrac{1}{16} = \dfrac{1}{4}$.

b Outcomes for a product of 4:
(1, 4), (2, 2), (4, 1)
Therefore, the probability of a product of 4 is $\dfrac{3}{16}$.

> **Note**
> The variables T_1 and T_2 each follow a uniform distribution.

> **Note**
> Because of the independence of the outcomes on the two dice, the probability of each of these outcomes is $\frac{1}{4} \times \frac{1}{4} = \frac{1}{16}$.

> **Note**
> It is probably best to think of the two rolls occurring one after the other. Then, outcome 1 and 4 is different from outcome 4 and 1.

> **Note**
> Axiom 3, $P(A \cup B) = P(A) + P(B)$; $A \cap B = \varnothing$

Mean and variance

The mean and variance of a discrete uniform distribution, U, can be derived using the formulae you learned in your A-level Mathematics course:

$$\mu = E[X] = \sum_x xP(X = x), \quad \sigma^2 = \sum_x x^2 P(X = x) - \mu^2$$

$$\mu_U = E[U] = \sum_i x_i P(U = x_i) = \sum_{i=1}^{a} \frac{x_i}{a} = \frac{1}{a}\sum_{i=1}^{a} x_i$$

$$\sigma_U^2 = \sum_i x_i^2 P(U = x_i) - \mu_U^2 = \sum_{i=1}^{a} \frac{x_i^2}{a} - \left(\mu_U\right)^2 = \frac{1}{a}\sum_{i=1}^{a} x_i^2 - \left(\mu_U\right)^2$$

> **Tip**
> $P(U = x_i) = \dfrac{1}{a}$

A discrete uniform random variable, U, with probability distribution $P(U = u_i) = \frac{1}{a}$, $i = 1, 2, \ldots, a$, has a mean and variance given

by $\mu_U = \frac{1}{a}\sum_{i=1}^{a} u_i$ and $\sigma_U^2 = \frac{1}{a}\sum_{i=1}^{a} u_i^2 - \left(\frac{1}{a}\sum_{i=1}^{a} u_i\right)^2$.

For the special case where u_i is a set of consecutive integers $1 - n$ (for example, the score on a fair dice, $n = 6$), these results can be proved.

A discrete uniform random variable, U, with probability distribution $P(U = i) = \frac{1}{n}$, $i = 1, 2, \ldots, n$, has a mean and variance given by

$$\mu_U = \sum_{i=1}^{n} \frac{i}{n} = \frac{n+1}{2} \qquad [1]$$

$$\sigma_U^2 = \sum_{i=1}^{n} \frac{i^2}{n} - \mu_U^2 = \frac{n^2-1}{12} \qquad [2]$$

You should learn the proofs for the mean and variance of a discrete uniform distribution as it is possible for them to be requested in the examination.

Tip

You will be asked to prove these results in question **1** of the practice examination questions at the end of the chapter.

Formulae booklet

Equations [1] and [2] are given in the *Formulae and Statistical Tables* booklet. Unless you are specifically asked to prove them, you can quote them without proof.

Example 2

Question

An ordinary dice is rolled once. Calculate the mean and variance of the number that is scored.

Answer

If X is the face number scored, then $P(X = x) = \frac{1}{6}$; $x = 1, 2, \ldots, 6$.

$$\mu_X = \frac{6+1}{2} = \frac{7}{2}$$

$$\sigma_X^2 = \frac{6^2-1}{12} = \frac{35}{12}$$

Note

As proof has not been requested, you can simply quote the equations.

Note

Use equation $\mu_U = \sum_{i=1}^{n} \frac{i}{n} = \frac{n+1}{2}$.

Note

Use equation $\sigma_U^2 = \frac{n^2-1}{12}$.

Example 3

Question

The random variable, X, takes values 2, 4, 6, ..., 20 with equal probabilities.

Find the mean and variance of X.

Answer

X has probability distribution given by $P(X = 2c) = \frac{1}{10}$; $c = 1, 2, \ldots, 10$.

$$\mu_X = \sum_{c=1}^{10} \frac{2c}{10}$$

$$= \frac{1}{5}\sum_{c=1}^{10} c$$

$$= \frac{1}{5} \times \frac{10 \times 11}{2} = 11$$

Note

Use $\mu = \sum_{\forall x} x P(X = x)$.

Note

Use $\sigma^2 = \sum_{\forall x} x^2 P(X = x) - \mu^2$

Note

Expressing the distribution in this form means that the **summation** can be over *consecutive* integers even though X takes only even values.

Note

Use $\sum_{i=1}^{n} i = \frac{n(n+1)}{2}$.

(continued)

(continued)

$$\sigma_X^2 = \sum_{c=1}^{10} \frac{(2c)^2}{10} - 11^2 = \frac{2}{5}\sum_{c=1}^{10} c^2 - 11^2$$

$$= \frac{2}{5} \times \frac{1}{6} \times 10 \times 11 \times 21 - 11^2 = 154 - 121 = 33$$

Even if the values of the random variable are not given explicitly, its mean and variance can still be found.

Example 4

A random variable X has the probability function

$$P(X = x) = \begin{cases} \dfrac{1}{2n}; & x = 1, 2, 3, \ldots, 2n \\ 0; & \text{otherwise} \end{cases}$$

where n is a positive integer.

a Determine, in terms of n, an expression for $E[X]$.

b Prove that $\text{Var}[X] = \dfrac{4n^2 - 1}{12}$.

a $P(X = x) = \begin{cases} \dfrac{1}{2n}; & x = 1, 2, 3, \ldots, 2n \\ 0; & \text{otherwise} \end{cases}$

$$E[X] = \sum_{x=1}^{2n} x\frac{1}{2n} = \frac{1}{2n}\sum_{x=1}^{2n} x$$

$$= \frac{1}{2n} \times \frac{2n(2n+1)}{2} = \frac{2n+1}{2}$$

b $\text{Var}[X] = E[X^2] - (E[X])^2$

$$= \sum_{x=1}^{2n} \frac{1}{2n}x^2 - \left(\frac{2n+1}{2}\right)^2$$

$$= \frac{1}{2n} \times \frac{1}{6} \times 2n(2n+1)(4n+1) - \frac{4n^2+4n+1}{4}$$

$$= \frac{(4n+2)(4n+1) - 3(4n^2+4n+1)}{12}$$

$$= \frac{4n^2-1}{12}$$

Note

$$\sum_{x=1}^{n} x^2 = \frac{1}{6}n(n+1)(2n+1)$$

Exercise 1

1 A fair tetrahedral dice numbered 1–4 is rolled once. Calculate the mean and variance of the number that is scored.

2 The random variable Y takes odd values 1–9 inclusive with equal probabilities. Find the mean and variance of X.

3 X is a random variable for the score on a fair, 10-sided dice.

State the distributions of these random variables and find their means and variances.

a $2X$ b $X - 1$

4 A random number generator is programmed to produce numbers from 1 to 400 with equal probabilities. X and Y are random variables for the first and last digits of the number generated. The number generator is then reprogrammed to produce numbers from 1 to 415 and S is a variable for the final digit.

    a Which of these variables, if any, is discrete uniform?

    b For any uniform variable, give the value of its parameter p.

5 A random experiment involves rolling a fair, tetrahedral dice with sides numbered 1–4, followed by a card being chosen from a pack of eight cards numbered 1–8. By considering two uniformly distributed random variables, find the probability of obtaining a total score that is a prime number.

6 Two fair, eight-sided dice are each numbered 1, 3, 5, …, 15. Find the probability that, when both dice are rolled, the total score is less than 9.

7 A digital clock shows the time in hours and minutes but not seconds. You look at the clock at some randomly chosen time of day. What is the probability that the clock shows a time between 02:10 hours and 04:40 hours?

Hint

Be careful: the time for which the clock shows those minutes may not be exactly what you at first think.

8 A number is chosen at random from the set of all prime numbers between 30 and 50 inclusive. What is the probability that its value is greater than 40?

9 A fair 10-sided dice with faces numbered 1–10 is thrown once. From first principles (that is, from the probability distribution and the general definitions of mean and variance), prove that the mean and variance of the score obtained are $\frac{11}{2}$ and $\frac{33}{4}$.

10 A random number generator shows numbers from 1 to 20 with equal probabilities.

    a Find the expected value obtained when one number is generated.

    The generator then malfunctions by failing to produce sixes, with the other numbers being produced with equal probability.

    b Find the new expected value.

11 A fairground game consists of throwing darts at a set of cards pinned to a board. The cards are numbered 0–50 and the player wins, in fairground tokens, the number on the card on which the dart lands. For a player whose dart lands on the set of cards at random, what is the expected number of tokens won per game? (A fair game is one where, *in the long run*, the player can expect to win or lose nothing.)

12 A random variable X has the probability function

$$P(X = x) = \begin{cases} \dfrac{1}{4n}; & x = 1, 2, 3, …, 4n \\ 0; & \text{otherwise} \end{cases}$$

where n is a positive integer.

    a Determine, in terms of n, an expression for $E[X]$.

    b Prove that
$$\text{Var}[X] = \frac{16n^2 - 1}{12}$$

    c If $n = 8$, calculate the exact value of
$$P(E[X] - \sqrt{\text{Var}[X]} < X < E[X] + \sqrt{\text{Var}[X]})$$

11.2 Geometric distribution

The geometric distribution is used when a set of independent and identical random experiments is performed and the random variable of interest, X, is the number of experiments up to and including a particular outcome. For example, X could be the number of rolls of a dice up to and including the first 6.

Probability model and probability distribution

To see the real-life significance of this distribution, consider this situation.

As a market researcher, you have been asked to interview adults who say that they are the main earner in their household. You stand at the entrance to a shopping mall and approach adults until you find one who says that they are the main earner.

It is possible to consider this situation mathematically by modelling each approach to a shopper as a Bernoulli trial where p is the probability of success (that is, finding a main earner).

> **Tip**
>
> Bernoulli trials are defined in the Recap section if you need a reminder.

Denote the number of approaches *up to and including* the main earner by the random variable X. Once this person has been identified and questioned and you have the first value of X, counting starts from 1 again to get the next value.

The distribution of X is $P(X = x) = (1 - p)^{x-1}p$, $x = 1, 2, 3, \ldots,$ because exactly x approaches requires $x - 1$ failures followed by one success and it is assumed that there is independence between outcomes of these approaches.

The random variable X is said to follow a **geometric distribution**.

> A random variable X follows a geometric distribution if it denotes the number of independent and identical Bernoulli trials up to and including the first success. Its probability distribution is given by
>
> $$P(X = x) = (1 - p)^{x-1}p, \ x = 1, 2, 3, \ldots$$
>
> where p is a constant and is the probability of success in any given trial.

> **Tip**
>
> The distribution is called geometric because the successive probabilities, p, $(1 - p)p$, $(1 - p)^2 p \ldots$, form a geometric sequence, first term p, common ratio $(1 - p)$.

This probability distribution applies when the random variable X is the number of trials *up to and including* the first success.

Occasionally, the variable is the number of trials *up to but not including the success*. Using identical principles, the probability distribution becomes $P(Y = y) = (1 - p)^y p$, $y = 0, 1, 2, 3, \ldots$, where in this case the variable values start from 0 rather than 1.

Example 5

A random variable, X, has a geometric probability distribution with $p = 0.7$.

Find

a $P(X = 2)$ b $P(X = 4)$ c $P(X < 3)$ d $P(X \geq 3)$

$P(X = x) = (1-p)^{x-1}p$

a $P(X = 2) = 0.3 \times 0.7 = 0.21$

b $P(X = 4) = 0.3^3 \times 0.7 = 0.0189$

c $P(X < 3) = P(X = 1) + P(X = 2)$

         $= 0.7 + 0.3 \times 0.7 = 0.91$

d $P(X \geq 3) = 1 - P(X < 3) = 1 - 0.91 = 0.09$

> **Note**
>
> State the probability distribution.

> **Note**
>
> For X to equal 2, there must be one failure and one success.

> **Note**
>
> Use Axiom 3, $P(A \cup B) = P(A) + P(B)$; $A \cap B = \emptyset$. Remember, X cannot equal 0.

> **Note**
>
> Axioms 1, $P(S) = 1$ and 3.

Example 6

A fair dice is rolled until a 1 is scored. If X is a random variable for the number of rolls required, find the probability that

a $X = 4$ b $X \geq 3$

> **Note**
>
> As the dice is fair, $p = \dfrac{1}{6}$.

$P(X = x) = (1-p)^{x-1}p$, $x = 1, 2, 3, \ldots$

a $P(X = 4) = \left(1 - \dfrac{1}{6}\right)^3 \dfrac{1}{6} = \dfrac{125}{1296}$

b $P(X \geq 3) = 1 - P(X \leq 2) = 1 - \left(\dfrac{1}{6} + \dfrac{5}{6} \times \dfrac{1}{6}\right) = \dfrac{25}{36}$

Mean and variance

The mean and variance of a geometric distribution can be derived using the general formulae you learned in your AS-level Mathematics course and which are given in the Recap section at the start of this chapter,

$$\mu = \mathrm{E}[X] = \sum_{\forall x} xP(X = x) = p\sum_{x=1}^{\infty} x(1-p)^{x-1}$$

$$= p(1 + 2q + 3q^2 + \cdots) \text{ where } q = 1 - p$$

$$= p(1-q)^{-2} \qquad\qquad [3]$$

$$\text{Set } p = 1 - q$$

$$= \dfrac{1}{p}$$

$$\sigma^2 = \mathrm{Var}[X] = \mathrm{E}[X^2] - \mu^2$$

Now, $\mathrm{E}[X^2] = p\sum_{x=1}^{\infty} x^2 q^{x-1}$

$$= (1 - q)(1 + 4q + 9q^2 + \cdots)$$

$$= (1 + 4q + 9q^2 + \cdots) - (q + 4q^2 + 9q^3 + \cdots)$$

$$= 1 + 3q + 5q^2 + 7q^3 + \cdots$$

This is not as easy to sum as it looks but it can be summed using this method:

Adding and subtracting the series $1 + q + q^2 + q^3 + \cdots$

$$\mathrm{E}[X^2] = (2 + 4q + 6q^2 + \cdots) - (1 + q + q^2 + \cdots)$$

$$= 2(1 + 2q + 3q^2 + \cdots) - (1 + q + q^2 + \cdots)$$

> **Tip**
>
> The proof of this result is beyond the AS specification and will therefore have to be memorized.

> **Tip**
>
> When reproducing this proof in an exam, you might have to remember this line.

The first of these series has been summed using a binomial expansion earlier in the proof. The second is an infinite geometric series, first term 1, common ratio q, and can be summed using the formula given in the Recap section of this chapter.

$$E[X^2] = \frac{2}{(1-q)^2} - \frac{1}{1-q}$$

$$= \frac{2}{p^2} - \frac{1}{p}$$

Now, $\text{Var}[X] = E[X^2] - \mu^2$.

Therefore, $\text{Var}[X] = \frac{2}{p^2} - \frac{1}{p} - \left(\frac{1}{p}\right)^2 = \frac{1-p}{p^2}$.

> **A geometric random variable, X, with probability distribution**
> $P(X = x) = (1-p)^{x-1}p$, $x = 1, 2, 3, \ldots$, **has a mean and variance**
> **given by** $\mu = E[X] = \frac{1}{p}$ **and** $\sigma^2 = \text{Var}[X] = \frac{1-p}{p^2}$.

You should learn these proofs for the mean and variance of a geometric distribution as it is possible for them to be requested in the examination. If the proofs are not requested, you can use the formulae without proof.

Example 7

R is a random variable for the number of Bernoulli trials up to and including the first success. Given that $P(R = 1) = 0.2$

a   show that $P(R = 3) = 0.128$

b   find the mean and variance of R.

Note

Use $P(X = x) = (1-p)^{x-1}p$, $x = 1, 2, 3, \ldots$

a   Let the probability of success in any given trial be p.

$P(R = r) = (1-p)^{r-1}p$

Make $r = 1$, then

$P(R = 1) = p$

If $p = 0.2$, then $(1-p)^2 p = 0.128$.

b $\mu = \frac{1}{0.2} = 5$

$\sigma^2 = \frac{0.8}{0.2^2} = 20$

Note

Use $\mu = \frac{1}{p}$.

Note

Use $\sigma^2 = \frac{1-p}{p^2}$.

When modelling real situations, one of the skills required is to be able to choose the correct probability model out of all of the ones you have met. Make sure that you understand why, in a question, a particular model is the correct one. In the next worked question, for example, be careful as only part is geometric.

Example 8

As part of their archery practice, Robin and William are playing a game consisting of a number of rounds. For each round of the game, they each shoot one arrow at the gold inner circle of a target. The probability that Robin hits the gold with any one arrow is $\frac{1}{5}$, independently of all previous shots. This probability for William is $\frac{1}{6}$. In each round, Robin shoots first. If they both hit the gold, then the game ends with a draw. If one of them hits the gold and the other misses, the one who hits the gold wins. If they both miss the gold, then the game continues to the next round.

Find the probability that

a   the game is drawn after no more than three rounds have been completed

b   the game is drawn

c   Robin wins the game.

The rules are now changed so that the only way the game can finish is by either Robin or William winning a round. Otherwise the game continues to another round.

d   Find the probability that the game lasts exactly three rounds.

Let X be the number of rounds up to and including the first $G_R G_W$ with only $G'_R G'_W$ occurring before it. Let p be the probability of a draw in any given round.

$$p = P(G_R G_W) = \frac{1}{5} \times \frac{1}{6} = \frac{1}{30}$$

$$P(G'_R G'_W) = \frac{4}{5} \times \frac{5}{6} = \frac{2}{3}$$

> **Note**
>
> Use the obvious notation.

a $P(X \leq 3) = P(X = 1) + P(X = 2) + P(X = 3)$

$$= \frac{1}{30} + \frac{2}{3} \times \frac{1}{30} + \left(\frac{2}{3}\right)^2 \times \frac{1}{30} = \frac{19}{270}$$

> **Note**
>
> The game is drawn at the first $G_R G_W$ and therefore if $X = 1$ or $X = 2$ or ...

b $P(\text{game drawn}) = P(X \geq 1)$

$$= \frac{1}{30} + \frac{2}{3} \times \frac{1}{30} + \left(\frac{2}{3}\right)^2 \times \frac{1}{30} + \cdots$$

$$= \frac{\frac{1}{30}}{1 - \frac{2}{3}} = \frac{1}{10}$$

> **Note**
>
> An infinite geometric series.

c   In any given round, Robin wins when $G_R G'_W$ occurs, probability $\frac{1}{6}$.

$$P(\text{Robin wins the game}) = \frac{1}{6} + \frac{2}{3} \times \frac{1}{6} + \left(\frac{2}{3}\right)^2 \times \frac{1}{6} + \cdots$$

$$= \frac{\frac{1}{6}}{1 - \frac{2}{3}} = \frac{1}{2}$$

> **Note**
>
> $$= \frac{1}{5} \times \frac{5}{6}$$

> **Note**
>
> Sum of an infinite geometric series.

d $P(\text{a game is won}) = P(G_R G'_W \text{ or } G'_R G_W)$

$$= \frac{1}{5} \times \frac{5}{6} + \frac{4}{5} \times \frac{1}{6} = \frac{3}{10}$$

Let Y denote the number of games until a win.

$$P(Y = 3) = \left(\frac{7}{10}\right)^2 \left(\frac{3}{10}\right) = 0.147$$

> **Note**
>
> With the new rules, the probability of any round being won is still $\frac{3}{10}$ but the game continues until this happens.

> **Note**
>
> Remember, the game continues after each $G'_R G'_W$ round and ends with a draw after a $G_R G_W$ round as long as $G'_R G_W$ or $G_R G'_W$ does not occur. X is not a geometric random variable because the probabilities are not of the form $(1-p)p^{n-1}$.

> **Note**
>
> Y is geometric, $p = \frac{3}{10}$.

Exercise 2

1 A fair coin is flipped until a head is scored. If X is a random variable for the number of flips required, find the probability that **a** $X = 4$ **b** $X \geq 4$

2 A random variable, G, has probability distribution given by

$P(G = g) = (1 - p)^{g-1}p$, $g = 1, 2, 3, \ldots$

If $p = 0.9$, find **a** $P(G = 2)$ **b** $P(G = 6)$ **c** $P(G < 3)$ **d** $P(G \geq 2)$

3 A random variable, X, has probability distribution given by

$P(X = x) = (1 - p)^{x-1}p$, $x = 1, 2, 3, \ldots$

Find the mean and variance of X if $p = 0.5$.

4 An ordinary dice is rolled until a prime number appears. Find the mean and variance of the number of rolls required.

5 A random variable, Y, has probability distribution given by

$P(Y = y) = (1 - p)^{y-1}p$, $y = 1, 2, 3, \ldots$

Given that the mean of Y is 8, find the variance of Y.

6 Prove that the cumulative probability distribution, $P(X \leq x)$, for a geometric random variable X, parameter p, is given by $P(X \leq x) = 1 - q^x$, where $q = 1 - p$.

7 Y is a random variable for the number of Bernoulli trials up to, *but not including*, the first success.

a Write the probability distribution of Y and find its expected value.

b Check your answer to the expected value of Y by using the relationship $Y = X - 1$ where X is the number of trials *up to and including* the first success.

8 An ordinary dice is repeatedly rolled. What is the probability that a 6 occurs for the second time on the fourth throw?

9 Prove that the probability distribution

$P(X = x) = (1 - p)^{x-1}p$, $x = 1, 2, 3, \ldots$

gives probabilities for $P(S) = 1$ where S is the sample space.

Summary

▶ A discrete uniform random variable, U, taking values u_i, $i = 1, 2, \ldots, a$, has
  • A probability distribution given by
  $$P(U = u_i) = \frac{1}{m}; i = 1, 2, \ldots, m; u_i \in \mathbb{R}$$
  • Mean and variance given by
  $$\mu_U = \frac{1}{a}\sum_{i=1}^{a} u_i \text{ and } \sigma_U^2 = \frac{1}{a}\sum_{i=1}^{a} u_i^2 - \mu_U^2$$

This distribution is used when the random variable of interest takes one of a finite number of values each with an equal probability.

- For the special case of a discrete uniform random variable, U, with probability distribution given by $P(U = i) = \frac{1}{n}$, $i = 1, 2, \ldots, n$, the mean and variance are given by

$$\mu_U = \frac{n+1}{2} \text{ and } \sigma_U^2 = \frac{n^2-1}{12}$$

- A random variable X follows a geometric distribution if its probability distribution is given by

$$P(X = x) = (1-p)^{x-1}p, \; x = 1, 2, 3, \ldots$$

where p is a constant and is the probability of success in any given trial.

This distribution is used when you can identify a number of independent Bernoulli trials and the random variable is the number of trials up to and including the first success.

- The mean and variance of a random variable with a geometric distribution are given by

$$\mu = \frac{1}{p} \text{ and } \sigma^2 = \frac{1-p}{p^2}$$

Review exercises

1. Two fair, six-sided dice are each numbered with integers 1–6. Find the probability that, when both dice are rolled, the total score is less than six.

2. Pairs of random digits are chosen so that all values from 00 to 99 are equally likely. Using the general definitions of mean and variance, find the probability that, when a pair of digits is chosen at random, the number obtained is more than one standard deviation away from its mean value.

3. R is a random variable for the number of Bernoulli trials up to and including the first success. Given that $P(R = 2) = \frac{3}{16}$, find two possible values of $P(R = 1)$.

4. Recall the scenario earlier in the chapter, where you are a market researcher interviewing adults who say that they are the main earner in their household. Using the same scenario, this time you find that, on average, you have to question five people until you meet one person who fits the description of 'main earner' (the fifth person).

Find the probability that

a any given person in the shopping mall fits this description

b that you have to approach at least three people in order to find someone to question.

Practice examination questions

1 Using the results for the mean and variance of a discrete uniform distribution:

$$\mu_U = \frac{1}{a}\sum_{i=1}^{a} u_i \text{ and } \sigma_U^2 = \frac{1}{a}\sum_{i=1}^{a} u_i^2 - \left(\frac{1}{a}\sum_{i=1}^{a} u_i\right)^2,$$

prove that, for the special case where $u_i = i; i = 1, 2, \ldots, n$

$$\mu_U = \frac{n+1}{2}$$
(2 marks)

and

$$\sigma_U^2 = \frac{n^2 - 1}{12}$$
(3 marks)

2 The random variable x has a geometric distribution with parameter p.

a Prove, from first principles, that $E(X^2) = \dfrac{1}{p} + \dfrac{2(1-p)}{p^2}$. (4 marks)

b Hence, given that $E(X) = \dfrac{1}{p}$, deduce that $\text{Var}(X) = \dfrac{(1-p)}{p^2}$. (1 mark)

c Given that $p = \dfrac{1}{2}$, calculate $P(X > \text{Var}(X))$. (3 marks)

AQA MS04 June 2014

3 a For a geometric random variable, X, with probability distribution

$$P(X = x) = (1-p)^{x-1}p; x = 1, 2, 3, \ldots$$

prove that the mean is given by (3 marks)

$$\mu = \frac{1}{p}$$

b In a school fair raffle, tickets are drawn in front of a group of parents. The tickets have been on sale for some weeks and only 12% of the tickets are with the parents present. The rules state that you have to be present to win a prize.

i What is the probability that, of the first 5 tickets drawn, exactly one prize is won? (2 marks)

ii What is the probability that the first prize is won on the 5th draw? (2 marks)

4 a The random variable X has a geometric distribution with parameter p.

i Prove from first principles that $E(X(X-1)) = \dfrac{2(1-p)}{p^2}$. (3 marks)

ii Hence, given that $E(X) = \dfrac{1}{p}$, prove that $\text{Var}(X) = \dfrac{1-p}{p^2}$. (2 marks)

b The independent random variables X_1 and X_2 have geometric distributions with parameters p_1 and p_2 respectively.

It is given that $\dfrac{E(X_1)}{E(X_2)} = \dfrac{2}{3}$ and that $\dfrac{\text{Var}(X_1)}{\text{Var}(X_2)} = \dfrac{1}{3}$.

i Show that $p_1 = \dfrac{1}{2}$. (5 marks)

ii Hence find the least value of N such that $P(X_1 > N) < 10^{-5}$. (4 marks)

AQA MS04 June 2010

5 Andy plays a game with Bea by throwing an unbiased six-sided die until a 5 is obtained. When a 5 is obtained, Bea pays Andy £10.

Find, to the nearest penny, the amount that Bea should charge Andy **per throw** so that, in the long run, Bea makes a profit of £1 **per game**. (4 marks)

AQA MS04 June 2011

6 a The random variable X has a geometric distribution with parameter p.

　i Prove that $E(X) = \dfrac{1}{p}$. (3 marks)

　ii Given that $E(X^2) = \dfrac{2-p}{p^2}$, show that $\text{Var}(X) = \dfrac{1-p}{p^2}$. (2 marks)

b An unbiased tetrahedral die has faces marked 1 to 4. When it is thrown on a table, the score is the number on the face that is in contact with the table.

　i Calculate the probability that it takes more than two throws to obtain a score of 4. (2 marks)

　ii The number of throws, Y, that it takes for **two different** scores to occur at least once is given by $Y = 1 + X$, where X has a geometric distribution with parameter $\dfrac{3}{4}$.

　　Determine values for $E(Y)$ and $\text{Var}(Y)$. (3 marks)

AQA MS04 June 2012

7 The random variable X has a geometric distribution with parameter p. It is given that $\text{Var}(X)$ is four times $E(X)$.

a Show that the non-zero value of p is 0.20. (3 marks)

b Hence find:

　i $P(X > 7 \mid X > 4)$; (3 marks)

　ii the least integer n such that $P(X > n) < 0.0001$. (4 marks)

AQA MS04 June 2013

Introduction

Functions suggest a dependency or mapping of one variable onto another. A **probability generating function**, however, is a function that is used slightly differently. Instead of defining a mapping, the probability generating function is used to generate probabilities of the various values of the corresponding random variable in a mathematically convenient way. It provides a neat way of summarising the whole probability distribution. It can also be used to calculate the mean and variance of the corresponding probability distribution.

Objectives

By the end of this chapter, you should know how to:

▶ Find probability generating functions for well-known discrete probability distributions.

▶ Use probability generating functions to calculate probabilities and the mean and variance of the corresponding random variables.

▶ Find probability generating functions for sums of independent random variables.

Recap

You will need to remember...

▶ The binomial expansion
$$(a+bx)^n = a^n + na^{n-1} bx + \frac{n(n-1)a^{n-2} b^2}{2!} x^2 +$$

▶ The formula for the sum of squares of integers 1 to n
$$\sum_{x=1}^{n} x^2 = \frac{1}{6}n(n+1)(2n+1).$$

▶ The formula for the sum to infinity of a geometric series
$$S_\infty = \frac{a}{1-r}; |r| < 1.$$

▶ The distribution function of a binomial random variable, X, and parameters n and p
$$P(X = x) = \binom{n}{x} p^x q^{n-x}; x = 0, 1, 2, \ldots, n \text{ where } q = 1-p.$$

▶ Rules of probability; for an experiment, E, with associated events A and B defined on a sample space S,
1. $P(S) = 1$
2. $0 \le P(A) \le 1$ for $A \subseteq S$
3. $P(A \cup B) = P(A) + P(B); A \cap B = \emptyset.$

Formulae booklet

The expectation of functions of random variables is
$$E[g(X)] = \sum_x g(x)P(X = x)$$
where $g(X)$ is a function of the random variable X.

12.1 Probability generating functions and their properties

For a random variable taking integer values and with a given probability distribution, it is possible to define a **probability generating function**. Consider a random variable X with probability distribution

$$P(X = x_i) = p_i; i = 1, 2, \ldots, n$$

The probability generating function, $G_X(t)$, of X is defined as $G_X(t) = E[t^X]$.

Now, the expectation of a function of a random variable is given by

$$E[g(X)] = \sum_{\forall x} g(x)P(X = x)$$

Therefore, substituting t^X for $g(x)$ and p_i for $P(X = x)$ gives

$$G_X(t) = E[t^X] = \sum_{i=1}^{n} p_i t^{x_i}$$

> **Tip**
>
> Axiom 3 gives $G_X(1) = 1$ because this is the sum of the probabilities of all possible outcomes of an experiment.

> The probability generating function, $G_X(t)$, of a random variable X,
>
> defined by the equation $G_X(t) = E[t^X] = \sum_{i=1}^{n} p_i t^{x_i}$, is a function of t where
>
> the coefficient of t^{x_i} is the probability that X takes the value x_i.

In general, $P(X = x)$ is the coefficient of t^x in $G_X(t)$. For example, the coefficient of x^2 in a probability generating function is $P(X = 2)$.

Example 1

An ordinary fair dice is rolled once and the random variable X denotes the score obtained. Write the probability generating function, $G_X(t)$, of X.

$$G_X(t) = E[t^X] = \sum_{i=1}^{n} p_i t^{x_i}$$

$$G_X(t) = \frac{t}{6} + \frac{t^2}{6} + \cdots + \frac{t^6}{6}$$

> **Note**
>
> The probability of each value 1–6, is $\frac{1}{6}$.

Example 2

A random variable Y has probability distribution given by

$$P(Y = y) = \begin{cases} p; & y = 1 \\ 1 - p; & y = 0 \end{cases}$$

Find its probability generating function.

> **Note**
>
> When a probability distribution is given in this form, it means $P(Y = 1) = p$ and $P(Y = 0) = 1 - p$.

> **Note**
>
> $G_Y(t)$ is the probability generating function of Y and so, by definition of a probability generating function, this is equivalent to the statements $P(Y = 1) = p$ and $P(Y = 0) = 1 - p$.

$$G_X(t) = \sum_{i=1}^{n} p_i t^{x_i}$$

$$G_Y(t) = t^0(1 - p) + t^1 p$$

$$= 1 - p + pt$$

Mean and variance using probability generating functions

Probability generating functions provide an efficient way to calculate the mean and variance of random variables.

Suppose $G_X(t)$ is the probability generating function of a random variable X with mean μ and variance σ^2. $G_X(t)$ can be differentiated with respect to t in order to find general formulae for calculating the mean and variance of X.

This can be differentiated with respect to t as follows.

$$G_X'(t) = \frac{d}{dt}G_X(t) = \frac{d}{dt}\sum_{i=1}^{n}p_i t^{x_i}$$

$$= \sum_{i=1}^{n}p_i x_i t^{x_i-1} \qquad [1]$$

Tip

Since $G_X(t) = \sum_{i=1}^{n}p_i t^{x_i}$.

Also

$$G_X''(t) = \frac{d}{dt}G_X'(t) = \sum_{i=1}^{n}p_i x_i(x_i-1)t^{x_i-2} \qquad [2]$$

Setting $t = 1$ in equations [1] and [2] gives

$$G_X'(1) = \sum_{i=1}^{n}p_i x_i = \mu \qquad [3]$$

$$G_X''(1) = \sum_{i=1}^{n}p_i x_i(x_i-1)$$

$$= \sum_{i=1}^{n}(p_i x_i^2 - p_i x_i)$$

$$= \sum_{i=1}^{n}p_i x_i^2 - \sum_{i=1}^{n}p_i x_i$$

$$= \sum_{i=1}^{n}p_i x_i^2 - \mu^2 + \mu^2 - \mu$$

$$= \sigma^2 + \mu^2 - \mu$$

Therefore,

$$\sigma^2 = G_X''(1) + \mu - \mu^2 \qquad [4]$$

Tip

Note that t can be taken as any value but setting it equal to 1 in the expressions for $G_X'(t)$ and $G_X''(t)$ gives, respectively, the mean and an expression that can be used to calculate the variance.

Tip

$\sigma^2 = \sum_{i=1}^{n}p_i x_i^2 - \mu^2$

If $G_X(t)$ is the probability generating function of a random variable X, and μ and σ^2 are the mean and variance of X, respectively, then $\mu = G_X'(1)$ and $\sigma^2 = G_X''(1) + \mu - \mu^2$.

Example 3

A random variable X has probability distribution $P(X = x) = \begin{cases} 0.3; & x = 1 \\ 0.4; & x = 2 \\ 0.3; & x = 3 \end{cases}$

Write its probability generating function, $G_X(t)$, and hence find the mean and variance of X.

$$G_X(t) = 0.3t + 0.4t^2 + 0.3t^3$$

Therefore,

$$G'_X(t) = 0.3 + 0.8t + 0.9t^2$$
$$G''_X(t) = 0.8 + 1.8t$$

Therefore,

$$\mu = G'_X(1) = 2$$
$$\sigma^2 = G''_X(1) + \mu - \mu^2 = 2.6 + 2 - 4 = 0.6$$

> **Note**
>
> $G_X(t)$ is the probability generating function of X.

> **Note**
>
> Recall setting $t = 1$ gives the value of the mean.

> **Note**
>
> Setting $t = 1$ gives expressions that can be solved for the variance.

If you are not given the probability distribution fully, for example it may be given in terms of some unknown constant, you may have to do some preliminary calculations. In the next example, remember that $G_X(1) = 1$ because this is the sum of the probabilities of all possible outcomes of an experiment.

Example 4

A random variable X has probability distribution
$$P(X = x) = k(x+1)^2; \ x = 2, 3, 4$$

a   Using the total probability axiom, find the value of k and hence the generating function, $G_X(t)$, of X.

b   Use $G_X(t)$ to find the mean and variance of X.

a
$$P(X = x) = k(x+1)^2$$
$$9k + 16k + 25k = 1$$
$$k = \frac{1}{50}$$

b $$G_X(t) = \frac{9}{50}t^2 + \frac{16}{50}t^3 + \frac{25}{50}t^4$$

$$G'_X(t) = \frac{9}{25}t + \frac{24}{25}t^2 + 2t^3$$

$$G''_X(t) = \frac{9}{25} + \frac{48}{25}t + 6t^2$$

Therefore,

$$\mu = G'_X(1) = \frac{83}{50}$$

$$\sigma^2 = G''_X(1) + \mu - \mu^2 = \frac{207}{25} + \frac{83}{50} - \left(\frac{83}{50}\right)^2 = 7.18 \ \text{(2dp)}$$

> **Note**
>
> Set x equal to 2, 3 and 4 in the probability distribution, sum these probabilities and, using Axiom 1, equate to 1.

> **Note**
>
> Differentiate $G_X(t)$ twice to find $G'_X(t)$ and $G''_X(t)$.

Sometimes you might be given the probability generating function and be asked to find the probability distribution from it.

Exercise 1

1   A random variable R has probability distribution
$$P(R = r) = \begin{cases} 0.4; \ r = 1 \\ 0.4; \ r = 2 \\ 0.2; \ r = 3 \end{cases}$$
Write its probability generating function.

Bernoulli distribution

The probability distribution function of a Bernoulli random variable, B, parameter p, is

$$P(B = b) = \begin{cases} p; & b = 1 \\ 1 - p; & b = 0 \end{cases}$$

Using the definition of a probability generating function, it can be shown that

> **A Bernoulli random variable, B, parameter p, has a probability generating function $G_B(t) = 1 - p + pt$ and mean and variance given by $\mu_B = p$ and $\sigma_B^2 = p(1 - p)$.**

The proof of this result follows the same method as that for the geometric distribution.

> **Tip**
>
> You will be asked to prove this result in Exercise 2, question **6**.

Example 7

Using probability generating functions, find the parameter p and the variance of a Bernoulli random variable with mean 0.4.

Let X be a Bernoulli random variable, parameter p.

$G_X(t) = t^0(1 - p) + t^1 p = 1 - p + pt$

$G'_X(t) = p$

$G''_X(t) = 0$

Therefore, $\mu = G'_X(1) = p \implies p = 0.4$

$\sigma^2 = G''_X(1) + \mu - \mu^2 = 0.24$

> **Note**
>
> Probability generating function of a Bernoulli random variable, parameter p.

> **Note**
>
> Differentiate the probability generating function twice.

> **Note**
>
> No need to set $t = 1$ because the derivatives are not functions of t.

> **Note**
>
> Since $\mu = 0.4$.

Binomial distribution

A binomial random variable, X, parameters n and p, has a probability distribution

$$P(X = x) = \binom{n}{x} p^x q^{n-x}; \ x = 0, 1, 2, \ldots, n$$

where $q = 1 - p$.

This gives a probability generating function

$$G_X(t) = E[t^X] = \sum_{x=0}^{n} \binom{n}{x} p^x q^{n-x} t^x$$

$$= \sum_{x=0}^{n} \binom{n}{x} (pt)^x q^{n-x}$$

Expanding gives $G_X(t) = q^n + \binom{n}{1}(pt)q^{n-1} + \cdots + (pt)^n$ which you should recognise as the binomial expansion of $(q + pt)^n$.

Therefore, $G_X(t) = (q + pt)^n$.

A binomial random variable, X, parameters n and p, with probability distribution

$$P(X = x) = \binom{n}{x} p^x q^{n-x}; \; x = 0, 1, 2, \ldots, n$$

has a probability generating function $G_B(t) = (q + pt)^n$.

Example 8

Using its probability distribution function, prove that the probability generating function of a binomial random variable, X, with $n = 3$ and $p = \frac{3}{4}$, is given by $G_X(t) = \frac{1}{64}(1 + 3t)^3$.

$$P(X = x) = \left(\frac{3}{4}\right)^x \left(\frac{1}{4}\right)^{3-x}; \; x = 0, 1, 2, 3$$

$$G_X(t) = E[t^X] = \sum_{x=0}^{n} \left(\binom{3}{x} \left(\frac{3}{4}\right)^x \left(\frac{1}{4}\right)^{3-x} t^x \right)$$

$$= \sum_{x=0}^{n} \left(\binom{3}{x} \left(\frac{3t}{4}\right)^x \left(\frac{1}{4}\right)^{3-x} \right)$$

$$G_X(t) = \left(\frac{1}{4}\right)^3 + \binom{3}{1} \left(\frac{3t}{4}\right) \left(\frac{1}{4}\right)^2 + \cdots + \left(\frac{3t}{4}\right)^3$$

Therefore, $G_X(t) = \left(\frac{1}{4} + \frac{3t}{4}\right)^3 = \frac{1}{64}(1 + 3t)^3$.

> **Note**
>
> State the probability distribution function.

> **Note**
>
> The probability generating function of X.

> **Note**
>
> Write out a few terms. You should recognise this as the binomial expansion of $\left(\frac{1}{4} + \frac{3t}{4}\right)^3$.

Exercise 2

1. For each of these probability generating functions, write the distribution of the corresponding random variable.

   a $G_P(t) = \dfrac{t}{10} + \dfrac{t^2}{10} + \cdots + \dfrac{t^{10}}{10}$

   b $G_Q(t) = 0.4 + 0.6t$

   c $G_R(t) = 0.3t \left(\dfrac{1}{1 - 0.7t} \right)$

   d $G_S(t) = (0.1 + 0.9t)^{13}$

2. Write the probability generating functions for these distributions.

   a Binomial $(n = 7, p = 0.4)$

   b Geometric $(p = 0.1)$

   c Bernoulli $(p = 0.4)$

3. A discrete uniform random variable, U, has a probability distribution given by.

   $$P(U = u_i) = \frac{1}{a}; \; i = 1, 2, \ldots, a$$

   Prove that its probability generating function is

   $$G_U(t) = \frac{t^{u_1}}{a} + \frac{t^{u_2}}{a} + \cdots + \frac{t^{u_a}}{a} = \sum_{i=1}^{a} \frac{t^{u_i}}{a}$$

4 A Bernoulli random variable, B, has probability distribution given by

$$P(B = b) = \begin{cases} p; & b = 1 \\ 1 - p; & b = 0 \end{cases}$$

Show that its probability generating function is

$$G_B(t) = 1 - p + pt$$

and use it to show that the mean and variance of B are given by

$$\mu_B = p \text{ and } \sigma_B^2 = p(1 - p)$$

5 A fair dice with six faces numbered 1, 3, 5, 7, 9 and 11 is rolled once. Using probability generating functions, find the mean and variance of the score obtained.

6 Prove that the probability generating function for a binomial-distributed random variable, parameters n and p, is $(q + pt)^n$ where $q = 1 - p$.

7 An urn contains 12 balls, four of which are red. One ball is taken at random and its colour noted.

a Write the probability generating function for the number, 0 or 1, of red balls drawn.

b Use this probability generating function to find the mean and variance for this random variable.

c Write the probability generating function for the number of balls which are not red and find the corresponding mean and variance.

d Explain the relationship between the variances in parts **b** and **c**.

8 A binomial random variable, R, has probability generating function given by

$$G_R(t) = \frac{1}{32}(1 + t)^5$$

a Prove that

$$P(R = r) = \binom{5}{r}\left(\frac{1}{2}\right)^5$$

b Using $G_R(t)$, find E[R] and Var[R].

12.3 Sums of independent random variables

Sums and mean values of random variables occur frequently in probability questions, especially in the context of sampling theory. Probability generating functions can often simplify solutions to these problems.

Two cases for the probability generating functions of sums of independent random variables are considered in this section, those where the variables are independent and those where they are also identical.

Consider two independent random variables X_1 and X_2 and let X_T be the sum of these variables, that is, $X_T = X_1 + X_2$.

The probability generating function of X_T is

$$G_{X_T}(t) = E[t^{X_T}] = E[t^{X_1+X_2}]$$

$$= E[t^{X_1} \times t^{X_2}]$$

$$= E[t^{X_1}] \times E[t^{X_1}]$$

because X_1 and X_2 are independent random variables.

But $G_X(t) = E[t^X]$,

therefore $G_{X_T}(t) = G_{X_1}(t) \times G_{X_2}(t)$.

> **For two independent random variables, X_1 and X_2, the probability generating function for the sum of the variables is the product of their generating functions; that is, $G_{X_T}(t) = G_{X_1}(t) \times G_{X_2}(t)$ where $X_T = X_1 + X_2$.**

Note: this result for two independent random variables can be extended to any number of variables as long as they are independent of each other.

Example 9

Question

A fair six-sided dice with faces numbered 1–6 and a fair 12-sided dice with faces numbered 1–12 are both rolled once. Find the probability generating function for the sum of the scores obtained.

Answer

Let X and Y be random variables for the scores on the 6- and 12-sided dice, respectively.

$$G_X(t) = \frac{t^1}{6} + \frac{t^2}{6} + \cdots + \frac{t^6}{6} = \frac{t(1-t^6)}{6(1-t)}$$

$$G_Y(t) = \frac{t^1}{12} + \frac{t^2}{12} + \cdots + \frac{t^{12}}{12} = \frac{t(1-t^{12})}{12(1-t)}$$

> **Note**
> The generating functions of X and Y.

If T is the total score on the two dice, then

$$G_T(t) = G_X(t) \times G_Y(t) = \frac{t(1-t^6)}{6(1-t)} \times \frac{t(1-t^{12})}{12(1-t)}$$

$$= \frac{t^2(1-t^6)(1-t^{12})}{72(1-t)^2}$$

> **Note**
> Use the result for the generating function of a sum of random variables.

Example 10

Question

You have two packs of cards, the first a pack of 10 cards numbered 1–10 and the second with eight cards numbered 1–8. You sample with replacement from the first pack until you get a six, then you repeat this with the second pack. Find the probability generating function of T, the total number of cards drawn including the two sixes.

Let R and S be random variables for the number of draws required from each of the two packs.

R Geometric$\left(p = \dfrac{1}{10}\right)$, S Geometric$\left(p = \dfrac{1}{8}\right)$

$G_T(t) = t^2 \dfrac{1}{10} \times \dfrac{1}{8}\left(\dfrac{1}{1 - \frac{9}{10}t}\right)\left(\dfrac{1}{1 - \frac{7}{8}t}\right) = t^2 \times \dfrac{1}{10 - 9t} \times \dfrac{1}{8 - 7t}$

$\qquad = \dfrac{t^2}{80 - 142t + 63t^2}$

Note

This is the probability generating function of T because $T = R + S$.

Independent and identically distributed random variables

With the further assumption that X_1 and X_2 are not only independent but also identically distributed, if $X_T = X_1 + X_2$ and $G_x(t)$ is the generating function of both X_1 and X_2, then

$G_{X_T}(t) = G_X(t) \times G_X(t) = (G_X(t))^2.$

> For two independent and identically distributed random variables, X_1 and X_2, each with probability generating function $G_x(t)$, the probability generating function for the sum of the variables is the product of their generating functions; that is
>
> $G_{X_T}(t) = (G_X(t))^2$ where $X_T = X_1 + X_2$.

As with the previous result, this can be extended to any number of independent and identically distributed random variables.

Using the generating function for X_T, its mean and variance can be found using the results discussed above.

This result for independent and identically distributed random variables is important because it helps to solve many problems in statistics. First, the X_i are independent and identically distributed random variables if X_i, $i = 1, 2, \ldots, n$, is an independent random sample from a large population (see Example 13 below). Further, the connection between the Bernoulli distribution and the binomial distribution is important; the sum of n independent and identical Bernoulli variables is the total number of successes in n Bernoulli trials, and therefore has a binomial distribution. (See Exercise 3, questions **5** and **6** below.)

Example 11

A fair dice is rolled three times. Treating this as an independent random sample of size three from a uniform distribution on the integers 1–6, show that the generating function of the sum of the scores is

$\dfrac{1}{216} \times \dfrac{t^3\left(1 - t^6\right)^3}{\left(1 - t\right)^3}$

Answer

Let U_i be a random variable for the score on the ith throw and X_T be the total score on the three throws.

$$G_{U_i}(t) = \sum_{i=1}^{6} \frac{t^i}{6} = \frac{1}{6}(t + t^2 + \cdots + t^6)$$

$$= \frac{t}{6} \times \frac{1 - t^6}{1 - t}$$

$$X_T = \sum_{i=1}^{3} U_i$$

$$G_{X_T}(t) = (G_{U_i}(t))^3$$

$$= \left(\frac{t}{6} \times \frac{1 - t^6}{1 - t} \right)^3$$

$$= \frac{1}{216} \times \frac{t^3 (1 - t^6)^3}{(1 - t)^3}$$

Note

The generating function of the U_i.

Note

Apply the results for the probability generating function of a uniform random variable taking consecutive integers and the probability generating function for the sum of independent and identically distributed random variables.

Example 12

Note

$G_{X_T}(t)$ is the probability generating function of X_T.

Question

A fair coin is flipped until heads shows for the third time. Using probability generating functions, find the probability generating function of the number of rolls required.

Answer

Let X_i be the number of rolls since the last heads, or the beginning of the experiment, up to and including the ith heads.

$$X_i \sim \text{Geometric}\left(\frac{1}{2}\right)$$

$$G_X(t) = \frac{t}{2 - t}$$

The total number of rolls required is given by $X_T = \sum_{i=1}^{3} X_i$.

$$G_{X_T}(t) = ((G_{X_i}(t))^3 = \left(\frac{t}{2 - t} \right)^3$$

Therefore, $\mu_{X_T} = G'_{X_T}(1) = 3 \times 2 = 6$.

Note

The X_i have a geometric distribution, $p = \frac{1}{2}$.

Note

Geometric probability generating function, $p = \frac{1}{2}$.

Note

Apply the result for the probability generating function of independent and identically distributed random variables.

Exercise 3

1 Two independent observations are taken from a geometric distribution, $p = 0.8$. Write the probability generating function for the sum of these observations.

2 X and Y are two independent Poisson-distributed random variables with means 3 and 4, respectively. Given that the probability generating function of a Poisson random variable parameter μ is given by $G(t) = e^{-\mu(1-t)}$, write and simplify the probability generating function for the random variable $X + Y$. Hence state the distribution of $X + Y$.

3 A fair six-sided dice with faces numbered 1–6 and a fair eight-sided dice with faces numbered 1–8 are each thrown once. Find the probability generating function for the sum of the scores obtained.

4 X and Y are two binomial random variables
$X \quad B(n, 0.25); Y \quad B(n, 0.5)$
Find the probability generating function of the variable $X + Y$. Use the probability generating function to prove that the mean of $X + Y$ is $0.75n$.

5 X_1, X_2, X_3 are three independent random variables, each following a Bernoulli distribution with parameter 0.4. Find the probability generating function for the variable T, defined by
$T = X_1 + X_2 + X_3$
Hence find the mean and variance of T.

6 X_i, $i = 1, 2, \ldots, n$, are independent and identically distributed random variables, each following a Bernoulli distribution, parameter p. If
$X_T = \sum_{i=1}^{n} X_i$ find the mean and variance of X_T and comment on the distribution of X_T. (This is a generalisation of the result in question **5**, above.)

7 Two six-sided dice and an eight-sided dice are thrown. If all three dice are fair, prove that the probability generating function of X_T, the total score, is given by $G_{X_T}(t) = \dfrac{t^3(1-t^6)^2(1-t^8)}{288(1-t)^3}$.

8 Two bags contain counters. Bag A contains one red and three black counters; bag B contains three red and three black ones. Three counters are taken from each bag with replacement and their colour noted. Find the probability generating function and the mean of the total number of red counters drawn.

9 A random variable X takes values 1, 2 and 4 with probabilities $\frac{1}{2}, \frac{1}{4}$ and $\frac{1}{4}$. Find the generating function of X. If T is the total of the first three independent observations of X, use this generating function to find $E[T]$.

Summary

▶ Probability generating functions give probabilities of the various values of the corresponding random variable. They can also be used to calculate the expected value and variance of the random variable.

▶ The generating function, $G_X(t)$, of a discrete random variable, X, defined by the equation $G_X(t) = E[t^X]$, is a function of t where the coefficient of t^{x_i} is the probability that X takes the value x_i.

▶ If $G_X(t)$ is the probability generating function of a random variable X and μ and σ^2 are the mean and variance of X, respectively, then $\mu = G'_X(1)$ and $\sigma^2 = G''_X(1) + \mu - \mu^2$.

▶ A geometric random variable, X, parameter p, has a probability generating function $G_X(t) = tp\left(\dfrac{1}{1 - qt}\right)$ and mean and variance given by $\mu_X = \dfrac{1}{p}$ and $\sigma_X^2 = \dfrac{1 - p}{p^2}$.

▶ A Bernoulli random variable, B, parameter p, has a probability generating function $G_B(t) = 1 - p + pt$ and mean and variance given by $\mu_B = p$ and $\sigma_B^2 = p(1 - p)$.

▶ A binomial random variable, X, parameters n and p, has a probability generating function $G_B(t) = (q + pt)^n$ and mean and variance given by $\mu_X = np$ and $\sigma_X^2 = np(1 - p)$.

▶ For a set of n independent random variables X_i, $i = 1, 2, \ldots, n$, each with generating function $G_{X_i}(t)$, if $X_T = \displaystyle\sum_{i=1}^{n} X_i$, then $G_{X_T}(t) = \displaystyle\prod_{i=1}^{n} G_{X_i}(t)$.

▶ For a set of n independent and identically distributed random variables X_i, $i = 1, 2, \ldots, n$, each with generating function $G_X(t)$, if $X_T = \displaystyle\sum_{i=1}^{n} X_i$, then $G_{X_T}(t) = G_X(t))^n$.

Review exercises

1 A random variable X has probability distribution

$$P(X = x) = \begin{cases} 0.4; & x = 0 \\ 0.2; & x = 2 \\ 0.4; & x = 5 \end{cases}$$

Write its probability generating function and hence find the mean and variance of X.

2 A random variable, U, has a probability distribution given by
$$P(U = u_i) = \frac{1}{10}; i = 1, 2, \ldots, 10$$

a State the distribution of U.

b By finding the probability generating function of U, express its mean and variance in terms of the u_i.

3 An ordinary dice is rolled once and the random variable X denotes the score obtained. Using probability generating functions, find the mean and variance of X.

4 A random variable R has probability generating function given by

$$G_R(t) = 0.2 + 0.8t$$

  a Write the probability distribution function of R.

  b Use the probability generating function to find the mean and variance.

5 A biased dice with faces numbered 1–6 is rolled until a 3 appears for the first time. The probability of obtaining a 3 in any given throw is 0.2 and X is a random variable for the number of rolls up to and including the first 3.

  a Give the distribution function of X.

  b Use your answer to part **a** to find the generating function for X and hence find $E[X]$.

6 A random variable, X, has probability generating function given by

$$G_X(t) = \frac{1}{3125}(2+3t)^5$$

  a State the distribution of X with the values of its parameters.

  b Use $G_X(t)$ to find the mean and variance of X.

7 A fair four-sided dice is thrown three times. Show that the generating function for the sum of the scores is

$$\frac{1}{64} \times \frac{t^3(1-t^4)^3}{(1-t)^3}$$

and find the probability that the total score is less than 7.

Practice examination questions

1 **a** A Bernoulli random variable B has probability distribution given by

$$P(B = b) = \begin{cases} p; & b = 1 \\ 1-p; & b = 0 \end{cases}$$

    Find its probability generating function.   (3 marks)

  b A random variable S has probability generating function

$$G_S(t) = \frac{1}{5}(t + 2t^3 + at^5)$$

    Find the value of a and the probability distribution of S.   (4 marks)

2 A random variable X has probability distribution

$$P(X = x) = \begin{cases} 0.3; & x = 1 \\ 0.3; & x = 3 \\ 0.4; & x = 4 \end{cases}$$

Write down its probability generating function and hence find the mean and variance of X.   (7 marks)

3 A random variable X has a probability distribution given by

$$P(X = i) = \frac{1}{12}; i = 1, 2, \dots, 12$$

By finding the probability generating function of X, calculate its mean and variance. (7 marks)

4 A random variable, X, has a geometric distribution with $p = \frac{1}{3}$. Find the probability generating function of X and hence show that $E[X] = 3$ and $\text{Var}[X] = 6$. (7 marks)

5 Prove from first principles (that is, starting with the distribution function) that the probability generating function of a binomial distribution, X, with $n = 4$ and $p = \frac{2}{3}$, is given by

$$G_X(t) = \frac{1}{81}(1 + 2t)^4 \qquad \text{(4 marks)}$$

Prove also, using the probability generating function, that the mean and variance of X are $\frac{8}{3}$ and $\frac{8}{9}$, resp. (6 marks)

6 X and Y are two geometric random variables:

$X \sim \text{Geo}(0.35); Y \sim \text{Geo}(0.6)$

Find the probability generating function of the variable $X + Y$. (3 marks)

13 Linear Combinations of Discrete Random Variables

Introduction

In your study of statistics in AS-level Mathematics, you met functions of single random variables such as $aX + b$. However, many real-life problems deal with more than one variable that are combined to form a single variable. For example, the weight of a pack of beverage cans is the sum of the weights of the drink, the can and the packaging. In this chapter, you will learn how to deal with functions of more than one random variable.

Recap

You will need to remember…

▶ The mean and variance of the random variable, X: $\mu = E[X]$ and $\sigma^2 = E[X^2] - E^2[X]$

▶ AS-level expectation algebra of $aX + b$:
  • $E[aX + b] = aE[X] + b$
  • $\text{Var}[aX + b] = a^2 \text{Var}[X]$

Objectives

By the end of this chapter, you should know how to:

▶ Calculate the mean and variance of linear combinations of two random variables.

▶ Define and calculate the covariance and the product-moment correlation coefficient of two random variables.

▶ Apply these results to linear combinations of independent discrete random variables.

- -

Linear combinations of discrete random variables

In this section you will meet linear combinations of more than one random variable. No assumption will be made about the independence or otherwise of X and Y; this case will be dealt with in the second part of this chapter.

Mean of $aX \pm bY$

Suppose that X and Y are two discrete random variables. They can be combined into a single variable by the single variable T, where $T = aX \pm bY$, where a and b are both constants.

> For two discrete random variables, X and Y, the mean of the linear
> function $aX \pm bY$ is given by $E[aX \pm bY] = aE[X] \pm bE[Y]$.   [1]

The proof of this deceptively simple result requires the use of joint distributions, which are beyond the scope of this course. However, the example below shows that it is plausible.

Example 1

The table gives the bivariate probability distribution for the variables X and Y.

Y X	1	2	3	$P(X=x)$
1	0.15	0.1	0.05	0.3
2	0.2	0.1	0.4	0.7
$P(Y=y)$	0.35	0.2	0.45	1

Note

For example,
$P(X=1 \text{ and } Y=1) = 0.15$.

a Write the probability distribution of the random variable T given by
$T = 4X + 2Y$.

b By finding the expected values of the three variables, show that
$E[T] = 4E[X] + 2E[Y]$.

Note

For example, $P(T=10) =$
$P(X=1, Y=3 \text{ or } X=2, Y=1)$
$= P(X=1, Y=3) + P(X=2,$
$Y=1) = 0.05 + 0.2 = 0.25$.

a

t	6	8	10	12	14
$P(T=t)$	0.15	0.1	0.25	0.1	0.4

b $E[X] = 1 \times 0.3 + 2 \times 0.7 = 1.7$

$E[Y] = 1 \times 0.35 + 2 \times 0.2 + 3 \times 0.45 = 2.1$

$4E[X] + 2E[Y] = 4 \times 1.7 + 2 \times 2.1 = 11$

$E[T] = 6 \times 0.15 + 8 \times 0.1 + 10 \times 0.25 + 12 \times 0.1 + 14 \times 0.4 = 11$

Therefore, $E[T] = 4E[X] + 2E[Y]$.

A similar result also applies to more general combinations of X and Y, for example $f(X)$ and $g(Y)$.

> **For two discrete random variables, X and Y, the mean**
> **of $f(X) \pm g(Y)$ is given by $E[f(X) \pm g(Y)] = E[f(X)] \pm E[g(Y)]$.** [2]

Tip

This result will be used to calculate the variance of linear combinations of variables.

The variance of $aX \pm bY$ and the covariance of X and Y

Consider again the linear function of two random variables, X and Y, given by $aX + bY$, where a and b are constants. The variance of this function is

$$\text{Var}[aX + bY] = E[(aX + bY)^2] - E^2[aX + bY]$$
$$= a^2 E[X^2] + b^2 E[Y^2] + 2abE[XY] - (a^2 E^2[X] + b^2 E^2[Y]$$
$$+ 2ab\, E[X]E[Y])$$
$$= a^2(E[X^2] - E^2[X]) + b^2(E[Y^2] - E^2[Y])$$
$$+ 2ab(E[XY] - E[X]E[Y])$$
$$= a^2\, \text{Var}[X] + b^2\, \text{Var}[Y] + 2ab(E[XY] - E[X]E[Y])$$

Tip

Use the familiar result
$\text{Var}[X] = E[X^2] - (E[X])^2$.
$E^2[X]$ is a shorthand way of writing the square of the expected value of X, $(E[X])^2$.

Tip

Use equation [2], the formula for the mean of $f(X) \pm g(Y)$.

Defining the covariance of X and Y by the equation $\text{Cov}(X, Y) = E[XY] - E[X]\,E[Y]$ gives $\text{Var}[aX + bY] = a^2\,\text{Var}[X] + b^2\,\text{Var}[Y] + 2ab\,\text{Cov}(X, Y)$.
A similar proof for $aX - bY$ gives
$\text{Var}[aX - bY] = a^2\,\text{Var}[X] + b^2\,\text{Var}[Y] - 2ab\,\text{Cov}(X, Y)$.
Hence,

> **For two discrete random variables, X and Y,**
> **the variance of the linear function $aX \pm bY$ is given by**
> $\text{Var}\,[aX \pm bY] = a^2\,\text{Var}\,[X] + b^2\,\text{Var}\,[Y] \pm 2ab\,\text{Cov}\,(X, Y)$.

Note that setting $Y = X$ in the expression for $\text{Cov}(X, Y)$ gives $\text{Cov}(X, X) = E[X^2] - E^2[X]$. In other words, the covariance of X *with itself* is the variance of X, which explains the word 'covariance'.

Tip

Remember that the **covariance** is a statistical measure of the degree of linear dependence between the two variables.

Tip

The formula for covariance is written
$\text{Cov}(X, Y) = E((X - \mu_x)(Y - \mu_y)) = E(XY) - \mu_x\,\mu_y$
in the *Formulae and Statistical Tables* booklet.

Formulae booklet

Where
$\text{Cov}(X, Y) = E[XY] - E[X]\,E[Y]$

Example 2

Question

Two discrete random variables, X and Y, have means 12 and 9 and variances 4 and 3, respectively. Find the mean and variance of the linear combination $4X - 3Y$ if the covariance of X and Y is 3.

Answer

$E[4X - 3Y] = 4E[X] - 3E[Y]$
$\qquad\qquad = 4 \times 12 - 3 \times 9 = 21$
$\text{Var}[4X - 3Y] = 16\,\text{Var}[X] + 9\,\text{Var}[Y] - 24\,\text{Cov}(X, Y)$
$\qquad\qquad = 16 \times 4 + 9 \times 3 - 24 \times 3$
$\qquad\qquad = 19$

Note

Use [1]
$E[aX \pm bY] = a\,E[X] \pm b\,E[Y]$

Note

Use
$\text{Var}[aX \pm bY] = a^2\text{Var}[X] + b^2\text{Var}[Y] \pm 2ab\,\text{Cov}(X,Y)$

You can use these relationships to solve more complicated problems.

Example 3

Question

Two random variables, X and Y, are related by the equation $Y = 2X$. Prove that $\text{Cov}(X, Y) = 2\,\text{Var}(X)$.

Answer

$\text{Cov}(X, Y) = E[XY] - E[X]E[Y]$
$\text{Cov}(X, 2X) = E[2X^2] - E[X]E[2X]$
$\qquad\qquad = 2E[X^2] - 2E^2[X]$
$\qquad\qquad = 2(E[X^2] - E^2[X])$
$\qquad\qquad = 2\text{Var}[X]$

Note

This is the definition of covariance.

Note

Set $Y = 2X$.

Product-moment correlation coefficient

The covariance is useful for demonstrating the degree of statistical dependence between two random variables. However, it has the significant defect that its value is dependent on the units in which the variables are measured. For example, changing the units of measurement for two variables from centimetres to millimetres does not change the degree of dependence between the variables but will change the covariance.

This dependence on scale is removed by dividing the covariance by the product of the variables' standard deviations, $\sigma_X\sigma_Y$. This then gives the well-known product-moment correlation coefficient, ρ.

> For two random variables, X and Y, the product-moment correlation coefficient is given by $\rho = \frac{\text{Cov}(X, Y)}{\sigma_X\sigma_Y}$ where σ_X and σ_Y are the standard deviations of X and Y, respectively.

Example 4

Two variables, X and Y, have the expectations and variances below.

$E[X] = 1.7; \text{Var}[X] = 0.21; \; E[Y] = 2.1; \text{Var}[Y] = 0.79; E[XY] = 3.7$

Find   a $\text{Cov}(X, Y)$ b ρ_{XY} c   the mean and variance of $X + Y$.

a $\text{Cov}(X, Y) = E[XY] - E[X]\,E[Y]$
$= 3.7 - 1.7 \times 2.1 = 0.13$

b $\rho_{XY} = \dfrac{\text{Cov}(X, Y)}{\sigma_X\sigma_Y}$

$= \dfrac{0.13}{\sqrt{0.21 \times 0.79}} = 0.32 \text{ (2dp)}$

c $E[X+Y] = E[X] + E[Y] = 1.7 + 2.1 = 3.8$
$\text{Var}[X+Y] = \text{Var}[X] + \text{Var}[Y] + 2\text{Cov}(X, Y)$
$= 0.21 + 0.79 + 2 \times 0.13 = 1.26$

Example 5

Two discrete random variables, X and Y, have the bivariate probability distribution shown in the table.

Y \ X	1	2	3	$P(X=x)$
1	0.3	0.1	0.1	0.5
2	0.1	0.1	0.3	0.5
$P(Y=y)$	0.4	0.2	0.4	1

a   Show that $E[XY] \neq E[X]E[Y]$.
b   Find the covariance between X and Y.
c   By finding the variances of X and Y, calculate the population correlation coefficient, ρ_{XY}.

a $E[X] = 1 \times 0.5 + 2 \times 0.5 = 1.5$
$E[Y] = 1 \times 0.4 + 2 \times 0.2 + 3 \times 0.4 = 2$
$E[XY] = 1 \times 0.3 + 2 \times 0.2 + 3 \times 0.1 + 4 \times 0.1 + 6 \times 0.3 = 3.2$
$E[X]E[Y] = 1.5 \times 2 = 3$
Therefore, $E[XY] \neq E[X]E[Y]$.
b $\text{Cov}(X, Y) = E[XY] - E[X]\,E[Y] = 3.2 - 3 = 0.2$

(continued)

(continued)

c $\text{Var}[X] = 1^2 \times 0.5 + 2^2 \times 0.5 - 1.5^2 = 0.25$

$\text{Var}[Y] = 1^2 \times 0.4 + 2^2 \times 0.2 + 3^2 \times 0.4 - 2^2 = 0.8$

Therefore, $\rho_{XY} = \dfrac{\text{Cov}(X,Y)}{\sigma_X \sigma_Y}$

$= \dfrac{0.2}{\sqrt{0.25 \times 0.8}} = 0.45 \, (\text{2dp})$

Exercise 1

1. Two discrete random variables, X and Y, have means 10 and 12 and variances 3 and 4, respectively. Find the mean and variance of the linear combination $5X - 4Y$ if the covariance of X and Y is 2.

2. Given that $E[R] = 4.1$; $\text{Var}[R] = 0.8$; $E[S] = 14$; $\text{Var}[S] = 2.1$; $\text{Cov}(R, S) = 1.1$, find   **a** $E[RS]$ **b** ρ_{RS} **c** the mean and variance of $R + S$.

3. Prove the formula $E[2X - 3Y] = 2E[X] - 3E[Y]$ for this bivariate distribution.

Y \ X	1	2	3	$P(X=x)$
4	0.05	0.1	0.35	0.5
5	0.3	0.1	0.1	0.5
$P(Y=y)$	0.35	0.2	0.45	1

4. Two variables, R and S, have a bivariate probability distribution as shown in this table.

S \ R	1	2	3	$P(R=r)$
2	0.05	0.1	0.15	0.3
4	0.1	0.1	0.1	0.3
5	0.25	0.1	0.05	0.4
$P(S=s)$	0.4	0.3	0.3	1

   a Find the covariance of R and S.

   b By finding the variances of R and S, calculate the population correlation coefficient, ρ_{RS}.

5. By expanding the brackets in the expression $(X - \mu_X)(Y - \mu_Y)$ and using the result $E[f(X) \pm g(Y)] = E[f(X)] \pm E[g(Y)]$ show that

$E[(X - \mu_X)(Y - \mu_Y)] = E[XY] - E[X]E[Y]$.

> **Tip**
> This proves the alternative form for the covariance given in the *Formulae and Statistical Tables* booklet.

6. X, Y and Z are three random variables with means and standard deviations as follows:

$\mu_X = 10, \sigma_X = 2$; $\mu_Y = 12, \sigma_Y = 3$; $\mu_Z = 13, \sigma_Z = 4$

The values of the product-moment correlation coefficient for the above variables are $\rho_{YZ} = -0.8$ and $\rho_{XY} = \rho_{XZ} = 0$. Find the mean and variance of

   a $X + Z$ **b** $Y + Z$

7 Given that E[X] = 7, Var[X] = 9 and Y = 2X − 1, find the covariance and the correlation coefficient of X and Y.

8 Given that E[X] = 5, Var[X] = 8 and Y = aX + b where a and b are constants, find the covariance and the correlation coefficient of X and Y. Comment on your result.

9 Ben either cycles or takes the bus to school. He is also quite often late. As part of a project on transport, over many weeks he notes his method of transport to school and whether or not he is late. This gives the probability data in the table, where C is the number of times he cycles in a week and L is the number of times he is late in that week.

L \ C	1	2	3	P(C = c)
3	0.15	0.1	0.05	0.3
4	0.1	0.1	0.1	0.3
5	0.05	0.1	0.25	0.4
P(L = l)	0.3	0.3	0.4	1

a Find the covariance of C and L.

b By finding the variances of C and L, calculate the population correlation coefficient, ρ_{CL}.

c On a day when he has an exam and it is particularly important that he is not late, would you recommend that he cycles or that he takes the bus? Give a reason for your answer.

13.2 Independent discrete random variables

The discussion has so far been about two random variables with no assumptions being made about their independence. However, sets of independent random variables occur frequently in problems. These include cases where there can be no causal connection between the variables (for example, the scores when two dice are rolled) and any case of independent random sampling.

You will now consider the implications of adding the assumption of independence to the theory of linear combinations of random variables.

Mean, covariance and variance

The covariance between any two random variables, X and Y, is given by $\text{Cov}(X, Y) = E[XY] - E[X]E[Y]$. The covariance, a measure of the dependence of X and Y, equals zero if the variables are **independent**. Therefore, for independent X and Y, $E[XY] = E[X]E[Y]$ and $\text{Var}[aX + bY] = a^2\,\text{Var}[X] + b^2\,\text{Var}[Y]$.

> **Tip**
>
> This is the case because, in general,
> $\text{Var}[aX + bY] = a^2\,\text{Var}[X] + b^2\,\text{Var}[Y] + 2ab\,\text{Cov}(X, Y)$.

Since the product-moment correlation coefficient between X and Y is given by $\rho = \dfrac{\text{Cov}(X, Y)}{\sigma_X \sigma_Y}$, it follows that ρ is also zero.

If X and Y are two independent random variables, then
- $E[aX \pm bY] = aE[X] \pm bE[Y]$
- $\text{Cov}(X, Y) = 0$
- $E[XY] = E[X]E[Y]$
- $\text{Var}[aX \pm bY] = a^2\,\text{Var}[X] + b^2\,\text{Var}[Y]$
- The product-moment correlation coefficient $\rho_{XY} = 0$.

Take care! Independent X and Y implies that $E[XY] = E[X]E[Y]$ and $\rho = 0$. However, if you know that $E[XY] = E[X]E[Y]$ or that $\rho = 0$, you cannot assume that X and Y are necessarily independent. Question **5** in Exercise 2 illustrates these ideas.

Example 6

Question

R and S are two random variables. Given that

$E[R] = 4.3$; $\text{Var}[R] = 1.8$; $E[S] = 16$; $\text{Var}[S] = 1.2$; $\text{Cov}(R, S) = 0$

find   **a** $E[RS]$ **b** ρ_{RS} **c** the mean and variance of $R + S$.

Answer

a $E[RS] = E[R] \times E[S]$

       $= 4.3 \times 16 = 68.8$

b $\rho_{RS} = 0$

c $E[R + S] = E[R] + E[S] = 4.3 + 16 = 20.3$

   $\text{Var}[R + S] = \text{Var}[R] + \text{Var}[S] = 1.8 + 1.2$

> **Note**
>
> $\rho_{RS} = \dfrac{\text{Cov}(R, S)}{\sigma_R \sigma_S}$ and
>
> $\text{Cov}(R, S) = 0$

> **Note**
>
> Since $\text{Cov}(R, S) = 0$.

> **Note**
>
> $\text{Cov}[R, S] = E[RS] - E[R]E[S]$ and $\text{Cov}(R, S) = 0$

> **Note**
>
> This applies to any R and S.

For problems involving the mean and variance of **normally distributed** independent variables, it is necessary to assume that a linear combination of the variables is also normally distributed.

Exercise 2

1 X and Y are two independent random variables. Given that

    $E[X] = 7.4$; $\text{Var}[X] = 9.8$; $E[Y] = 21$; $\text{Var}[Y] = 11$

    find   **a** $E[XY]$ **b** the mean and variance of **i** $X + Y$ **ii** $2X - 3Y$

2 There are two independent random variables: P has a mean of 32 and a variance of 21, and Q has a mean of 20 and a variance of 9. Find the mean and variance of

    a $P - Q$ **b** $2P + 3Q$ **c** $5P - 3Q + 4$

3 X_i, $i = 1, 2, 3$, is an independent random sample from a population with mean 12 and variance 6. Write the mean and variance of these statistics.

   a $\displaystyle\sum_{i=1}^{3} X_i$ b $\displaystyle\frac{1}{3}\sum_{i=1}^{3} X_i$ c $\dfrac{X_1 + 2X_2 + 3X_3}{6}$

4 X, Y and Z are three random variables with means 13, 15 and 12, and variances 1.2, 1.4 and 0.9, respectively. Find the mean and variance of the weighted mean, R, given by $R = \frac{X + 2Y + 5Z}{8}$.

> **Tip**
>
> Remember that a weighted mean is one where some data points contribute more to the final average than other points.

5 A random variable, X, has the distribution in the table.

x	-2	-1	0	1	2
$P(X=x)$	0.2	0.2	0.2	0.2	0.2

   a Show that the covariance and the correlation of X with X^2 are both equal to zero.

   b Comment on the result that independence of two variables implies zero covariance and zero correlation.

Summary

▶ For two discrete random variables X and Y, the mean and variance of the linear function $aX \pm bY$ are as follows:
- Mean, $E[aX \pm bY] = aE[X] \pm bE[Y]$
- Variance, $\mathrm{Var}[aX \pm bY] = a^2\,\mathrm{Var}[X] + b^2\,\mathrm{Var}[Y] \pm 2ab\,\mathrm{Cov}(X,Y)$
  where $\mathrm{Cov}(X,Y) = E[XY] - E[X]E[Y]$.

▶ For two random variables X and Y, the product-moment correlation coefficient is given by $\rho = \frac{\mathrm{Cov}(X,Y)}{\sigma_X \sigma_Y}$ where σ_X and σ_Y are the standard deviations of X and Y, respectively.

▶ If X and Y are two independent random variables, then
- $E[aX \pm bY] = aE[X] \pm bE[Y]$
- $\mathrm{Cov}(X, Y) = 0$
- $E[XY] = E[X]E[Y]$
- $\mathrm{Var}[aX \pm bY] = a^2\,\mathrm{Var}[X] + b^2\,\mathrm{Var}[Y]$
- The product-moment correlation coefficient $\rho_{XY} = 0$.

Review exercises

1 X and Y are two discrete random variables with means 9.4 and 11.3 and variances 0.8 and 1.1, respectively. Given that $T = 5X - 11Y$, find the mean of T and covariance of X and Y if the variance of T is 11.

2 Given that $E[X] = 5.1$; $\mathrm{Var}[X] = 1.2$; $E[Y] = 12$; $\mathrm{Var}[Y] = 2.8$; $\mathrm{Cov}(X,Y) = 1.1$
   find   a $E[XY]$ b ρ_{XY} c the mean and variance of $X + 2Y$.

3 X, Y and Z are three random variables with means and standard deviations

$\mu_X = 12, \sigma_X = 2.1; \mu_Y = 17, \sigma_Y = 2.8; \mu_Z = 13, \sigma_Z = 4.1$.

The only pair of non-zero correlated variables among these is Y and Z where $\rho_{YZ} = 0.6$.

Find the mean and variance of $\qquad$ **a** $2X + Z$ $\qquad$ **b** $Y - Z$

Practice examination questions

1 In the manufacture of desk drawer fronts, a machine cuts sheets of veneered chipboard into rectangular pieces of width W millimetres and height H millimetres. The 4 edges of each of these pieces are then covered with matching veneered tape.

The distributions of W and H are such that

$E(W) = 350 \qquad Var(W) = 5 \qquad E(H) = 210 \qquad Var(H) = 4 \qquad \rho_{WH} = 0.75$

Calculate the mean and the variance of the length of tape, $T = 2W + 2H$, needed for the edges of a drawer front. $\qquad$ (5 marks)

AQA MS03 June 2010

2 Alyssa lives in the countryside but works in a city centre. Her journey to work each morning involves a car journey, a walk and wait, a train journey, and a walk.

Her car journey time, U minutes, from home to the village car park has a mean of 13 and a standard deviation of 3. Her time, V minutes, to walk from the village carpark to the village railway station and wait for a train to depart has a mean of 15 and a standard deviation of 6.

Her train journey time, W minutes, from the village railway station to the city centre railway station has a mean of 24 and a standard deviation of 4.

Her time, X minutes, to walk from the city centre railway station to her office has a mean of 9 and a standard deviation of 2.

The values of the product moment correlation coefficient for the above 4 variables are $\rho_{UV} = -0.6$ and $\rho_{UW} = \rho_{UX} = \rho_{VW} = \rho_{VX} = \rho_{WX} = 0$.

Determine values for the mean and the variance of:

$\qquad$ **i** $M = U + V$ $\qquad\qquad$ (4 marks)

$\qquad$ **ii** $D = W - 2U$ $\qquad\qquad$ (3 marks)

$\qquad$ **iii** $T = M + W + X$ given that $\rho_{MW} = \rho_{MX} = 0$ $\qquad$ (2 marks)

AQA MS03 June 2012

3 The schedule for an organisation's afternoon meeting is as follows.

Session A (Speaker 1) 2.00 pm to 3.15 pm

Session B (Discussion) 3.15 pm to 3.45 pm

Session C (Speaker 2) 3.45 pm to 5.00 pm

Records show that:

the duration, X, of Session A has mean 68 minutes and standard deviation 10 minutes;

the duration, Y, of Session B has mean 25 minutes and standard deviation 5 minutes;

the duration, Z, of Session C has mean 73 minutes and standard deviation 15 minutes; and that:

$\rho_{XZ}=0$ $\qquad$ $\rho_{XY}=-0.8$ $\qquad$ $\rho_{YZ}=0$

Determine the means and the variances of:

i $\quad L=X+Z$; (2 marks)

ii $\quad M=X+Y$. (3 marks)

AQA MS03 June 2013

4 The numbers of daily morning operations, X, and daily afternoon operations, Y, in an operating theatre of a small private hospital can be modelled by the following bivariate probability distribution.

			Number of morning operations (X)				
		2	3	4	5	6	$P(Y=y)$
Number of	3	0.00	0.05	0.20	0.20	0.05	0.50
afternoon	4	0.00	0.15	0.10	0.05	0.00	0.30
operations (Y)	5	0.05	0.05	0.10	0.00	0.00	0.20
	$P(X=x)$	0.05	0.25	0.40	0.25	0.05	1.00

a i State why $E(X)=4$ and show that $Var(X)=0.9$. (4 marks)

ii Given that $E(Y)=3.7$, $Var(Y)=0.61$ and $E(XY)=14.4$ calculate values for $Cov(X, Y)$ and ρ_{XY}. (4 marks)

b Calculate values for the mean and the variance of:

i $\quad T=X+Y$;

ii $\quad D=X-Y$. (4 marks)

AQA MS03 June 2014

14 Constant Velocity in Two Dimensions

Introduction

If the pilot of a helicopter sets a course to meet up with a ship at sea, the course will depend on the present position and velocity of the ship, and must allow for the speed and direction of the wind. This chapter shows how to analyse situations like this, where an object can be affected by more than one velocity.

Objectives

By the end of this chapter, you should know how to:

▶ Use vectors to represent displacement and velocity in two dimensions.

▶ Find the resultant of two velocities.

▶ Find the velocity of one object relative to another.

▶ Use relative velocity to find an interception or closest-approach course.

▶ Find how near to each other two moving objects will pass.

Recap

You will need to remember...

▶ The sine rule and cosine rule for a triangle ABC.

▶ How to differentiate y in terms of x.

▶ How to solve a quadratic equation by completing the square.

▶ A vector, written for example as **a** or $\overrightarrow{AB}$, has magnitude and direction.

▶ The magnitude of **a** is written $|\mathbf{a}|$.

▶ Vectors are equal if they have the same magnitude and direction.

▶ $k\mathbf{a}$ is parallel to **a**, and $|k\mathbf{a}| = k|\mathbf{a}|$.

▶ In the diagram, $\overrightarrow{AC} = \overrightarrow{AB} + \overrightarrow{BC}$; $\overrightarrow{AC}$ is the **resultant** of $\overrightarrow{AB}$ and $\overrightarrow{BC}$.

▶ Subtracting a vector is the same as adding its negative: $\mathbf{p} - \mathbf{q} = -\mathbf{q} + \mathbf{p}$.

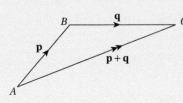

14.1 Vectors in component form

A vector with a magnitude of 1 is called a **unit vector**.

Vectors **i** and **j** are unit vectors in the x- and y-directions.

The vector $\overrightarrow{OP}$ shown has an x-**component** of 3 and a y-**component** of 2. You can write it in **component form** in terms of **i** and **j**, or as a **column vector**:

$$\overrightarrow{OP} = 3\mathbf{i} + 2\mathbf{j} = \begin{pmatrix} 3 \\ 2 \end{pmatrix}$$

> **Note**
>
> You can also use column vector notation: $\overrightarrow{OP} = x\mathbf{i} + y\mathbf{j}$.

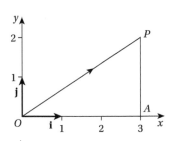

Finding magnitude and direction

Suppose that the vector $\overrightarrow{OP}$ has magnitude r, direction θ and components x and y, as shown.

Given r and θ, you can **resolve** the vector into components:
$x = r\cos\theta$ and $y = r\sin\theta$

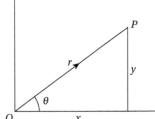

Given x and y, you can find the magnitude and direction:

$r = \sqrt{x^2 + y^2}$ and $\tan\theta = \dfrac{y}{x}$

It is a good idea to sketch a diagram when finding θ because, for example, $(3\mathbf{i} - 2\mathbf{j})$ and $(-3\mathbf{i} + 2\mathbf{j})$ would give the same value of $\tan\theta$.

Example 1

Express these vectors in component form.

a

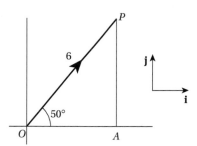

b

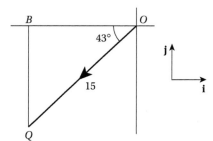

a $OA = 6\cos 50° = 3.86$ and $AP = 6\sin 50° = 4.60$

$\overrightarrow{OP} = 3.86\mathbf{i} + 4.60\mathbf{j}$ or $\begin{pmatrix} 3.86 \\ 4.60 \end{pmatrix}$

b $OB = 15\cos 43° = 11.0$ and $BQ = 15\sin 43° = 10.2$

$\overrightarrow{OQ} = -11.0\mathbf{i} - 10.2\mathbf{j}$ or $\begin{pmatrix} -11.0 \\ -10.2 \end{pmatrix}$

Example 2

Find the magnitude and direction of these vectors.

a $\mathbf{p} = 5\mathbf{i} + 2\mathbf{j}$

b $\mathbf{q} = -\mathbf{i} - 2\mathbf{j}$

a

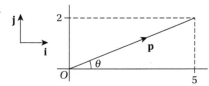

b

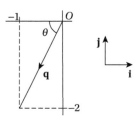

> **Note**
>
> You should always make the direction clear. You can mark the angle in a diagram, as in this case, or you can state the rotation from the positive x-direction (the answer to part **b** would then be given as $-116.6°$). In some questions you will need to give the direction as a bearing.

$|\mathbf{p}| = \sqrt{5^2 + 2^2} = 5.39$

$|\mathbf{q}| = \sqrt{(-1)^2 + (-2)^2} = 2.24$

$\tan\theta = \dfrac{2}{5}$ giving $\theta = 21.8°$

$\tan\theta = 2$ giving $\theta = 63.4°$

14.2 Position, displacement, distance, velocity and speed

You should already know from earlier studies that the position of an object, its displacement and its velocity as it moves are all vector quantities.

Position, displacement and distance

A point has a **position vector** relative to a given origin, O. The position vector of a point A is the vector $\overrightarrow{OA}$, also written as $\mathbf{a}$ or $\mathbf{r}_A$. Similarly, the position vector of B is $\overrightarrow{OB}$, $\mathbf{b}$ or $\mathbf{r}_B$, and so on.

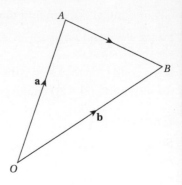

Displacement is a vector whose magnitude and direction correspond to a change of position. The displacement of point B from point A is $\overrightarrow{AB} = \mathbf{b} - \mathbf{a}$.

The **distance** from A to B is the magnitude, $|\mathbf{b} - \mathbf{a}|$, of the displacement. Distance is a scalar.

Example 3

Question

East and north are the x- and y-directions. An object moves from P, position vector $(4\mathbf{i} + 6\mathbf{j})$ km, to Q, position vector $(14\mathbf{i} + 8\mathbf{j})$ km, and then 8 km on a bearing of $045°$ to R. Find the distance and bearing of R from P.

Answer

$\mathbf{p} = (4\mathbf{i} + 6\mathbf{j})$ and $\mathbf{q} = (14\mathbf{i} + 8\mathbf{j})$

$\overrightarrow{PR} = \overrightarrow{PQ} + \overrightarrow{QR}$

$\overrightarrow{PQ} = \mathbf{q} - \mathbf{p} = 10\mathbf{i} + 2\mathbf{j}$

$\overrightarrow{QR} = 8\cos 45°\mathbf{i} + 8\sin 45°\mathbf{j} = 5.657\mathbf{i} + 5.657\mathbf{j}$

$\overrightarrow{PR} = \overrightarrow{PQ} + \overrightarrow{QR} = 15.657\mathbf{i} + 7.657\mathbf{j}$

Distance $PR = \left|\overrightarrow{PR}\right| = \sqrt{15.657^2 + 7.657^2} = 17.4$ km

Let PR make angle θ with the x-direction.

$\theta = \tan^{-1}\left(\dfrac{7.657}{15.657}\right) = 26.1°$

The bearing is $90° - 26.1° = 063.9°$.

> **Note**
> Sketch a diagram.

> **Note**
> Resolve $\overrightarrow{QR}$ into components.

> **Note**
> Use $r = \sqrt{x^2 + y^2}$.

> **Note**
> Use $\tan\theta = \dfrac{y}{x}$.

> **Note**
> You could solve this using the cosine and sine rules. However, if there were several stages to the journey, that approach would be much harder.

Velocity and speed

For a point moving in two dimensions with constant velocity, its **velocity vector** is the displacement it undergoes in each unit of time, and its **speed** is the magnitude of its velocity.

If a particle has constant velocity $\mathbf{v}$ and travels for a time t, it will undergo a displacement of $\mathbf{v}t$. If it starts at the point with position vector $\mathbf{a}$, its position vector at time t is given by $\mathbf{r} = \mathbf{a} + \mathbf{v}t$.

Example 4

Question

A particle travels from point A, position vector $(3\mathbf{i} + 7\mathbf{j})$ m, with constant velocity $(2\mathbf{i} - \mathbf{j})$ ms^{-1}.

Find   **a** its speed   **b** its position, B, after 3 s.

a Speed $= |2\mathbf{i} - \mathbf{j}| = \sqrt{2^2 + (-1)^2} = 2.24$ ms^{-1}

> **Note**
>
> Speed is the magnitude of velocity.

b Initial position $\mathbf{a} = 3\mathbf{i} + 7\mathbf{j}$

Displacement $\mathbf{v}t = (2\mathbf{i} - \mathbf{j}) \times 3 = 6\mathbf{i} - 3\mathbf{j}$

Final position $\mathbf{r} = \mathbf{a} + \mathbf{v}t = (3\mathbf{i} + 7\mathbf{j}) + (6\mathbf{i} - 3\mathbf{j}) = 9\mathbf{i} + 4\mathbf{j}$

B has position vector $(9\mathbf{i} + 4\mathbf{j})$ m.

> **Note**
>
> Its position changes by $(2\mathbf{i} - \mathbf{j})$ m every second.

Some problems might involve situations with two or more moving objects. You can use what you know about calculating the position vector of each object using their velocity, to determine if they will collide.

Example 5

Coastguard radar shows a ship, A, at $(4\mathbf{i} + 7\mathbf{j})$ km and another ship, B, at $(7\mathbf{i} + \mathbf{j})$ km. One hour later the ships are at $(8\mathbf{i} + 9\mathbf{j})$ km and $(10\mathbf{i} + 5\mathbf{j})$ km respectively. Show that without a change of speed or course there will be a collision.

Velocity of A is $\quad \mathbf{v}_A = (8\mathbf{i} + 9\mathbf{j}) - (4\mathbf{i} + 7\mathbf{j}) = (4\mathbf{i} + 2\mathbf{j})$ km/h

Velocity of B is $\quad \mathbf{v}_B = (10\mathbf{i} + 5\mathbf{j}) - (7\mathbf{i} + \mathbf{j}) = (3\mathbf{i} + 4\mathbf{j})$ km/h

> **Note**
>
> Velocity is displacement in 1 h.

At time t s the positions of A and B are

$$\mathbf{a} = (4\mathbf{i} + 7\mathbf{j}) + t(4\mathbf{i} + 2\mathbf{j}) = (4t + 4)\mathbf{i} + (2t + 7)\mathbf{j}$$

and $\quad \mathbf{b} = (7\mathbf{i} + \mathbf{j}) + t(3\mathbf{i} + 4\mathbf{j}) = (3t + 7)\mathbf{i} + (4t + 1)\mathbf{j}$

They collide if $\mathbf{a} = \mathbf{b}$.

$$(4t + 4)\mathbf{i} + (2t + 7)\mathbf{j} = (3t + 7)\mathbf{i} + (4t + 1)\mathbf{j}$$

Equate components: $4t + 4 = 3t + 7 \quad \Rightarrow \quad t = 3$

$\qquad\qquad\qquad\quad 2t + 7 = 4t + 1 \quad \Rightarrow \quad t = 3$

> **Note**
>
> The $\mathbf{i}$ and $\mathbf{j}$ components must both match for the same value of t.

So the ships will collide after 3 h.

Exercise 1

1 Points A and B have position vectors $\mathbf{a} = 2\mathbf{i} + \mathbf{j}$ and $\mathbf{b} = 5\mathbf{i} - 6\mathbf{j}$ respectively. Find the distance AB.

2 A particle starts from position vector $(3\mathbf{i} - 2\mathbf{j})$ m. It moves with constant velocity, and 3 s later its position vector is $(12\mathbf{i} + 10\mathbf{j})$ m. Find

    a its velocity   **b** its speed   **c** its position 5 s after starting.

3 Vectors $\mathbf{p}$ and $\mathbf{q}$ have magnitudes p and q and make angles θ and ϕ, respectively, with the x-direction. In each of these cases, find

    i vectors $\mathbf{p}$ and $\mathbf{q}$ in component form

    ii the magnitude and direction of $\mathbf{p} + \mathbf{q}$.

      a $p = 4, q = 5, \theta = 25°, \phi = 50°$

      b $p = 3, q = 6, \theta = 110°, \phi = -20°$

> **Tip**
>
> Remember negative angles rotate clockwise from the x-direction.

4 Find the magnitude and direction of the resultant of a displacement of 3.5 km on a bearing of 050° and a displacement of 5.4 km on a bearing of 128°.

5 Yuri and Zak start walking at the same time from a point O in a field. Yuri has velocity $(\mathbf{i} + 2\mathbf{j})$ ms^{-1} and Zak has velocity $(3\mathbf{i} + \mathbf{j})$ ms^{-1}.

  a Find Yuri's speed.

  b Find the angle between their paths.

  c Find the vector $\overrightarrow{YZ}$ at time t s.

  d Find the value of t when they are 90 m apart.

6 At a certain time, particle A is at $(\mathbf{i} + 4\mathbf{j})$ m and moving with constant velocity $(3\mathbf{i} + 3\mathbf{j})$ ms^{-1}. At the same time, particle B is at $(5\mathbf{i} + 2\mathbf{j})$ m and moving with constant velocity $(2\mathbf{i} + 3.5\mathbf{j})$ ms^{-1}.

  a Find the vector $\overrightarrow{AB}$ at time t s.

  b Show that the particles collide, and find the position of the point of collision.

7 An aircraft leaves an airfield at 0900 h and flies on a bearing of 030° at an average speed of 450 km/h. A second aircraft leaves the same airfield at 0930 h and flies on a bearing of 110° at an average speed of 400 km/h.

  a Find their distance apart at 1100 h.

  b Find the bearing of the second aircraft from the first at that time.

14.3 Resultant velocity

An object can be affected by more than one velocity. For example, the actual velocity of an aircraft is a combination of the velocity with which it is flown and the velocity of the wind. Its actual velocity is therefore the vector sum of the two velocities, and is called its **resultant velocity**.

> An object affected by two velocities v_1 and v_2 moves with a resultant velocity V given by $V = v_1 + v_2$.

When solving problems involving resultant velocity, it can help to sketch a diagram. There are generally two key approaches: using trigonometry, or expressing the velocities in component form and then adding.

Example 6

A swimmer, who can swim at 0.8 ms^{-1} in still water, wants to cross a river flowing at 0.5 ms^{-1}.

  a If she aims straight across the river, what will be her actual velocity?

  b If she wants to travel straight across,

  i in what direction should she aim

  ii what will be her actual speed?

> **Note**
>
> Sketch a velocity diagram.

  a $v = \sqrt{0.8^2 + 0.5^2} = 0.943$ ms^{-1}

  $\tan\theta = \dfrac{0.8}{0.5} \Rightarrow \theta = 58°$

  She travels at 0.943 ms^{-1} at 58° to the direction of flow of the river.

> **Note**
>
> Remember, for velocity, you need to state both speed and direction.

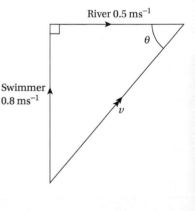

(continued)

(continued)

b i $\cos\phi=\dfrac{0.5}{0.8}\Rightarrow \phi=51.3°$

> **Note**
> Sketch a diagram.

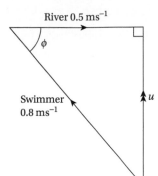

She should aim upstream at 51.3° to the bank.

ii $u=\sqrt{0.8^2-0.5^2}=0.624\text{ ms}^{-1}$

She travels straight across the river at 0.624 ms⁻¹.

Example 7

A boat is being steered due north. It can travel at 10 ms⁻¹ in still water but the current is making it travel at 18 ms⁻¹ on a bearing of 060°. Find the speed and direction of the current.

East and north are the **i**- and **j**-directions.

$\mathbf{v_B}$ is the still-water velocity of the boat, $\mathbf{v_C}$ is the velocity of the current and $\mathbf{V}$ is the resultant velocity.

> **Note**
> Draw the vector diagram.

> **Note**
> State the meaning of your notation.

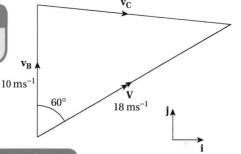

$\mathbf{v_B}=10\mathbf{j}$

$\mathbf{V}=18\cos30°\mathbf{i}+18\sin30°\mathbf{j}=9\sqrt{3}\mathbf{i}+9\mathbf{j}$

$9\sqrt{3}\mathbf{i}+9\mathbf{j}=10\mathbf{j}+\mathbf{v_C}$

> **Note**
> $\mathbf{V}=\mathbf{v_B}+\mathbf{v_C}$

> **Note**
> Write the given velocities in component form.

$\mathbf{v_C}=9\sqrt{3}\mathbf{i}-\mathbf{j}$

The speed of the current is $|\mathbf{v_C}|=\sqrt{(9\sqrt{3})^2+(-1)^2}=15.6\text{ ms}^{-1}$.

$\tan^{-1}\left(\dfrac{-1}{9\sqrt{3}}\right)=-3.7°$

> **Note**
> Calculate the angle between $\mathbf{v_C}$ and the **i**-direction.

The current has bearing 93.7°.

Example 8

An aircraft has a still-air speed of 200 km/h. It needs to travel north-east. A wind of 40 km/h is blowing from the east. In what direction should the pilot steer?

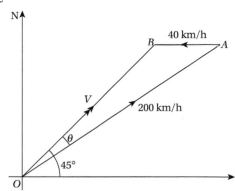

Angle $\widehat{OBA}=135°$

$\dfrac{\sin\theta}{40}=\dfrac{\sin135°}{200}$

> **Note**
> Use the sine rule.

$\sin\theta=\dfrac{40\sin135°}{200}=0.1414\Rightarrow \theta=8.13°$

The pilot must steer on a bearing of 053.13°.

Exercise 2

1 In each of these cases, find the magnitude and direction of the resultant of the given velocities.

    a   20 ms^{-1} due east and 15 ms^{-1} on a bearing of 030°

    b   45 km/h on a bearing of 125° and 30 km/h on a bearing of 320°

    c   10 ms^{-1} due north, 12 ms^{-1} due west and 18 ms^{-1} on a bearing of 200°.

2 Rain is falling at a speed of 24 ms^{-1}. It is being blown by a wind so that it falls at an angle of 64° to the horizontal ground. What is the speed of the wind?

3 A bird which can fly at 30 km/h heads due north but is blown by a 15 km/h wind from the south-west. Find the bird's resultant velocity.

4 A ship is being steered due east. A current flows from north to south causing the ship to travel at 12 km/h on a bearing of 120°. Find

    a   the speed of the current   **b**   the still water speed of the ship.

5 A boat can travel at 5 ms^{-1} in still water. It is crossing a river 200 m wide, flowing at 2 ms^{-1}. Points A and B are on the banks, directly opposite each other.

    a   If the boat leaves A and steers towards B, at what speed will it travel and how far downstream will it reach the other bank?

    b   If the boat needs to travel directly from A to B, in what direction should it be steered and at what speed will it travel?

6 A canoeist paddles north at 5 ms^{-1}, but is affected by a current of 3 ms^{-1} from the south-east and a wind of 6 ms^{-1} from a bearing of 290°. Find the canoeist's actual velocity.

7 An aircraft has a speed in still air of 400 km/h. A wind is blowing from the south at 80 km/h.

    a   The pilot steers the aircraft due east. Find the actual speed and direction of travel.

    b   The pilot wishes the plane to travel due east. Find the direction in which the aircraft should be steered and the speed at which it will travel.

8 A boat P, capable of 6 ms^{-1} in still water, travels from a point A around a course ABC, which is an equilateral triangle of side length 500 m. A uniform current of 4 ms^{-1} flows in the direction $\overrightarrow{AB}$. An identical boat Q starts from A at the same time and travels the other way (ACB) around the course. Which boat gets back to A first, and by how much does the winner win?

9 A river is D m wide and flows at u ms^{-1}. A man can swim at v ms^{-1} in still water, where $v > u$. Find the ratio between the time it would take him to swim directly across the river and back, and the time it would take him to swim D m upstream and back.

10 Malik and Ibrahim intend to cross a river, flowing at 3 ms^{-1}, to a point Q directly opposite their start point P. They can both swim at 5 ms^{-1}. Malik swims from P so he travels directly to Q. Ibrahim runs upstream until he can aim straight across the river and be carried by the current down to Q.

    a   How fast must Ibrahim run to arrive at Q at the same time as Malik?

    b   Show that by running at the same speed but only half as far before entering the water, Ibrahim can arrive at Q before Malik.

14.4 Relative velocity, closest approach and interception

All velocities are measured relative to some frame of reference in which the observer is stationary.

Imagine you are driving car A at 60 km/h when your friend in car B, travelling at 80 km/h, overtakes you and drives on ahead. It appears to you that your friend is travelling forwards at 20 km/h. This is her velocity relative to you. It seems to your friend that you are travelling backwards at 20 km/h. This is your velocity relative to her. The 60 km/h and 80 km/h are your respective velocities relative to an observer on the side of the road, that is, relative to the Earth.

You use $_A\mathbf{v}_B$ to mean 'the velocity of A relative to B'.

So in this case, $_A\mathbf{v}_B = -20$ km/h and $_B\mathbf{v}_A = 20$ km/h.

You can see that the relative velocities are just the differences of the two velocities:

$$_A\mathbf{v}_B = \mathbf{v}_A - \mathbf{v}_B = 60 - 80 = -20 \text{ km/h}$$

$$_B\mathbf{v}_A = \mathbf{v}_B - \mathbf{v}_A = 80 - 60 = 20 \text{ km/h}$$

The truth of this relationship, even if the velocities are not in the same direction, can be shown as follows:

The velocities, $\mathbf{v}_A$ and $\mathbf{v}_B$, of objects A and B is the rate at which their position vectors, $\mathbf{a}$ and $\mathbf{b}$, are changing,

$$\mathbf{v}_A = \frac{d\mathbf{a}}{dt} \quad \text{and} \quad \mathbf{v}_B = \frac{d\mathbf{b}}{dt}$$

To an observer on B, object B appears to be stationary. The apparent velocity of A is the rate at which the displacement $\overrightarrow{BA}$ is changing. This is $_A\mathbf{v}_B$, the velocity of A **relative to** B.

$$_A\mathbf{v}_B = \frac{d\overrightarrow{BA}}{dt} = \frac{d(\mathbf{a} - \mathbf{b})}{dt} = \frac{d\mathbf{a}}{dt} - \frac{d\mathbf{b}}{dt} = \mathbf{v}_A - \mathbf{v}_B$$

This confirms that

> If observers A and B have velocities $\mathbf{v}_A$ and $\mathbf{v}_B$, the velocity $_A\mathbf{v}_B$ of A relative to B is given by $_A\mathbf{v}_B = \mathbf{v}_A - \mathbf{v}_B$.

Example 9

Anita is skating due east on a pond at 12 km/h. Bruno skates nearby at 10 km/h on a bearing of 330°. How fast, and in what direction, does it seem to Anita that Bruno is moving?

$$\mathbf{v}_A = 12\mathbf{i}$$

$$\mathbf{v}_B = -5\mathbf{i} + 5\sqrt{3}\mathbf{j}$$

$$_B\mathbf{v}_A = (-5\mathbf{i} + 5\sqrt{3}\mathbf{j}) - 12\mathbf{i}$$

$$= -17\mathbf{i} + 5\sqrt{3}\mathbf{j}$$

$$\left|_B\mathbf{v}_A\right| = \sqrt{(-17)^2 + (5\sqrt{3})^2} = 19.1 \text{ km/h}$$

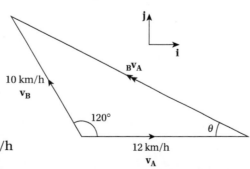

> **Tip**
>
> Strictly, the velocity of A relative to the Earth would be $_A\mathbf{v}_E$ but it is usual to just label this $\mathbf{v}_A$ (the velocity of A).

> **Note**
>
> Sketch a diagram. East and north are the $\mathbf{i}$- and $\mathbf{j}$-directions.

(continued)

(continued)

$\tan\theta = \dfrac{5\sqrt{3}}{17}$ which gives $\theta = 27°$.

To Anita, Bruno appears to be moving at 19.1 km/h on a bearing of 297°.

Closest approach

You can use relative velocity to decide how close two moving objects will be as they pass one another. Effectively, you imagine that one object is stationary and plot the relative course of the other object.

Example 10

A ferry is travelling due south at 24 km/h. The captain spots a tanker, 8 km away on a bearing of 210°. The tanker signals that it is travelling due east at 16 km/h. How far apart will they be on their closest approach? Assume that wind and currents affect both vessels equally and so can be ignored.

Let F and T represent the ferry and the tanker.

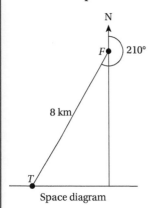

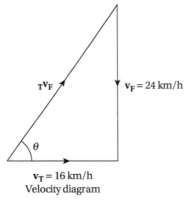

$\mathbf{v_T} = 16$ km/h due east and $\mathbf{v_F} = 24$ km/h due south

$_T\mathbf{v_F} = \mathbf{v_T} - \mathbf{v_F}$

$\tan\theta = \dfrac{24}{16} = 1.5 \quad\Rightarrow\quad \theta = 56.3°$

To the captain the tanker appears to be travelling on a bearing of 33.7°.

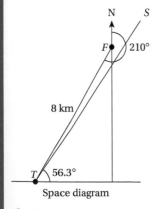

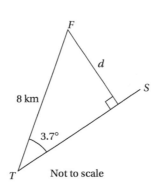

$d = 8\sin 3.7° = 0.516$ km.

So the ships pass within 516 m of each other.

Example 11

Model cars A and B have initial positions $(5\mathbf{i} + 8\mathbf{j})$ m and $(2\mathbf{i} - 4\mathbf{j})$ m respectively. They are moving with constant velocities $(2\mathbf{i} + \mathbf{j})$ ms^{-1} and $(\mathbf{i} + 2\mathbf{j})$ ms^{-1} respectively. Find when they are closest together, and the distance that then separates them.

At time t $\mathbf{r}_A = (5\mathbf{i} + 8\mathbf{j}) + (2\mathbf{i} + \mathbf{j})t = (5 + 2t)\mathbf{i} + (8 + t)\mathbf{j}$

and $\mathbf{r}_B = (2\mathbf{i} - 4\mathbf{j}) + (\mathbf{i} + 2\mathbf{j})t = ((2 + t)\mathbf{i} + (2t - 4)\mathbf{j}$

$\overrightarrow{BA} = \mathbf{r}_A - \mathbf{r}_B = (3 + t)\mathbf{i} + (12 - t)\mathbf{j}$

$\left|\overrightarrow{BA}\right|^2 = (3 + t)^2 + (12 - t)^2 = 2t^2 - 18t + 153$

$\left|\overrightarrow{BA}\right|^2 = 2(t - 4.5)^2 + 112.5$

This is a minimum when $t = 4.5$ s.

Minimum $BA = \sqrt{112.5} = 10.6$ m

> **Note**
> Find the position of A relative to B.

> **Note**
> Closest approach occurs when this is a minimum. You can find this by completing the square.
> Find their distance apart.

> **Note**
> You could also find the minimum by differentiation: $\dfrac{d\left(|BA|^2\right)}{dt} = 4t - 18 = 0$ for a minimum. This gives $t = 4.5$ s. Substituting back gives $BA = 10.6$ m.

Interception

It is often necessary for an object, for example ship A, to set a course to meet another object, ship B. This is known as an **interception course**. When on such a course, it will appear to ship A that ship B is heading straight towards it. In other words, the direction of the relative velocity $_B\mathbf{v}_A$ must be the same as the vector $\overrightarrow{BA}$.

Example 12

A patrol boat, P, which can travel at 50 km/h, spots a smuggler, S, 800 m away on a bearing of $060°$, as shown. Radar shows that S is travelling at 30 km/h due west.

a What course should P set to intercept S in the shortest time?

b How long before they meet?

a $\mathbf{v}_S = 30$ km/h due west

$|\mathbf{v}_P| = 50$ km/h

The direction of $_S\mathbf{v}_P$ must be $240°$.

$_S\mathbf{v}_P = \mathbf{v}_S - \mathbf{v}_P$

$\sin\theta = \dfrac{30\sin 30°}{50} = 0.3$

$\theta = 17.5°$ (or $\theta = 162.5°$)

> **Note**
> For interception, it must appear to P that S is heading straight for them. Sketch a velocity diagram.

> **Note**
> Use the sine rule.

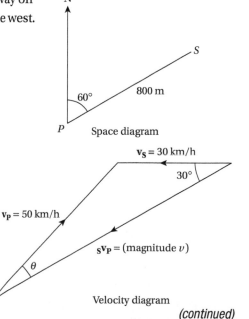

Space diagram

Velocity diagram

(continued)

(continued)

So *P* must set a course on a bearing of $060° - 017.5° = 042.5°$ to intercept *S*.

Answer

> **Note**
>
> The second value of θ does not give a viable triangle. If both values were viable, there would be two possible interception courses. If it is not possible to draw the velocity triangle, then interception is *impossible* (see Example 13).

b $|_S\mathbf{v}_P| = v$

$$v = \frac{50\sin 132.5°}{\sin 30°} = 73.7 \text{ km/h}$$

Initial distance was 800 m = 0.8 km.

Time to interception in seconds is $\dfrac{0.8}{73.7} \times 3600 = 39.1$ s.

> **Note**
>
> You need v to find the time to interception. The third angle in the velocity triangle is 132.5°.

Example 13

Question

A speedboat, *S*, is travelling at 16 km/h. A second boat, *B*, which can travel at 10 km/h, is initially 500 m from *S* on a bearing of 120°, as shown.

a If *S* is travelling east, find the course *B* should set to intercept it.

b If *S* is travelling south, show that *B* cannot intercept it.

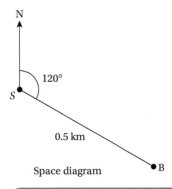

0.5 km

Space diagram

Answer

a $\mathbf{v}_S = 16$ km/h

$|\mathbf{v}_B| = 10$ km/h

$_S\mathbf{v}_B = \mathbf{v}_S - \mathbf{v}_B$

> **Note**
>
> If *B* is to intercept *S*, $_S\mathbf{v}_B$ must be in the direction $\overrightarrow{SB}$ on the diagram.

> **Note**
>
> Sketch the velocity diagram.

There are two possible triangles, ABC_1 and ABC_2, that fit the facts.

$$\frac{\sin\theta}{16} = \frac{\sin 30°}{10} \implies \theta = 53.1° \text{ or } 126.9°$$

B can set a course on a bearing of 353.1° or 066.9°.

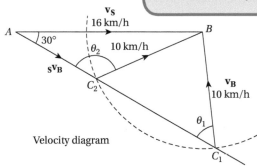

Velocity diagram

b The velocity $\mathbf{v}_B = 10$ km/h is not large enough to form a triangle in this case, so no interception course is possible.

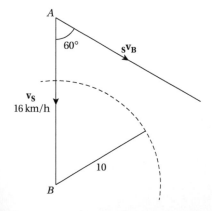

> **Note**
>
> For any course between these, *B* could reach a meeting point and then wait for *S* to arrive.

> **Note**
>
> Try to sketch the velocity triangle.

Setting a closest-approach course

If interception is not possible, the best you can achieve is to set a course to come as close as possible.

Suppose, for example, that ship B is travelling due west at 20 km/h and that ship A is initially 10 km due south of B. If ship A has a maximum speed of 16 km/h, it cannot intercept ship B.

The initial situation is shown in the space diagram.

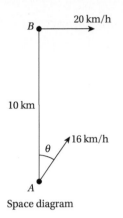

Space diagram

You need to choose the value of θ so that A passes as close as possible to B.

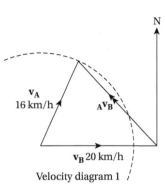

Velocity diagram 1

To an observer on B, ship B appears to be stationary and ship A travelling with velocity $_A\mathbf{v_B} = \mathbf{v_A} - \mathbf{v_B}$. You need the direction of $_A\mathbf{v_B}$ to be as close as possible to the direction AB – that is, due north.

Velocity diagram 1 shows one possible situation.

As the direction of $\mathbf{v_A}$ varies, the vertex of the velocity triangle could be anywhere on the dotted circle.

If the direction of $_A\mathbf{v_B}$ is to be as close as possible to north, it will be tangential to the dotted circle. This is shown in Velocity diagram 2.

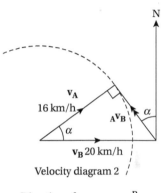

Velocity diagram 2

So the closest-approach course happens when $\mathbf{v_A}$ is perpendicular to $_A\mathbf{v_B}$.

From the diagram, $\cos \alpha = \dfrac{16}{20} = 0.8$
$$\alpha = 36.1°$$

The course that ship A should set is 053.1°.

You can find how close the ships will pass by drawing another space diagram, as shown.

The closest approach is d, where $\dfrac{d}{10} = \sin 36.9°$
$$d = 6$$

So ship A can come within 6 km of ship B.

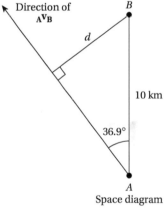

Space diagram

Exercise 3

1. Particle A has velocity $(10\mathbf{i} + 3\mathbf{j})$ ms^{-1} and particle B has velocity $(3\mathbf{i} + 5\mathbf{j})$ ms^{-1}. Find

   a   the velocity of A relative to B b   the velocity of B relative to A.

2. Two roads leave a town, one heading due west, the other on a bearing of 240°. Car A travels on the first road at 50 km/h, while car B takes the second road at 30 km/h. Find the magnitude and direction of the velocity of car B relative to car A.

3. A horse rider feels that the wind is blowing at 12 km/h directly across her path from left to right. If she is travelling from west to east at 16 km/h, what is the magnitude and direction of the wind?

4. At midday, ships P and Q are 12 km apart, with Q due west of P. P has velocity $(-2\mathbf{i} + 2\mathbf{j})$ km/h and Q has velocity $(\mathbf{i} + 3\mathbf{j})$ km/h. Find

 a the minimum distance between the ships in the subsequent motion

 b the time at which they are closest.

5. A boat, A, is 10 km south of a second boat, B. Boat A is travelling at 24 km/h due east. Boat B has a maximum speed of 26 km/h. B needs to intercept A in the shortest possible time.

 a What course should B set?

 b How long will it take before the boats meet?

6. A ferry, travelling at 35 km/h, is about to cross a sea lane perpendicularly. It observes a tanker in the lane travelling at 20 km/h. The tanker is 5 km away from the ferry in a direction making an angle of $35°$ with the direction of the ferry. Will the two ships collide? If not, how far apart are the ships when they are at their point of closest approach?

7. A motor cyclist is travelling at a steady speed of 60 km/h due south. The wind appears to be blowing in a direction with bearing $040°$. On her return journey, at the same speed and with the same wind conditions, the wind appears to be blowing in from a direction with bearing $150°$. What is the velocity of the wind?

8. Particles A and B start at points with position vectors $(2\mathbf{i} + \mathbf{j})$ m and $(3\mathbf{i} + 4\mathbf{j})$ m respectively. They have constant velocities $(-\mathbf{i} + 2\mathbf{j})$ ms^{-1} and $(p\mathbf{i} + q\mathbf{j})$ ms^{-1} respectively.

 a Show that, if the particles are to collide, $p < -1$, $q < 2$ and $q = 3p + 5$.

 b Show that, if the particles have their closest approach to one another sometime after the start, $p + 3q < 5$.

 c If $p = -2$ and $q = 2$, find the time taken for the particles to reach their point of closest approach, and find the distance that then separates them.

9. Ship A is initially 8 km north-west of ship B. Ship B is travelling north at 12 km/h. Ship A has a maximum speed of 6 km/h.

 a What course should ship A set in order to pass as close as possible to ship B?

 b What will be the minimum distance between the ships?

Summary

For motion in two dimensions:

▶ The displacement from A to B is the vector $\overrightarrow{AB} = \mathbf{b} - \mathbf{a}$.

▶ The distance AB is the magnitude, $\left|\overrightarrow{AB}\right|$, of the displacement.

▶ Velocity is a vector. Constant velocity is displacement per unit time.

▶ Speed is the magnitude of velocity.

▶ For an object affected by two velocities, $\mathbf{v}_1$ and $\mathbf{v}_2$, the resultant velocity $\mathbf{V}$ is given by $\mathbf{V} = \mathbf{v}_1 + \mathbf{v}_2$.

- If observers A and B have velocities $\mathbf{v_A}$ and $\mathbf{v_B}$ (relative to the Earth) the velocity $_A\mathbf{v_B}$ of A relative to B is given by $_A\mathbf{v_B} = \mathbf{v_A} - \mathbf{v_B}$.
- To find the closest approach (minimum distance) of object B to object A, find the relative velocity $_B\mathbf{v_A}$ and plot this relative path.
- To intercept B, A must set a course so that the relative velocity $_B\mathbf{v_A}$ is in the direction BA (so B appears to be heading straight for A).
- If A cannot intercept B, it can set a closest-approach course by making $\mathbf{v_A}$ perpendicular to $_A\mathbf{v_B}$.

Review exercises

1 The road from P to Q makes a detour round a mountain. It first goes 6 km from P on a bearing of 080°, then 7 km on a bearing of 020° and finally 5 km on a bearing of 295° to reach Q. There is a plan to bore a tunnel through the mountain from P to Q. It will be cost-effective if it reduces the journey by more than 10 km. Decide if the road should be built.

2 In order to travel due north at 25 km/h, a ship with a cruising speed of 20 km/h has to steer a course of 320°. What is the speed of the current?

3 An aircraft capable of 300 km/h in still air must fly to an airport 500 km east. A wind of 50 km/h is blowing from the south-west.
  a Show that the journey will take $1\frac{1}{2}$ hours.
  b Find how long the return journey will take if the wind continues unchanged.

4 A sheep, S, is running across a field at 8 ms⁻¹ on a bearing of 110°. The shepherd's dog, D, is initially 200 m south-west of the sheep. The dog can run at 12 ms⁻¹. Assuming that both animals run in a straight line, find
  a the direction in which the dog should run to intercept the sheep
  b the time it will take to reach the sheep.

5 Aircraft P is initially 10 km due north of aircraft Q. P is travelling east at 200 km/h and Q is flying north-east at 300 km/h. Find the shortest distance between the two aircraft if they do not change their speed or direction.

6 Two boats, A and B, are initially 10 km apart, with A due west of B. B is travelling due north at 15 km/h. A has a maximum speed of 12 km/h. What course should A set to pass as close as possible to B?

Practice examination questions

1 An aeroplane is travelling due north at 180 ms⁻¹ relative to the air.
  The air is moving north-west at 50 ms⁻¹.
  a Find the magnitude of the resultant velocity of the aeroplane. (4 marks)
  b Find the direction of the resultant velocity, giving your answer as a three-figure bearing to the nearest degree. (4 marks)

AQA MM1B June 2008

2 A fishing boat is travelling between two ports, A and B, on the shore of a lake. The bearing of B from A is 130°. The fishing boat leaves A and travels directly towards B with speed 2 ms⁻¹. A patrol boat on the lake is travelling with speed 4 ms⁻¹ on a bearing of 040°.

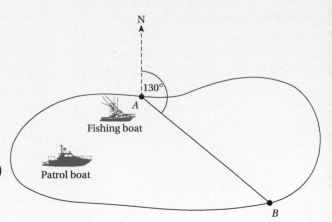

a Find the velocity of the fishing boat relative to the patrol boat, giving your answer as a speed together with a bearing. (5 marks)

b When the patrol boat is 1500 m due west of the fishing boat, it changes direction in order to intercept the fishing boat in the shortest possible time.

   i Find the bearing on which the patrol boat should travel in order to intercept the fishing boat. (4 marks)

   ii Given that the patrol boat intercepts the fishing boat before it reaches B, find the time, in seconds, that it takes the patrol boat to intercept the fishing boat after changing direction. (4 marks)

   iii State a modelling assumption necessary for answering this question, other than the boats being particles. (1 mark)

AQA MM03 June 2009

3 Two boats, A and B, are moving on straight courses with constant speeds. At noon, A and B have position vectors $(\mathbf{i} + 2\mathbf{j})$ km and $(-\mathbf{i} + \mathbf{j})$ km respectively relative to a lighthouse. Thirty minutes later, the position vectors of A and B are $(-\mathbf{i} + 3\mathbf{j})$ km and $(2\mathbf{i} - \mathbf{j})$ km respectively relative to the lighthouse.

a Find the velocity of A relative to B in the form $(m\mathbf{i} + n\mathbf{j})$ km/h, where m and n are integers. (4 marks)

b The position vector of A relative to B at time t hours after noon is $\mathbf{r}$ km. Show that $\mathbf{r} = (2 - 10t)\mathbf{i} + (1 + 6t)\mathbf{j}$. (3 marks)

c Determine the value of t when A and B are closest together. (5 marks)

d Find the shortest distance between A and B. (2 marks)

AQA MM03 June 2014

4 At noon, two ships, A and B, are a distance of 12 km apart, with B on a bearing of 065° from A. The ship B travels due north at a constant speed of 10 km/h. The ship A travels at a constant speed of 18 km/h.

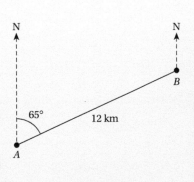

a Find the direction in which A should travel in order to intercept B. Give your answer as a bearing. (4 marks)

b In fact, the ship A actually travels on a bearing of 065°.

   i Find the distance between the ships when they are closest together. (7 marks)

   ii Find the time when the ships are closest together. (3 marks)

AQA MS03 June 2012

5 The unit vectors **i** and **j** are directed due east and due north respectively. Two runners, Albina and Brian, are running on level parkland with constant velocities of $(5\mathbf{i} - \mathbf{j})$ ms^{-1} and $(3\mathbf{i} + 4\mathbf{j})$ ms^{-1} respectively.

Initially, the position vectors of Albina and Brian are $(-60\mathbf{i} + 160\mathbf{j})$ m and $(40\mathbf{i} - 90\mathbf{j})$ m respectively, relative to a fixed origin in the parkland.

a Write down the velocity of Brian relative to Albina. (2 marks)

b Find the position vector of Brian relative to Albina t seconds after they leave their initial positions. (3 marks)

c Hence determine whether Albina and Brian will collide if they continue running with the same velocities. (3 marks)

AQA MM03 June 2008

15 Dimensional Analysis

Introduction

When you write a formula in mechanics, it only makes sense if you are equating like with like. For example, you would not create a formula that puts a length equal to a mass. Dimensional analysis can be used to help you avoid such problems.

Recap

There is no specific prior knowledge required for this chapter.

Objectives

By the end of this chapter, you should know how to:

► Find the dimensions of physical quantities in terms of mass, length and time.

► Check a formula for dimensional consistency.

► Predict the form that a formula will take.

15.1 Dimensions and dimensional consistency

All quantities in mechanics are defined in terms of three basic measures: **mass**, **length** and **time**.

Dimensions

The basic **dimensions** mass (M), length (L) and time (T) are used to describe **physical** quantities. They are **independent**, which means, for example, that you cannot define mass in terms of length and time.

You use square brackets to refer to the dimensions of a quantity. For instance, [force] means 'the dimensions of force'.

So

► If h is the height of a cylinder $[h] \equiv L$
► If m is the mass of a block $[m] \equiv M$

Take note that dimensions are more fundamental than units. For example, depth could be measured in kilometres, miles, inches or fathoms but it always has dimension L.

Most quantities are **compound**, which means they are a combination of two or more other quantities. For example, consider area. To find an area you always multiply two lengths. Familiar examples include: area of a rectangle = length × width and area of a circle = π × radius × radius. It follows that $[\text{area}] \equiv L^2$.

Using this approach for other quantities, you get several general results.

▶ Area is found by multiplying two lengths:
[area] ≡ [length] × [length] ≡ L^2

▶ Velocity is found by dividing distance by time:
[velocity] ≡ [distance] ÷ [time] ≡ LT^{-1}

▶ Acceleration is found by dividing (change of) velocity by time:
[acceleration] ≡ [velocity] ÷ [time] ≡ LT^{-2}

▶ Force is found by multiplying mass by acceleration:
[force] ≡ [mass] × [acceleration] ≡ MLT^{-2}

▶ Work is found by multiplying force by distance:
[work] ≡ [force] × [distance] ≡ ML^2T^{-2}

Some quantities are **dimensionless**. These include constants such as π but also any quantity defined as the ratio of two quantities that have the same dimensions, since the dimensions effectively cancel out.

Examples of dimensionless quantities include:

▶ Angles; in radians an angle is found by dividing the arc length by the radius:
[angle] ≡ [arc length] ÷ [radius] ≡ $L \times L^{-1} \equiv L^0$. So an angle has no dimensions.

▶ Coefficient of friction (μ); this is defined as friction force ÷ normal reaction force: $[\mu] \equiv$ [force] ÷ [force], which is therefore dimensionless.

However, some constants do have dimensions as you will see in Example 1, where you will find the dimensions of the gravitational constant, G.

Example 1

Newton's law of gravitation states that the attractive force (F) between two bodies of mass m_1 and m_2 a distance r apart is given by $F = \frac{Gm_1m_2}{r^2}$, where G is the gravitational constant. Find the dimensions of G.

$[F] \equiv MLT^{-2}$; $[m_1] \equiv [m_2] \equiv M$ and $[r] \equiv L$

$MLT^{-2} \equiv \dfrac{[G] \times M^2}{L^2}$

$[G] \equiv M^{-1}L^3T^{-2}$

Example 2

Power can be defined as force × velocity. State possible units for power.

[power] = [force] × [velocity]

= $MLT^{-2} \times LT^{-1}$

so [power] = ML^2T^{-3}

So, possible units for power are kgm^2s^{-3}.

Note
Units often reflect the dimensions involved. For example, the SI units for velocity (ms⁻¹) and acceleration (ms⁻²) indicate the dimensions LT⁻¹ and LT⁻². This is not always the case though; for example, the SI units of force and work are the newton (N) and the joule (J) respectively.

Tip
Expressing an angle in degrees is just a change of units, so its dimensionality does not change.

Note
State the known dimensions and substitute in the given formula.

Note
In fact, the SI unit for power is the watt (W).

Dimensional consistency

When you work through a problem or write a formula, each statement you make must be dimensionally consistent. That is, the dimensions of both sides of the equation must be the same. You can use this to check your working or to confirm that formulae and statements are consistent.

For example, the formula area $= \pi r^2$ is consistent, since πr^2 has the correct dimensions (L^2). However, if b and c are lengths, then the expression $5b^2c$ cannot represent an area, because its dimensions (L^3) are wrong. (Of course, it could represent a volume.)

Similarly, the expression $(3d^2e + 2f)$, where d, e and f are lengths, cannot represent anything sensible, since the two parts have different dimensions. It is equivalent to adding a volume to a length, which does not give a meaningful result.

> In any equation or formula relating to physical quantities, you cannot equate or add terms which are not dimensionally the same. That is, if $a \equiv b + c$ then $[a] \equiv [b] \equiv [c]$.

Example 3

Question

Gina finds a formula for the range, R, of a projectile fired at initial speed v up a slope inclined at θ to the horizontal. She obtains $R = \frac{v^2(1-\sin\theta)}{g\cos\theta}$. Is this dimensionally consistent?

Answer

$[R] \equiv L$

$[v] \equiv LT^{-1}$

$[g] \equiv LT^{-2}$

$\sin\theta$ and $\cos\theta$ are both dimensionless

For Gina's formula   [Left-hand side] $= L$

$$[\text{Right-hand side}] = \left[\frac{v^2(1-\sin\theta)}{g\cos\theta}\right] \equiv \frac{(LT^{-1})^2}{LT^{-2}} \equiv L$$

So the formula is dimensionally consistent.

> **Note**
>
> State the known dimensions. Remember g is an acceleration, and trigonometric functions are the ratio of two lengths.

Exercise 1

1 If m_1 and m_2 are masses, x is a length, v is a velocity, F is a force and t is a time, write the dimensions of

a mv^2 **b** Fv **c** $\dfrac{m_1v}{m_1+m_2}$

d $\dfrac{v^2t}{x}$ **e** $Fx\sin t$

2 Write the dimensions of

a pressure (force per unit area)

b density (mass per unit volume)

c speed of rotation (revolutions per second)

d angle of elevation (degrees)

3 Two forms of mechanical energy are kinetic energy, $\frac{1}{2}mv^2$, and gravitational potential energy, mgh. Do these two formulae have the same dimensions?

4 Momentum is defined as mass $\times$ velocity. Impulse is defined as force $\times$ time. Are momentum and impulse dimensionally equivalent?

5 Show that the formula $s = ut + \frac{1}{2}at^2$ is dimensionally consistent.

6 If a system is working at a constant rate, its power is the work done per unit time.

  a Find the dimensions of power.

  b Show that the formula

$$\text{power} = \text{force} \times \text{velocity}$$

  is dimensionally consistent.

7 For a body of mass m to move in a circle of radius r with angular speed ω, there must be a force, F, acting towards the centre of the circle. F is given by the formula $F \equiv mr\omega^2$. What are the dimensions of ω?

8 Check these formulae for dimensional consistency.

  a The time taken for a particle to slide from rest down a smooth slope of length l and inclination α is

$$t = \sqrt{\frac{2l}{g\sin\alpha}}$$

  b The highest point on a wall reachable by a projectile with initial speed u fired from a distance a from the base of the wall is $h = \frac{u^4 - g^2a^2}{2gu}$.

  c Two particles of mass m_1 and m_2 are connected by an elastic string of length l and modulus λ. One particle is set in motion with speed u.

  The maximum extension, x, reached by the string is $x = \left[\frac{m_1 m_2 l u^2}{(m_1 + m_2)\lambda}\right]^{\frac{1}{2}}$ (the dimensions of λ are those of a force).

9 A metal plate of area A slides across a second plate. The plates are a distance y apart and are separated by a layer of liquid. A force F is needed to keep the plate moving at a constant speed u. The dynamic viscosity of the liquid, μ, is defined by $\mu = \frac{Fy}{Au}$. Find the dimensions of μ and hence state possible units for dynamic viscosity.

15.2 Finding a formula

If you can decide which quantities should appear in a formula, you can use dimensional consistency to decide on a possible form for the expression. You can do this when the expression involves up to three quantities.

If you assume that A is related to b, c and d by a formula $A \equiv Kb^\alpha c^\beta d^\gamma$, where K is a dimensionless constant, you can use dimensional consistency to find the values of α, β and γ, and so find a *possible* formula for A.

Tip

The result will include an arbitrary, dimensionless constant. To have the complete formula you would need to find the value of this constant, perhaps by experimental means.

Example 4

It is thought that the speed, c, of sound in a gas depends only on the density, ρ, and the pressure, p, of the gas. That is, $c \equiv K\rho^\alpha p^\beta$, where K is a dimensionless constant.

Find the values of α and β required for the formula to be dimensionally consistent.

$[c] \equiv LT^{-1}$

$[\rho] \equiv \dfrac{[\text{mass}]}{[\text{volume}]} \equiv ML^{-3}$

$[p] \equiv \dfrac{[\text{force}]}{[\text{area}]} \equiv ML^{-1}T^{-2}$

$[c] \equiv [K] \times [\rho]^\alpha \times [p]^\beta$

$\Rightarrow \quad LT^{-1} \equiv (ML^{-3})^\alpha \times (ML^{-1}T^{-2})^\beta$

$\Rightarrow \quad LT^{-1} \equiv M^{(\alpha+\beta)}L^{(-3\alpha-\beta)}T^{-2\beta}$

> **Note**
>
> State the known dimensions.

> **Note**
>
> Substitute into the proposed formula.

$0 = \alpha + \beta$	[1]
$1 = -3\alpha - \beta$	[2]
$-1 = -2\beta$	[3]

From [3] $\qquad \beta = \dfrac{1}{2}$ $\qquad$ [4]

From [1] and [4] $\quad \alpha = -\dfrac{1}{2}$

> **Note**
>
> For dimensional consistency, the powers of M, L and T must be the same on both sides.

Check these values in [2]: $-3\alpha - \beta = -3 \times \left(-\dfrac{1}{2}\right) - \dfrac{1}{2} = 1$

Equation [2] is satisfied.

The required formula is $\quad c \equiv K\rho^{-\frac{1}{2}}p^{\frac{1}{2}} \quad$ or $\quad c \equiv K\sqrt{\dfrac{p}{\rho}}$

> **Note**
>
> With two unknowns it may not be possible to satisfy all three equations. This would tell you that your assumptions about the nature of the formula were incorrect. Although you can satisfy all three equations here, it does not guarantee that your assumptions were correct, though it is good evidence for it.

Example 5

A body falling through the air experiences a force, F, due to air resistance. Assuming that F depends on the mass, m, of the body, its velocity, v, and its cross-sectional area, A, find a dimensionally consistent formula for F.

Assume that $\quad F \equiv Km^\alpha v^\beta A^\gamma$

$[F] \equiv MLT^{-2}$

$[m] \equiv M$

$[v] \equiv LT^{-1}$

$[A] \equiv L^2$

$[F] \equiv [K] \times [m]^\alpha \times [v]^\beta \times [A]^\gamma$

$\Rightarrow \quad MLT^{-2} \equiv M^\alpha \times (LT^{-1})^\beta \times (L^2)^\gamma$

$\Rightarrow \quad MLT^{-2} \equiv M^\alpha L^{(\beta+2\gamma)}T^{-\beta}$

> **Note**
>
> State the known dimensions.

> **Note**
>
> Substitute into the proposed formula.

$1 = \alpha$	[1]
$1 = \beta + 2\gamma$	[2]
$-2 = -\beta$	[3]

> **Note**
>
> Equate the powers of each dimension.

(continued)

(continued)

Solving gives $\alpha = 1$, $\beta = 2$, $\gamma = -\dfrac{1}{2}$.

The required formula is $F \equiv Kmv^2 A^{-\frac{1}{2}}$ or $F \equiv \dfrac{Kmv^2}{\sqrt{A}}$

If there were four factors, you could not use the method in Example 5 because you could not find four variables, α, β, γ and δ, from only three equations.

Exercise 2

1 In the formula $A \equiv Kb^\alpha c^\beta$, K is a dimensionless constant, $[A] = ML^2T^{-2}$, $[b] = ML^{-1}T$ and $[c] = MLT^{-1}$. Find the values of α and β, and hence state the formula.

2 In the formula $A \equiv Kb^\alpha c^\beta d^\gamma$, K is a dimensionless constant, $[A] = LT^{-2}$, $[b] = MT^{-1}$, $[c] = ML^{-1}$ and $[d] = L^2T$. Find the values of α, β and γ, and hence state the formula.

3 A simple pendulum consists of a particle of mass m suspended on a light string of length l. The time of oscillation, T, is believed to be given by the formula $T \equiv Km^\alpha l^\beta g^\gamma$. Use dimensional analysis to find the values of α, β and γ, and hence write a possible formula.

4 The velocity, v, of waves on an ocean is believed to be related to the water density, ρ, the wavelength, λ, of the waves, and g. Use dimensional analysis to find a possible formula for v.

5 A skateboarder is skating on a ramp in the form of half a cylinder. Assuming that the force between the skateboard and the surface of the ramp depends on the mass m kg of the skateboarder, his velocity v ms^{-1} and the radius r m of the cylinder, find a possible formula for that force.

6 Liquid flows through a pipe because there is a pressure difference between the ends of the pipe. The rate of flow, V m^{3}s^{-1}, depends upon this pressure difference, p Nm^{-2}, the viscosity of the liquid, η kgm^{-1}s^{-1}, the length of the pipe, l m, and the radius of the pipe, r m.

a Explain why it would be impossible using dimensional analysis to find a formula for V of the form $V \equiv K\eta^\alpha r^\beta p^\gamma l^\delta$.

b It is decided that the important thing is the pressure gradient p/l. Use dimensional analysis to obtain a possible formula relating the rate of flow, V, to the viscosity of the liquid, the radius of the pipe and the pressure gradient.

Summary

▶ All quantities in mechanics are defined in terms of three dimensions: mass M, length L and time T.

▶ Square brackets denote dimension, for example height, h, has dimension $[h] \equiv L$.

- Most quantities are compound, that is, a combination of two or more other quantities. For example,
  - $[\text{area}] \equiv [\text{length}] \times [\text{length}] \equiv L^2$
  - $[\text{velocity}] \equiv [\text{distance}] \div [\text{time}] \equiv LT^{-1}$
  - $[\text{acceleration}] \equiv [\text{velocity}] \div [\text{time}] \equiv LT^{-2}$
  - $[\text{force}] \equiv [\text{mass}] \times [\text{acceleration}] \equiv MLT^{-2}$
  - $[\text{work}] \equiv [\text{force}] \times [\text{distance}] \equiv ML^2T^{-2}$
- Constants are usually dimensionless but some, such as the modulus of elasticity and the gravitational constant, G, do have dimensions.
- Some quantities, for example angle and the coefficient of friction, are dimensionless. Dimensionless quantities are often created when the quantity is defined as the ratio of two quantities that have the same dimensions, since the dimensions effectively cancel out.
- Formulae must be dimensionally consistent, that is if $a \equiv b + c$ then $[a] \equiv [b] \equiv [c]$. This gives a check on the validity (but not the truth) of a proposed formula.
- If you assume that A is related to b, c and d by a formula $A \equiv Kb^{\alpha}c^{\beta}d^{\gamma}$, where K is a dimensionless constant, you can use dimensional consistency to find the values of α, β and γ, and so find a *possible* formula for A.

Review exercises

1 The time, T, taken for a satellite to complete a circular orbit of radius r around the Earth (radius R) is given by

$$T^2 = \frac{4\pi^2 r^3}{gR^2}$$

a Identify each dimension in the formula.

b Hence state if this formula is dimensionally consistent.

2 A particle is placed at a distance r from the centre of a rough horizontal disc. The coefficient of friction between the disc and the particle is μ. The disc rotates at n revolutions per second. The maximum value of n for the particle to remain in place is given by

$$n = \frac{\mu}{2\pi}\sqrt{\frac{g}{r}}$$

Show that this formula is dimensionally consistent.

3 A particle is projected vertically into the air. Assuming that the greatest height, h, reached by the particle depends only on its initial velocity, v, and gravity, g, use dimensional analysis to predict the formula for h in terms of v, g and a dimensionless constant k.

4 A rocket fired vertically upwards will escape from the Earth's gravitational field if its velocity reaches a value V, the escape velocity. It is believed that V depends on the mass, m, of the rocket, the acceleration due to gravity, g, and the radius, r, of the Earth. Use dimensional analysis to find the form which the formula for V should take.

5 The frequency, f vibrations per second, with which a guitar string vibrates is thought to depend on the length, l, of the string, the density, ρ, of the steel and the tension, T, applied to it. Assuming this to be the case, find a possible formula for f in terms of l, ρ, T and a dimensionless constant k.

Practice examination questions

1 The time, t, for a single vibration of a piece of taut string is believed to depend on the length of the taut string, l, the tension in the string, F, the mass per unit length of the string, q, and a dimensionless constant, k, such that $t = kl^\alpha F^\beta q^\gamma$, where α, β and γ are constants.

By using dimensional analysis, find the values of α, β and γ. (5 marks)

AQA MM03 June 2011

2 A tank containing a liquid has a small hole in the bottom through which the liquid escapes. The speed, u ms^{-1}, at which the liquid escapes is given by $u = CV\rho g$, where V m^3 is the volume of the liquid in the tank, ρ kgm^{-3} is the density of the liquid, g is the acceleration due to gravity and C is a constant.

By using dimensional analysis, find the dimensions of C. (5 marks)

AQA MM03 June 2010

3 The magnitude of the gravitational force, F, between two planets of masses m_1 and m_2 with centres at a distance x apart is given by $F = \frac{Gm_1m_2}{x^2}$, where G is a constant.

a By using dimensional analysis, find the dimensions of G. (3 marks)

b The lifetime, t, of a planet is thought to depend on its mass, m, its initial radius, R, the constant G and a dimensionless constant, k, so that $T = km^\alpha R^\beta G^\gamma$, where α, β and γ are constants.

Find the values of α, β and γ. (5 marks)

AQA MM03 June 2007

4 A ball of mass m is travelling vertically downwards with speed u when it hits a horizontal floor. The ball bounces vertically upwards to a height h. It is thought that h depends on m, u, the acceleration due to gravity g, and a dimensionless constant k, such that $h = km^\alpha u^\beta g^\gamma$, where α, β and γ are constants.

By using dimensional analysis, find the values of α, β and γ. (5 marks)

AQA MM03 June 2009

5 The time T taken for a simple pendulum to make a single small oscillation is thought to depend only on its length l, its mass m and the acceleration due to gravity g. By using dimensional analysis:

a Show that T does not depend on m (3 marks)

b Express T in terms of l, g and k, where k is a dimensionless constant. (4 marks)

AQA MM03 June 2006

16 Collisions in One Dimension

Introduction

From planning a shot in snooker, to investigating a road crash, there is a need to understand what happens when objects collide. This chapter looks at the mathematics behind head-on collisions, that is, collisions in one dimension.

Recap

There is no specific prior learning required for this chapter.

Objectives

By the end of this chapter, you should know how to:

► Define impulse and momentum, and understand the relationship between them.

► Find the impulse of a variable force.

► Use the principle of conservation of momentum.

► Use Newton's coefficient of restitution to find the result of collisions between elastic particles.

16.1 Impulse and momentum

If you apply a force to an object, the effect of the force depends on the magnitude of the force and the length of time for which you apply it. Together, these form the **impulse** of the force. The result of the impulse will be a change in the object's velocity but the size of the change will also depend on its mass. Together, mass and velocity form the **momentum** of the object.

Suppose a constant force, F N, acts for a time t s in the direction of motion of a particle of mass m kg. The particle accelerates at a constant $\mathbf{a}$ ms^{-2}. Its velocity changes from $\mathbf{u}$ ms^{-1} to $\mathbf{v}$ ms^{-1}:

From $\mathbf{v} = \mathbf{u} + \mathbf{a}t$ $\qquad \mathbf{a} = \dfrac{\mathbf{v} - \mathbf{u}}{t}$ $\qquad\qquad$ [1]

From Newton's second law $\quad \mathbf{F} = m\mathbf{a}$ $\qquad\qquad\qquad$ [2]

Combining [1] and [2] $\qquad \mathbf{F} = \dfrac{m(\mathbf{v} - \mathbf{u})}{t}$

$$\mathbf{F}t = m\mathbf{v} - m\mathbf{u} \qquad\qquad\qquad [3]$$

The equation $\mathbf{F}t = m\mathbf{v} - m\mathbf{u}$ defines two quantities:

► The left-hand side, $\mathbf{F}t$, defines the **impulse** of the force.
   The impulse of a constant force, $\mathbf{F}$, applied for a time, t, is $\mathbf{F}t$:
   impulse = force × time.
   The SI unit of impulse is the **newton second** (Ns). Impulse is a vector quantity because force is a vector.

► The right-hand side shows the change in the **momentum**.
   Momentum is the quantity 'mass × velocity'. Its units must be the same as the impulse.
   For a particle of mass m moving with velocity $\mathbf{v}$, momentum = $m\mathbf{v}$.
   The SI unit of momentum is the **newton second** (Ns). Momentum is a vector quantity because velocity is a vector.

In summary,

> **Impulse = change of momentum**
>
> $Ft = mv - mu$

Example 1

A spacecraft of mass 120 kg is travelling in a straight line at 4 ms⁻¹. Its rocket is fired for 5 s, exerting a force of 150 N. Find the new velocity, v, of the spacecraft if the force was directed

a in the direction of motion   **b** in the opposite direction.

(Assume that the loss of mass when firing the rocket is negligible.)

a Impulse $= 150 \times 5 = 750$ Ns

Initial momentum $= 120 \times 4 = 480$ Ns

Final momentum $= 120v$

$750 = 120\,v - 480$

$v = 10.25$ ms⁻¹

b $-750 = 120v - 480$

$v = -2.25$ ms⁻¹

The spacecraft travels at 2.25 ms⁻¹ in the opposite direction.

Note

Momentum = mass × velocity

Note

Impulse = force × time

Note

Impulse = change of momentum

Note

The situation in **b** is the same as in **a** except that the impulse is −750 Ns.

Impulse of a variable force

In the case above, the force is constant. Imagine that there is a variable force, F, acting on a particle of mass m in its direction of motion. If F is a function of time, you can use calculus to find the impulse it exerts.

The acceleration at any instant is $\dfrac{dv}{dt}$.

From Newton's second law, $m\dfrac{dv}{dt} = F$.

If the velocity changes from u to v as t changes from 0 to t

$$m\int_u^v dv = \int_0^t F\,dt$$

$$\Rightarrow \int_0^t F\,dt = mv - mu$$

So, $\displaystyle\int_0^t F\,dt$ gives the change of momentum and is therefore the impulse of the force.

> If a variable force, F, acts for a time, t, the impulse of the force is given by impulse $= \displaystyle\int_0^t F\,dt$.

Example 2

A particle of mass 3 kg is acted on by a force $F = 3t^2$ for a period of 3 s. The particle is initially travelling at 2 ms⁻¹ in the same direction as the force. Find

a the impulse of the force   **b** the final velocity, v, of the particle.

a The impulse, J, is given by $J = \int_0^3 3t^2 \, dt = \left[t^3\right]_0^3 = 27$ Ns.

b Initial momentum $= 3 \times 2 = 6$ Ns

Final momentum $= 3v$

$3v - 6 = 27$

$v = 11$ ms^{-1}

The concept of impulse is most useful when the period of time is very short. For example, if you hit a ball with a racquet, you cannot easily measure either the force (which is probably not constant) or the time for which the force acted. However, you *can* calculate the change of momentum and then use this to calculate the impulse. The average force in this situation is the constant force that would produce the same effect. If you do in fact know the time involved, you can then find the average force since impulse = average force × time.

Example 3

A ball of mass 0.15 kg is travelling at 40 ms^{-1}. It strikes some netting and comes to rest in 0.5 s. Find

a the impulse the ball receives

b the average force exerted by the netting on the ball.

a

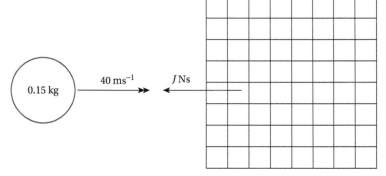

In the diagram, take left to right as the positive direction, so the net exerts an impulse $-J$ Ns on the ball.

Initial momentum $= 0.15 \times 40 = 6$ Ns

Final momentum $= 0$ Ns

$-J = 0 - 6$

$J = 6$ Ns

The netting exerts an impulse of magnitude 6 Ns from right to left in the diagram.

b Let the average force be F.

$6 = F \times 0.5$

$F = 12$ N

The netting exerts an average force of 12 N from right to left in the diagram.

Exercise 1

1 In each of the cases illustrated, state the momentum of particles *A* and *B*.

a

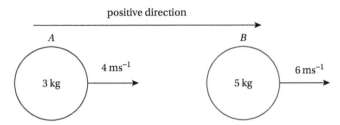

positive direction

A   3 kg   4 ms⁻¹   *B*   5 kg   6 ms⁻¹

b

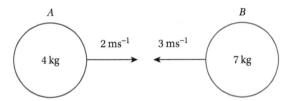

A   4 kg   2 ms⁻¹   3 ms⁻¹   *B*   7 kg

c

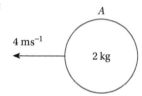 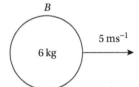

4 ms⁻¹   *A*   2 kg   *B*   6 kg   5 ms⁻¹

2 A particle *A*, of mass *m*, receives an impulse *I* from a force *F* applied for a time *t*.

   a Find *I* if $F = 12$ N and $t = 7$ s.

   b Find *I* if $m = 2$ kg and the velocity of *A* changes from 3 ms⁻¹ to 8 ms⁻¹.

   c Find *I* if $m = 5$ kg and the velocity of *A* changes from 4 ms⁻¹ to −3 ms⁻¹.

   d Find the average value of *F* if $I = 24$ Ns and $t = 4$ s.

3 An object of mass 4 kg is at rest. It receives an impulse of magnitude 28 Ns. With what speed will it start to move?

4 An object of mass 7 kg is travelling in a straight line at a speed of 4 ms⁻¹. It is acted on by a constant force in the direction of the line, which makes its speed increase to 10 ms⁻¹.

   a Find the impulse exerted on the object.

   b Find the force involved if the process took 0.35 s.

5 A particle, of mass 8 kg, travelling with velocity 2 ms⁻¹, is acted upon for a period of 4 s by a forward force $F = (3t^2 + 1)$ N. Find

   a the impulse of the force

   b the final velocity of the particle.

6 A tennis player strikes a ball so that its path is exactly reversed. The ball approaches the racket at 35 ms⁻¹ and leaves at 45 ms⁻¹. The mass of the ball is 90 g. Find the magnitude of the impulse exerted on the ball.

7. An engine of mass 20 tonnes is travelling at 54 km/h. Its brakes are applied for 3 s, after which it is travelling at 45 km/h. Find the change in momentum of the engine and hence the average braking force applied.

8. A particle of mass 3 kg travelling at 8 ms⁻¹ strikes a wall at right angles and receives an impulse of magnitude 39 Ns. With what speed does it rebound from the wall?

9. A particle of mass 2 kg is travelling at 8 ms⁻¹. A variable braking force is applied, so that after t s the magnitude of the force is $2t$ N. Find the length of time before the particle is brought to rest.

16.2 The principle of conservation of linear momentum

Forces that are *internal* to a system do not change the total momentum of the system. Suppose a system comprises particles A and B, which are separately free to move, and which have momentum M_A and M_B respectively.

The total momentum of the system $= (M_A + M_B)$ Ns.

If particle A now exerts a force F N on particle B for a time t s, B receives an impulse $J = Ft$ Ns.

By Newton's third law, B exerts a force $-F$ N on A, also for time t s. So, A receives an impulse $-J$ Ns.

Impulse = change of momentum, so:

The new momentum of $A = (M_A - J)$ Ns

The new momentum of $B = (M_B + J)$ Ns

The total momentum of the system $= (M_A - J) + (M_B + J) = (M_A + M_B)$ Ns

The total momentum has not been changed.

Even if they are not constant, the two forces are equal and opposite at all times, so the changes in momentum are also equal and opposite.

This leads to the **principle of conservation of linear momentum**.

> The principle of conservation of linear momentum states that the total momentum of a system in a particular direction remains constant unless an external force is applied in that direction.

So, for example, if two particles have masses m_1 and m_2, initial velocities u_1 and u_2, and final velocities v_1 and v_2, then $m_1 u_1 + m_2 u_2 = m_1 v_1 + m_2 v_2$.

Example 4

A particle of mass 4 kg, travelling at 6 ms⁻¹, collides with a second particle of mass 3 kg, travelling in the opposite direction at 2 ms⁻¹. After the collision, the first particle continues in the same direction at 1 ms⁻¹. Find the velocity, v, of the second particle after the collision.

Question

2 A ball of mass 0.2 kg is hit directly by a bat. Just before the impact, the ball is travelling horizontally with speed 18 ms⁻¹. Just after the impact, the ball is travelling horizontally with speed 32 ms⁻¹ in the opposite direction.

   a Find the magnitude of the impulse exerted on the ball.   (2 marks)

   b At time t seconds after the ball first comes into contact with the bat, the force exerted by the bat on the ball is $k(0.9t - 10t^2)$ newtons, where k is a constant and $0 \leq t \leq 0.09$. The bat stays in contact with the ball for 0.09 seconds. Find the value of k.   (4 marks)

AQA MM03 June 2011

3 Three smooth spheres A, B and C of equal radii and masses m, m and $2m$ respectively lie at rest on a smooth horizontal table. The centres of the spheres lie in a straight line with B between A and C. The coefficient of restitution between any two spheres is e. The sphere A is projected directly towards B with speed u and collides with B.

   a Find, in terms of u and e, the speed of B immediately after the impact between A and B.   (5 marks)

   b The sphere B subsequently collides with C. The speed of C immediately after this collision is $\frac{3}{8}u$. Find the value of e.   (7 marks)

AQA MM03 June 2006

4 A smooth sphere A, of mass m, is moving with speed $4u$ in a straight line on a smooth horizontal table. A smooth sphere B, of mass $3m$, has the same radius as A and is moving on the table with speed $2u$ in the same direction as A.

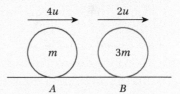

The sphere A collides directly with sphere B. The coefficient of restitution between A and B is e.

   a Find, in terms of u and e, the speeds of A and B immediately after the collision.   (6 marks)

   b Show that the speed of B after the collision cannot be greater than $3u$.   (2 marks)

   c Given that $e = \frac{2}{3}$, find, in terms of m and u, the magnitude of the impulse exerted on B in the collision.   (3 marks)

AQA MM03 June 2013

5 An ice-hockey player has mass 60 kg. He slides in a straight line at a constant speed of 5 ms⁻¹ on the horizontal smooth surface of an ice rink towards the vertical perimeter wall of the rink, as shown in the diagram.

The player collides directly with the wall, and remains in contact with the wall for 0.5 seconds.

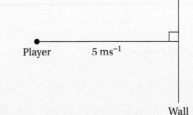

At time t seconds after coming into contact with the wall, the force exerted by the wall on the player is $4 \times 10^4 t^2 (1 - 2t)$ newtons, where $0 \leq t \leq 0.5$.

   a Find the magnitude of the impulse exerted by the wall on the player.   (4 marks)

   b The player rebounds from the wall. Find the player's speed immediately after the collision.   (3 marks)

AQA MM03 June 2012

Introduction

Understanding and taking advantage of the relationship between the roots of a polynomial and its coefficients is useful in a variety of mathematical topics and is used widely throughout the study of pure mathematics, including Complex Numbers.

Objectives

By the end of this chapter, you should know how to:

▶ Identify the sum and product of real roots of cubic functions.

▶ Identify the sum and product of real roots of polynomials of degree n.

▶ Find real and complex roots of polynomials of a higher degree.

▶ Identify the sum and product of real and complex roots of polynomials of a higher degree.

Recap

You will need to remember…

▶ How to write a complex number in Cartestian form.

▶ How to find the complex roots of quadratics.

▶ How to find the sum, difference, product and quotient of two complex numbers.

See Chapter 2 Complex Numbers.

▶ That if $ax^2 + bx + c = 0$, then the two roots α and β have sums and products:

$$\alpha + \beta = -\frac{b}{a} \quad \text{and} \quad \alpha\beta = \frac{c}{a}$$

Roots and Coefficients of a Quadratic Function.

▶ How to find equations with roots that are a function of existing roots.

▶ That if $f(x)$ is continuous on an interval $a \le x \le b$, and changes sign between $x = a$ and $x = b$, then there is a root of $f(x)$ between a and b.

See Chapter 9 Numerical Methods.

17.1 Roots of higher-order equations

You saw in Chapter *3 Roots and Coefficients of a Quadratic Function,* that you can find the equation of a quadratic function given the roots α and β using the sum and product of the roots.

The **sum** of the roots is $-\dfrac{b}{a}$ and the **product** of the roots is $\dfrac{c}{a}$.

You can use the roots and coefficients of a higher-order function in a similar way.

Roots of cubic equations

If α, β and γ are the roots of a cubic equation, $ax^3 + bx^2 + cx + d = 0$, then the equation is in the form $a(x - \alpha)(x - \beta)(x - \gamma) = 0$.

$$ax^3 + bx^2 + cx + d = a(x - \alpha)(x - \beta)(x - \gamma)$$
$$= a[x^3 - (\alpha + \beta + \gamma)x^2 + (\alpha\beta + \beta\gamma + \gamma\alpha)x - \alpha\beta\gamma] = 0$$

▶ Equating coefficients of x^2 gives: $\alpha + \beta + \gamma = -\dfrac{b}{a}$

- Equating coefficients of x gives: $\alpha\beta + \beta\gamma + \gamma\alpha = \dfrac{c}{a}$

- Equating the constants gives: $\alpha\beta\gamma = -\dfrac{d}{a}$

   Given a cubic equation, $ax^3 + bx^2 + cx + d = 0$, the sum of the roots is $-\dfrac{b}{a}$,

   the sum of all the possible products of the roots taken two at a time is $\dfrac{c}{a}$

   and the product of the roots is $-\dfrac{d}{a}$.

Therefore, if you know the roots of a cubic, you can use them to find the equation of the cubic. Similarly, if you have the equation of a cubic, you can use it to find the values of the roots.

Example 1

Find the cubic equation in x which has roots 4, 3 and -2.

$\alpha + \beta + \gamma = 4 + 3 + (-2) = 5 \Rightarrow b = -5$

$\alpha\beta + \beta\gamma + \gamma\alpha = 4 \times 3 + 3 \times -2 + (-2 \times 4) = -2$
$\qquad \Rightarrow c = -2$

$\alpha\beta\gamma = 4 \times 3 \times -2 = -24 \Rightarrow d = 24$

The equation in x is $x^3 - 5x^2 - 2x + 24 = 0$

Note Using $\alpha + \beta + \gamma = -\dfrac{b}{a}$.

Note Using $\alpha\beta + \beta\gamma + \gamma\alpha = \dfrac{c}{a}$.

Note Using $\alpha\beta\gamma = -\dfrac{d}{a}$.

Example 2

The cubic equation $x^3 + 3x^2 - 7x + 2 = 0$ has roots α, β, γ.
Find the value of $\alpha^2 + \beta^2 + \gamma^2$.

$\alpha + \beta + \gamma = -\dfrac{b}{a} = -3 \qquad \alpha\beta + \beta\gamma + \gamma\alpha = \dfrac{c}{a} = -7$

$\alpha\beta\gamma = -\dfrac{d}{a} = -2$

$\alpha^2 + \beta^2 + \gamma^2 = (\alpha + \beta + \gamma)^2 - 2(\alpha\beta + \beta\gamma + \gamma\alpha)$

$\alpha^2 + \beta^2 + \gamma^2 = (-3)^2 - 2 \times -7 = 23$

$\alpha^2 + \beta^2 + \gamma^2 = 23$

Note From expanding $(\alpha + \beta + \gamma)^2$.

Note Substitute in the values of $(\alpha + \beta + \gamma)$ and $(\alpha\beta + \beta\gamma + \gamma\alpha)$.

You learned how to find an equation with roots that are a function of existing roots in Chapter *3 Roots and Coefficients of a Quadratic Function*.

In example 2, you were able to calculate $\alpha^2 + \beta^2 + \gamma^2$ by expressing it in terms of the known values $(\alpha + \beta + \gamma)$ and $(\alpha\beta + \beta\gamma + \gamma\alpha)$ by expanding $(\alpha + \beta + \gamma)^2$. In doing so, you made use of a very important result:

$$\sum \alpha^2 = \left(\sum \alpha\right)^2 - 2\sum \alpha\beta,$$ **where $\sum \alpha$ is the sum of the roots taken one at a time and $\sum \alpha\beta$ is the sum of the products of all possible pairs of roots.**

Roots of a polynomial equation of degree n

You know that a polynomial of degree n has n roots.

From the properties of the roots of quadratic and cubic equations, it follows that in a polynomial equation of degree n, $ax^n + bx^{n-1} + cx^{n-2} + \cdots = 0$, then you have

▶ the sum of the roots (taken one at a time): $\sum \alpha = -\dfrac{b}{a}$

▶ the sum of the products of all possible pairs of roots: $\sum \alpha\beta = \dfrac{c}{a}$

▶ the sum of the products of all possible combinations of roots taken three at a time: $\sum \alpha\beta\gamma = -\dfrac{d}{a}$

▶ the sum of the products of all possible combinations of roots taken four at a time ... and so on

▶ finally, the product of the n roots: $\alpha\beta\gamma\ldots = (-1)^n \dfrac{\text{Last term}}{a}$.

> **Note**
> Since the last term is the product of $-\alpha, -\beta, -\gamma, -\delta, \ldots$

Example 3

Question

The roots of $f(x) = 4x^5 + 6x^4 - 3x^3 + 7x^2 - 11x - 3 = 0$ are α, β, γ, δ and ε.

a Find the sum of the five roots.

b **i** Show that $x = 1$ is a root of the equation.

 ii Hence show that the sum of the roots other than 1 is $-\dfrac{5}{2}$.

Answer

a $-\dfrac{b}{a} = -\dfrac{6}{4} = -\dfrac{3}{2}$

> **Note**
> Using sum of roots = $\alpha + \beta + \gamma + \delta + \varepsilon = -\dfrac{b}{a}$.

b **i** When $x = 1$,

 $f(1) = 4 + 6 - 3 + 7 - 11 - 3 = 0$

> **Note**
> Substitute 1 into the polynomial.

 Using the factor theorem, $x = 1$ is one root of the equation.

> **Note**
> From part **a**.

 ii $\alpha + \beta + \gamma + \delta + \varepsilon = -\dfrac{3}{2}$

 If $\varepsilon = 1$, then

 $\alpha + \beta + \gamma + \delta + 1 = -\dfrac{3}{2}$

 $\Rightarrow \alpha + \beta + \gamma + \delta = -\dfrac{5}{2}$

> **Note**
> Any of the roots could be made to equal 1.

Exercise 1

1 Write down the sum of the roots of each of these equations.

 a $x^3 + 3x - 7 = 0$ **b** $x^3 - 11x^2 + 5 = 0$

 c $x^3 + 5x - 4 = 0$ **d** $3x^3 + 7x^2 + 2 = 0$

2 Write down the product of the roots of each of these equations.

 a $3x^3 + 7x^2 + 2 = 0$ **b** $x^2 + 2 = \dfrac{7}{x}$ **c** $2x^3 = 7 - 4x$

3 For the equation, $f(x) = 8x^5 + 7x^4 - 5x^3 - 2x^2 + 8x + 2 = 0$, find

 a the sum of the roots **b** the product of the roots

4. The roots of $f(x) = 3x^4 + 5x^3 - 4x^2 - 19x - 12 = 0$ are α, β, γ and δ.

   a Find the sum of the four roots.   b Find the product of the four roots.

5. If α, β, γ are the roots of the equation $x^3 - 5x + 3 = 0$, find the values of

   a $\alpha + \beta + \gamma$ b $\alpha^2 + \beta^2 + \gamma^2$ c $\alpha^3 + \beta^3 + \gamma^3$

6. Given that α, β, γ are the roots of the equation $x^3 + 2x^2 - 4x + 2 = 0$, find the cubic equation whose roots are $\alpha\beta$, $\beta\gamma$, and $\gamma\alpha$.

7. Given the cubic equation $x^3 - 13x + q = 0$ has roots α, β and 3β, find the possible values of q.

17.2 Complex roots of a polynomial equation

When a polynomial function with **real coefficients** has no real roots, two of the roots will be a complex number. The complex roots of a polynomial with real coefficients always occur in **conjugate pairs**.

You learned about complex conjugate pairs in Chapter 2 Complex Numbers.

> If $z = x + iy$ is a root of a polynomial equation with real coefficients, then $z^* = x - iy$ is also a root of the polynomial equation, where z^* is the conjugate of z.

Suppose z is a root of the polynomial $a_n z^n + a_{n-1} z^{n-1} + a_{n-2} z^{n-2} + \cdots + a_0 = 0$

Then, taking the conjugate of both sides,

$$\overline{a_n z^n + a_{n-1} z^{n-1} + a_{n-2} z^{n-2} + \cdots + a_0} = 0$$

Using $\overline{z_1 + z_2} = \overline{z_1} + \overline{z_2}$, gives

$$\overline{a_n z^n} + \overline{a_{n-1} z^{n-1}} + \overline{a_{n-2} z^{n-2}} + \cdots + \overline{a_0} = 0$$

which gives

$$\overline{a_n}(z^*)^n + \overline{a_{n-1}}(z^*)^{n-1} + \overline{a_{n-2}}(z^*)^{n-2} + \cdots + \overline{a_0} = 0$$

Since all the a_i are real, $\overline{a_i} = a_i$. Therefore, you have

$$a_n(z^*)^n + a_{n-1}(z^*)^{n-1} + a_{n-2}(z^*)^{n-2} + \cdots + a_0 = 0$$

Hence, z^* is also a root of the polynomial.

In general, if the coefficients of a polynomial function are not real, then there is no expectation that roots will occur in complex conjugate pairs. You can see this in Example 4, where the quadratic has a b coefficient of $3 - i$ and the roots are not a complex conjugate pair.

Example 4

The equation $z^2 + (3 + i)z + p = 0$ has a root of $2 - i$. Find the value of p and the other root of the equation.

Using, $z = 2 - i$

$(2 - i)^2 + (3 + i)(2 - i) + p = 0$

$\Rightarrow \qquad p = -10 + 5i$

So the equation is: $z^2 + (3 + i)z - 10 + 5i = 0$

$\alpha + \beta = -\dfrac{b}{a} \Rightarrow -(3 + i)$

Taking $\alpha = (2 - i)$ (given), then

$(2 - i) + \beta = -(3 + i)$

$\beta = -(3 + i) - (2 - i) = -5$

Note

Since $2 - i$ is a root, $z = 2 - i$ satisfies the equation.

Note

Solve for β.

Note

The sum of the roots is $-(3 + i)$.

Note

The solutions in this case are not complex conjugate pairs because the coefficients of the quadratic are not real.

Example 5

Show that $4 - i$ is a root of the polynomial equation $f(z) = z^3 - 6z^2 + z + 34 = 0$.

Hence find the other roots.

Substituting $z = 4 - i$ in $f(z) = z^3 - 6z^2 + z + 34 = 0$

$$f(4 - i) = (4 - i)^3 - 6(4 - i)^2 + (4 - i) + 34$$

$$= 52 - 47i - 90 + 48i + 4 - i + 34$$

$$= 0$$

Therefore, $4 - i$ is a root of $f(z) = z^3 - 6z^2 + z + 34 = 0$.

Hence, $4 + i$ is also a root (complex conjugate pairs).

$$[z - (4 + i)][z - (4 - i)] = z^2 - 8z + 17$$

Dividing $z^3 - 6z^2 + z + 34 = 0$ by $z^2 - 8z + 17$:

$$f(z) = (z^2 - 8z + 17)(z + 2)$$

Therefore, the three roots of $f(z) = z^3 - 6z^2 + z + 34 = 0$ are $4 + i$, $4 - i$ and -2.

> **Note**
>
> To prove that $z = 4 - i$ is a root, you prove that $f(4 - i) = 0$. If $z = 4 - i$ is a root, then $z = 4 + i$ is also a root, since the roots occur as complex conjugate pairs, so you will have found one of the other two roots.

> **Note**
>
> Clearly state your reasoning.

> **Note**
>
> The function is cubic, so you know it has three roots; there is still one root to find.

> **Note**
>
> Next, you need to find the quadratic with **real** coefficients that is a factor, then divide $f(z)$ by this quadratic to find the third factor.

> **Note**
>
> If $z - (4 + i)$ and $z - (4 - i)$ are factors of the polynomial, then it follows that the quadratic with real coefficients that results from multiplying them together is also a factor.

Example 6

Show that $2 + i$ is a root of the polynomial equation $z^4 - 12z^3 + kz^2 - 140z + 125 = 0$.

Find the remaining roots of the equation.

Substituting $z = 2 + i$ into $f(z) = z^4 - 12z^3 + kz^2 - 140z + 125$

$$f(2 + i) = (2 + i)^4 - 12(2 + i)^3 + k(2 + i)^2 - 140(2 + i) + 125$$

$$= -7 + 24i - 24 - 132i + k(3 + 4i) - 280 - 140i + 125$$

$$= 0$$

Therefore, $(2 + i)$ is a root of $f(z) = z^4 - 12z^3 + 62z^2 - 140z + 125 = 0$.

Hence, $z = 2 - i$ is also a root (complex conjugate pairs).

If $z - (2 + i)$ and $z - (2 - i)$ are factors of the polynomial, so is

$$[z - (2 + i)][z - (2 - i)] = z^2 - 4z + 5$$

Dividing $z^4 - 12z^3 + 62z^2 - 140z + 125$ by $z^2 - 4z + 5$

$$f(z) = (z^2 - 4z + 5)(z^2 - 8z + 25)$$

The roots of $z^2 - 8z + 25 = 0$ are $4 \pm 3i$ (using quadratic formula).

Therefore, the four roots of $f(z) = z^4 - 12z^3 + 62z^2 - 140z + 125 = 0$ are $2 + i$, $2 - i$, $4 + 3i$ and $4 - 3i$.

> **Note**
>
> By proving that $f(2 + i) = 0$, you prove that $z = 2 + i$ is a root. Correspondingly, $z = 2 - i$ is also a root.

> **Note**
>
> Clearly state your reasoning.

> **Note**
>
> Find the quadratic with real coefficients that is a factor.

> **Note**
>
> Divide $f(z)$ by the quadratic to find the other factors.

> **Note**
>
> State how you found the roots.

To solve problems involving polynomials, it is important that you understand the properties of the various different types of function. For instance, a cubic function has a degree of three and will therefore have up to three roots. When $f(x)$ has no stationary points, there is only one real solution to $f(x) = 0$. When the values of $f(x)$ at its stationary points are of opposite sign, $f(x) = 0$ has three real solutions since it means the curve must cross the x-axis three times. Therefore, the curve of a cubic function could take the form of one of the following:

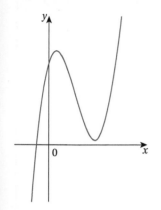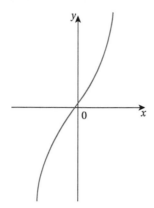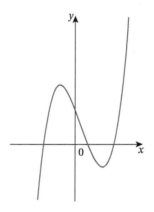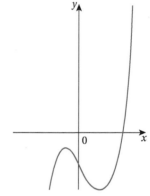

Example 7

The roots of the equation $f(x) \equiv 2x^3 - 3x^2 + 7x - 19 = 0$ are α, β and γ. Show that

a there is only one real root

b the real root lies between $x = 2$ and $x = 3$

c the real part of the two complex roots
  lies between $-\dfrac{1}{4}$ and $-\dfrac{3}{4}$.

a $f(x) \equiv 2x^3 - 3x^2 + 7x - 19$

   $f'(x) = 6x^2 - 6x + 7$

   $6x^2 - 6x + 7 = 0$

   $\Rightarrow x = \dfrac{6 \pm \sqrt{36 - 168}}{12}$

   $f'(x) = 0$ has no real roots.

   Hence, the cubic $f(x)$ has no stationary points, which means that $f(x) = 0$ has only one real root.

> **Note**
> To show that a cubic equation has only one real root, find the values of $f(x)$ at its stationary points.

> **Note**
> To find the values of $f(x)$ at its stationary points, differentiate $f(x)$ and find the roots of the quadratic $f'(x) = 0$.

b $f(2) = -1$ and $f(3) = +29$

   $f(x)$ has opposite signs at $x = 2$ and $x = 3$ and is continuous for $2 \le x \le 3$. Therefore, the real root of $f(x) = 0$ lies between $x = 2$ and $x = 3$.

> **Note**
> Substitute the given values of x into $f(x)$.

See Chapter 9 *Numerical Methods* for a reminder if you need to.

c Let the three roots of the equation be α, β, γ, where α is a real number between 2 and 3, and β and γ are complex numbers.

(continued)

(continued)

Let the roots be represented by $p + iq$ and $p - iq$.

Using the sum of roots $\alpha + \beta + \gamma = -\dfrac{b}{a}$,

$$\alpha + \beta + \gamma = \frac{3}{2}$$

$$\alpha + p + iq + p - iq = \frac{3}{2}$$

$$\Rightarrow \qquad 2p = \frac{3}{2} - \alpha$$

Since $2 < \alpha < 3$,

$$\frac{3}{2} - 3 < 2p < \frac{3}{2} - 2$$

$$\Rightarrow \quad -\frac{3}{2} < 2p < -\frac{1}{2}$$

$$\Rightarrow \quad -\frac{3}{4} < p < -\frac{1}{4}$$

Hence, the real part of each complex root, p, lies between $-\dfrac{1}{4}$ and $-\dfrac{3}{4}$.

Note

Since the roots of a polynomial with real coefficients occur in conjugate complex pairs, β and γ are conjugate complex numbers.

Exercise 2

1. Solve the equation $x^4 - 5x^3 + 2x^2 - 5x + 1 = 0$, given that i is a root.

2. Solve the equation $3x^4 - x^3 + 2x^2 - 4x - 40 = 0$, given that 2i is a root.

3. The cubic equation $4z^3 + kz^2 + 25 = 0$ has a root $z = 2 - i$.

   a If $\theta = 2 - i$, calculate θ^2 and θ^3.

   b Hence find the value of k.

4. Determine the number of real roots of the equation $2x^3 - 7x + 2 = 0$.

5. Determine the range of possible values of k if the equation $x^3 + 3x^2 = k$ has three real roots.

6. One root of the equation $z^4 - 5z^3 + 13z^2 - 16z + 10 = 0$ is $1 + i$. Find the other roots.

7. a Show that one root of the equation $z^3 + 5z^2 - 56z + 110 = 0$ is $3 + i$.

   b Find the other roots of the equation.

Summary

▶ Given a cubic equation, $ax^3 + bx^2 + cx + d = 0$

- the sum of the roots is $-\dfrac{b}{a}$

- the sum of all the possible products of the roots taken two at a time is $\dfrac{c}{a}$

- the product of the roots is $-\dfrac{d}{a}$.

Therefore, if you know the roots of a cubic, you can use them to find the equation of the cubic. Similarly, if you have the equation of a cubic, you can use it to find the values of the roots.

Answer

▶ $\displaystyle\sum\alpha^2 = \left(\sum\alpha\right)^2 - 2\sum\alpha\beta$, where $\displaystyle\sum\alpha$ is the sum of the roots taken one at a time and $\displaystyle\sum\alpha\beta$ is the sum of the products of all possible pairs of roots. This is an important result.

▶ In a polynomial equation of degree n, $ax^n + bx^{n-1} + cx^{n-2} + \cdots = 0$, then the

- sum of the roots (taken one at a time): $\displaystyle\sum\alpha = -\frac{b}{a}$

- sum of the products of all possible pairs of roots: $\displaystyle\sum\alpha\beta = \frac{c}{a}$

- sum of the products of all possible combinations of roots taken three at a time is $\displaystyle\sum\alpha\beta\gamma = -\frac{d}{a}$

- sum of the products of all possible combinations of roots taken four at a time ... and so on

- product of the n roots: $\alpha\beta\gamma\cdots = (-1)^n \dfrac{\text{Last term}}{\text{First term}}$.

▶ If $z = x + iy$ is a root of a polynomial equation with real coefficients, then $z^* = x - iy$ is also a root of the polynomial equation, where z^* is the conjugate of z. The complex roots are said to occur in conjugate pairs.

▶ In general, if the coefficients of a polynomial function are *not* real, then there is no expectation that roots will occur in complex conjugate pairs.

Review exercises

1 Find the sum and product of the roots of the equation.
$$7(x+3)^2 = \frac{5}{x+1}$$

2 The numbers α, β and γ satisfy the equations
$$\alpha^2 + \beta^2 + \gamma^2 = -4 + 6i$$
$$\alpha\beta + \beta\gamma + \gamma\alpha = 2 - 3i$$

a Show that $\alpha + \beta + \gamma = 0$.

b The numbers α, β and γ are also the roots of the equation
$$z^3 + pz^2 + qz + r = 0$$
Write down the values of p and q.

c Given that $\alpha = 2i$, find the value of r.

3 The cubic equation $z^3 + pz^2 + qz + 7 - 2i = 0$, where p and q are constants, has three complex roots α, β and γ. It is known that $\beta = 2 + i$ and $\gamma = 3 - i$.

a Write down the value of $\alpha\beta\gamma$, and hence show that $(7+i)\alpha = -7 + 2i$.

b Hence find the value of α, given in the form $x + iy$.

c Find the value of p.

4 **a** Show that one root of the equation $z^4 - 2z^3 + 6z^2 + 22z + 13 = 0$ is $2 - 3i$.

b **i** Find the other roots of the equation.

ii Hence factorise $z^4 - 2z^3 + 6z^2 + 22z + 13$ into two quadratics, each of which has real coefficients.

Practice examination questions

1 The cubic equation

$$z^3 + pz^2 + 25z + q = 0$$

where p and q are real, has a root $\alpha = 2 - 3i$.

a Write down another non-real root, β, of this equation. (1 mark)

b Find

    i the value of $\alpha\beta$ (1 mark)

    ii the third root, γ, of the equation (3 marks)

    iii the values of p and q. (3 marks)

AQA MFP2 June 2009

2 The cubic equation

$$z^3 + qz + (18 - 12i) = 0$$

where q is a complex number, has roots α, β, and γ.

a Write down the value of

    i $\alpha\beta\gamma$ (1 mark)

    ii $\alpha + \beta + \gamma$. (1 mark)

b Given that $\beta + \gamma = 2$, find the value of

    i α (1 mark)

    ii $\beta\gamma$ (2 marks)

    iii q. (3 marks)

c Given that β is of the form ki, where k is real, find β and γ. (4 marks)

AQA MFP2 June 2008

3 The cubic equation

$$z^3 + pz^2 + 6z + q = 0$$

has roots α, β and γ.

a Write down the value of $\alpha\beta + \beta\gamma + \gamma\alpha$. (1 mark)

b Given that p and q are real and that $\alpha^2 + \beta^2 + \gamma^2 = -12$

    i explain why the cubic equation has two non-real roots and one real root (2 marks)

    ii find the value of p. (4 marks)

c One root of the cubic equation is $-1 + 3i$.

    Find

    i the other two roots (3 marks)

    ii the value of q. (2 marks)

AQA MFP2 June 2007

4 The cubic equation

$$x^3 + px^2 + qx + r = 0$$

where p, q and r are real, has roots α, β and γ.

a Given that

$$\alpha + \beta + \gamma = 4 \quad \text{and} \quad \alpha^2 + \beta^2 + \gamma^2 = 20$$

find the values of p and q. (5 marks)

b Given further that one root is $3 + i$, find the value of r. (5 marks)

AQA MFP2 January 2006

Proof by Induction and Finite Series

Introduction

If you are the member of a club and your membership automatically renews on payment each year, then you can prove that you are a member simply by showing that you were a member once, and that you paid your dues each year thereafter. A similar concept occurs in mathematics. For instance, if you can show that a formula is true for $k = 1$ and you can establish that the formula must be true for $k + 1$, provided it was true for k, then that formula *must* also be true for *all* values of k. In this chapter, this idea is used to prove several formulae to be true including the formula for $\sum r^2$ that you saw in Chapter *4 Series*.

Objectives

By the end of this chapter, you should know how to:
▶ Use proof by induction to prove divisibility and De Moivre's theorem.
▶ Sum various finite series, using techniques such as partial fractions, induction and differencing.

Recap

You will need to remember...
▶ How to manipulate expressions given in the summation notation, such as $\sum r^2 - 3r$.
▶ How to manipulate partial fractions.

For a reminder, see Chapter *4 Series*

18.1 Proof by induction

Mathematical induction, or **proof by induction**, can be used to prove that a given statement applies to any natural number, n.

> **To prove a statement by induction, you proceed in two steps:**
>
> 1 **You assume that the statement is true for $n = k$ and use this assumption to prove that it is true for $n = k + 1$.**
>
> 2 **You then prove the statement for $n = 1$.**

Step 2 tells you that the statement is true for $n = 1$. Step 1 then tells us that, when $k = 1$, the statement is true for $n = 2$. Using step 1 again, when $k = 2$, the statement must be true for $n = 3$. Using step 1 yet again, the statement is true for $n = 4$. Similarly, step 1 can be repeated for $n = 5$, $n = 6$, and so on. Therefore, the statement is true for *all* integers n (≥ 1).

The processes used in step 1 to prove that the statement is true for $n = k + 1$, will vary according to the nature of the statement. In general, you start with $n = k$ then show that that this expression is true when $n = k + 1$.

Is it important to note that when proving a result by induction, it is *essential* that you state what it is you are trying to show and that you write *all* of the explanatory statements in full as part of your working.

Some examples of proof by induction for different types of statement are covered in this section.

Series

When proving $n = k + 1$ for a statement about the sum of a series, it is important to note the following relationship:

for a sequence, (u_n), with sum $\sum S_n = u_1 + u_2 + \cdots + u_n$, it follows that

$$\sum S_{n+1} = S_n + u_{n+1}$$

Example 1

a Prove that $\displaystyle\sum_{r=1}^{n} r = \frac{1}{2}n(n+1)$.

b Hence find $\displaystyle\sum_{10}^{20} r$.

a Assuming formula is true for $n = k$, then $\displaystyle\sum_{r=1}^{k} r = \frac{1}{2}k(k+1)$.

To prove that $\displaystyle\sum_{r=1}^{n} r = \frac{1}{2}n(n+1)$ is true for $n = k + 1$, need to prove

$$\sum_{r=1}^{k+1} r = \frac{1}{2}(k+1)(k+2).$$

$$\sum_{r=1}^{k+1} r = \sum_{r=1}^{k} r + (k+1)\text{th term}$$

$$\sum_{r=1}^{k+1} r = \frac{1}{2}k(k+1) + k + 1$$

$$= \frac{1}{2}[k(k+1) + 2(k+1)]$$

$$= \frac{1}{2}(k+1)(k+2)$$

Therefore, $\displaystyle\sum_{r=1}^{n} r = \frac{1}{2}n(n+1)$ is true for $n = k + 1$, if it is true for $n = k$.

When $n = 1$, LHS of the formula $= 1$; RHS of the formula $= \frac{1}{2} \times 1 \times 2 = 1$

Therefore, the formula is true for $n = 1$.

Therefore, $\displaystyle\sum_{r=1}^{n} r = \frac{1}{2}n(n+1)$ is true for all integers $n \geq 1$.

b $\displaystyle\sum_{10}^{20} r = \sum_{1}^{20} r - \sum_{1}^{9} r$

$$= \frac{1}{2} \times 20 \times 21 - \frac{1}{2} \times 9 \times 10 = 210 - 45 = 165$$

Note
Step 1, assume the formula is true for $n = k$ and write down the result of $n = k$.

Note
State what you need to show.

Note
Using $\sum S_{n+1} = S_n + u_{n+1}$

Note
Simplify the right-hand side.

Note
Step 2, prove it is true for $n = k + 1$.

Note
Step 3, prove the statement is true for $n = 1$.

Note
The sum of a series is given from 1 to n, so you must convert the series to a series starting at 1.

Example 2

Prove that $\displaystyle\sum_{r=1}^{n} r.r! = (n+1)! - 1.$

To prove formula is true for $k + 1$, need to prove $\displaystyle\sum_{r=1}^{k+1} r.r! = (k+2)! - 1.$

Assuming that the formula is true for $n = k$, then

$$\sum_{r=1}^{k} r.r! = (k+1)! - 1$$

$$\sum_{r=1}^{k+1} r.r! = (k+1)! - 1 + (k+1)\text{th term}$$

$$= (k+1)! - 1 + (k+1)(k+1)!$$
$$= (k+1)!(1 + k + 1) - 1$$
$$= (k+2)(k+1)! - 1$$
$$= (k+2)! - 1$$

Therefore, the formula is true for $n = k + 1$.

When $n = 1$: LHS of $\displaystyle\sum_{r=1}^{n} r.r! = 1$; RHS of $\displaystyle\sum_{r=1}^{n} r.r! = (n+1)! - 1 = 2! - 1 = 1$

Therefore, the formula is true for $n = 1$.

Therefore, $\displaystyle\sum_{r=1}^{n} r.r! = (n+1)! - 1$ is true for all $n \geq 1$.

Sequences

When a sequence is defined by a recurrence relation, you can use proof by induction to prove that a given formula for the sequence is true for all values. As before, in order to express the statement in terms of $n = k + 1$ you need to use the result of $n = k$.

Example 3

The sequence $a_{n+1} = a_n + 2n + 1$, $a_1 = 2$. Use induction to prove that $a_n = n^2 + 1$ for $n \geq 1$.

Need to prove that $a_{k+1} = (k+1)^2 + 1$.

Assuming that the statement is true for $n = k$, then

$a_{k+1} = a_k + 2k + 1 = (k^2 + 1) + 2k + 1 = k^2 + 2k + 2 = (k+1)^2 + 1.$

Therefore the result is also true for $n = k + 1$.

Now check that the hypothesis is true for $n = 1$.

LHS $= a_1 = 2$

RHS $= a_1 = 1^2 + 1 = 2$

Therefore the result is true for $n = 1$, and for all positive integers n.

If each term in a recurrence sequence is based on two previous terms, you would have to assume that the formula works for both $n = k$ and $n = k + 1$, then prove it works for $k = n + 2$ and that it is true for $n = 1$ and $n = 2$.

De Moivre's theorem

De Moivre's theorem is an important formula in mathematics because it connects complex numbers with trigonometry. You will learn more about this formula in Chapter *20 De Moivre's Theorem*. It is included in this chapter as an example of proof by induction on trigonometric equations.

Example 4

De Moivre's theorem states that, for all integral values of n,
$(\cos \theta + i \sin \theta)^n = \cos n\theta + i \sin n\theta.$

Use proof by induction to prove that this is true for all positive integers.

Assuming that the statement is true when $n = k,$ then

$(\cos \theta + i \sin \theta)^k = (\cos k\theta + i \sin k\theta)$

$\Rightarrow (\cos \theta + i \sin \theta)^{k+1} = (\cos k\theta + i \sin k\theta)(\cos \theta + i \sin \theta)$

Using $(\cos \theta + i \sin \theta)(\cos \phi + i \sin \phi) = \cos(\theta + \phi) + i \sin(\theta + \phi),$

$(\cos \theta + i \sin \theta)^{k+1} = \cos(k+1)\theta + i \sin(k+1)\theta$

Therefore, the statement is true for $n = k + 1$.

When $n = 1$, you have $(\cos \theta + i \sin \theta)^n = \cos \theta + i \sin \theta$

And $\cos n\theta + i \sin n\theta = \cos \theta + i \sin \theta$

Therefore, the statement is true for $n = 1$.

Therefore, de Moivre's theorem is true for all positive integers.

> **Note**
>
> $\cos(\theta + \varphi) = \cos \theta \cos \varphi - \sin \theta \sin \varphi$
> and $\sin(\theta + \varphi) = \sin \theta \cos \varphi + \cos \theta \sin \varphi$

Some other examples

Sometimes you might have a more complicated looking series but the same principles apply.

In the case of a differential equation, in order to create your expression in terms of $n = k + 1$, you will need to use $\dfrac{d^{n+1}y}{dx^{n+1}} = \dfrac{d}{dx}\left(\dfrac{d^n y}{dx^n}\right).$ You then simplify the RHS as usual.

Example 5

Prove that $\dfrac{d^n}{dx^n}(e^x \sin x) = 2^{\frac{n}{2}} e^x \sin\left(x + \frac{1}{4}n\pi\right).$

Assuming the formula is true of $n = k$, then $\dfrac{d^k}{dx^k}(e^x \sin x) = 2^{\frac{k}{2}} e^x \sin\left(x + \frac{1}{4}k\pi\right).$

Therefore,

$\dfrac{d^{k+1}}{dx^{k+1}}(e^x \sin x) = \dfrac{d}{dx}\left(\dfrac{d^k}{dx^k}(e^x \sin x)\right) = \dfrac{d}{dx}\left[2^{\frac{k}{2}} e^x \sin\left(x + \frac{1}{4}k\pi\right)\right]$

$= 2^{\frac{k}{2}} e^x \left[\sin\left(x + \frac{1}{4}k\pi\right) + \cos\left(x + \frac{1}{4}k\pi\right)\right]$

> **Note**
>
> Using product rule.

(continued)

When $x = 0$, $e^0 = a_1 \Rightarrow a_1 = 1$.

Differentiating again, you obtain $e^x = 2a_2 + 3 \times 2a_3 x + 4 \times 3a_4 x^2 + 5 \times 4a_5 x^3 + \cdots$

When $x = 0$, $e^0 = 2a_2 \Rightarrow a_2 = \dfrac{1}{2}$.

Differentiating yet again, you obtain $e^x = 3 \times 2a_3 + 4 \times 3 \times 2a_4 x + 5 \times 4 \times 3a_5 x^2 + \cdots$

When $x = 0$, $e^0 = 3 \times 2a_3 \Rightarrow a_3 = \dfrac{1}{3 \times 2 \times 1} = \dfrac{1}{3!}$.

Repeating the differentiation, you obtain $a_4 = \dfrac{1}{4!}$, $a_5 = \dfrac{1}{5!}$, $a_6 = \dfrac{1}{6!}$, $a_7 = \dfrac{1}{7!}$

Therefore,

$$e^x = 1 + x + \frac{x^2}{2!} + \frac{x^3}{3!} + \frac{x^4}{4!} + \frac{x^5}{5!} + \cdots + \frac{x^n}{n!} + \cdots$$

This series **converges** for all real x.

Power series for $(1 + x)^n$ for rational values of n

Let $(1 + x)^n = a_0 + a_1 x + a_2 x^2 + a_3 x^3 + \cdots$ Setting $x = 0$, you discover that $a_0 = 1$.

Differentiating, you obtain $n(1 + x)^{n-1} = a_1 + 2a_2 x + 3a_3 x^2 + \cdots$

When $x = 0$, you discover that $a_1 = n$.

If n is a positive integer (or zero), the series will eventually terminate, giving the normal expansion for $(1 + x)^n$. However, in all other cases you can find the general value of a_k by repeated differentiation, which is $a_k = \dfrac{n \times (n-1) \times \cdots \times (n-k+1)}{k!}$

This series converges when $|x| < 1$.

$$(1 + x)^n = 1 + \frac{nx}{1!} + \frac{n(n-1)x^2}{2!} + \cdots \quad \text{for} -1 < x < 1 \text{ when } n \text{ is an integer.}$$

You will recognise this series as the binomial theorem expansion which you met in your A-level mathematics studies. Since n is not a natural number, this expansion is not a finite series.

Power series for $\ln(1 + x)$

Since $\ln 0$ is not finite, you cannot have a power series for $\ln x$. Instead, you use a power series for $\ln(1 + x)$.

Let $\ln(1 + x) = a_0 + a_1 x + a_2 x^2 + a_3 x^3 + \cdots$ When $x = 0$, $\ln 1 = a_0$. But $\log 1 = 0$, therefore $a_0 = 0$.

Differentiating $\ln(1 + x) = a_1 x + a_2 x^2 + a_3 x^3 + \cdots$, you obtain

$\dfrac{1}{1+x} = a_1 + 2a_2 x + 3a_3 x^2 + 4a_4 x^3 + \cdots$

However, using the power series expansion of $(1+x)^n$ above, you can expand $\frac{1}{1+x}$ as $(1+x)^{-1}$ to give $1-x+x^2-x^3+x^4-x^5+\cdots$. Hence, you have

$$1-x+x^2-x^3+x^4-x^5+\cdots \equiv a_1+2a_2x+3a_3x^2+4a_4x^3+\cdots$$

Equating coefficients, you obtain $a_1=1$, $a_2=-\frac{1}{2}$, $a_3=\frac{1}{3}$, $a_4=-\frac{1}{4}$, $\cdots$

Therefore, you have

$$\ln(1+x)=x-\frac{x^2}{2}+\frac{x^3}{3}-\frac{x^4}{4}+\frac{x^5}{5}-\cdots$$

When $|x|<1$, the series **converges**. By inspection, you notice that the expansion is valid when $x=1$, but not when $x=-1$. Hence, you have

$$\ln(1+x)=x-\frac{x^2}{2}+\frac{x^3}{3}-\frac{x^4}{4}+\frac{x^5}{5}-\cdots+(-1)^{n+1}\frac{x^n}{n}+\cdots \text{ for } -1<x\leq 1$$

and,

$$\ln(1-x)=-x-\frac{x^2}{2}-\frac{x^3}{3}-\frac{x^4}{4}-\frac{x^5}{5}-\cdots-(-1)^{n+1}\frac{x^n}{n}+\cdots \text{ for } -1\leq x<1$$

You can use these techniques to expand a power series of related functions and to determine the range of values for which the expansions are valid.

Power series for more complicated functions

You can combine power series for simple functions to make power series for more complicated functions.

Example 2

Question

Find the power series for $\cos x^2$.

Answer

The power series for $\cos x$ is

$$\cos x=1-\frac{x^2}{2!}+\frac{x^4}{4!}-\cdots+\frac{(-1)^n}{(2n)!}x^{2n}+\cdots$$

$$\cos x^2=1-\frac{(x^2)^2}{2!}+\frac{(x^2)^4}{4!}-\cdots+\frac{(-1)^n}{(2n)!}(x^2)^{2n}+\cdots$$

$$=1-\frac{x^4}{2!}+\frac{x^8}{4!}-\cdots+\frac{(-1)^n}{(2n)!}x^{4n}+\cdots$$

Since the power series for $\cos x$ is valid for all values of x, the power series for $\cos x^2$ is valid for all values of x^2, that is, for all values of x.

> **Note**
>
> To obtain the power series for $\cos x^2$, replace every x in the $\cos x$ series with x^2.

Example 3

Find the power series for $\ln(1+3x)$, stating when the expansion is valid.

In the expansion for $\ln(1+x)$, $\ln(1+x)=x-\dfrac{x^2}{2}+\dfrac{x^3}{3}-\cdots$

$\ln(1+3x)=(3x)-\dfrac{(3x)^2}{2}+\dfrac{(3x)^3}{3}-\cdots$

$=3x-\dfrac{9}{2}x^2+9x^3-\cdots$

Since the expansion for $\ln(1+x)$ is valid for $-1<x\le1$, the expansion for

$\ln(1+3x)$ is valid for $-1<3x\le1$, so $-\dfrac{1}{3}<x\le\dfrac{1}{3}$.

Therefore,

$\ln(1+3x)=3x-\dfrac{9}{2}x^2+9x^3-\cdots$ for $-\dfrac{1}{3}<x\le\dfrac{1}{3}$

Example 4

Find the power series for $e^{4x}\sin 3x$, up to and including the term in x^4.

The power series for e^x is $e^x=1+x+\dfrac{x^2}{2!}+\dfrac{x^3}{3!}+\dfrac{x^4}{4!}+\cdots$

Therefore, the power series for e^{4x} is $e^{4x}=1+(4x)+\dfrac{(4x)^2}{2!}+\dfrac{(4x)^3}{3!}+\dfrac{(4x)^4}{4!}+\cdots$

$\sin 3x=(3x)-\dfrac{(3x)^3}{3!}+\dfrac{(3x)^5}{5!}-\cdots$

Therefore, the power series for $e^{4x}\sin 3x$ is

$e^{4x}\sin 3x=\left[1+(4x)+\dfrac{(4x)^2}{2!}+\dfrac{(4x)^3}{3!}+\dfrac{(4x)^4}{4!}+\cdots\right]\left[(3x)-\dfrac{(3x)^3}{3!}+\dfrac{(3x)^5}{5!}-\cdots\right]$

$=\left(1+4x+8x^2+\dfrac{32}{3}x^3+\dfrac{32}{3}x^4+\cdots\right)\left(3x-\dfrac{9}{2}x^3+\cdots\right)$

Ignoring terms in x^5 and higher powers,

$e^{4x}\sin 3x=3x+12x^2+24x^3-\dfrac{9}{2}x^3+32x^4-18x^4$

Therefore,

$e^{4x}\sin 3x=3x+12x^2+\dfrac{39}{2}x^3+14x^4$

Example 5

Find all the terms up to and including x^4 in the power series for $e^{\sin x}$.

Using the power series for e^x, $e^{\sin x} = 1 + \dfrac{\sin x}{1!} + \dfrac{\sin^2 x}{2!} + \dfrac{\sin^3 x}{3!} + \cdots$

Therefore,

$$e^{\sin x} = 1 + \frac{x - \dfrac{x^3}{3!} + \cdots}{1!} + \frac{\left(x - \dfrac{x^3}{3!} + \cdots\right)^2}{2!} + \frac{\left(x - \dfrac{x^3}{3!} + \cdots\right)^3}{3!} + \frac{\left(x - \dfrac{x^3}{3!} + \cdots\right)^4}{4!} + \cdots$$

$$\Rightarrow\ e^{\sin x} = 1 + x - \frac{x^3}{3!} + \frac{x^2 - \dfrac{2x^4}{3!}}{2!} + \frac{x^3}{3!} + \frac{x^4}{4!} + \cdots$$

which gives

$$e^{\sin x} = 1 + x + \frac{x^2}{2} - \frac{x^4}{8}$$

> **Note**
>
> You now apply the power series for $\sin x$. Since you are asked for terms only up to x^4, you can ignore terms in higher powers of x.

Exercise 1

1. Find out whether these infinite series converge or diverge.

   a $\displaystyle\sum_{n=1}^{\infty} \frac{5^n}{n!}$ **b** $\displaystyle\sum_{n=2}^{\infty} \frac{1}{2^n - 1}$ **c** $\displaystyle\sum_{n=1}^{\infty} \frac{n^2}{2^n}$

2. **a** Find the binomial expansion of $(1+8x)^{-\frac{1}{4}}$ up to and including the term in x^3.

   b Find the binomial expansion of $(16+8x)^{-\frac{1}{4}}$ up to and including the term in x^3.

3. **a** Find the binomial expansion of $\left(1 - \dfrac{7}{2}x\right)^{-3}$ up to and including the term in x^3.

   b Find the range of values for which the binomial expansion of $\left(1 - \dfrac{7}{2}x\right)^{-3}$ is valid.

   c Given that x is small, show that $\left(\dfrac{8}{2-7x}\right)^3 \approx a + bx + cx^2$, where a, b and c are to be found.

4. Find the power series for

   a e^{3x} **b** $\cos x^2$

5. Find the power series for

   a $\sin 2x$ **b** $\cos 5x$ **c** e^{8x} **d** $\ln(1 + x^2)$ **e** $\ln(1 - 2x)$

6. Find the power series of each of these functions, up to and including the term in x^4.

   a $\sin x^2$ **b** $(1+x)e^{3x}$ **c** $(2+x^2)\cos 3x$ **d** $e^{\cos x}$ **e** $\ln(1 + \cos x)$

7. Find the power series expansion of e^{2x^2}.

19.3 Finding the limits of a series

You already know that the limit of the sum of an infinite series gives the number a series heads towards. The limit of the sum of an infinite series is approximately equal to the sum when enough terms have been included. We can ensure that the sum is as close as we'd like to the limit by making sure that we include enough terms.

Finding the limit of a series requires a **limiting process**, whereby you solve the limit to find the value of the function at its limit. You can evaluate a limit expression of a series, f(x) directly by substituting in the appropriate value of x.

When it is not easily possible to evaluate a limit expression directly, for example when the limit is a quotient such that at $x = 0$ it is undefined, you can use series expansion to find the limit.

Using power series

Power series are useful to find the limits of a series as x tends to zero. For example, to find limit of $\dfrac{f(x)}{g(x)}$ as $x \to 0$, when $f(0) = g(0) = 0$. To find such a limit, you expand both the numerator and the denominator of the expression as a power series in x and then divide both by the lowest power of x present. Then you use $x = 0$.

If you simply use $x = 0$ to begin with, you would obtain $\dfrac{f(0)}{g(0)} = \dfrac{0}{0}$, which means that you have proceeded incorrectly.

Example 6

Find the limit of $\dfrac{x - \sin x}{x^2(e^x - 1)}$ as $x \to 0$.

$$\frac{x - \sin x}{x^2(e^x - 1)} = \frac{x - \left(x - \dfrac{x^3}{3!} + \dfrac{x^5}{5!} - \cdots\right)}{x^2\left(1 + x + \dfrac{x^2}{2!} + \cdots - 1\right)} = \frac{\dfrac{x^3}{3!} - \dfrac{x^5}{5!} + \cdots}{x^3 + \dfrac{x^4}{2!} + \cdots}$$

> **Note**
> Expand using power series.

$$\frac{x - \sin x}{x^2(e^x - 1)} = \frac{\dfrac{1}{3!} - \dfrac{x^2}{5!} + \cdots}{1 + \dfrac{x}{2!} + \dfrac{x^2}{3!} \cdots}$$

> **Note**
> Divide the numerator and the denominator by x^3.

Therefore,

$$\lim_{x \to 0} \frac{x - \sin x}{x^2(e^x - 1)} = \frac{\dfrac{1}{3!}}{1} = \frac{1}{6}$$

> **Note**
> Let $x \to 0$.

Example 7

Find the limit of $\dfrac{1-\cos x}{\sin^2 x}$ as $x \to 0$.

$$\frac{1-\cos x}{\sin^2 x} = \frac{1-\left(1-\dfrac{x^2}{2!}+\dfrac{x^4}{4!}-\cdots\right)}{\left(x-\dfrac{x^3}{3!}+\dfrac{x^5}{5!}-\cdots\right)^2} = \frac{\dfrac{x^2}{2!}-\dfrac{x^4}{4!}+\cdots}{x^2-\dfrac{2x^4}{3!}+\cdots}$$

$$\frac{1-\cos x}{\sin^2 x} = \frac{\dfrac{1}{2!}-\dfrac{x^2}{4!}+\cdots}{1-\dfrac{2x^2}{3!}+\cdots}$$

Therefore,

$$\lim_{x\to 0}\frac{1-\cos x}{\sin^2 x} = \frac{\dfrac{1}{2!}}{1} = \frac{1}{2}$$

> **Note**
>
> Expand using power series.

> **Note**
>
> Divide the numerator and the denominator by x^2.

Exercise 2

1. You are told that $y = \displaystyle\sum_{n=0}^{\infty}\frac{nx^n}{3^n}$. When does this series converge?

2. Find the value of $\displaystyle\lim_{x\to 0}\frac{e^x-1}{x}$.

3. Find the value of $\displaystyle\lim_{x\to 0}\frac{\sin 3x}{x}$.

4. Find the power series expansion of $\cos x^3$. Which values of x is this valid for?

5. Prove that the series $\displaystyle\sum_{n=1}^{\infty}\frac{4}{n}$ does not converge.

19.4 Improper integrals

You know from Chapter 6 *Calculus*, that an **improper integral** is one that has either

▶ a limit of integration of $\pm\infty$, or

▶ an integrand that is infinite at one or other of its limits of integration, or between these limits.

In the first case, if you replace $\pm\infty$ with n you can then find the limit of the integral as $n \to \pm\infty$. When this limit is **finite**, the integral **can be found** (see *Example 12*). When this limit is **not finite**, the integral **cannot be found**. In the first case, you replace one or other of the limits of integration with p, for example, and then find the limit of the integral as p tends to the value of the limit it has replaced.

Example 8

Question

Determine $\displaystyle\int_1^\infty \frac{1}{x^2}\,dx$.

Answer

$$\int_1^\infty \frac{1}{x^2}\,dx = \lim_{n\to\infty}\int_1^n \frac{1}{x^2}\,dx$$

$$= \lim_{n\to\infty}\left[-\frac{1}{x}\right]_1^n = \lim_{n\to\infty}\left(-\frac{1}{n}+1\right)$$

As $n\to\infty, \dfrac{1}{n}\to 0$ and $\lim_{n\to\infty}\left(-\dfrac{1}{n}+1\right)=1$

Therefore $\displaystyle\int_1^\infty \frac{1}{x^2}\,dx=1$

> **Note**
> The upper limit is ∞, so you replace it with n; this is the same method you used in Chapter 6 *Calculus*.

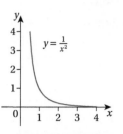

> **Note**
> This shows that the area under the curve $y=\dfrac{1}{x^2}$ is finite even though the boundary is of infinite length.

This section will extend the use of improper integrals from what you saw in Chapter 6 *Calculus*; you will be required to use a limiting process in order to evaluate the integral. You will need to explain in detail how you use the limiting process.

Example 9

Question

Determine $\displaystyle\int_0^e x\ln x\,dx$.

Answer

$$\int_0^e x\ln x\,dx = \lim_{p\to 0}\int_p^e x\ln x\,dx$$

$$= \lim_{p\to 0}\left(\left[\frac{x^2}{2}\ln x\right]_p^e - \int_p^e \frac{x}{2}dx\right)$$

$$= \lim_{p\to 0}\left(\frac{e^2}{2}-\frac{p^2}{2}\ln p-\frac{e^2}{4}+\frac{p^2}{4}\right)$$

$$\int_0^e x\ln x\,dx = \frac{e^2}{2}-\frac{e^2}{4}=\frac{e^2}{4}$$

> **Note**
> At $x = 0$, the integrand is infinite. Therefore, replace the lower limit with p and find the limit as $p \to 0$. Use integration by parts to perform the integration; this is the method you have seen before.

> **Note**
> You need to apply a limiting process, so use the generic statement that polynomials dominate logarithms.

Example 10

Question

Evaluate the improper integral $\displaystyle\int_0^\infty x\,e^{-x}\,dx$ showing the limiting process used.

Answer

Integral $\displaystyle\int_0^\infty x\,e^{-x}\,dx = \lim_{n\to\infty}\int_0^n x\,e^{-x}\,dx$

$$= \lim_{n\to\infty}\left(\left[-x\,e^{-x}\right]_0^n + \int_0^n e^{-x}\,dx\right)$$

$$= \lim_{n\to\infty}(-ne^{-n}-e^{-n})+e^0=1$$

> **Note**
> Since e^{-n} dominates the polynomial n.

> **Note**
> At $x = \infty$, the limits are not finite. Therefore, replace the upper limit with n and find the limit as $n \to \infty$. Use integration by parts to perform the integration.

Exercise 3

Find the value, where it exists, of each of the integrals in questions **1** to **4**.

1 $\displaystyle\int_{-a}^{\infty} \frac{1}{x^2 - a^2}\,\mathrm{d}x$ **2** $\displaystyle\int_{0}^{\infty} \frac{1}{x^2 + a^2}\,\mathrm{d}x$ **3** $\displaystyle\int_{-2}^{2} \frac{1}{x+2}\,\mathrm{d}x$ **4** $\displaystyle\int_{0}^{\frac{\pi}{2}} \tan x\,\mathrm{d}x$

5 Evaluate the improper integral $\displaystyle\int_{0}^{\infty} x\,\mathrm{e}^{-5x}\,\mathrm{d}x$ showing the limiting process used.

6 Evaluate the improper integral $\displaystyle\int_{0}^{\infty} x\,\mathrm{e}^{-7x}\,\mathrm{d}x$ showing the limiting process used.

7 Show that $\displaystyle\lim_{p\to\infty}\frac{4p+5}{3p+7}=\frac{4}{3}$. Hence evaluate $\displaystyle\int_{1}^{\infty}\left(\frac{4}{4p+5}-\frac{3}{3p+7}\right)$ and show the limiting process used.

8 Evaluate the improper integral $\displaystyle\int_{1}^{\infty}\left(\frac{1}{x}-\frac{5}{5x+7}\right)$ showing the limiting process used and giving your answer in the form $\ln k$.

Summary

▶ The Maclaurin theorem can be used to express some functions as a series in ascending positive integral powers of x. The Maclaurin theorem states that
$$\mathrm{f}(x) = \mathrm{f}(0) + x\mathrm{f}'(0) + \frac{x^2}{2!}\mathrm{f}''(0) + \cdots + \frac{x^r}{r!}\mathrm{f}^{(r)}(0) + \cdots \text{ provided that } \mathrm{f}(0),\ \mathrm{f}'(0) \dots$$
$\mathrm{f}^{(r)}(0)$ all have finite values.

▶ The Maclaurin theorem can be applied to a number of functions, some examples are given below and included in the *Formulae and Statistical Tables* booklet. You need to know when each of these Maclaurin's theorems converge.

• $\mathrm{e}^x = 1 + \dfrac{x}{1!} + \dfrac{x^2}{2!} + \dfrac{x^3}{3!} + \cdots$, converges for all real numbers.

• $\ln(1+x) = x - \dfrac{x^2}{2} + \dfrac{x^3}{3} - \cdots$, converges for $-1 < x \le 1$.

• $\sin x = x - \dfrac{x^3}{3!} + \dfrac{x^5}{5!} - \cdots$, converges for all real numbers.

• $\cos x = 1 - \dfrac{x^2}{2!} + \dfrac{x^4}{4!} - \cdots$, converges for all real numbers.

• $(1+x)^n = 1 + nx + \dfrac{n(n-1)}{2!}x^2 + \cdots$ converges if n is a positive integer, or if $|x| < 1$. (**Note** that this formula is in the *Formulae and Statistical Tables* booklet in the Binomial section.)

▶ You can combine power series for simple functions to make power series for more complicated functions.

• diverges when $\displaystyle\lim_{n\to\infty}\left|\frac{a_{n+1}}{a_n}\right| > 1$

• provides no information when $\displaystyle\lim_{n\to\infty}\left|\frac{a_{n+1}}{a_n}\right| = 1$.

▶ Understand that when a limit contains terms involving a polynomial and an exponential or logarithmic function, for example, $x^k e^{-x}$ as $x \to \infty$, and $x^k \ln x$ as $x \to 0$, that e^x dominates x^n, while x^n dominates $\log x$.

▶ It is possible to calculate various improper integrals involving e^x and $\ln x$, which often require a limiting process.

Review exercises

1 Find the limit $\displaystyle\lim_{x \to 0} \frac{x^2 e^x}{\cos 2x - 1}$.

2 Find the limit $\displaystyle\lim_{x \to 0} \frac{\sqrt{2+x} - \sqrt{2}}{x}$.

3 Find the first four terms of the power series expansions of the following, and state where they converge.

   a $\tan 7x$ **b** $\ln(1-2x)$

4 Using Maclaurin's theorem, find the first three non-zero terms in the expansion of $\ln(1 + 3\sin x)$. Hence find $\displaystyle\lim_{x \to 0} \frac{\ln(1 + 3\sin x)}{\ln(1 - x)}$.

5 By using the series expansion for $\cos x$, find the first three non-zero terms in the expansion of $\sec x$. Using Maclaurin's theorem, find the first two non-zero terms in the series expansion of $\tan x$. Hence find $\displaystyle\lim_{x \to 0} \frac{x \tan 3x}{\sec x - 1}$.

Practice examination questions

1 **a** Find the binomial expansion of $(1 + 4x)^{\frac{1}{2}}$ up to and including the term in x^2.   (2 marks)

   b **i** Find the binomial expansion of $(4 - x)^{-\frac{1}{2}}$ up to and including the term in x^2.   (3 marks)

      ii State the range of values of x for which the expansion in part **b i** is valid.   (1 mark)

   c Find the binomial expansion of $\sqrt{\dfrac{1 + 4x}{4 - x}}$ up to and including the term in x^2.   AQA MPC4 June 2012

2 **a** Write down the expansion of e^{3x} in ascending powers of x up to and including the term in x^2.   (1 mark)

   b Hence, or otherwise, find the term in x^2 in the expansion, in ascending powers of x, of $e^{3x}(1 + 2x)^{-\frac{3}{2}}$.   (4 marks)

                                       AQA MFP3 January 2013

3 It is given that $y = \ln(1 + \sin x)$.

   a Find $\dfrac{dy}{dx}$.   (2 marks)

   b Show that $\dfrac{d^2 y}{dx^2} = -e^{-y}$.   (3 marks)

c Express $\dfrac{d^4y}{dx^4}$ in terms of $\dfrac{dy}{dx}$ and e^{-y}. (3 marks)

d Hence, by using Maclaurin's theorem, find the first four non-zero terms in the expansion, in ascending powers of x, of $\ln(1 + \sin x)$. (3 marks)

AQA MFP3 June 2012

4 a Use integration by parts to show that $\displaystyle\int \ln x \, dx = x \ln x - x + c$, where c is an arbitrary constant. (2 marks)

b Hence evaluate $\displaystyle\int_0^1 \ln x \, dx$, showing the limiting process used. (4 marks)

AQA MFP3 January 2009

5 a Find $\displaystyle\int x^2 \ln x \, dx$. (3 marks)

b Explain why $\displaystyle\int_0^e x^2 \ln x \, dx$ is an improper integral. (1 mark)

c Evaluate $\displaystyle\int_0^e x^2 \ln x \, dx$, showing the limiting process used. (3 marks)

AQA MFP3 June 2011

6 a Write down the expansions in ascending powers of x up to and including the term in x^3 of

i $\cos x + \sin x$ (1 mark)

ii $\ln(1+3x)$ (1 mark)

b It is given that $y = e^{\tan x}$.

i Find $\dfrac{dy}{dx}$ and show that $\dfrac{d^2y}{dx^2} = (1 + \tan x)^2 \dfrac{dy}{dx}$. (5 marks)

ii Find the value of $\dfrac{d^3y}{dx^3}$ when $x = 0$. (2 marks)

iii Hence, by using Maclaurin's theorem, show that the first four terms in the expansion, in ascending powers of x, of $e^{\tan x}$ are

$$1 + x + \frac{1}{2}x^2 + \frac{1}{2}x^3.$$ (2 marks)

c Find

$$\lim_{x \to 0}\left[\frac{e^{\tan x} + (\cos x + \sin x)}{x \ln(1+3x)}\right]$$ (3 marks)

AQA MFP3 January 2011

7 a Explain why $\displaystyle\int_1^\infty x e^{-3x} \, dx$ is an improper integral. (1 mark)

b Find $\displaystyle\int x e^{-3x} \, dx$. (3 marks)

c Hence evaluate $\displaystyle\int_1^\infty x e^{-3x} \, dx$ showing the limiting process used. (3 marks)

AQA MFP3 January 2008

20 De Moivre's Theorem

Introduction

The definition $i = \sqrt{-1}$ in Chapter 2 *Complex Numbers*, helped us to understand the non-real roots of quadratic equations. The majority of this chapter will be about de Moivre's theorem, including its use to solve equations in the form $z^n = a + ib$. You will also use the theorem to express terms such as $\cos 6\theta$ in powers of $\cos\theta$, and conversely to express $\cos^6\theta$ in terms of multiple angles such as $\cos 6\theta$ without having to learn complicated multiple-angle formulae. You will also consider functions such as e^{ix} and use them to solve integrals such as $\int \sin^n\theta \, d\theta$.

Objectives

By the end of this chapter, you should know how to:
► Prove de Moivre's theorem.
► Solve equations of the form $z^n = a + ib$.
► Express $\cos^n\theta$ in terms of multiple angles, and express $\cos n\theta$ in terms of powers of $\cos\theta$; similarly for sin and tan.
► Use de Moivre's theorem to integrate functions like $\int \sin^n\theta \, d\theta$.
► Use the formula $e^{ix} = \cos x + i\sin x$.

Recap

You will need to remember...
► How to add, multiply and divide complex numbers.
► How to write complex numbers in polar (modulus-argument) form.
► How to manipulate sines and cosines using trigonometric identities.
► How to use induction to prove results.
► How to expand using the binomial theorem.

See Chapter 2 Complex Numbers.

See Chapter 18 Proof by Induction and Finite Series

- -

20.1 De Moivre's theorem

De Moivre's theorem has many applications in mathematics. For example, finding powers and roots of complex numbers when expressed in their polar form; expressing powers of trigonometric functions in terms of multiple angles and vice versa; and expressing a function as an exponential of a complex function and thus allowing you to integrate certain functions.

You saw in Chapter 2 *Complex Numbers* that a complex number given in the Cartesian form ($z = x + iy$) can be written in the polar form (r, θ; where r is the modulus and θ is the argument) as $z = r(\cos\theta + i\sin\theta)$.

> De Moivre's theorem states that $(\cos\theta + i\sin\theta)^n = \cos n\theta + i\sin n\theta$, for all integral values of n. When n is not an integer, then $\cos n\theta + i\sin n\theta$ is only one of the possible values.

Proof when n is a positive integer

You can use proof by induction to show that de Moivre's theorem is true for all positive integer values of n.

You assume that the statement is true when $n = k$. Hence, you have

$(\cos\theta + i\sin\theta)^k = (\cos k\theta + i\sin k\theta)$

$\Rightarrow (\cos\theta + i\sin\theta)^{k+1} = (\cos k\theta + i\sin k\theta)(\cos\theta + i\sin\theta)$

Using $(\cos\theta + i\sin\theta)(\cos\phi + i\sin\phi) = \cos\theta\cos\phi - \sin\theta\sin\phi + i(\cos\theta\sin\phi + \sin\theta\cos\phi)$

$= \cos(\theta + \phi) + i\sin(\theta + \phi)$, you obtain

$(\cos\theta + i\sin\theta)^{k+1} = \cos(k+1)\theta + i\sin(k+1)\theta$

Therefore, the statement is true for $n = k + 1$.

When $n = 1$, you have $(\cos\theta + i\sin\theta)^n = \cos\theta + i\sin\theta$

And $\cos n\theta + i\sin n\theta = \cos\theta + i\sin\theta$

Therefore, the statement is true for $n = 1$.

Therefore, de Moivre's theorem is true for all values of $n \geq 1$. That is, for all positive integers.

Proof when n is a negative integer

When n is a negative integer, $n = -p$, where p is a positive integer, you have

$$(\cos\theta + i\sin\theta)^n = (\cos\theta + i\sin\theta)^{-p} = \frac{1}{(\cos\theta + i\sin\theta)^p}$$

Using de Moivre's theorem for the positive integer p, you obtain

$$\frac{1}{(\cos\theta + i\sin\theta)^p} = \frac{1}{\cos p\theta + i\sin p\theta}$$

$$= \frac{\cos p\theta - i\sin p\theta}{(\cos p\theta + i\sin p\theta)(\cos p\theta - i\sin p\theta)}$$

which gives

$$\frac{1}{(\cos\theta + i\sin\theta)^p} = \cos p\theta - i\sin p\theta$$

But $n = -p$, hence you have

$\cos p\theta - i\sin p\theta = \cos(-n\theta) - i\sin(-n\theta) = \cos n\theta + i\sin n\theta$.

Therefore, you have $(\cos\theta + i\sin\theta)^n = \cos n\theta + i\sin n\theta$ for all negative integers.

Simple application to complex numbers

You must always make sure the complex number is in the form $z = r(\cos\theta + i\sin\theta)$ when applying de Moivre's theorem.

You can use the theorem to express a complex number, z^n, in multiple-angled form, that is, in terms of $\cos n\theta$ and $\sin n\theta$.

Example 1

Find the value of $(\cos\theta + i\sin\theta)^5$ in multiple-angle form.

$(\cos\theta + i\sin\theta)^5 \equiv \cos 5\theta + i\sin 5\theta$

The theorem can also be used to simplify an expression in the form $\cos n\theta + i\sin n\theta$.

> **Note**
>
> Using de Moivre's theorem:
> $(\cos\theta + i\sin\theta)^n = \cos n\theta + i\sin n\theta$.

Example 2

Simplify $(\cos 8\theta + i \sin 8\theta)(\cos 6\theta + i \sin 6\theta)$.

Answer

Using de Moivre's theorem:

$$(\cos 8\theta + i \sin 8\theta) = (\cos \theta + i \sin \theta)^8$$

$$(\cos 6\theta + i \sin 6\theta) = (\cos \theta + i \sin \theta)^6$$

$$\Rightarrow (\cos \theta + i \sin \theta)^8 (\cos \theta + i \sin \theta)^6 = (\cos \theta + i \sin \theta)^{14}$$

$$= \cos 14\theta + i \sin 14\theta$$

> **Note**
>
> Using the laws of indices.

> **Note**
>
> Use de Moivre's theorem to express the answer as $\cos n\theta + i \sin n\theta$.

It is possible to calculate positive powers of complex numbers, z^n, using de Moivre's theorem. First you express the complex number in its r, θ form, then using de Moivre's theorem you express it in multiple-angle form and evaluate.

Example 3

Question

Find $\left[\cos\left(\dfrac{\pi}{6}\right) + i \sin\left(\dfrac{\pi}{6}\right) \right]^3$.

Answer

$$\left[\cos\left(\frac{\pi}{6}\right) + i \sin\left(\frac{\pi}{6}\right) \right]^3 \equiv \cos\left(3 \times \frac{\pi}{6}\right) + i \sin\left(3 \times \frac{\pi}{6}\right)$$

$$\equiv \cos\left(\frac{\pi}{2}\right) + i \sin\left(\frac{\pi}{2}\right) = i$$

> **Note**
>
> Appling de Moivre's theorem:
> $(\cos \theta + i \sin \theta)^n = \cos n\theta + i \sin n\theta$.

Example 4

Question

Find $\left[\sin\left(\dfrac{\pi}{3}\right) + i \cos\left(\dfrac{\pi}{3}\right) \right]^6$.

Answer

Using $\cos\left(\dfrac{\pi}{2} - \theta\right) = \sin \theta$,

$$\left[\sin\left(\frac{\pi}{3}\right) + i \cos\left(\frac{\pi}{3}\right) \right]^6 \equiv \left[\cos\left(\frac{\pi}{6}\right) + i \sin\left(\frac{\pi}{6}\right) \right]^6$$

$$\equiv \cos \pi + i \sin \pi$$

$$= -1$$

Therefore, $\left[\sin\left(\dfrac{\pi}{3}\right) + i \cos\left(\dfrac{\pi}{3}\right) \right]^6 = -1$

> **Note**
>
> You need the complex number to be in the '$\cos \theta + i \sin \theta$' form before you can apply the theorem.

> **Note**
>
> $\sin \dfrac{\pi}{3} = \cos\left(\dfrac{\pi}{2} - \dfrac{\pi}{3}\right) = \cos \dfrac{\pi}{6}$
>
> and similarly for $\cos \dfrac{\pi}{3}$.

Example 5 demonstrates an alternative method to get the complex number in Example 4 into the correct form to apply the theorem.

Example 5

Find $\left[\sin\left(\dfrac{\pi}{3}\right)+i\cos\left(\dfrac{\pi}{3}\right)\right]^6$.

$\left[\sin\left(\dfrac{\pi}{3}\right)+i\cos\left(\dfrac{\pi}{3}\right)\right]^6 \equiv \left(i\left[\cos\left(\dfrac{\pi}{3}\right)-i\sin\left(\dfrac{\pi}{3}\right)\right]\right)^6$

Note
Factor out the i.

$\equiv \left(i\left[\cos\left(-\dfrac{\pi}{3}\right)+i\sin\left(-\dfrac{\pi}{3}\right)\right]\right)^6$

Note
Using $\cos(x) = \cos(-x)$ and $\sin(-x) = -\sin(x)$.

$\left[\sin\left(\dfrac{\pi}{3}\right)+i\cos\left(\dfrac{\pi}{3}\right)\right]^6 \equiv i^6[\cos(-2\pi)+i\sin(-2\pi)] = -1\times1 = -1$

Therefore, $\left[\sin\left(\dfrac{\pi}{3}\right)+i\cos\left(\dfrac{\pi}{3}\right)\right]^6 = -1$

Note
Applying de Moivre's theorem to the RHS.

Note that in *Example 5*, $\left[\cos\left(\dfrac{\pi}{3}\right)-i\sin\left(\dfrac{\pi}{3}\right)\right]^6 \equiv \cos 2\pi - i\sin 2\pi$ and hence you might have deduced that $(\cos\theta - i\sin\theta)^n \equiv \cos n\theta - i\sin n\theta$. However, this **cannot** be used as a correct version of de Moivre's theorem, which is only applicable to $(\cos\theta + i\sin\theta)^n$. Therefore, if you are asked to use de Moivre's theorem to find the value of $\left[\cos\left(\dfrac{\pi}{3}\right)-i\sin\left(\dfrac{\pi}{3}\right)\right]^6$ for example, you *must* change this to $\left[\cos\left(-\dfrac{\pi}{3}\right)+i\sin\left(-\dfrac{\pi}{3}\right)\right]^6$ as shown in *Example 5*.

Example 6

Find the value of $(1+i)^4$.

$z = (1+i) \Rightarrow z = \sqrt{2}\cos\left(\dfrac{\pi}{4}\right)+i\sqrt{2}\sin\left(\dfrac{\pi}{4}\right)$

$\Rightarrow (1+i)^4 = \left(\sqrt{2}\left[\cos\left(\dfrac{\pi}{4}\right)+i\sin\left(\dfrac{\pi}{4}\right)\right]\right)^4$

Note
Factor out $\sqrt{2}$.

Note
Using $z \equiv x+iy = r\cos\theta + i r\sin\theta$, where $r = |z| = \sqrt{x^2+y^2}$ and $\tan\theta = \dfrac{y}{x}$.

$= (\sqrt{2})^4\left[\cos\left(\dfrac{\pi}{4}\right)+i\sin\left(\dfrac{\pi}{4}\right)\right]^4$

$= (\sqrt{2})^4\left[\cos\left(4\times\dfrac{\pi}{4}\right)+i\sin\left(4\times\dfrac{\pi}{4}\right)\right]$

Note
Apply de Moivre's theorem.

$= 4(\cos\pi + i\sin\pi)$

Therefore, $(1+i)^4 = -4$.

You learned how to create the modulus-argument form of a complex number in Chapter 2 *Complex Numbers*.

Example 7

Find the value of $\dfrac{1}{(4-4i)^3}$.

$$4-4i = 4\sqrt{2}\left[\cos\left(-\frac{\pi}{4}\right)+i\sin\left(-\frac{\pi}{4}\right)\right]$$

$$\Rightarrow \frac{1}{(4-4i)^3} = \frac{1}{\left(4\sqrt{2}\left[\cos\left(-\frac{\pi}{4}\right)+i\sin\left(-\frac{\pi}{4}\right)\right]\right)^3}$$

$$= \frac{1}{128\sqrt{2}\left[\cos\left(-\frac{\pi}{4}\right)+i\sin\left(-\frac{\pi}{4}\right)\right]^3}$$

$$= \frac{1}{128\sqrt{2}}\left[\cos\left(-\frac{\pi}{4}\right)+i\sin\left(-\frac{\pi}{4}\right)\right]^{-3}$$

$$\Rightarrow \frac{1}{(4-4i)^3} = \frac{1}{128\sqrt{2}}\left[\cos\left(\frac{3\pi}{4}\right)+i\sin\left(\frac{3\pi}{4}\right)\right]$$

$$= \frac{1}{128\sqrt{2}}\left(-\frac{1}{\sqrt{2}}+\frac{1}{\sqrt{2}}i\right) = \frac{1}{128\sqrt{2}}\times-\frac{1}{\sqrt{2}}(1-i)$$

Therefore,

$$\frac{1}{(4-4i)^3} = \frac{1}{256}(-1+i)$$

> **Note**
>
> Convert 4 − 4i into its (r, θ) form.

> **Note**
>
> Use de Moivre's theorem.

Exercise 1

1. Simplify $(\cos 11\theta + i\sin 11\theta)(\cos 3\theta + i\sin 3\theta)$.

2. Using de Moivre's theorem, find the value of

    a $(\cos\theta + i\sin\theta)^6$

    b $(\cos 2\theta + i\sin 2\theta)^4$

    c $\left[\cos\left(\frac{\pi}{3}\right)+i\sin\left(\frac{\pi}{3}\right)\right]^9$

    d $\left[\cos\left(\frac{\pi}{4}\right)+i\sin\left(\frac{\pi}{4}\right)\right]^6$

    e $\dfrac{1}{(\cos 2\theta + i\sin 2\theta)^4}$

    f $\dfrac{1}{\left[\cos\left(\frac{\pi}{6}\right)+i\sin\left(\frac{\pi}{6}\right)\right]^6}$

    g $\left[\cos\left(\frac{2\pi}{5}\right)+i\sin\left(\frac{2\pi}{5}\right)\right]^{10}$

    h $\left[\cos\left(-\frac{\pi}{18}\right)+i\sin\left(-\frac{\pi}{18}\right)\right]^9$

3. Simplify.

    a $(\cos 3\theta + i\sin 3\theta)(\cos 7\theta + i\sin 7\theta)$

    b $(\cos 5\theta + i\sin 5\theta)(\cos 6\theta - i\sin 6\theta)$

    c $\dfrac{\left[\cos\left(\frac{\pi}{3}\right)+i\sin\left(\frac{\pi}{3}\right)\right]^5}{\left[\cos\left(\frac{\pi}{3}\right)-i\sin\left(\frac{\pi}{3}\right)\right]^4}$

    d $(1+i)^4 + (1-i)^4$

4 Simplify.

   a $(1+i)^8$ **b** $(3-\sqrt{3}\,i)^6$

   c $(1-i)^4$ **d** $(2+2\sqrt{3}\,i)^6$

5 Simplify.

   a $(\cos\theta-i\sin\theta)^5$ **b** $(\sin\theta+i\cos\theta)^4$

   c $\dfrac{1}{(\sin\theta+i\cos\theta)^6}$ **d** $\dfrac{1}{\left[\sin\left(\dfrac{\pi}{5}\right)-i\cos\left(\dfrac{\pi}{5}\right)\right]^{10}}$

6 Show that $\dfrac{\cos4x+i\sin4x}{\cos5x-i\sin5x}$ can be expressed in the form $\cos nx+i\sin nx$, where n is an integer to be found.

20.2 Further applications of de Moivre's theorem to complex numbers

*n*th roots of unity

When n is not an integer, de Moivre's theorem gives **only one** of the possible values for $(\cos\theta+i\sin\theta)^n$,

that is, it gives $\cos n\theta+i\sin n\theta$.

However, $(\cos\theta+i\sin\theta)^{\frac{1}{n}}$ can take n different values.

Let $(\cos\theta+i\sin\theta)^{\frac{1}{n}}=r(\cos\phi+i\sin\phi)$

Comparing the moduli of both sides, you have $r=1$.

Raising both sides to the nth power, and using $[(\cos\theta+i\sin\theta)^{\frac{1}{n}}]^n=\cos\theta+i\sin\theta$

you obtain

$$\cos\theta+i\sin\theta=[(\cos\theta+i\sin\theta)^{\frac{1}{n}}]^n=(\cos\phi+i\sin\phi)^n$$

$\Rightarrow$ $\cos\theta+i\sin\theta=\cos n\phi+i\sin n\phi$

Therefore, you have

$$\cos\theta=\cos n\phi \text{ and } \sin\theta=\sin n\phi$$

which gives

$$n\phi=\theta,\ \theta+2\pi,\theta+4\pi,\theta+6\pi,\ldots$$

since $\cos(\theta+2\pi)=\cos\theta$, and $\sin(\theta+2\pi)=\sin\theta$.

That is, you have

$$\phi=\frac{\theta}{n},\frac{\theta+2\pi}{n},\frac{\theta+4\pi}{n},\ldots$$

which means that $(\cos\theta+i\sin\theta)^{\frac{1}{n}}$ is equal to

$$\cos\left(\frac{\theta}{n}\right)+i\sin\left(\frac{\theta}{n}\right)$$

or $\cos\left(\dfrac{\theta+2\pi}{n}\right)+i\sin\left(\dfrac{\theta+2\pi}{n}\right)$

or $\cos\left(\dfrac{\theta+4\pi}{n}\right)+i\sin\left(\dfrac{\theta+4\pi}{n}\right)$

and so on, adding $\dfrac{2\pi}{n}$ each time until you obtain

$$(\cos\theta + i\sin\theta)^{\frac{1}{n}} \equiv \cos\left[\frac{\theta+(n-1)2\pi}{n}\right] + i\sin\left[\frac{\theta+(n-1)2\pi}{n}\right]$$

All subsequent values are repeats of the n different values given above.

Therefore, $(\cos\theta + i\sin\theta)^{\frac{1}{n}}$ has n different values.

Note that these n solutions are symmetrically placed on a circle drawn on an Argand diagram.

Solving equations in the form $z^n = a + ib$

The method explained above for finding the nth roots of unity, can be used in general, to find the solutions to equations of the form $z^n = a + ib$.

The general method is to convert $z = a + ib$ to the r, θ form, find one root using de Moivre's theorem and then sketch an Argand diagram and use symmetry to find the other values.

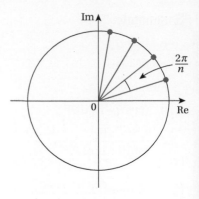

You learned how to draw Argand diagrams in Chapter 2 *Complex Numbers*.

Example 8

Find all the roots of $z^6 = -64$.

$-64 = 64(\cos\pi + i\sin\pi)$

$\Rightarrow (-64)^{\frac{1}{6}} = 64^{\frac{1}{6}}(\cos\pi + i\sin\pi)^{\frac{1}{6}}$

$\qquad = 2\left[\cos\left(\dfrac{\pi}{6}\right) + i\sin\left(\dfrac{\pi}{6}\right)\right]$

Using symmetry, the other roots are as shown in the diagram.

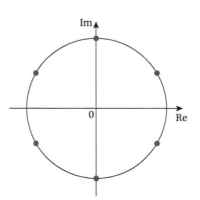

$2\left[\cos\left(\dfrac{\pi}{2}\right) + i\sin\left(\dfrac{\pi}{2}\right)\right]$

$2\left[\cos\left(\dfrac{5\pi}{6}\right) + i\sin\left(\dfrac{5\pi}{6}\right)\right]$

$2\left[\cos\left(-\dfrac{5\pi}{6}\right) + i\sin\left(-\dfrac{5\pi}{6}\right)\right]$

$2\left[\cos\left(-\dfrac{\pi}{2}\right) + i\sin\left(-\dfrac{\pi}{2}\right)\right]$

$2\left[\cos\left(-\dfrac{\pi}{6}\right) + i\sin\left(-\dfrac{\pi}{6}\right)\right]$

$(-64)^{\frac{1}{6}} = \pm\left(\dfrac{\sqrt{3}}{2} \pm \dfrac{i}{2}\right), \pm i$

Note

Expressing $z = -64$ in the form $z = r(\cos\theta + i\sin\theta)$.

Note

Using de Moivre's theorem.

Note

Sketch an Argand diagram and use symmetry to find the other roots.

Note

Since all of these values can be expressed simply in the form $a + ib$, it is common to give these answers in the form shown here.

Example 9

Find the values of the roots of $z^2 = -1 - \sqrt{3}\,i$

$$-1 - \sqrt{3}i = 2\left[\cos\left(-\frac{2\pi}{3}\right) + i\sin\left(-\frac{2\pi}{3}\right)\right]$$

From de Moivre's theorem, one root is

$$\left(2\left[\cos\left(-\frac{2\pi}{3}\right) + i\sin\left(-\frac{2\pi}{3}\right)\right]\right)^{\frac{1}{2}} = 2^{\frac{1}{2}}\left[\cos\left(-\frac{\pi}{3}\right) + i\sin\left(-\frac{\pi}{3}\right)\right]$$

$$= \sqrt{2}\left(+\frac{1}{2} - \frac{\sqrt{3}}{2}i\right)$$

$$= \frac{\sqrt{2}}{2} - \frac{\sqrt{6}}{2}i$$

From the diagram, by symmetry, the other root is

$$-\frac{\sqrt{2}}{2} + \frac{\sqrt{6}}{2}i$$

Therefore,

$$(-1 - \sqrt{3}\,i)^{\frac{1}{2}} = \pm\left(\frac{\sqrt{2}}{2} - \frac{\sqrt{6}}{2}i\right)$$

> **Note**
>
> Express $-1 - \sqrt{3}i$ in the form $\cos\theta + i\sin\theta$.

> **Note**
>
> Sketch an Argand diagram.

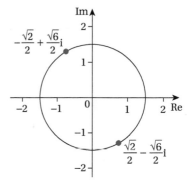

Example 10

Find the solutions of $27z^3 = 8$.

$27z^3 = 8 \Rightarrow 3z = \sqrt[3]{1} \times 2$

$\sqrt[3]{1}$ has the following values: $1, -\frac{1}{2} \pm \frac{\sqrt{3}}{2}$

Using $\sqrt[3]{1} = 1$,

$3z = 2 \Rightarrow z = \frac{2}{3}$

Using $\sqrt[3]{1} = -\frac{1}{2} + \frac{\sqrt{3}}{2}i$,

$3z = -1 + \sqrt{3}i \Rightarrow z = -\frac{1}{3} + \frac{\sqrt{3}}{3}i$

Using $\sqrt[3]{1} = -\frac{1}{2} - \frac{\sqrt{3}}{2}i$,

$3z = -1 - \sqrt{3}i \Rightarrow z = -\frac{1}{3} - \frac{\sqrt{3}}{3}i$

> **Note**
>
> Take the cube root of both sides, remembering to multiply one side of the resulting equation by each of the three cube roots of unity, taken one at a time. In this case, it is simpler to multiply $\sqrt[3]{8}$ by the three cube roots.

Example 11

Find the solutions of $16z^4 = (z-1)^4$.

$16z^4 = (z-1)^4$

$\Rightarrow 2z = \sqrt[4]{1}(z-1)$

$\sqrt[4]{1} = 1, -1, i, -i$

Using $\sqrt[4]{1} = 1$,

$2z = z - 1 \quad \Rightarrow \quad z = -1$

Using $\sqrt[4]{1} = -1$,

$2z = -(z-1) \quad \Rightarrow \quad 3z = 1 \quad \Rightarrow \quad z = \dfrac{1}{3}$

Using $\sqrt[4]{1} = i$,

$2z = i(z-1)$

$\Rightarrow z = -\dfrac{i}{2-i}$

$\Rightarrow z = -\dfrac{i(2+i)}{(2-i)(2+i)}$

which gives, $z = \dfrac{1}{5}(1-2i)$

Using $\sqrt[4]{1} = -i$, $\quad 2z = -i(z-1)$

$\Rightarrow z = \dfrac{i}{2+i}$

$\Rightarrow z = \dfrac{i(2-i)}{5}$

which gives, $z = \dfrac{1}{5}(1+2i)$

Therefore, the four solutions of $16z^4 = (z-1)^4$ are $-1, \dfrac{1}{3}, \dfrac{1}{5}(1 \pm 2i)$.

Exponential form of a complex number

Using the Maclaurin theorem for e^x, $\cos x$ and $\sin x$, you can show that

$e^{ix} = \cos x + i \sin x$.

This is the **exponential form** of a complex number.

Expressed generally, you have

$$z = r(\cos \theta + i \sin \theta) \quad \Rightarrow \quad z = re^{i\theta}$$

You can use the exponential form to simplify many types of problem, including finding certain integrals.

Note that using the exponential form of $(\cos \theta + i \sin \theta)^n$, you have

$$(\cos \theta + i \sin \theta)^n = (e^{i\theta})^n = e^{i(n\theta)} \equiv \cos n\theta + i \sin n\theta$$

which is another way of proving de Moivre's theorem.

You learned about the Maclaurin theorem in Chapter *19 Series and Limits*.

Tip

You can use the identity $e^{ix} = \cos x + i \sin x$ without justification.

Example 12

Express $2 + 2i$ in $re^{i\theta}$ form.

The modulus of $2 + 2i$ is $2\sqrt{2}$ and its argument is $\dfrac{\pi}{4}$.

$\Rightarrow 2 + 2i = 2\sqrt{2}e^{\frac{i\pi}{4}}$

Note

Using $z = r(\cos \theta + i \sin \theta) \Rightarrow z = re^{i\theta}$.

Example 13

Express $1 - i\sqrt{3}$ in $re^{i\theta}$ form.

The modulus of $1 - i\sqrt{3}$ is 2 and its argument is $-\dfrac{\pi}{3}$.

Hence,

$1 - i\sqrt{3} = 2e^{\frac{-i\pi}{3}}$

You can express the roots of a complex number using the exponential form.

Example 14

Find the values of $(-2 + 2i)^{\frac{1}{3}}$ and show their positions on an Argand diagram.

$$(-2 + 2i)^{\frac{1}{3}} = \left(2\sqrt{2}\left[\cos\left(\frac{3\pi}{4} \right) + i\sin\left(\frac{3\pi}{4} \right) \right] \right)^{\frac{1}{3}}$$

$$= \left(2^{\frac{3}{2}}\left[\cos\left(\frac{3\pi}{4} \right) + i\sin\left(\frac{3\pi}{4} \right) \right] \right)^{\frac{1}{3}}$$

> **Note**
> Expressing the complex numbers in the exponential form.

> **Note**
> Draw an Argand diagram.

Therefore, from de Moivre's theorem,

one value of $(-2 + 2i)^{\frac{1}{3}}$ is $2^{\frac{1}{2}}\left[\cos\left(\frac{\pi}{4} \right) + i\sin\left(\frac{\pi}{4} \right) \right]$.

> **Note**
> Find one value of $(-2 + 2i)^{\frac{1}{3}}$.

By symmetry, the other roots are

$$2^{\frac{1}{2}}\left[\cos\left(\frac{\pi}{4} + \frac{2\pi}{3} \right) + i\sin\left(\frac{\pi}{4} + \frac{2\pi}{3} \right) \right]$$

and $\quad 2^{\frac{1}{2}}\left[\cos\left(\frac{\pi}{4} + \frac{4\pi}{3} \right) + i\sin\left(\frac{\pi}{4} + \frac{4\pi}{3} \right) \right]$.

These three roots are therefore:

$2^{\frac{1}{2}} e^{\frac{i\pi}{4}}, 2^{\frac{1}{2}} e^{\frac{11i\pi}{12}}, 2^{\frac{1}{2}} e^{\frac{19i\pi}{12}}$

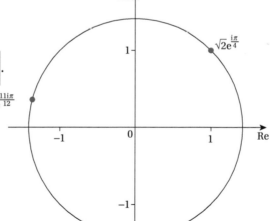

Multiplying one complex number by another

Expressing the two numbers, z_1 and z_2, in their exponential form, you have

$$z_1 z_2 = r_1 e^{i\theta_1} \times r_2 e^{i\theta_2}$$

which gives $z_1 z_2 = r_1 r_2 e^{i(\theta_1 \theta_2)}$

This is a very simple way of proving that to find the product of two complex numbers, you multiply the moduli and add the arguments.

You saw this rule in Chapter 2 Complex Numbers.

Simplifying certain integrals

You can simplify integrals of the type $\int e^{ax} \cos bx \, dx$.

Example 15

Find $\int e^{2x} \sin x \, dx$.

$\int e^{2x} \sin x \, dx = \text{Im} \int e^{2x} (\cos x + i \sin x) \, dx$ where $\text{Im} \int$ is the imaginary part of the given integral.

Using the exponential form of $\cos x + i \sin x$,

$\int e^{2x} \sin x \, dx = \text{Im} \int e^{2x} \times e^{ix} \, dx$

$= \text{Im} \int e^{(2+i)x} dx$

$\Rightarrow \int e^{2x} \sin x \, dx = \text{Im} \left(\dfrac{1}{2+i} e^{(2+i)x} \right)$

$= \text{Im} \left[\dfrac{2-i}{(2+i)(2-i)} e^{2x} (\cos x + i \sin x) + c \right]$

$= \text{Im} \left[e^{2x} \dfrac{(2\cos x + \sin x + 2i \sin x - i \cos x)}{2^2 - (i)^2} \right] + c$

Therefore, $\int e^{2x} \sin x \, dx = \dfrac{e^{2x}}{5} (2 \sin x - \cos x) + c$.

> **Note**
>
> Express this in terms of the exponential form.

Example 16

Find $\int e^{4x} \cos 3x \, dx$.

$\int e^{4x} \cos 3x \, dx = \text{Re} \int e^{4x} (\cos 3x + i \sin 3x) \, dx$ where $\text{Re} \int$ is the real part of the given integral.

Using the exponential form of $\cos 3x + i \sin 3x$,

$\int e^{4x} \cos 3x \, dx = \text{Re} \int e^{(4+3i)x} dx$

$= \text{Re} \left(\dfrac{1}{4+3i} e^{(4+3i)x} + c \right)$

$= \text{Re} \left[\dfrac{4-3i}{(4+3i)(4-3i)} e^{4x} (\cos 3x + i \sin 3x) + c \right]$

Therefore, $\int e^{4x} \cos 3x \, dx = \dfrac{e^{4x}}{25} \left(\dfrac{4\cos 3x + 3\sin 3x}{25} \right) + c$.

Exercise 2

1. Find the possible values of z, giving your answers in

   i $a + ib$ form

   ii $re^{i\theta}$ form

   a $z^4 = -16$ **b** $z^3 = -8 + 8i$ **c** $z^3 = 27i$

   d $z^2 = 16i$ **e** $z^5 = -32$

2 Find the six sixth roots of unity.

3 Solve each of these.

  a $(z+2i)^2 = 4$ **b** $(z-1)^3 = 8$ **c** $z^2 = (z+1)^2$ **d** $(z+3i)^2 = (2z-1)^2$

4 Find the seven seventh roots of unity in the $e^{i\theta}$ form.

5 Solve $z^5 = 32i$. Give your answers in the $r^{i\theta}$ form, and show them on an Argand diagram.

6 By considering the ninth roots of unity, show that

$$\cos\left(\tfrac{2\pi}{9}\right) + \cos\left(\tfrac{4\pi}{9}\right) + \cos\left(\tfrac{6\pi}{9}\right) + \cos\left(\tfrac{8\pi}{9}\right) = -\tfrac{1}{2}.$$

7 By considering the seventh roots of unity, show that

$$\cos\left(\tfrac{\pi}{7}\right) + \cos\left(\tfrac{3\pi}{7}\right) + \cos\left(\tfrac{5\pi}{7}\right) = \tfrac{1}{2}.$$

8 Evaluate each of these.

  a $\displaystyle\int e^{4x}\cos 5x \, dx$ **b** $\displaystyle\int e^{3x}\sin 7x \, dx$

  c $\displaystyle\int e^{-2x}\sin 4x \, dx$ **d** $\displaystyle\int e^{-4x}\cos 3x \, dx$

20.3 Trigonometric identities

You can also use de Moivre's theorem to express multiple-angled trigonometric terms as powers of the related single-angle trigonometric term. For example, you can use the theorem to express $\cos 6\theta$, in powers of $\cos \theta$, and conversely to express $\cos^6\theta$ in terms of multiple angles such as $\cos 6\theta$.

Expressions for $\cos^n \theta$ and $\sin^n \theta$ in terms of $\cos k\theta$ and $\sin k\theta$

It is possible to express powers of trigonometric functions in terms of multiple angles without having to learn complicated multiple-angle formulae.

Let $z = \cos\theta + i\sin\theta$.

You then have $\dfrac{1}{z} = (\cos\theta + i\sin\theta)^{-1} = \cos\theta - i\sin\theta$,

which gives $z + \dfrac{1}{z} = 2\cos\theta$ and $z - \dfrac{1}{z} = 2i\sin\theta$.

You also have $z^n = (\cos\theta + i\sin\theta)^n = \cos n\theta + i\sin n\theta$ (de Moivre's theorem) so

$$\dfrac{1}{z^n} = \dfrac{1}{\cos n\theta + i\sin n\theta}$$

$$\dfrac{1}{z^n} = \cos n\theta - i\sin n\theta$$

which gives

$$z^n + \dfrac{1}{z^n} = 2\cos n\theta$$

$$z^n - \dfrac{1}{z^n} = 2i\sin n\theta$$

With the help of these four identities for $z \pm \dfrac{1}{z}$ and $z^n \pm \dfrac{1}{z^n}$, you can write any power of $\cos\theta$ or $\sin\theta$ in terms of multiples of θ.

If $z = \cos\theta + i\sin\theta$, to express

▶ $\cos n\theta$ in terms of z, use the identity $z^n + \dfrac{1}{z^n} = 2\cos n\theta$

▶ $\sin n\theta$ in terms of z, use the identity $z^n - \dfrac{1}{z^n} = 2i\sin n\theta$.

For example, to express $\cos^n\theta$ in terms of $\cos r\theta$:

Set $z = \cos\theta + i\sin\theta$ and use the identity $\cos\theta = \dfrac{1}{2}\left(z + \dfrac{1}{z}\right)$, where $z + \dfrac{1}{z} = 2\cos\theta$.

Raise both sides of the identity to the nth power and collect terms of the form $z^r + \dfrac{1}{z^r}$.

Then use the identity $z^r + \dfrac{1}{z^r} = 2\cos r\theta$ to write the expression in the required form.

Similarly, to express $\sin^n\theta$ in terms of $\sin r\theta$ and $\cos r\theta$:

Set $z = \cos\theta + i\sin\theta$ and use the identity $\sin\theta = \dfrac{1}{2i}\left(z - \dfrac{1}{z}\right)$, where $z - \dfrac{1}{z} = 2i\sin\theta$.

Raise both sides of the identity to the nth power and collect terms of the form $z^r + \dfrac{1}{z^r}$ or $z^r - \dfrac{1}{z^r}$.

Use the identity $z^r + \dfrac{1}{z^r} = 2\cos r\theta$ (if n is even) or $z^r - \dfrac{1}{z^r} = 2i\sin r\theta$ (if n is odd) to write the expression in the required form.

Example 17

Question

If $z = \cos\theta + i\sin\theta$, express the following in terms of θ.

a z^4 **b** z^{-3}

Answer

a $z^4 = \cos 4\theta + i\sin 4\theta$

b $z^{-3} = \cos(-3\theta) + i\sin(-3\theta)$

$\Rightarrow z^{-3} = \cos 3\theta - i\sin 3\theta$

> **Note**
>
> You know that, when $z = \cos\theta + i\sin\theta$, $z^n = \cos n\theta + i\sin n\theta$ for all integer values of n.

Example 18

Question

Given that $z = \cos\theta + i\sin\theta$, express the following in terms of z.

a $\cos 6\theta$ **b** $\sin 3\theta$

> **Note**
>
> It can help to write down the appropriate identities.

Answer

When $z = \cos\theta + i\sin\theta$,

$z^n + \dfrac{1}{z^n} \equiv 2\cos n\theta$ and $z^n - \dfrac{1}{z^n} \equiv 2i\sin n\theta$

a $2\cos 6\theta \equiv z^6 + \dfrac{1}{z^6}$

$\Rightarrow \cos 6\theta \equiv \dfrac{1}{2}\left(z^6 + \dfrac{1}{z^6}\right)$

> **Note**
>
> Using the identity, $z^n + \dfrac{1}{z^n} \equiv 2\cos n\theta$.

b $2i\sin 3\theta \equiv z^3 - \dfrac{1}{z^3}$

> **Note**
>
> Using the identity, $z^n - \dfrac{1}{z^n} \equiv 2i\sin n\theta$.

$\Rightarrow \sin 3\theta \equiv \dfrac{1}{2i}\left(z^3 - \dfrac{1}{z^3}\right) = -\dfrac{i}{2}\left(z^3 - \dfrac{1}{z^3}\right)$

Example 19

Question

Express $\cos^3 \theta$ as the cosines of multiples of θ.

Answer

$$\cos\theta \equiv \frac{1}{2}\left(z+\frac{1}{z}\right)$$

$$\cos^3\theta \equiv \left[\frac{1}{2}\left(z+\frac{1}{z}\right)\right]^3$$

$$\equiv \frac{1}{2^3}\left(z+\frac{1}{z}\right)^3$$

$$\equiv \frac{1}{8}\left(z^3+3z^2\times\frac{1}{z}+3z\times\frac{1}{z^2}+\frac{1}{z^3}\right)$$

$$\equiv \frac{1}{8}\left(z^3+3z+\frac{3}{z}+\frac{1}{z^3}\right)$$

$$\cos^3\theta \equiv \frac{1}{8}\left[\left(z^3+\frac{1}{z^3}\right)+3\left(z+\frac{1}{z}\right)\right]$$

$$\cos^3\theta \equiv \frac{1}{8}(2\cos 3\theta+3\times 2\cos\theta)$$

$$\cos^3\theta \equiv \frac{1}{4}\cos 3\theta+\frac{3}{4}\cos\theta$$

Note

Express $\cos\theta$ in terms of z using identity $z+\frac{1}{z}=2\cos\theta$.

Note

Collect terms of the type $z^n+\frac{1}{z^n}$, according to the values of n (as you are required to give the answer as the cosines of multiples of θ).

Note

Hence express $\cos^3\theta$ in terms of z.

Note

Rearrange the terms on the RHS.

Note

Convert the RHS.

Example 20

Express $\cos^6 \theta$ as the cosines of multiples of θ.

$$\cos^6\theta \equiv \left[\frac{1}{2}\left(z+\frac{1}{z}\right)\right]^6 \text{ where } z=\cos\theta+i\sin\theta.$$

Using the binomial theorem, $\left(z+\frac{1}{z}\right)^6=z^6+6z^5\times\frac{1}{z}+15z^4\times\frac{1}{z^2}+\cdots+\frac{1}{z^6}$

Therefore,

$$\cos^6\theta \equiv \frac{1}{64}\left(z^6+6z^4+15z^2+20+\frac{15}{z^2}+\frac{6}{z^4}+\frac{1}{z^6}\right)$$

$$\equiv \frac{1}{64}\left[\left(z^6+\frac{1}{z^6}\right)+6\left(z^4+\frac{1}{z^4}\right)+15\left(z^2+\frac{1}{z^2}\right)+20\right]$$

$$\cos^6\theta \equiv \frac{1}{64}(2\cos 6\theta+6\times 2\cos 4\theta+15\times 2\cos 2\theta+20)$$

$$\Rightarrow \cos^6\theta \equiv \frac{1}{32}\cos 6\theta+\frac{3}{16}\cos 4\theta+\frac{15}{32}\cos 2\theta+\frac{5}{16}$$

Note

Using identity $z+\frac{1}{z}=2\cos\theta$.

Note

Collect terms in the form $z^r+\frac{1}{z^r}$.

Note

Converting the RHS using identity $z^n+\frac{1}{z^n}=2\cos n\theta$.

Example 21

Express $\sin^5 \theta$ as the sines of multiples of θ.

$$\sin^5 \theta \equiv \left[\frac{1}{2i} \left(z - \frac{1}{z} \right) \right]^5 \text{ where } z = \cos\theta + i\sin\theta.$$

Using the binomial theorem,

$$\sin^5 \theta = \frac{1}{32i^5} \left(z^5 - 5z^3 + 10z - \frac{10}{z} + \frac{5}{z^3} - \frac{1}{z^5} \right)$$

$$= \frac{1}{32i} \left[\left(z^5 - \frac{1}{z^5} \right) - 5\left(z^3 - \frac{1}{z^3} \right) + 10\left(z - \frac{1}{z} \right) \right]$$

$$\sin^5 \theta \equiv \frac{1}{32i} [2i\sin 5\theta - 10i\sin 3\theta + 20i\sin\theta]$$

$$\Rightarrow \sin^5 \theta \equiv \frac{1}{16}\sin 5\theta - \frac{5}{16}\sin 3\theta + \frac{5}{8}\sin\theta$$

Expansions of cos $n\theta$ and sin $n\theta$ as powers of cos θ and sin θ

It is also possible to express multiple-angle trigonometric functions in terms of powers of trigonometric functions using de Moivre's theorem.

To change the function $\cos n\theta$ into powers of $\cos\theta$, you express $\cos n\theta$ as the **real part** (Re) of $\cos n\theta + i\sin n\theta$:

$$\cos n\theta = \text{Re}(\cos n\theta + i\sin n\theta)$$

By de Moivre's threoem, you have:

$$\cos n\theta + i\sin n\theta = (\cos\theta + i\sin\theta)^n$$

$$\Rightarrow \cos n\theta = \text{Re}(\cos\theta + i\sin\theta)^n$$

You then expand the right-hand side by the binomial theorem and extract the real terms to find $\cos n\theta$ in terms of powers of $\cos\theta$.

Similarly, to change the function $\sin n\theta$ into powers of $\sin\theta$, express $\sin n\theta$ as the **imaginary part** (Im) of $\cos n\theta + i\sin n\theta$. You then expand the right-hand side of $\sin n\theta = \text{Im}(\cos\theta + i\sin\theta)^n$ and extract the imaginary terms.

A similar technique can be used to express $\tan n\theta$ as powers of $\tan\theta$: express $\cos n\theta$ and $\sin n\theta$ as powers of sin and cos and then divide the two expressions.

Example 22

Express $\sin 3\theta$ in terms of $\sin\theta$.

$\sin 3\theta = \text{Im}(\cos 3\theta + i\sin 3\theta)$, where Im (z) is the imaginary part of z.

$$\cos 3\theta + i\sin 3\theta = (\cos\theta + i\sin\theta)^3$$

$$\Rightarrow \sin 3\theta = \text{Im}(\cos\theta + i\sin\theta)^3$$

Expanding the RHS using the binomial theorem,

$$\sin 3\theta = \text{Im}[\cos^3\theta + 3\cos^2\theta(i\sin\theta) + 3\cos\theta(i\sin\theta)^2 + (i\sin\theta)^3]$$

$$= \text{Im}(\cos^3\theta + 3i\cos^2\theta\sin\theta - 3\cos\theta\sin^2\theta - i\sin^3\theta)$$

$$= 3\cos^2\theta\sin\theta - \sin^3\theta$$

(continued)

(continued)

Using $\cos^2\theta = 1 - \sin^2\theta$

$\sin 3\theta = 3(1 - \sin^2\theta)\sin\theta - \sin^3\theta$

$\qquad = 3\sin\theta - 3\sin^3\theta - \sin^3\theta$

Therefore,

$\sin 3\theta = 3\sin\theta - 4\sin^3\theta$

> **Note**
>
> The answer must be in terms of $\sin\theta$, so you need to choose an appropriate identity.

Example 23

Express the following in terms of powers of $\cos\theta$.

a $\cos 6\theta$ **b** $\dfrac{\sin 6\theta}{\sin\theta}$

a $\cos 6\theta = \mathrm{Re}(\cos 6\theta + i\sin 6\theta)$

Hence, $\cos 6\theta = \mathrm{Re}(\cos\theta + i\sin\theta)^6$

Expanding the RHS using the binomial theorem,

> **Note**
>
> Using de Moivre's theorem.

$\cos 6\theta = \mathrm{Re}[\cos^6\theta + 6\cos^5\theta(i\sin\theta) + \dfrac{6\times5}{2\times1}\cos^4\theta(i\sin\theta)^2 + \dfrac{6\times5\times4}{3\times2\times1}\cos^3\theta(i\sin\theta)^3$

$\qquad + \dfrac{6\times5\times4\times3}{4\times3\times2\times1}\cos^2\theta(i\sin\theta)^4 + \dfrac{6\times5\times4\times3\times2}{5\times4\times3\times2\times1}\cos\theta(i\sin\theta)^5 + (i\sin\theta)^6]$

$\Rightarrow \cos 6\theta = \cos^6\theta - 15\cos^4\theta\sin^2\theta + 15\cos^2\theta\sin^4\theta - \sin^6\theta$

Using $\sin^2\theta = 1 - \cos^2\theta$,

$\cos 6\theta = \cos^6\theta - 15\cos^4\theta(1 - \cos^2\theta) + 15\cos^2\theta(1 - \cos^2\theta)^2 - (1 - \cos^2\theta)^3$

$\qquad = \cos^6\theta - 15\cos^4\theta + 15\cos^6\theta + 15\cos^2\theta - 30\cos^4\theta + 15\cos^6\theta - 1$

$\qquad + 3\cos^2\theta - 3\cos^4\theta + \cos^6\theta$

Therefore,

$\cos 6\theta = 32\cos^6\theta - 48\cos^4\theta + 18\cos^2\theta - 1$

b $\sin 6\theta = \mathrm{Im}(\cos 6\theta + i\sin 6\theta)$

Hence, $\sin 6\theta = \mathrm{Im}(\cos\theta + i\sin\theta)^6$

Expanding the RHS using the binomial theorem,

$\sin 6\theta = \mathrm{Im}[\cos^6\theta + 6\cos^5\theta(i\sin\theta) + \dfrac{6\times5}{2\times1}\cos^4\theta(i\sin\theta)^2 + \dfrac{6\times5\times4}{3\times2\times1}\cos^3\theta(i\sin\theta)^3$

$\qquad + \dfrac{6\times5\times4\times3}{4\times3\times2\times1}\cos^2\theta(i\sin\theta)^4 + \dfrac{6\times5\times4\times3\times2}{5\times4\times3\times2\times1}\cos\theta(i\sin\theta)^5 + (i\sin\theta)^6]$

$\Rightarrow \sin 6\theta = 6\cos^5\theta\sin\theta - 20\cos^3\theta\sin^3\theta + 6\cos\theta\sin^5\theta$

Therefore,

$\dfrac{\sin 6\theta}{\sin\theta} = 6\cos^5\theta - 20\cos^3\theta(1 - \cos^2\theta) + 6\cos\theta(1 - \cos^2\theta)^2$

$\qquad = 6\cos^5\theta - 20\cos^3\theta + 20\cos^5\theta + 6\cos\theta - 12\cos^3\theta + 6\cos^5\theta$

Therefore,

$\dfrac{\sin 6\theta}{\sin\theta} = 32\cos^5\theta - 32\cos^3\theta + 6\cos\theta$

Example 24

a Express $\sin 5\theta$ in terms of $\sin \theta$.

b Hence, prove that $\sin\left(\dfrac{\pi}{5}\right)$, $\sin\left(\dfrac{2\pi}{5}\right)$, $\sin\left(\dfrac{6\pi}{5}\right)$ and $\sin\left(\dfrac{7\pi}{5}\right)$ are the roots of the equation $16x^4 - 20x^2 + 5 = 0$.

c Deduce that $\sin^2\left(\dfrac{\pi}{5}\right)$ and $\sin^2\left(\dfrac{2\pi}{5}\right)$ are roots of the equation $16y^2 - 20y + 5 = 0$, and hence find the exact value of

   i $\sin\left(\dfrac{\pi}{5}\right)\sin\left(\dfrac{2\pi}{5}\right)$ ii $\cos\left(\dfrac{2\pi}{5}\right)$

a $\sin 5\theta = \mathrm{Im}(\cos 5\theta + i \sin 5\theta)$

Hence,

$\sin 5\theta = \mathrm{Im}(\cos\theta + i\sin\theta)^5$

$\qquad = \mathrm{Im}(\cos^5\theta + 5i\cos^4\theta\sin\theta + 10i^2\cos^3\theta\sin^2\theta + 10i^3\cos^2\theta\sin^3\theta$
$\qquad\quad + 5i^4\cos\theta\sin^4\theta + i^5\sin^5\theta)$

Therefore,

$\sin 5\theta = 5\cos^4\theta\sin\theta - 10\cos^2\theta\sin^3\theta + \sin^5\theta$

Using $\cos^2\theta = 1 - \sin^2\theta$,

$\sin 5\theta = 5(1 - \sin^2\theta)^2 \sin\theta - 10(1 - \sin^2\theta)\sin^3\theta + \sin^5\theta$

$\Rightarrow \sin 5\theta = 16\sin^5\theta - 20\sin^3\theta + 5\sin\theta$

b From part **a**,

$\dfrac{\sin 5\theta}{\sin\theta} = 16\sin^4\theta - 20\sin^2\theta + 5$

When $\sin 5\theta = 0$, $16\sin^4\theta - 20\sin^2\theta + 5 = 0$, therefore,

$16x^4 - 20x^2 + 5 = 0$ on substituting $x = \sin\theta$.

$\sin 5\theta = 0 \quad \Rightarrow \quad \theta = 0$ (excluded), $\dfrac{\pi}{5}, \dfrac{2\pi}{5}, \dfrac{3\pi}{5}, \ldots$

which give the following values for $\sin\theta$:

$\sin\left(\dfrac{\pi}{5}\right)$, $\sin\left(\dfrac{2\pi}{5}\right)$, $\sin\left(\dfrac{3\pi}{5}\right)$ which is the same as $\sin\left(\dfrac{2\pi}{5}\right)$,

$\sin\left(\dfrac{4\pi}{5}\right)$ which is the same as $\sin\left(\dfrac{6\pi}{5}\right)$ and $\sin\left(\dfrac{7\pi}{5}\right)$,

$\sin\pi$ which is zero and hence excluded.

Therefore, the four *different* non-zero values of x for $16x^4 - 20x^2 + 5 = 0$

are:

$\sin\left(\dfrac{\pi}{5}\right)$ $\sin\left(\dfrac{2\pi}{5}\right)$ $\sin\left(\dfrac{6\pi}{5}\right)$ $\sin\left(\dfrac{7\pi}{5}\right)$

c Substitute $y = x^2$ to obtain the equation $16y^2 - 20y + 5 = 0$, whose roots are the two different values for y given by the substitution.

There are just two values of x^2, $\sin^2\left(\dfrac{\pi}{5}\right)$ and $\sin^2\left(\dfrac{2\pi}{5}\right)$, since

$\sin\left(\dfrac{6\pi}{5}\right) = -\sin\left(\dfrac{\pi}{5}\right)$ and $\sin\left(\dfrac{7\pi}{5}\right) = -\sin\left(\dfrac{2\pi}{5}\right)$, which give

$\sin^2\left(\dfrac{6\pi}{5}\right) = \sin^2\left(\dfrac{\pi}{5}\right)$ and $\sin^2\left(\dfrac{7\pi}{5}\right) = \sin^2\left(\dfrac{2\pi}{5}\right)$

Note

The solutions of $16x^4 - 20x^2 + 5 = 0$ are $x = \sin\theta$, where θ satisfies $\dfrac{\sin 5\theta}{\sin\theta} = 0$. All the x are different, and since $\sin 5\theta$ is divided by $\sin\theta$, you exclude the possible root $\sin\theta = 0$.

(continued)

(continued)

Therefore, the two different roots of the equation $16y^2 - 20y + 5 = 0$ are
$y = \sin^2\left(\dfrac{\pi}{5}\right)$ and $y = \sin^2\left(\dfrac{2\pi}{5}\right)$.

i Using the product of the roots of a polynomial, you have
for $16y^2 - 20y + 5 = 0$

$$\alpha\beta = \frac{5}{16}$$

$$\Rightarrow \quad \sin^2\left(\frac{\pi}{5}\right)\sin^2\left(\frac{2\pi}{5}\right) = \frac{5}{16}$$

$$\Rightarrow \quad \sin\left(\frac{\pi}{5}\right)\sin\left(\frac{2\pi}{5}\right) = \pm\sqrt{\frac{5}{16}}$$

Since both $\sin\left(\dfrac{\pi}{5}\right)$ and $\sin\left(\dfrac{2\pi}{5}\right)$ are positive, you obtain

$$\sin\left(\frac{\pi}{5}\right)\sin\left(\frac{2\pi}{5}\right) = \frac{\sqrt{5}}{4}$$

> You saw how to find the product and sum of roots of a polynomial in Chapter *17 Roots and Polynomials.*

ii Since $16y^2 - 20y + 5 = 0$ is a quadratic equation, its roots are

$$y = \frac{20 \pm \sqrt{400 - 320}}{32}$$

$$\Rightarrow \quad y = \frac{20 \pm \sqrt{80}}{32} = \frac{5 \pm \sqrt{5}}{8}$$

Since these two roots are $\sin^2\left(\dfrac{\pi}{5}\right)$ and $\sin^2\left(\dfrac{2\pi}{5}\right)$, and

$\sin\left(\dfrac{2\pi}{5}\right) > \sin\left(\dfrac{\pi}{5}\right) > 0$, you have $\sin^2\left(\dfrac{\pi}{5}\right) = \dfrac{5 - \sqrt{5}}{8}$

Using the identity $\cos\theta = 1 - \sin^2\left(\dfrac{\theta}{2}\right)$, you obtain

$$\cos\left(\frac{2\pi}{5}\right) = 1 - 2\sin^2\left(\frac{\pi}{5}\right).$$

$$= 1 - 2 \times \frac{5 - \sqrt{5}}{8}$$

$$\Rightarrow \quad \cos\left(\frac{2\pi}{5}\right) = \frac{\sqrt{5} - 1}{4}$$

Exercise 3

1 If $z = \cos\theta + i\sin\theta$, find the values of each of the following.

    a $z^2 - \dfrac{1}{z^2}$ **b** $z^4 + \dfrac{1}{z^4}$ **c** $z^5 + \dfrac{1}{z^5}$

2 Express each of the following in terms of z, where $z = \cos\theta + i\sin\theta$.

    a $\cos 6\theta$ **b** $\sin 5\theta$ **c** $\cos 4\theta$

    d $\sin^3\theta$ **e** $\cos^4 3\theta$

3 Express each of the following in terms of $\cos\theta$.

    a $\cos 6\theta$ **b** $\cos 4\theta$ **c** $\dfrac{\sin 4\theta}{\sin\theta}$

4 Express each of the following in terms of $\sin\theta$.

    a $\sin 3\theta$ **b** $\sin 5\theta$ **c** $\dfrac{\cos 7\theta}{\cos\theta}$ **d** $\dfrac{\cos 5\theta}{\cos\theta}$

5 Express each of the following in terms of sines or cosines of multiple angles.

    a $\sin^3\theta$ **b** $\cos^3\theta$ **c** $\cos^5\theta$ **d** $\sin^5\theta$

6 Prove that $\cos^4\theta = \dfrac{1}{8}(\cos 4\theta + 4\cos 2\theta + 3)$.

7 Prove that $\tan 3\theta = \dfrac{3\tan\theta - \tan^3\theta}{1 - 3\tan^2\theta}$. Hence solve $t^3 - 3t^2 - 3t + 1 = 0$.

8 When $\cos 4\theta = \cos 3\theta$, prove that $\theta = 0, \dfrac{2\pi}{7}, \dfrac{4\pi}{7}, \dfrac{6\pi}{7}$.

Hence prove that $\cos\left(\dfrac{2\pi}{7}\right), \cos\left(\dfrac{4\pi}{7}\right)$ and $\cos\left(\dfrac{6\pi}{7}\right)$ are the roots of $8x^3 + 4x^2 - 4x - 1 = 0$.

Summary

▶ De Moivre's theorem states that $(\cos\theta + \mathrm{i}\sin\theta)^n = \cos n\theta + \mathrm{i}\sin n\theta$ for any integer n.

▶ When n is not an integer, de Moivre's theorem gives only one of the possible values for $(\cos\theta + \mathrm{i}\sin\theta)^n$, that is, it gives $\cos n\theta + \mathrm{i}\sin n\theta$. The other values can be found using the symmetry on an Argand diagram.

▶ De Moivre's theorem allows you to express a complex number, z^n in multiple-angle form, for example, $(\cos\theta + \mathrm{i}\sin\theta)^5 \equiv \cos 5\theta + \mathrm{i}\sin 5\theta$.

▶ You can use de Moivre's theorem to simplify an expression in the form $\cos n\theta + \mathrm{i}\sin n\theta$, for example,
$(\cos n\theta + \mathrm{i}\sin n\theta)(\cos m\theta + \mathrm{i}\sin m\theta) = \cos(n + m)\theta + \mathrm{i}\sin(n+m)\theta$.

▶ It is possible to calculate positive powers of complex numbers, z^n, using de Moivre's theorem. First you express the complex number in its r, θ form, then use de Moivre's theorem to express it in multiple-angle form and evaluate.

▶ It is possible to solve equations of the form $z^n = a + \mathrm{i}b$ by expressing the complex number in its r, θ form, then applying de Moivre's theorem to find one root and using symmetry to find the other roots.

▶ The exponential form of a complex number is $\mathrm{e}^{\mathrm{i}x} = \cos x + \mathrm{i}\sin x$, expressed generally you have $z = r\mathrm{e}^{\mathrm{i}\theta}$.

▶ Using the exponential form of $(\cos\theta + \mathrm{i}\sin\theta)^n$, you have another proof of de Moivre's theorem: $(\cos\theta + \mathrm{i}\sin\theta)^n = (\mathrm{e}^{\mathrm{i}\theta})^n = \mathrm{e}^{\mathrm{i}(n\theta)} \equiv \cos n\theta + \mathrm{i}\sin n\theta$

▶ De Moivre's theorem can be used to express powers of trigonometric functions, $\cos n\theta$ and $\sin n\theta$, in terms of multiple-angle trigonometric function, for example $\cos k\theta$ and $\sin k\theta$, using the following identities:

   • $z + \dfrac{1}{z} = 2\cos\theta$

   • $z - \dfrac{1}{z} = 2\mathrm{i}\sin\theta$

   • $z^n + \dfrac{1}{z^n} = 2\cos n\theta$

   • $z^n - \dfrac{1}{z^n} = 2\mathrm{i}\sin n\theta$

▶ The formulae $z + \dfrac{1}{z} = 2\cos\theta$ and $z - \dfrac{1}{z} = 2\mathrm{i}\sin\theta$ can also be used to help to solve integrals like $\displaystyle\int \sin^5 x \, \mathrm{d}x$.

▶ De Moivre's theorem can be used to express multiple-angle trigonometric functions as powers of trigonometric functions.

▶ To change the function $\cos n\theta$ into powers of $\cos\theta$, you express $\cos n\theta$ as the real part (Re) of $\cos n\theta + \mathrm{i}\sin n\theta$, that is, $\cos n\theta = \mathrm{Re}(\cos\theta + \mathrm{i}\sin\theta)^n$. Then you expand the RHS using the binomial theorem and extract the real terms.

▶ Similarly, to change the function $\sin n\theta$ into powers of $\sin\theta$, express $\sin n\theta$ as the imaginary part (Im) of $\cos n\theta + \mathrm{i}\sin n\theta$. You then expand the RHS of $\sin n\theta = \mathrm{Im}(\cos\theta + \mathrm{i}\sin\theta)^n$ and extract the imaginary terms.

Review exercises

1 Find all solutions of $z^2 = -25\mathrm{i}$.

2 Solve $(z-\mathrm{i})^4 = 81(z+2)^4$.

3 Express $\sin^2 5\theta$ in terms of z, where $z = \cos\theta + \mathrm{i}\sin\theta$.

4 Express $\cos^6\theta$ in terms of sines and cosines of multiple angles.

Practice examination questions

1 a i Express each of the numbers $1 + \sqrt{3}\mathrm{i}$ and $1 - \mathrm{i}$ in the form $r\mathrm{e}^{\mathrm{i}\theta}$, where $r > 0$. *(3 marks)*

　　ii Hence express

$$(1 + \sqrt{3}\mathrm{i})^8(1-\mathrm{i})^5$$

　　in the form $r\mathrm{e}^{\mathrm{i}\theta}$, where $r > 0$. *(3 marks)*

b Solve the equation

$$z^3 = (1 + \sqrt{3}\mathrm{i})^8(1-\mathrm{i})^5$$

giving your answers in the form $a\sqrt{2}\mathrm{e}^{\mathrm{i}\theta}$, where a is a positive integer, and $-\pi < \theta \le \pi$. *(4 marks)*

AQA MFP2 June 2010

2 Use de Moivre's theorem to find the smallest positive angle θ for which

$$(\cos\theta + \mathrm{i}\sin\theta)^{15} = -\mathrm{i}$$ *(5 marks)*

AQA MFP2 June 2007

3 a i Use de Moivre's theorem to show that

$$\cos 5\theta = \cos^5\theta - 10\cos^3\theta\sin^2\theta + 5\cos\theta\sin^4\theta$$

　　and find a similar expression for $\sin 5\theta$. *(5 marks)*

　　ii Deduce that

$$\tan 5\theta = \frac{\tan\theta(5 - 10\tan^2\theta + \tan^4\theta)}{1 - 10\tan^2\theta + 5\tan^4\theta}$$ *(3 marks)*

b Explain why $t = \tan\dfrac{\pi}{5}$ is a root of the equation

$$t^4 - 10t^2 + 5 = 0$$

and write down the other three roots of this equation in trigonometrical form. *(3 marks)*

c Deduce that

$$\tan\frac{\pi}{5}\tan\left(\frac{2\pi}{5}\right) = \sqrt{5}.$$ *(5 marks)*

AQA MFP2 June 2011

4 a Express $-4 + 4\sqrt{3}i$ in the form $re^{i\theta}$, where $r > 0$ and $-\pi < \theta \le \pi$. (3 marks)

b i Solve the equation $z^3 = -4 + 4\sqrt{3}i$, giving your answers in the form $re^{i\theta}$, where $r > 0$ and $-\pi < \theta \le \pi$. (4 marks)

ii The roots of the equation $z^3 = -4 + 4\sqrt{3}i$ are represented by the points P, Q and R on an Argand diagram.

Find the area of the triangle PQR, giving your answer in the form $k\sqrt{3}$, where k is an integer. (3 marks)

c By considering the roots of the equation $z^3 = -4 + 4\sqrt{3}i$, show that

$$\cos\frac{2\pi}{9} + \cos\frac{4\pi}{9} + \cos\frac{8\pi}{9} = 0.$$ (4 marks)

AQA MFP2 January 2013

5 a i Show that $\omega = e^{\frac{2\pi i}{7}}$ is a root of the equation $z^7 = 1$. (1 mark)

ii Write down the five other non-real roots in terms of ω. (2 marks)

b Show that

$$1 + \omega + \omega^2 + \omega^3 + \omega^4 + \omega^5 + \omega^6 = 0.$$ (2 marks)

c Show that

i $\omega^2 + \omega^5 = 2\cos\frac{4\pi}{7}$ (3 marks)

ii $\cos\frac{2\pi}{7} + \cos\frac{4\pi}{7} + \cos\frac{6\pi}{7} = -\frac{1}{2}$ (4 marks)

AQA MFP2 January 2010

6 a i By applying de Moivre's theorem to $(\cos\theta + i\sin\theta)^3$, show that

$\cos 3\theta = \cos^3\theta - 3\cos\theta\sin^2\theta.$ (3 marks)

ii Find a similar expression for $\sin 3\theta$. (1 mark)

iii Deduce that

$$\tan 3\theta = \frac{\tan^3\theta - 3\tan\theta}{3\tan^2\theta - 1}.$$ (3 marks)

b i Hence show that $\tan\frac{\pi}{12}$ is a root of the cubic equation

$x^3 - 3x^2 - 3x + 1 = 0.$ (3 marks)

ii Find two other values of θ, where $0 < \theta < \pi$, for which $\tan\theta$ is a root of this cubic equation. (2 marks)

c Hence show that

$$\tan\frac{\pi}{12} + \tan\frac{5\pi}{12} = 4.$$ (6 marks)

AQA MFP2 January 2008

Introduction

Coordinates tell us the location of something. Imagine programming a robot that mops the floor. Cartesian coordinates tell the robot how far to go east and then how far north. Polar coordinates, in contrast, would tell the robot which direction to face (how much to rotate) and then how far to go in that direction.

Objectives

By the end of this chapter, you should know how to:

▶ Relate Cartesian and polar coordinates.

▶ Convert Cartesian coordinates into polar and vice versa.

▶ Sketch many curves given in the form $r = f(\theta)$.

▶ Use a formula to find the sector area under a curve given in terms of polar coordinates.

Recap

You will need to remember…

▶ Trigonometric identities such as the double-angle formula $\cos 2\theta = 1 - 2\sin^2\theta$, and how to use them to integrate functions such as $\sin^2\theta$

▶ How to use angles in radians.

▶ The area of a sector of a circle with central angle θ radians and radius r, is $\frac{1}{2}r^2\theta$.

21.1 Position of a point

The position of a point, P, in a plane can be given in terms of its distance from a fixed point, O, called the **pole**, and the angle that OP makes with a fixed line called the **initial line**. When the position of a point is given in this way, you have the **polar coordinates** of the point.

In the diagram, the **Cartesian** coordinates of point P would be given as (x, y). Its position in polar coordinates would be given as (r, θ), where r (≥ 0) is the distance of P from the origin, O, and θ is the **anticlockwise angle** that OP makes with the x-axis, which is normally taken as the initial line.

θ is normally measured in radians and its **principal value** is taken to be between $-\pi$ and π.

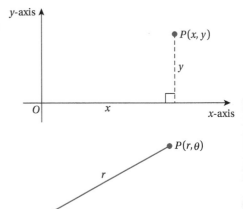

Example 1

Question

Plot the point P with polar coordinates $\left(4, \dfrac{\pi}{6}\right)$ and the point Q with coordinates $\left(2, -\dfrac{\pi}{3}\right)$.

You have already been introduced to the polar coordinates in Chapter *2 Complex Numbers*.

Connection between polar and Cartesian coordinates

You saw in Chapter *2 Complex Numbers* that there is a relationship between the polar and Cartesian coordinates. For example, from the diagram you can see that the point P is (x, y) in Cartesian coordinates and (r, θ) in polar coordinates, which leads to:

$$x = r \cos \theta \qquad y = r \sin \theta$$

$$r = \sqrt{x^2 + y^2} \qquad \tan \theta = \frac{y}{x}$$

If either x or y is negative, you should refer to the position of the point to determine the value of θ because the inverse tangent returns only values of θ between $-\dfrac{\pi}{2} < \theta < \dfrac{\pi}{2}$.

You can use the above equations to convert the equation of a curve from its Cartesian form to its polar form, or vice versa.

Example 2

Find the polar equation of the curve $x^2 + y^2 = 2x$.

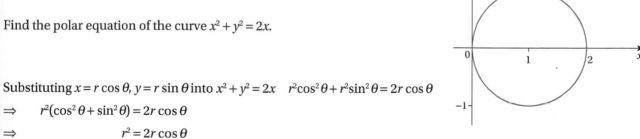

Substituting $x = r \cos \theta,\ y = r \sin \theta$ into $x^2 + y^2 = 2x$ $r^2\cos^2\theta + r^2\sin^2\theta = 2r \cos \theta$

$\Rightarrow \qquad r^2(\cos^2\theta + \sin^2\theta) = 2r \cos \theta$

$\Rightarrow \qquad\qquad\qquad r^2 = 2r \cos \theta$

$\Rightarrow \qquad\qquad\qquad r = 2 \cos \theta\ (\text{since } r \neq 0)$

Hence, the polar equation of the given curve is $r = 2 \cos \theta$.

Example 3

Convert the Cartesian coordinates $(-1, -\sqrt{3})$ into polar form.

$$r = \sqrt{1 + \left(\sqrt{3}\right)^2} = 2$$

$$\tan\theta = \frac{\sqrt{3}}{1}$$

$$\tan^{-1}\sqrt{3} = \frac{\pi}{3}$$

Therefore, $\theta = \dfrac{\pi}{3} - \pi = -\dfrac{2\pi}{3}$.

Note

Using $r = \sqrt{x^2 + y^2}$.

Note

To find the angle with the same tangent in the correct quadrant, add or subtract π to ensure that the value of θ is in the interval $-\pi < \theta \le \pi$.

Note

To find the value of θ, first find the angle using the formula $\tan\theta = \dfrac{y}{x}$ and then make sure you pick the corresponding value of θ in the correct quadrant.

Example 4

Find the Cartesian equation of the curve $r = a\cos\theta$.

$$r^2 = ar\cos\theta$$

Substituting $r^2 = x^2 + y^2$ and $x = r\cos\theta$,

$$x^2 + y^2 = ax$$

$$\Rightarrow \left(x - \frac{a}{2}\right)^2 + y^2 = \left(\frac{a}{2}\right)^2$$

Note

Multiply $r = a\cos\theta$ by r.

Note

This is a circle with centre $\left(\dfrac{a}{2}, 0\right)$ and radius $\dfrac{a}{2}$.

Exercise 1

1 Plot the points with the following polar coordinates.

  a $\left(3, \dfrac{\pi}{4}\right)$ **b** $\left(2, \dfrac{2\pi}{3}\right)$ **c** $\left(3, -\dfrac{\pi}{3}\right)$

  d $\left(2, \dfrac{3\pi}{2}\right)$ **e** $\left(4, -\dfrac{\pi}{4}\right)$

2 Find the Cartesian equation of each of these curves.

  a $r = 4$ **b** $r\cos\theta = 3$ **c** $r\sin\theta = 7$

  d $r = a(1 + \cos\theta)$ **e** $\dfrac{2}{r} = 1 + \cos\theta$

3 Find the polar equation of each of these curves.

  a $x^2 + y^2 = 9$ **b** $xy = 16$ **c** $\dfrac{x^2}{9} + \dfrac{y^2}{16} = 1$

  d $x^2 + y^2 = 6x$ **e** $(x^2 + y^2)^2 = x^2 - y^2$

21.2 Sketching curves given in polar coordinates

The normal way to sketch a curve expressed in polar coordinates is to plot points roughly using simple values of θ. Sometimes polar curves are closed; each area that is enclosed is called a **loop** of the curve.

Note that throughout this chapter, you will require that r is positive in order for a point to be on a curve.

To sketch a curve in polar coordinates, start by creating a table for some points on the curve. For example, let's sketch the curve of $r = a \cos 3\theta$.

θ	0	$\dfrac{\pi}{18}$	$\dfrac{\pi}{9}$	$\dfrac{\pi}{6}$	$\dfrac{\pi}{2}$	$\dfrac{5\pi}{9}$	$\dfrac{11\pi}{18}$	$\dfrac{2\pi}{3}$	$\dfrac{13\pi}{18}$	$\dfrac{7\pi}{9}$	$\dfrac{5\pi}{6}$
r	a	$\dfrac{\sqrt{3}}{2}a$	$\dfrac{1}{2}a$	0	0	$\dfrac{1}{2}a$	$\dfrac{\sqrt{3}}{2}a$	a	$\dfrac{\sqrt{3}}{2}a$	$\dfrac{1}{2}a$	0

Since r must always be positive, the curve does not exist for any values that make r negative, so these values should

not be plotted. For example, when $\theta = \dfrac{2\pi}{9}$, $r = a \cos\left(\dfrac{2\pi}{3}\right) = -\dfrac{1}{2}a$.

This means that the curve does not exist when $\theta = \dfrac{2\pi}{9}$. Similarly,

the curve does not exist for any value of θ between

$\dfrac{\pi}{6}$ and $\dfrac{\pi}{2}$, $\dfrac{5\pi}{6}$ and $\dfrac{7\pi}{6}$, $\dfrac{9\pi}{6}$ and $\dfrac{11\pi}{6}$, $\dfrac{13\pi}{6}$ and $\dfrac{15\pi}{6}$.

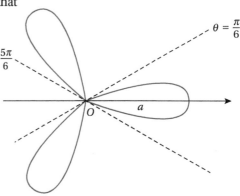

Plotting the values given in the table and joining the points gives a curve with three loops or lobes.

Notice that the half-lines $\theta = \dfrac{\pi}{6}$, $\theta = \dfrac{\pi}{2}$ and $\theta = \dfrac{5\pi}{6}$ are all

tangents to the loops. The tangents meet at the origin or pole. All three loops

are **congruent**.

In general, when sketching curves from polar coordinates,

▶ it is useful to look for any **symmetry**
  • if r is a function of $\cos\theta$ only, there is symmetry about the initial line
  • if r is a function of $\sin\theta$ only, there is symmetry about the line $\theta = \dfrac{\pi}{2}$

▶ the equations $r = a \sin\theta$ and $r = a \cos\theta$ are **circles**.

You should recognize $r = k$ as a circle for any positive constant k.

Example 5

Sketch $r = 1 + 2 \cos\theta$. Part of a table giving values for r is given.

θ	0	$\dfrac{\pi}{6}$	$\dfrac{\pi}{3}$	$\dfrac{\pi}{2}$	$\dfrac{2\pi}{3}$	$-\dfrac{2\pi}{3}$	$-\dfrac{\pi}{2}$	$-\dfrac{\pi}{3}$	$-\dfrac{\pi}{6}$	0
r	3	$1+\sqrt{3}$	2	1	0	0	1	2	$1+\sqrt{3}$	3

$$1 + 2\cos\theta = 0$$

$$\Rightarrow \quad \cos\theta = -\frac{1}{2}$$

$$\Rightarrow \quad \theta = \frac{2\pi}{3}, \frac{4\pi}{3}\left(\text{or} -\frac{2\pi}{3}\right)$$

Therefore, $1 + 2\cos\theta$ is negative for $\frac{2\pi}{3} < \theta < \frac{4\pi}{3}$.

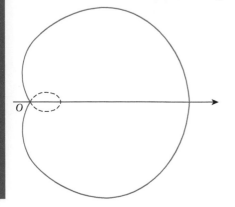

> **Note**
>
> To find when r is negative, solve $r = 0$.

> **Note**
>
> Since only positive values of r are relevant, the curve does not exist for values of θ between $\frac{2\pi}{3}$ and $\frac{4\pi}{3}$, and should not be shown on your sketch. Some graphing calculators include negative values of r, which you should ignore.

> **Note**
>
> Sketch $r = 1 + 2\cos\theta$.

> **Note**
>
> The dashed part represents the negative values of r, which are commonly displayed by graphics calculators, but you should *exclude* completely from any sketch you draw.

When a polar equation contains $\sec\theta$ or $\text{cosec }\theta$, it is often easier to use its Cartesian equation when sketching the curve.

Example 6

Sketch $r = a\sec\theta$.

$$r = a\sec\theta = \frac{a}{\cos\theta}$$

$$\Rightarrow r\cos\theta = a$$

$$\Rightarrow \qquad x = a$$

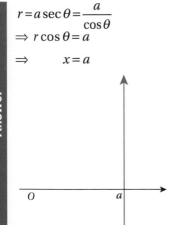

> **Note**
>
> Convert the polar form into Cartesian.

> **Note**
>
> Using trigonometric identity, $\sec\theta = \dfrac{1}{\cos\theta}$.

> **Note**
>
> $x = a$ is the straight line shown.

Example 7

Sketch $r = a\sec(\alpha - \theta)$.

$$r\cos(\alpha - \theta) = a$$
$$r\cos\theta\cos\alpha + r\sin\theta\sin\alpha = a$$
$$x\cos\alpha + y\sin\alpha = a$$

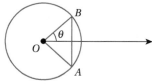

Note

Transpose terms using $\sec\theta = \dfrac{1}{\cos\theta}$, since $\cos\theta$ appears in many polar coordinate identities and $\sec\theta$ is less easy to work with.

Note

Replace $r\cos\theta$ with x and $r\sin\theta$ with y, using $x = r\cos\theta$ and $y = r\sin\theta$.

Note

Using the trigonometric identity, $\cos(u + v) = \cos u\cos v + \sin u\sin v$.

Note

Sketch the curve.

Example 8

The curve S is given by the formula $r = 8(1 - \cos\theta)$, $0 \le \theta \le 2\pi$.

The circle C with Cartesian equation $x^2 + y^2 = 16$ intersects the curve S at the points A and B.

a Find the polar coordinates of the points of intersection.

b Show that the length of AB is $4\sqrt{3}$.

a The polar equation of C is $r^2 = 16$, so $r = 4$.

At A and B, $4 = 8(1 - \cos\theta)$

Therefore $\cos\theta = \dfrac{1}{2}$.

$\theta = \dfrac{\pi}{3}$ or $\dfrac{5\pi}{3}$

The points of intersection are $\left(4, \dfrac{\pi}{3}\right)$ and $\left(4, \dfrac{5\pi}{3}\right)$.

Note

First ensure that both equations are in polar form.

Note

Find all possible points of intersection by finding all solutions of $\cos\theta = \dfrac{1}{2}$ with $0 < \theta < 2\pi$, and not only the principal value.

b

Using the diagram,

length $AB = 2r\sin\theta = 4\sqrt{3}$

Note

Sketch the graph.

Note

It would be possible to convert A and B into Cartesian coordinates to find the distance between them, but it's easier to think about where the points are.

Exercise 2

1 Sketch each of these curves.

   a $r = 4$ **b** $r\cos\theta = 3$ **c** $r\sin\theta = 7$

   d $r = a(1 + \cos\theta)$ **e** $\dfrac{2}{r} = 1 + \cos\theta$

2 Sketch each of these curves.

    a $r = a \sin 2\theta, \, 0 \le \theta < 2\pi$

    b $r = a \cos 4\theta, \, 0 \le \theta < 2\pi$

    c $r = 2 + 3 \cos \theta, \, -\pi \le \theta < \pi$

    d $r = a\theta, \, 0 \le \theta < 2\pi$

    e $r = 4 \sec \theta, \, -\dfrac{\pi}{2} \le \theta < \dfrac{\pi}{2}$

3 Sketch the curve $r = (4 + 2\cos\theta)\sqrt{\sin\theta}, \, 0 \le \theta \le \pi$.

4 Circle C is given by the equation $x^2 + (y-8)^2 = 64$.

    a   Prove that this locus can be written as $r = 16\sin\theta$.

    b   Curve S is given by $r = 8\sin\theta + 4, \, 0 \le \theta \le 2\pi$. The circle C meets the curve S at points P and Q.

The point M is the point so that $OPMQ$ is a rhombus. Find the area of the rhombus $OPMQ$.

21.3 Area of a sector of a curve

You already know from your A-level mathematics studies that integration can be used to find the area under a curve using the general expression $\int_{b}^{a} f(x)\, dx$, where a and b are the real number boundaries of the interval.

Now you will look at how to find the area of a sector of a curve. If you have the equation for part of a curve given in polar coordinates, there is a general formula that you can use.

Let A be the area bounded by the curve $r = f(\theta)$ and the two radii at α and at θ.

As θ increases by $\delta\theta$, the increase in area, δA, (shaded in the diagram) is given by

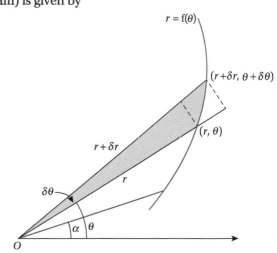

$$\frac{1}{2}r^2\delta\theta \le \delta A \le \frac{1}{2}(r+\delta r)^2\, \delta\theta \text{ (using areas of sectors)}$$

Dividing throughout by $\delta\theta$, you obtain

$$\frac{1}{2}r^2 \le \frac{\delta A}{\delta\theta} \le \frac{1}{2}(r+\delta r)^2$$

As $\delta\theta \to 0, \dfrac{\delta A}{\delta\theta} \to \dfrac{dA}{d\theta}$ and $\delta r \to 0$. Therefore, you have

$$\frac{dA}{d\theta} = \frac{1}{2}r^2$$

Integrating both sides with respect to θ, you obtain

$$\int \frac{dA}{d\theta}\, d\theta = \int \frac{1}{2}r^2\, d\theta$$

$$\Rightarrow \quad A = \int \frac{1}{2}r^2\, d\theta$$

The general equation for the area of a sector of a curve is $A = \int_{\alpha}^{\beta} \frac{1}{2} r^2 \, d\theta$, when the area is bounded by the radii $\theta = \alpha$ and $\theta = \beta$.

Tip

When applying this formula, it is important to remember that r must be defined and be non-negative throughout the interval $\alpha \le \theta \le \beta$.

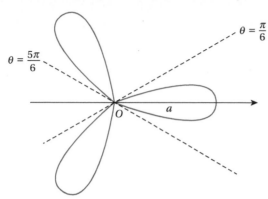

When finding the area using the general equation for the area of a sector of a curve, you should first sketch the curve as this makes it easier for you to find the boundaries.

Example 9

Find the area of one loop of the curve $r = a \cos 3\theta$.

$$A = \frac{1}{2} \int_{-\frac{\pi}{6}}^{\frac{\pi}{6}} r^2 \, d\theta$$

Note

You know from your sketch in *Example 4* that one loop is bounded by the tangent lines $\theta = \frac{\pi}{6}$ and $\theta = -\frac{\pi}{6}$.

Note

Using $A = \int_{\alpha}^{\beta} \frac{1}{2} r^2 \, d\theta$.

$$\Rightarrow \quad A = \frac{1}{2} \int_{-\frac{\pi}{6}}^{\frac{\pi}{6}} a^2 \cos^2 3\theta \, d\theta$$

Note

Using the double-angle formula, $\cos 2\theta = 2\cos^2\theta - 1$, to integrate.

$$A = \frac{1}{2} a^2 \int_{-\frac{\pi}{6}}^{\frac{\pi}{6}} \frac{1}{2}(\cos 6\theta + 1) \, d\theta$$

$$= \frac{a^2}{4} \left[\frac{\sin 6\theta}{6} + \theta \right]_{-\frac{\pi}{6}}^{\frac{\pi}{6}} = \frac{a^2}{4} \left(\frac{\pi}{6} + \frac{\pi}{6} \right) = \frac{a^2 \pi}{12}$$

So, the area of one loop of $r = a \cos 3\theta$ is $\dfrac{a^2 \pi}{12}$.

When a curve is symmetrical in other quadrants, it is often better to use only the area in the first quadrant. Therefore in *Example 9*, you could have used

$2 \times \int_{0}^{\frac{\pi}{6}} a^2 \cos^2 3\theta \, d\theta$ instead of using $\dfrac{1}{2} \int_{-\frac{\pi}{6}}^{\frac{\pi}{6}} a^2 \cos^2 3\theta \, d\theta$.

Example 10

The curve $r = k\theta$ and the half-lines $\theta = \dfrac{\pi}{2}$ and $\theta = \pi$ intersect.

a Sketch the curve of $r = k\theta$ for $\dfrac{\pi}{2} \leq \theta \leq \pi$.

b Find the area bounded by the curve $r = k\theta$ and the half-lines $\theta = \dfrac{\pi}{2}$ and $\theta = \pi$.

a

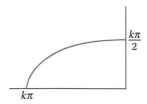

b $A = \dfrac{1}{2}\displaystyle\int_{\frac{\pi}{2}}^{\pi} k^2\theta^2\,d\theta$

$= \dfrac{k^2}{2}\left[\dfrac{\theta^3}{3}\right]_{\frac{\pi}{2}}^{\pi} = \dfrac{k^2}{2}\left(\dfrac{\pi^3}{3} - \dfrac{\pi^3}{24}\right) = \dfrac{7k^2\pi^3}{48}$

Hence, the required area is $\dfrac{7k^2\pi^3}{48}$.

> **Note**
>
> Using $A = \displaystyle\int_{\alpha}^{\beta} \dfrac{1}{2}r^2\,d\theta$.

Example 11

Sketch the curves $r = 1 + \cos\theta$ and $r = \sqrt{3}\sin\theta$. Find

a the points where the curves meet

b the area of the region enclosed by the two curves.

> **Note**
>
> Before sketching the two curves, you should note that $r = 1 + \cos\theta$ is similar to $r = 1 + 2\cos\theta$ (*Example 5*) and $r = \sqrt{3}\sin\theta$ is similar to $r = a\cos\theta$ (*Example 4*); curves of the form $r = a + b\cos n\theta$ have a similar shape.

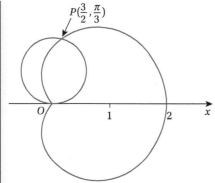

a $\sin\theta \equiv \sin\left(\dfrac{\theta}{2} + \dfrac{\theta}{2}\right)$

$\equiv 2\sin\left(\dfrac{\theta}{2}\right)\cos\left(\dfrac{\theta}{2}\right)$

$\cos\theta \equiv 2\cos^2\left(\dfrac{\theta}{2}\right) - 1$

> **Note**
>
> The two curves meet when $1 + \cos\theta = \sqrt{3}\sin\theta$. To solve the equation, there are two possible approaches: you can use double angle formulae; or, use the harmonic form to write $A\sin\theta + B\cos\theta$ in the form $R\cos(\theta - \alpha)$. Here, double angle formulae have been used.

> **Note**
>
> Using $\sin 2x = 2\sin x \cos x$.

> **Note**
>
> Using $\cos 2x = 2\cos^2 x - 1$.

(continued)

(continued)

Express $1 + \cos\theta = \sqrt{3}\sin\theta$ as

$$1 + 2\cos^2\left(\frac{\theta}{2}\right) - 1 = 2\sqrt{3}\sin\left(\frac{\theta}{2}\right)\cos\left(\frac{\theta}{2}\right)$$

$$2\cos^2\left(\frac{\theta}{2}\right) = 2\sqrt{3}\sin\left(\frac{\theta}{2}\right)\cos\left(\frac{\theta}{2}\right)$$

$$\cos\left(\frac{\theta}{2}\right) = \sqrt{3}\sin\left(\frac{\theta}{2}\right) \quad \text{or} \quad \cos\left(\frac{\theta}{2}\right) = 0$$

$$\Rightarrow \qquad \tan\left(\frac{\theta}{2}\right) = \frac{1}{\sqrt{3}} \qquad \text{or} \quad \theta = \pi$$

$$\Rightarrow \qquad \frac{\theta}{2} = \frac{\pi}{6} \Rightarrow \theta = \frac{\pi}{3}$$

Therefore, the curves meet at $\left(\dfrac{3}{2}, \dfrac{\pi}{3}\right)$ and $(0, \pi)$.

b

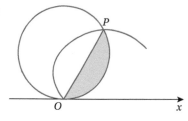

$$\text{Area} = \frac{1}{2}\int_0^{\frac{\pi}{3}} \left(\sqrt{3}\sin\theta\right)^2 d\theta$$

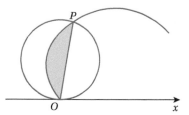

$$\text{Area} = \frac{1}{2}\int_{\frac{\pi}{3}}^{\pi} \left(1 + \cos\theta\right)^2 d\theta$$

Therefore,

$$\frac{1}{2}\int_0^{\frac{\pi}{3}} \left(\sqrt{3}\sin\theta\right)^2 d\theta + \frac{1}{2}\int_{\frac{\pi}{3}}^{\pi} (1 + \cos\theta)^2 d\theta$$

$$= \frac{1}{2}\int_0^{\frac{\pi}{3}} 3\sin^2\theta\, d\theta + \frac{1}{2}\int_{\frac{\pi}{3}}^{\pi} (1 + 2\cos\theta + \cos^2\theta)d\theta$$

Note

These are polar coordinates (r, θ). To find the area contained between the curves, you draw the line OP and consider separately the two areas so formed.

Note

The shaded region in the diagram is bounded by the curve $r = \sqrt{3}\sin\theta$ and the two radii $\theta = 0$ and $\theta = \dfrac{\pi}{3}$.

Note

The shaded region is bounded by the curve $r = 1 + \cos\theta$ and the two radii $\theta = \dfrac{\pi}{3}$ and π.

Note

Add the two areas together.

(continued)

$$= \frac{3}{4}\int_0^{\frac{\pi}{3}} \frac{1}{2}(1-\cos 2\theta)\,d\theta + \frac{1}{2}\int_{\frac{\pi}{3}}^{\pi} \left[1+2\cos\theta + \frac{1}{2}(\cos 2\theta+1) \right] d\theta$$

$$= \frac{3}{2}\left[\theta - \frac{1}{2}\sin 2\theta \right]_0^{\frac{\pi}{3}} + \frac{1}{2}\left[\frac{3}{2}\theta + 2\sin\theta + \frac{1}{4}\sin 2\theta \right]_{\frac{\pi}{3}}^{\pi}$$

$$= \frac{3}{4}\left(\frac{\pi}{3} - \frac{\sqrt{3}}{4} \right) + \frac{1}{2}\left(\frac{3\pi}{2} - \frac{3\pi}{6} - \sqrt{3} - \frac{\sqrt{3}}{8} \right)$$

$$= \frac{3\pi}{4} - \frac{3\sqrt{3}}{4}$$

Therefore, the area contained within the curves is $\dfrac{3\pi}{4} - \dfrac{3\sqrt{3}}{4}$.

Exercise 3

1 Find the area bounded by the curve $r = a\theta$ and the radii $\theta = \dfrac{\pi}{2}, \theta = \pi$.

2 For each of these curves, find the area enclosed by one loop.

   a $r = a\cos 2\theta$

   b $r = a\sin 2\theta$

   c $r = a\cos 4\theta$

3 Find the area enclosed by the curve $r = a\cos\theta$.

4 Find the area enclosed by the curve $r = 2 + 3\cos\theta$.

5 **a** Find the polar equation of the curve $(x^2 + y^2)^3 = y^4$.

   b Hence

     i sketch the curve

     ii find the area enclosed by the curve.

6 Find where these two curves intersect.

$$r = 2\sin\theta \qquad 0 \le \theta < \pi$$

   and $\quad r = 2(1 - \sin\theta) \qquad -\pi < \theta < \pi$

   Hence, find the area which is between the two curves.

7 The curve, S, is given by $r = 1 + \sin 2\theta$ for $-\pi \le \theta \le \pi$. Find θ when $r = 0$ and hence find the area of one loop.

Summary

▶ The position of a point in polar coordinates is given as (r, θ), where $r\,(\ge 0)$ is the distance of P from the pole, O, and θ is the anticlockwise angle that OP makes with the initial line (usually the x-axis).

▶ θ is normally measured in radians and its principal value is taken to be between $-\pi$ and π.

▶ You can convert Cartesian coordinates to polar coordinates, and vice versa, using the formulae

$$x = r\cos\theta \qquad y = r\sin\theta \qquad r = \sqrt{x^2 + y^2} \qquad \tan\theta = \frac{y}{x}$$

- To sketch a curve given in polar coordinates, plot some points roughly using simple values of θ and join using a smooth curve. Since r must always be positive, you should not sketch points where r is negative. In general,
  - it is useful to look for any symmetry
    - if r is a function of $\cos\theta$ only, there is symmetry about the initial line
    - if r is a function of $\sin\theta$ only, there is symmetry about the line $\theta = \dfrac{\pi}{2}$
  - the equations $r = a\sin\theta$ and $r = a\cos\theta$ are circles.
- You should recognize $r = k$ as a circle for any positive constant k.
- When a polar equation contains $\sec\theta$ or $\operatorname{cosec}\theta$, it is often easier to use its Cartesian equation when sketching the curve.
- When you are given part of a curve in polar coordinates, you can find the area of the sector using the formula $A = \dfrac{1}{2}\displaystyle\int_{\alpha}^{\beta} r^2 \, d\theta$.

Review exercises

1 Find the Cartesian equation of $r = a(1 - \cos\theta)$.

2 Find the polar equation of $x^2 + y^2 + 8y = 16$.

3 **a** Show that $y^2 = 1 - 2x$ can be written as $x^2 + y^2 = (1-x)^2$.

    b A curve has Cartesian equation $y^2 = 1 - 2x$. Find its polar equation.

4 A curve has polar equation $r = 8 - 5\cos\theta$ for $-\pi \le \theta \le \pi$.

    a Sketch the curve.

    b Find the area of the region bounded by the curve.

Practice examination questions

1 The diagram shows a sketch of a loop, the pole O and the initial line.

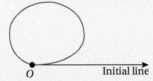

The polar equation of the loop is $r = (2 + \cos\theta)\sqrt{\sin\theta}$, $0 \le \theta \le \pi$.

Find the area enclosed by the loop.   (6 marks)

AQA MFP3 January 2009

2 A curve has polar equation $r(4 - 3\cos\theta) = 4$. Find its Cartesian equation in the form $y^2 = f(x)$.   (4 marks)

AQA MFP3 June 2014

3 The Cartesian equation of a circle is $(x + 8)^2 + (y - 6)^2 = 100$.

Using the origin O as the pole and the positive x-axis as the initial line, find the polar equation of this circle, giving your answer in the form $r = p\sin\theta + q\cos\theta$.   (4 marks)

AQA MFP3 June 2013

4 The diagram shows a sketch of a curve C, the pole O and the initial line.

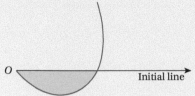

The polar equation of C is $r = 2\sqrt{1 + \tan\theta}$, $-\dfrac{\pi}{4} \leq \theta \leq \dfrac{\pi}{4}$.

Show that the area of the shaded region, bounded by the curve C and the

initial line, is $\dfrac{\pi}{2} - \ln 2$. **(4 marks)**

AQA MFP3 June 2012

5 The diagram shows a sketch of the curve C with polar equation
$r = 3 + 2\cos\theta$, $0 \leq \theta \leq 2\pi$.

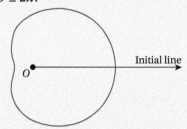

a Find the area of the region bounded by the curve C. **(6 marks)**

b A circle, whose Cartesian equation is $(x - 4)^2 + y^2 = 16$, intersects the curve C at the points A and B.

    i Find, in surd form, the length of AB. **(6 marks)**

    ii Find the perimeter of the segment AOB of the circle, where O is the pole. **(3 marks)**

AQA MFP3 January 2012

6 A curve C has polar equation $r(1 + \cos\theta) = 2$.

a Find the Cartesian equation of C, giving your answer in the form $y^2 = f(x)$. **(5 marks)**

b The straight line with polar equation $4r = 3\sec\theta$ intersects the curve C at the points P and Q. Find the length of PQ. **(4 marks)**

AQA MFP3 January 2011

7 The curve C_1 is defined by $r = 2\sin\theta$, $0 \leq \theta \leq \dfrac{\pi}{2}$.

The curve C_2 is defined by $r = \tan\theta$, $0 \leq \theta \leq \dfrac{\pi}{2}$.

a Find a Cartesian equation of C_1. **(3 marks)**

b **i** Prove that the curves C_1 and C_2 meet at the pole O and at one other point, P, in the given domain. State the polar coordinates of P. **(4 marks)**

    ii The point A is the point on the curve C_1 at which $\theta = \dfrac{\pi}{4}$.

    The point B is the point on the curve C_2 at which $\theta = \dfrac{\pi}{4}$.

Determine which of the points A or B is further away from the pole O, justifying your answer. (2 marks)

iii Show that the area of the region bounded by the arc OP of C_1 and the arc OP of C_2 is $a\pi + b\sqrt{3}$, where a and b are rational numbers. (10 marks)

AQA MFP3 June 2011

8 The polar equation of a curve C_1 is
$$r = 2(\cos\theta - \sin\theta),\ 0 \leq \theta \leq 2\pi$$

a **i** Find the Cartesian equation of C_1. (4 marks)

ii Deduce that C_1 is a circle and find its radius and the Cartesian coordinates of its centre. (3 marks)

b The diagram shows the curve C_2 with polar equation
$$r = 4 + \sin\theta,\ 0 \leq \theta \leq 2\pi$$

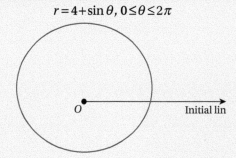

i Find the area of the region that is bounded by C_2. (6 marks)

ii Prove that the curves C_1 and C_2 do not intersect. (4 marks)

iii Find the area of the region that is outside C_1 but inside C_2. (2 marks)

AQA MFP3 June 2010

22 The Calculus of Inverse Trigonometrical Functions

Introduction

You know from previous studies how to differentiate the trigonometric functions sine, cosine and tangent. It is also possible to differentiate the inverses of these functions; the results are familiar rational functions. The inverse functions are very useful in calculus. A useful application is to integrate functions of the form $\dfrac{1}{a^2+x^2}$ or $\dfrac{1}{\sqrt{a^2-x^2}}$, for a given constant a.

Objectives

By the end of this chapter, you should know how to:

▶ Sketch the graphs of inverse trigonometric functions.

▶ Use the derivatives of $\sin^{-1}x$, $\cos^{-1}x$ and $\tan^{-1}x$.

▶ Integrate functions in the form $\dfrac{1}{ax^2+bx+c}$, and $\dfrac{1}{a^2+x^2}$

▶ Integrate functions in the form $\dfrac{1}{\sqrt{a^2-x^2}}$.

Recap

You will need to remember...

▶ Differentiate and integrate standard trigonometrical functions, for example $\sin x$, with x in radians.

▶ Find integrals such as $\int \sin^2 x \, dx$ using appropriate identities.

▶ Simplify trigonometric equations using appropriate trigonometric identities and formulae.

· ·

22.1 Inverse trigonometric functions

The inverse function $\sin^{-1}x$, is defined as the angle whose sine is x. For example,

$$\sin\left(\frac{\pi}{6}\right) = \frac{1}{2} \implies \sin^{-1}\left(\frac{1}{2}\right) = \frac{\pi}{6}$$

Hence, if $\theta = \sin^{-1}x$, then $\sin\theta = x$.

If $\sin\theta = \dfrac{1}{2}$, then there are infinitely many possible values of θ. If you use only values of $\sin^{-1}x$ that return a value of θ in the interval $-\dfrac{\pi}{2} \le \theta \le \dfrac{\pi}{2}$ then it ensures the function $\sin^{-1}x$ has only one value for any given input x.

Occasionally, **arcsin$x$** is used instead of $\sin^{-1}x$ and **arctan$x$** instead of $\tan^{-1}x$. Don't be confused by the notation $\sin^{-1}x$. In the past you have used positive powers of $\sin x$ to mean $\sin^3 x = (\sin x)^3$ and this is still correct. However, $\sin^{-1}x$ is **not** the same as $\dfrac{1}{\sin x}$; instead, it is a special notation to mean the inverse of $\sin x$.

Recall from your A-level mathematics studies that $\dfrac{d}{dx}\sin x = \cos x$ only because x is measured in radians. If you tried the same in degrees, the formula would be more complicated. The same applies here; any mention of angles in this chapter assumes that they are measured in **radians**.

Example 1

Question

If $\sin^{-1}x = \dfrac{2\pi}{5}$ find $\cos^{-1}x$.

Answer

$$\sin^{-1}x = \frac{2\pi}{5} \rightarrow \sin\frac{2\pi}{5} = x$$

$$\sin\frac{2\pi}{5} = \cos\left(\frac{\pi}{2} - \frac{2\pi}{5}\right) = \cos\frac{\pi}{10}$$

$$\therefore \cos^{-1}x = \frac{\pi}{10}$$

> **Note**
>
> Using the identity
> $$\sin\theta = \cos\left(\frac{\pi}{2} - \theta\right).$$

Sketching inverse trigonometric functions

You should already be very familiar with the shape of some trigonometric functions.

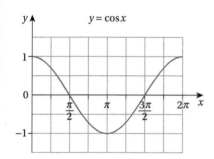

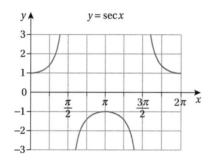

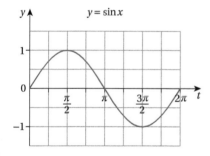

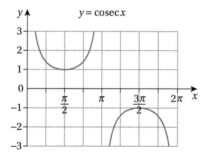

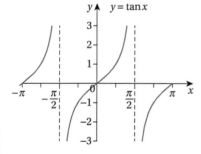

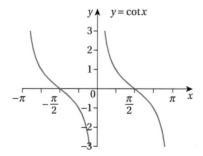

When sketching these functions, it is useful to remember the key coordinates:

▶ $\sin x$: $\sin 0 = 0$, $\sin\dfrac{\pi}{2} = 1$, $\sin\pi = 0$, $\sin\dfrac{3\pi}{2} = -1$ and $\sin 2\pi = 0$.

▶ $\cos x$: $\cos 0 = 1$, $\cos\dfrac{\pi}{2} = 0$, $\cos\pi = -1$, $\cos\dfrac{3\pi}{2} = 0$ and $\cos 2\pi = 1$.

▶ $\tan x$: $\tan 0 = 0$, $\tan\pi = 0$ and $\tan 2\pi = 0$; the following are vertical asymptotes: $\tan\dfrac{\pi}{2}$, $\tan\dfrac{3\pi}{2}$ etc.

Inverse sine graph

Before sketching an inverse trigonometric function it is important to define its domain.

The domain (possible x-values) of $y = \sin^{-1} x$ is $-1 \leq x \leq 1$ and the range (y-values) is $-\dfrac{\pi}{2} \leq \sin^{-1} x \leq \dfrac{\pi}{2}$. The graph of $y = \sin^{-1} x$ is obtained by reflecting the graph of $y = \sin x$ in the line $y = x$ for $-1 \leq x \leq 1$.

To draw the sketch to an acceptable degree of accuracy, you need to find the gradient of the sine curve at the origin. To do this, you differentiate $y = \sin x$, which gives $\dfrac{dy}{dx} = \cos x$.

At the origin, where $x = 0$, you have $\dfrac{dy}{dx} = \cos 0 = 1$.

So, the gradient of $y = \sin x$ at the origin is 1.

You then proceed as follows:

▶ Draw the line $y = x$. Show this as a dashed line.

▶ Next, carefully sketch the graph of $y = \sin x$ across the domain of $y = \sin^{-1} x$, remembering that $y = x$ is a tangent to $y = \sin x$ at the origin.

▶ Finally, carefully sketch the reflection of $y = \sin x$ in the line $y = x$, to give the graph shown below.

> **Note**
>
> Restrict the domain so that the graph of $\sin^{-1} x$ is a function.

> **Note**
>
> We restrict the domain so that $\sin^{-1} x$ has only one value. For example, we want to make sure that $\sin^{-1} 0$ is not simultaneously defined to be $n\pi$ for every integer n. By convention, the values of $\sin^{-1} x$ are all required to be in the interval $-\dfrac{\pi}{2} < \sin^{-1} x < \dfrac{\pi}{2}$.

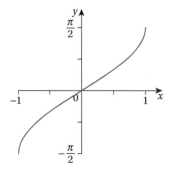

The graphs of other inverse trigonometric functions are found similarly: that is, by reflecting the graph of the relevant trigonometric function in the line $y = x$ across its domain. If the curve of the function passes through the origin, start by finding its gradient at that point.

Inverse tangent graph

$y = \tan^{-1} x$ has all values of x as its domain and $-\dfrac{\pi}{2} < \tan^{-1} x < \dfrac{\pi}{2}$ as its range.
The graphs of $y = \tan x$ and $y = \tan^{-1} x$ (or arctan x) are shown below.

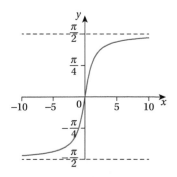

Inverse cosine graph

$y = \cos^{-1}x$ has $-1 \le x \le 1$ as its domain and $0 \le \cos^{-1}x \le \pi$ as its range. The graphs of $y = \cos x$ and $y = \cos^{-1}x$ (or $y = \arccos x$) are shown below.

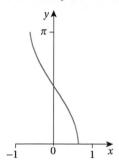

Exercise 1

1 Find the value of each of these inverse functions.

    a $\sin^{-1}0.5$ **b** $\sin^{-1}\left(-\dfrac{1}{2}\right)$ **c** $\cos^{-1}\left(-\dfrac{\sqrt{3}}{2}\right)$

    d $\tan^{-1}1$ **e** $\sec^{-1}\sqrt{2}$ **f** $\cot^{-1}3$

> **Tip**
>
> From your earlier studies, you should know how to use your calculator to work these out.

2 Sketch the graph of each of these inverse functions.

    a $\sec^{-1}x$ **b** $\operatorname{cosec}^{-1}x$ **c** $\cot^{-1}x$

3 If $\cos^{-1}x = \dfrac{2\pi}{5}$, find $\sin^{-1}x$.

4 Prove that $\tan^{-1}\left(\dfrac{1+x}{1-x}\right) = \dfrac{\pi}{4} + \tan^{-1}x$.

22.2 Differentiation and integration of inverse trigonometric functions

You should already know from earlier studies how to differentiate and integrate trigonometric functions.

► If $f(x) = \sin x$, then $f'(x) = \cos x$.
► If $f(x) = \cos x$ then $f'(x) = -\sin x$
► If $f(x) = \tan x$ then $f'(x) = \sec^2 x$.
► $\displaystyle\int \sin ax \, dx = -\dfrac{1}{a}\cos ax + C$
► $\displaystyle\int \cos ax \, dx = \dfrac{1}{a}\sin ax + C$
► $\displaystyle\int \tan ax \, dx = -\dfrac{1}{a}\ln|\cos ax| + C = \dfrac{1}{a}\ln|\sec ax| + C$

You will now look at how to differentiate and integrate inverse trigonometric functions. To differentiate, convert the trigonometric function into the form $\sin/\cos/\tan y = x$ first, and then differentiate. You then use the derivative to help you identify the associated standard integral.

$\sin^{-1}x$

If $y = \sin^{-1}x$, then $\sin y = x$.

- $S = \int\limits_{x_1}^{x_2} 2\pi y \sqrt{1 + \left(\dfrac{dy}{dx}\right)^2}\, dx$ (Cartesian form), where the rotation occurs between the lines x_1 and x_2

- $S = 2\pi \int\limits_{t_1}^{t_2} y \sqrt{\left(\dfrac{dx}{dt}\right)^2 + \left(\dfrac{dy}{dt}\right)^2}\, dt$ or $S = \int 2\pi y \sqrt{x^2 + y^2}\, dt$ (parametric form), where the rotation occurs between the points defined by t_1 and t_2.

Review exercises

1. If $y = \dfrac{1}{6}x^3 + \dfrac{1}{2x}$, find the arc length of f(x) between $x = 2$ and $x = 3$.
2. Find the arc length of the curve $y = \dfrac{1}{8}x^2 - \ln x$ between $x = 3$ and $x = 7$.
3. Find the area of the surface of revolution of $y = r\dfrac{x}{h}$ about the x-axis between $x = 0$ and $x = h$, and use this together with the area of a circle, $A = \pi r^2$, to find the surface area of a cone of radius r and height h.

Practice examination questions

1. The arc of the curve with equation $y = \dfrac{1}{2}\cos 4x$ between the points where $x = 0$ and $x = \dfrac{\pi}{8}$ is rotated by 2π radians about the x-axis. Show that the area S of the curved surfaced formed is given by

   $$S = \pi \int\limits_0^{\frac{\pi}{8}} \cos 4x \sqrt{1 + 4\sin^2 4x}\, dx$$
   (2 marks)
   AQA MFP2 June 2013

2. **a** Given that $x = \ln(\sec t + \tan t) - \sin t$ show that $\dfrac{dx}{dt} = \sin t \tan t$. (4 marks)

   b A curve is given parametrically by the equations
   $x = \ln(\sec t + \tan t) - \sin t,\ y = \cos t$
   The length of the arc of the curve between the points where $t = 0$ and $t = \dfrac{\pi}{3}$ is denoted by s.
   Show that $s = \ln p$, where p is an integer. (6 marks)
   AQA MFP2 January 2011

   > **Tip**
   > You can find the derivatives of trigonometric functions in the formulae booklet.

3. A curve has equation $y = 4\sqrt{x}$.
   Show that the length of arc s of the curve between the points where $x = 0$ and $x = 1$ is given by
   $$s = \int\limits_0^1 \sqrt{\dfrac{x+4}{x}}\, dx$$
   (4 marks)
   AQA MFP2 June 2007

4. A curve has parametric equations
   $x = t - \dfrac{1}{3}t^3,\ y = t^2$
   a Show that
   $$\left(\dfrac{dx}{dt}\right)^2 + \left(\dfrac{dy}{dt}\right)^2 = (1+t^2)^2$$
   (3 marks)
   b The arc of the curve between $t = 1$ and $t = 2$ is rotated through 2π radians about the x-axis.
   Show that S, the surface area generated, is given by $S = k\pi$, where k is a rational number to be found. (5 marks)
   AQA MFP2 June 2006

Introduction

In this chapter, you will be introduced to two new functions, cosh x and sinh x. The most common real-world example is the catenary curve, which describes how a wire hangs between two telegraph poles, and can be represented by the curve $y = a \cosh \dfrac{x}{a}$.

Recap

You will need to remember...

▶ How to use and algebraically manipulate parametric equations, exponential functions and logarithmic functions.

▶ How to use and algebraically manipulate parametric forms of equations.

▶ How to differentiate exponential and logarithmic functions, including using the

• quotient rule, $\dfrac{\mathrm{d}}{\mathrm{d}x}\left(\dfrac{\mathrm{f}(x)}{\mathrm{g}(x)}\right) = \dfrac{\frac{\mathrm{df}}{\mathrm{d}x}\mathrm{g}(x) - \mathrm{f}(x)\frac{\mathrm{dg}}{\mathrm{d}x}}{\left[\mathrm{g}(x)\right]^2}$

• chain rule, $\dfrac{\mathrm{d}}{\mathrm{d}x}\left[\mathrm{g}(\mathrm{f}(x))\right] = \dfrac{\mathrm{dg}}{\mathrm{d}x}(\mathrm{f}(x)) \times \dfrac{\mathrm{df}}{\mathrm{d}x}$

Objectives

By the end of this chapter, you should know how to:

▶ Define hyperbolic functions.

▶ Solve equations such as $a \sinh x + b \cosh x = c$.

▶ Prove identities such as $\sinh(x + y) = \sinh x \cosh y + \cosh y \sinh x$.

▶ Express inverse functions, such as $\sinh^{-1} x$, in logarithmic form.

▶ Prove differentiation formulae, including $\dfrac{\mathrm{d}}{\mathrm{d}x}(\cosh x) = \sinh x$

▶ Prove and use identities, such as $1 - \tanh^2 x = \mathrm{sech}^2 x$.

24.1 Hyperbolic functions

There are six **hyperbolic functions**. They are called 'hyperbolic' because they are related to the parametric equations for a hyperbola. Hyperbolic sine of x and hyperbolic cosine of x are written **sinh x** and **cosh x**. By convention, you pronounce sinh as 'shine' and cosh as 'cosh'.

They are defined by the following relationships:

$$\sinh x = \frac{1}{2}(\mathrm{e}^x - \mathrm{e}^{-x}) \qquad \cosh x = \frac{1}{2}(\mathrm{e}^x + \mathrm{e}^{-x})$$

Hyperbolic functions such as cosh x and sinh x are analogous to the regular (circular) trigonometrical functions in the sense that while $(x = \cos t, x = \sin t)$ is a parametric equation that represents a circle of radius 1, $(\cosh t, \sinh t)$ is a parametric equation that represents the hyperbola $x^2 - y^2 = 1$.

In a similar manner to the more familiar trigonometric functions, there are also four other hyperbolic functions.

See Chapter *1 Loci, Graphs and Algebra* if you need a reminder of what a hyperbolic function is.

The definitions of the six hyperbolic functions are as follows:

$$\sinh x = \frac{1}{2}(e^x - e^{-x}) \qquad \cosh x = \frac{1}{2}(e^x + e^{-x})$$

$$\tanh x = \frac{\sinh x}{\cosh x} = \frac{e^x - e^{-x}}{e^x + e^{-x}} \qquad \operatorname{cosech} x = \frac{1}{\sinh x}$$

$$\operatorname{sech} x = \frac{1}{\cosh x} \qquad \coth x = \frac{1}{\tanh x}$$

You pronounce tanh as 'than', (co)sech as '(co)sheck' and coth as 'coth'.

To evaluate a hyperbolic function, you can either consult your calculator instructions, or you can use their definition.

Example 1

Find, to 4 decimal places,

a $\sinh 2$ b $\operatorname{sech} 3$

a $\sinh 2 = \frac{1}{2}(e^2 - e^{-2}) = 3.6268$ (4 dp)

b $\operatorname{sech} 3 = \frac{1}{\cosh 3} = \frac{1}{10.0677...}$

$\operatorname{sech} 3 = 0.0993$ (4 dp)

Note

Substitute 2 in place of x in $\sinh x = \frac{1}{2}(e^x - e^{-x})$.

Note

Use your calculator to find cosh 3, then use $\operatorname{sech} x = \frac{1}{\cosh x}$.

Graphs of cosh x, sinh x and tanh x

You need to be familiar with, and recognise, the graphs of the following hyperbolic functions: $\sinh x$, $\cosh x$ and $\tanh x$.

$y = \cosh x$
You obtain the graph of $y = \cosh x$ by finding the mean values of a few corresponding pairs of values of $y = e^x$ and $y = e^{-x}$, and then plotting these mean values.

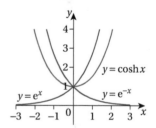

$y = \sinh x$
To produce the graph of $y = \sinh x$, you find half the difference between a few corresponding pairs of values of $y = e^x$ and $y = e^{-x}$, and then plot these values.

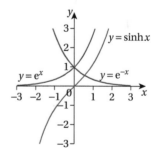

$y = \tanh x$
The tanh curve is obtained by dividing the values of sinh x by the values of cosh x:
you have $\tanh x = \frac{\sinh x}{\cosh x}$, which gives $\tanh x = \frac{(e^x - e^{-x})}{(e^x + e^{-x})}$

$$\Rightarrow \tanh x = \frac{1 - e^{-2x}}{1 + e^{-2x}}$$

Therefore, $\tanh x < 1$ for all values of x, and as $x \to +\infty$, $\tanh x \to 1$.

Since $\tanh x = -\frac{1 - e^{2x}}{1 + e^{2x}}$, $\tanh x > -1$ for all values of x, and as $x \to -\infty$, $\tanh x \to -1$.

Hence, the graph of $y = \tanh x$ lies between the asymptotes $y = 1$ and $y = -1$.

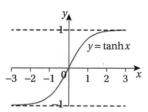

Standard hyperbolic identities

There are a number of standard hyperbolic identities that you will be expected to know, and should also be able to prove.

Standard hyperbolic identities:

► $\cosh^2 x - \sinh^2 x = 1$

► $\coth^2 x - 1 = \operatorname{cosech}^2 x$

► $1 - \tanh^2 x = \operatorname{sech}^2 x$

$\cosh^2 x - \sinh^2 x = 1$

From the exponential definitions of $\cosh x$ and $\sinh x$, you have

$$\cosh^2 x = \left[\frac{1}{2}(e^x + e^{-x})\right]^2 = \frac{1}{4}(e^{2x} + 2 + e^{-2x}) \qquad [1]$$

and $\qquad \sinh^2 x = \left[\frac{1}{2}(e^x - e^{-x})\right]^2 = \frac{1}{4}(e^{2x} - 2 + e^{-2x}) \qquad [2]$

Subtracting [2] from [1], you obtain

$$\cosh^2 x - \sinh^2 x = \frac{1}{4}(e^{2x} + 2 + e^{-2x}) - \frac{1}{4}(e^{2x} - 2 + e^{-2x}) = 1$$

Therefore, you have

$$\cosh^2 x - \sinh^2 x = 1$$

Notice the similarity of this hyperbolic identity with the usual trigonometric identity, $\cos^2 x + \sin^2 x \equiv 1$.

$\coth^2 x - 1 = \operatorname{cosech}^2 x$

Dividing the identity $\cosh^2 x - \sinh^2 x = 1$ by $\sinh^2 x$, you obtain

$$\frac{\cosh^2 x}{\sinh^2 x} - \frac{\sinh^2 x}{\sinh^2 x} \equiv \frac{1}{\sinh^2 x}$$

which gives

$$\coth^2 x - 1 = \operatorname{cosech}^2 x$$

$1 - \tanh^2 x = \operatorname{sech}^2 x$

Similarly, dividing the identity $\cosh^2 x - \sinh^2 x = 1$ by $\cosh^2 x$, you obtain

$$\frac{\cosh^2 x}{\cosh^2 x} - \frac{\sinh^2 x}{\cosh^2 x} \equiv \frac{1}{\cosh^2 x}$$

which gives

$$1 - \tanh^2 x = \operatorname{sech}^2 x$$

Some other useful hyperbolic identities

You will be asked to prove the following identities in *Exercise 1*; you will need to make sure you can prove them.

► $\cosh(A + B) = \cosh A \cosh B + \sinh A \sinh B$

► $\sinh(A - B) = \sinh A \cosh B - \cosh A \sinh B$

► $\sinh A + \sinh B = $ be able to $2 \sinh\left(\dfrac{A+B}{2}\right)\cosh\left(\dfrac{A-B}{2}\right)$

Solving equations involving hyperbolic functions

Equations involving hyperbolic functions are solved by either using the basic hyperbolic definitions (as shown in *Example 2*) or using the standard methods that you use to solve trigonometric equations (as shown in *Example 3*).

Example 2

Find the values of e^x for which $8 \cosh x - 12 \sinh x = 1$.

$$8 \cosh x - 12 \sinh x = 1 \rightarrow 4e^x + 4e^{-x} - (6e^x - 6e^{-x}) = 1$$
$$-2e^x + 10e^{-x} = 1$$
$$-2e^{2x} + 10 = e^x$$
$$2e^{2x} + e^x - 10 = 0$$
$$(e^x - 2)(2e^x + 5) = 0$$
$$e^x = 2 \text{ or } -\frac{5}{2}$$

> **Note**
>
> Using $\cosh x = \dfrac{1}{2}(e^x + e^{-x})$
>
> and $\sinh x = \dfrac{1}{2}(e^x - e^{-x})$.

> **Note**
>
> Solve each of the equations.

Example 3

Solve the equation $2 \cosh^2 x - \sinh x = 3$. Give your answer to four significant figures.

$$2(1 + \sinh^2 x) - \sinh x - 3 = 0$$
$$2 \sinh^2 x - \sinh x - 1 = 0$$
$$(2 \sinh x + 1)(\sinh x - 1) = 0$$
$$\Rightarrow \qquad \sinh x = 1 \text{ or } -\frac{1}{2}$$
$$\Rightarrow \qquad x = 0.8814 \text{ or } -0.4812$$

> **Note**
>
> Using the identity
> $\cosh^2 x - \sinh^2 x \equiv 1$.

> **Note**
>
> By factorising.

Exercise 1

1. Evaluate each of the following, giving your answer
   i in terms of e   **ii** to three significant figures.
   a $\cosh 2$ **b** $\sinh 3$ **c** $\tanh 4$
   d $\text{sech } 2$ **e** $\text{cosech } 4$ **f** $\coth 6$

2. Prove that $\cosh^2 x - \sinh^2 x = 1$.

3. Starting with the definitions of $\sinh x$ and $\cosh x$, prove each of the following identities.
   a $\cosh(A + B) = \cosh A \cosh B + \sinh A \sinh B$
   b $\sinh(A - B) = \sinh A \cosh B - \cosh A \sinh B$
   c $\sinh A + \sinh B \equiv 2 \sinh\left(\dfrac{A+B}{2}\right) \cosh\left(\dfrac{A-B}{2}\right)$

4. Solve each of these equations, giving your answer to three significant figures.
   a $3 \sinh x + 2 \cosh x = 4$ **b** $4 \cosh x - 8 \sinh x + 1 = 0$
   c $\cosh x + 4 \sinh x = 3$

5. Find the values of e^x for which $8 \cosh x - 4 \sinh x = 7$.
   Hence find the values of x giving your answers as natural logarithms.

6. Show that $\dfrac{1}{6 \cosh x + 8 \sinh x} = \dfrac{e^{-x}}{p - e^{-2x}}$ where p is a constant to be found.

24.2 Differentiation of hyperbolic functions

You need to be able to prove the results of differentiating hyperbolic functions.

Differentiating sinh x and cosh x

To differentiate sinh x and cosh x, you use their exponential definitions.

Hence, for sinh x, you have

$$\frac{d}{dx}\sinh x = \frac{d}{dx}\left[\frac{1}{2}(e^x - e^{-x})\right] = \frac{1}{2}(e^x + e^{-x})$$

From the definitions, you know that $\frac{1}{2}(e^x + e^{-x}) = \cosh x$

Therefore, you have

$$\frac{d}{dx}\sinh x = \cosh x$$

If you were then to differentiate cosh x:

$$\frac{d}{dx}\cosh x = \frac{d}{dx}\left[\frac{1}{2}(e^x + e^{-x})\right] = \frac{1}{2}(e^x - e^{-x})$$

From the definition of sinh x, you conclude

$$\frac{d}{dx}\cosh x = \sinh x$$

Differentiating tanh x

To differentiate tanh x, you use $\tanh x = \dfrac{\sinh x}{\cosh x}$

which gives

$$\frac{d}{dx}\tanh x = \frac{d}{dx}\frac{\sinh x}{\cosh x} = \frac{\cosh x \cosh x - \sinh x \sinh x}{\cosh^2 x} \quad \text{(Using the quotient rule.)}$$

$$= \frac{\cosh^2 x - \sinh^2 x}{\cosh^2 x} = \frac{1}{\cosh^2 x} = \text{sech}^2 x$$

Therefore, you have

$$\frac{d}{dx}\tanh x = \text{sech}^2 x$$

Differentiating functions of the form cosh ax

To differentiate functions such as cosh ax, again you use the exponential definitions. Hence, you have

$$\frac{d}{dx}\cosh ax = \frac{d}{dx}\left[\frac{1}{2}(e^{ax} + e^{-ax})\right]$$

$$= \frac{1}{2}(ae^{ax} - ae^{-ax})$$

From the exponential definitions, note that $a\left[\frac{1}{2}(e^{ax} - e^{-ax})\right] = a\sinh ax$

Therefore,

$$\frac{d}{dx}\cosh ax = a \sinh ax$$

$$\frac{d}{dx}\sinh ax = a \cosh ax$$

$$\frac{d}{dx}\tanh ax = a \,\text{sech}^2 ax$$

Differentiating functions of the form cosh²x

To differentiate functions such as $\cosh^2 x$, you express it as $(\cosh x)^2$ and then apply the chain rule (as demonstrated in part b of *Example 4*).

Example 4

Find $\dfrac{\mathrm{d}y}{\mathrm{d}x}$ when

a $y = 3\cosh 3x$ **b** $y = 3\cosh 3x + 5\sinh 4x + 2\cosh^4 7x$

> **Note**
>
> Use your answer from part a in part b then to differentiate $\cosh^4 7x$, express it as $(\cosh 7x)^4$ and apply the chain rule.

a $\dfrac{\mathrm{d}y}{\mathrm{d}x} = 3 \times 3\sinh 3x$

$= 9\sinh 3x$

> **Note**
>
> Using $\dfrac{\mathrm{d}}{\mathrm{d}x}\cosh ax = a\sinh ax$.

b $\dfrac{\mathrm{d}y}{\mathrm{d}x} = 9\sinh 3x + 20\cosh 4x + \dfrac{\mathrm{d}}{\mathrm{d}x}\,2\,(\cosh 7x)^4$

$= 9\sinh 3x + 20\cosh 4x + 2 \times 4 \times 7\sinh 7x\cosh^3 7x$

$= 9\sinh 3x + 20\cosh 4x + 56\sinh 7x\cosh^3 7x$

> **Note**
>
> Using the chain rule
> $\dfrac{\mathrm{d}}{\mathrm{d}x}\big[g(f(x))\big] = \dfrac{\mathrm{d}g}{\mathrm{d}x}(f(x)) \times \dfrac{\mathrm{d}f}{\mathrm{d}x}$

Integration of hyperbolic functions

It is possible to apply the differentiation formulae given above to deduce the following results of integration:

$$\int \cosh ax\,\mathrm{d}x = \frac{1}{a}\sinh ax + c$$

$$\int \sinh ax\,\mathrm{d}x = \frac{1}{a}\cosh ax + c$$

$$\int \tanh ax\,\mathrm{d}x = \int \frac{\sinh ax}{\cosh ax}\mathrm{d}x = \frac{1}{a}\ln\cosh ax + c$$

$$\int \operatorname{sech}^2 ax\,\mathrm{d}x = \frac{1}{a}\tanh ax + c$$

Example 5

Find $\displaystyle\int (2\sinh 4x + 9\operatorname{sech}^2 3x)\,\mathrm{d}x$.

$\displaystyle\int 2\sinh 4x\,\mathrm{d}x + \int 9\operatorname{sech}^2 3x\,\mathrm{d}x$

$= \dfrac{2}{4}\cosh 4x + \dfrac{9}{3}\tanh 3x + c$

$= \dfrac{1}{2}\cosh 4x + 3\tanh 3x + c$

> **Note**
>
> Using
> $\displaystyle\int \sinh ax\,\mathrm{d}x = \frac{1}{a}\cosh ax + c$
> and
> $\displaystyle\int \operatorname{sech}^2 ax\,\mathrm{d}x = \frac{1}{a}\tanh ax + c.$

Exercise 2

1 Differentiate, with respect to x, each of the following.

   a $\cosh 2x$ **b** $\sinh 5x$ **c** $\tanh 3x$

   d $3\cosh^5 3x$ **e** $2\sinh^4 8x$

2 Integrate, with respect to x, each of the following.

   a $\sinh 3x$ **b** $\cosh 4x$

   c $\sinh\left(\dfrac{x}{3}\right)$ **d** $\tanh 4x$

3 Differentiate, with respect to x, each of the following.

    a $\coth x$ **b** $\operatorname{sech} x$ **c** $\ln \tanh 5x$

24.3 Inverse hyperbolic functions

You can define the inverse functions of the hyperbolic functions in a similar way to the inverses of the ordinary trigonometric functions.

Hence, for example, if $y = \sinh^{-1}x$, then $\sinh y = x$. Likewise for $\cosh^{-1}x$, $\tanh^{-1}x$, $\operatorname{cosech}^{-1}x$, $\operatorname{sech}^{-1}x$ and $\coth^{-1}x$.

Sometimes these functions are written as arcsinh x, arccosh x etc.

Sketching inverse hyperbolic functions

You need to be familiar with, and recognise, the graphs of the following inverse hyperbolic functions: $\sinh^{-1}x$, $\cosh^{-1}x$ and $\tanh^{-1}x$.

See Chapter 22 The Calculus of Inverse Trigonometrical Functions if you need a reminder.

The curve of $y = \sinh^{-1}x$ is obtained by reflecting the curve of $y = \sinh x$ in the line $y = x$.

To draw the curve with reasonable accuracy, you need to find the gradient of $y = \sinh x$ at the origin. Accordingly, you differentiate $y = \sinh x$, to obtain

$$\frac{dy}{dx} = \cosh x$$

Thus, at the origin, where $x = 0$, you have

$$\frac{dy}{dx} = \cosh 0 = 1$$

That is, the gradient of $y = \sinh x$ at the origin is 1.

You now proceed as follows:

▶   Draw the line $y = x$ as a dashed line.
▶   Sketch carefully the graph of $y = \sinh x$, remembering that $y = x$ is a tangent to $y = \sinh x$ at the origin.
▶   Reflect this sinh curve in the line $y = x$.

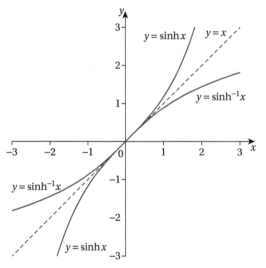

You can sketch the other inverse hyperbolic functions in a similar way; that is, by reflecting the curve of the relevant hyperbolic function in the line $y = x$. In each case, you must find the gradient of the hyperbolic curve at the origin.

Take for example, $y = \tanh x$, which gives

$$\frac{dy}{dx} = \operatorname{sech}^2 x$$

At the origin, where $x = 0$, you have

$$\frac{dy}{dx} = \operatorname{sech}^2 0 = \frac{1}{\cosh^2 0} = 1$$

That is, the gradient of $y = \tanh x$ at the origin is 1.

Also, you know that $y = \tanh x$ has asymptotes $y = 1$ and $y = -1$. Therefore, because $y = \tanh^{-1}x$ is the reflection of $y = \tanh x$ in $y = x$, $y = \tanh^{-1}x$ has asymptotes $x = 1$ and $x = -1$.

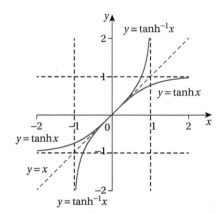

Logarithmic form of inverse hyperbolic functions

The inverse hyperbolic functions $\cosh^{-1}x$, $\sinh^{-1}x$ and $\tanh^{-1}x$ can all be expressed as logarithmic functions; doing so gives an exact answer. You need to know these forms and the proofs of these results.

Expressing $\cosh^{-1}x$ as a logarithmic function

Let $\cosh^{-1}x = y$.

You then have $\qquad x = \cosh y$

$\Rightarrow \quad x = \dfrac{1}{2}(e^y + e^{-y})$ (Using the exponential definition, $\cosh x = \dfrac{1}{2}(e^x + e^{-x})$.)

Multiplying throughout by $2e^y$, you obtain

$\qquad 2xe^y = e^{2y} + 1$

$\Rightarrow \quad e^{2y} - 2xe^y + 1 = 0$

To solve this equation, you treat it as a quadratic in e^y, which gives

$\qquad e^y = \dfrac{2x \pm \sqrt{4x^2 - 4}}{2}$

$\Rightarrow \quad e^y = x \pm \sqrt{x^2 - 1}$

Taking the logarithms of both sides, you obtain

$\qquad y = \ln(x \pm \sqrt{x^2 - 1})$

Since $\cosh x = y$ has two solutions for every $y > 1$, and because $\cosh^{-1}y$ must be a function, it is important to choose which of the two values should be defined to be $\cosh^{-1}y$. By convention, $\cosh^{-1}y$ is defined to be the solution that is not negative. To ensure that $\ln(x \pm \sqrt{(x^2-1)}) > 0$, you need $x \pm \sqrt{(x^2-1)} > 1$, and therefore it is important to choose the *larger* value, $x + \sqrt{(x^2+1)}$ for the value of $\cosh^{-1}y$.

> **The value of $\cosh^{-1}x$ is $\ln(x + \sqrt{x^2 - 1})$.**

The other solution of $\cosh y$ is $y = \ln(x - \sqrt{x^2 - 1})$. This other solution can be written:

$\ln(x - \sqrt{x^2 - 1}) = \ln\left[\dfrac{(x - \sqrt{x^2 - 1})(x + \sqrt{x^2 - 1})}{x + \sqrt{x^2 - 1}}\right]$

$\qquad\qquad\qquad\quad = \ln\left[\dfrac{x^2 - (x^2 - 1)}{x + \sqrt{x^2 - 1}}\right]$

$\qquad\qquad\qquad\quad = \ln\left[\dfrac{1}{x + \sqrt{x^2 - 1}}\right]$

$\qquad\qquad\qquad\quad = -\ln(x + \sqrt{x^2 - 1})$

Hence, you have

$\qquad \ln(x \pm \sqrt{x^2 - 1}) = \pm\ln(x + \sqrt{x^2 - 1})$

which matches the symmetry of the graph of $\cosh x$.

Example 6

Find the value of $\cosh^{-1} 2$ in logarithmic form.

$\cosh^{-1} 2 = \ln(2 + \sqrt{3})$

> **Note**
>
> Using
> $$\cosh^{-1} x = \ln(x + \sqrt{x^2 - 1})$$
> and substituting $x = 2$.

Example 7

Find the exact coordinates of the points where the line $y = 3$ cuts the graph of $y = \cosh x$.

When $y = 3$,

$$x = \cosh^{-1} 3$$

$$\Rightarrow \quad x = \ln(3 + \sqrt{8}) = \ln(3 + 2\sqrt{2})$$

By symmetry, the other value of x is $-\ln(3 + 2\sqrt{2})$.

Therefore, the two points are $(\ln(3 + 2\sqrt{2}), 3)$ and $(-\ln(3 + 2\sqrt{2}), 3)$

Expressing $\sinh^{-1} x$ as a logarithmic function

Let $y = \sinh^{-1} x$. You then have

$$x = \sinh y \quad \Rightarrow \quad x = \frac{1}{2}(e^y - e^{-y}) \text{ (Using the exponential definition,}$$
$$\sinh x = \frac{1}{2}(e^x - e^{-x}).)$$

Multiplying throughout by $2e^y$, you obtain

$$2xe^y = e^{2y} - 1$$

$$\Rightarrow \quad e^{2y} - 2xe^y - 1 = 0$$

Treating this equation as a quadratic in e^y, you have

$$e^y = \frac{2x \pm \sqrt{4x^2 + 4}}{2} \quad \Rightarrow \quad e^y = x \pm \sqrt{x^2 + 1}$$

Taking the logarithms of both sides, you obtain $y = \ln(x \pm \sqrt{x^2 + 1})$.

The value of $\sinh^{-1} x$ can only be $\ln(x + \sqrt{x^2 + 1})$. You cannot have $\sinh^{-1} x = \ln(x - \sqrt{x^2 + 1})$ because $x < \sqrt{x^2 + 1}$, which would give the logarithm of a negative number, which is a complex number.

Hence, you have

> **The logarithmic form of the inverse of sinh is, $\sinh^{-1} x = \ln(x + \sqrt{x^2 + 1})$.**

Example 8

Find the value, in logarithmic form, of $\sinh^{-1} 3$.

> **Note**
>
> Using $\sinh^{-1} x = \ln(x + \sqrt{x^2 + 1})$.

$$\sinh^{-1} 3 = \ln(3 + \sqrt{10})$$

Expressing $\tanh^{-1} x$ as a logarithmic function

Let $y = \tanh^{-1} x$. You then have

$$x = \tanh y = \frac{\sinh y}{\cosh y}$$

$$\Rightarrow \quad x = \frac{\frac{1}{2}(e^y - e^{-y})}{\frac{1}{2}(e^y + e^{-y})}$$

Multiplying the numerator and the denominator by $2e^y$, you obtain

$$x = \frac{e^{2y} - 1}{e^{2y} + 1}$$

$$\Rightarrow \quad e^{2y}x + x = e^{2y} - 1$$

Therefore, you have

$$e^{2y} = \frac{1+x}{1-x} \quad \Rightarrow \quad y = \frac{1}{2}\ln\left(\frac{1+x}{1-x}\right)$$

Hence,

The logarithmic value of $\tanh^{-1} x$ is $\frac{1}{2}\ln\left(\frac{1+x}{1-x}\right)$, for $|x| < 1$.

In summary,

The logarithmic forms of some of the hyperbolic functions are,

▶ $\cosh^{-1} x = \ln(x + \sqrt{x^2 - 1}) \quad x \geq 1$

▶ $\sinh^{-1} x = \ln(x + \sqrt{x^2 + 1})$

▶ $\tanh^{-1} x = \frac{1}{2}\ln\left(\frac{1+x}{1-x}\right) \quad -1 < x < 1$

Example 9

Find the value, in logarithmic form, of $\tanh^{-1}\frac{1}{2}$.

Note

Using $\tanh^{-1} x = \frac{1}{2}\ln\left(\frac{1+x}{1-x}\right)$.

$$\tanh^{-1}\frac{1}{2} = \frac{1}{2}\ln\left(\frac{\frac{3}{2}}{\frac{1}{2}}\right)$$

$$= \frac{1}{2}\ln 3$$

If you are asked to find the value of $\operatorname{cosech}^{-1} x$, $\operatorname{sech}^{-1} x$ or $\coth^{-1} x$ in logarithmic form, use the definition of the relevant hyperbolic function to express the equation in terms of $\sinh x$, $\cosh x$ or $\tanh x$, then express the term as the associated inverse function and then finally apply the appropriate logarithmic value.

Example 10

Find the value, in logarithmic form, of $\operatorname{sech}^{-1}\dfrac{1}{2}$.

$y = \operatorname{sech}^{-1}\dfrac{1}{2}$

$\Rightarrow \operatorname{sech} y = \dfrac{1}{2}$

$\Rightarrow \dfrac{1}{\cosh y} = \dfrac{1}{2}$

$\Rightarrow \cosh y = 2$

$\Rightarrow \qquad y = \cosh^{-1}2$

$\cosh^{-1}2 = \ln(2+\sqrt{3})$

Therefore, $\operatorname{sech}^{-1}\dfrac{1}{2} = \ln(2+\sqrt{3})$.

> **Note**
>
> Using $\operatorname{sech} x = \dfrac{1}{\cosh x}$.

> **Note**
>
> Using that fact that if $y = \cosh^{-1}x$, then $\cosh y = x$.

> **Note**
>
> Using the logarithmic form of $\cosh x$:
>
> $\cosh^{-1}x = \ln(x+\sqrt{x^2-1})$.

Differentiation and integration of inverse hyperbolic functions

You need to be able to prove the results of differentiating inverse hyperbolic functions.

$\sinh^{-1}x$

If $y = \sinh^{-1}x$, then $\sinh y = x$.

Differentiating $\sinh y = x$, you obtain

$\cosh y \dfrac{dy}{dx} = 1$

$\Rightarrow \qquad \dfrac{dy}{dx} = \dfrac{1}{\cosh y}$

You know that $\cosh y = \pm\sqrt{(1+\sinh^2 y)}$. But $\sinh^{-1}x$ is always increasing and therefore you need to ensure that $\dfrac{dy}{dx} > 0$.

Therefore

$\Rightarrow \qquad \dfrac{dy}{dx} = \dfrac{1}{\cosh y} = \dfrac{1}{\sqrt{1+\sinh^2 y}} = \dfrac{1}{\sqrt{1+x^2}}$

Therefore,

$\dfrac{d}{dx}\sinh^{-1}x = \dfrac{1}{\sqrt{1+x^2}}$ and it follows that, $\displaystyle\int \dfrac{1}{\sqrt{1+x^2}}\,dx = \sinh^{-1}x.$

If $y = \sinh^{-1}\left(\dfrac{x}{a}\right)$, then $\sinh y = \dfrac{x}{a}$.

Differentiating $\sinh y = \dfrac{x}{a}$, you obtain

$\cosh y \dfrac{dy}{dx} = \dfrac{1}{a}$

$\Rightarrow \qquad \dfrac{dy}{dx} = \dfrac{1}{a\cosh y} = \dfrac{1}{a\sqrt{1+\sinh^2 y}}$

which gives

$$\frac{dy}{dx} = \frac{1}{a\sqrt{1+\left(\frac{x}{a}\right)^2}} = \frac{1}{\sqrt{a^2 + x^2}}$$

Therefore,

$$\frac{d}{dx}\sinh^{-1}\left(\frac{x}{a}\right) = \frac{1}{\sqrt{a^2 + x^2}} \text{ and it follows that, } \int \frac{1}{\sqrt{a^2 + x^2}}\,dx = \sinh^{-1}\left(\frac{x}{a}\right).$$

$\cosh^{-1} x$

If $y = \cosh^{-1} x$, then $\cosh y = x$.

Differentiating $\cosh y = x$, you obtain

$$\sinh y \frac{dy}{dx} = 1$$

$$\Rightarrow \quad \frac{dy}{dx} = \frac{1}{\sinh y}$$

Since $\cosh^{-1} x$ is always increasing, again you choose the positive square root. Therefore

$$\Rightarrow \quad \frac{dy}{dx} = \frac{1}{\sinh y} = \frac{1}{\sqrt{\cosh^2 y - 1}} = \frac{1}{\sqrt{x^2 - 1}}$$

which gives

$$\frac{d}{dx}\cosh^{-1} x = \frac{1}{\sqrt{x^2 - 1}}$$

Therefore,

$$\frac{d}{dx}\cosh^{-1} x = \frac{1}{\sqrt{x^2 - 1}} \text{ and it follows that, } \int \frac{1}{\sqrt{x^2 - 1}}\,dx = \cosh^{-1} x.$$

If $y = \cosh^{-1}\left(\frac{x}{a}\right)$, then $\cosh y = \frac{x}{a}$.

Differentiating $\cosh y = \frac{x}{a}$, you obtain

$$\sinh y \frac{dy}{dx} = \frac{1}{a}$$

$$\Rightarrow \quad \frac{dy}{dx} = \frac{1}{a \sinh y} = \frac{1}{a\sqrt{\cosh^2 y - 1}}$$

which gives

$$\frac{dy}{dx} = \frac{1}{a\sqrt{\left(\frac{x}{a}\right)^2 - 1}} = \frac{1}{\sqrt{x^2 - a^2}}$$

Therefore,

$$\frac{d}{dx}\cosh^{-1}\left(\frac{x}{a}\right) = \frac{1}{\sqrt{x^2 - a^2}} \text{ and it follows that, } \int \frac{1}{\sqrt{x^2 - a^2}}\,dx = \cosh^{-1}\left(\frac{x}{a}\right).$$

$\tanh^{-1} x$

If $y = \tanh^{-1} x$, then $\tanh y = x$.

Differentiating, $\text{sech}^2 y \frac{dy}{dx} = 1$

$$\frac{dy}{dx} = \frac{1}{\text{sech}^2 y} = \frac{1}{(1 - \tanh^2 y)} = \frac{1}{(1 - x^2)}$$

Therefore,

$$\frac{d}{dx}\tanh^{-1} x = \frac{1}{1 - x^2} \text{ and it follows that, } \int \frac{1}{1 - x^2}\,dx = \tanh^{-1} x.$$

Formulae booklet

In the *Formulae and Statistical Tables* booklet you will see that
$$\int \frac{1}{\sqrt{a^2 + x^2}}\,dx = \sinh^{-1}\left(\frac{x}{a}\right) \text{ or}$$
$$\ln\left(x + \sqrt{x^2 + a^2}\right).$$

Formulae booklet

In the *Formulae and Statistical Tables* booklet you will see that
$$\int \frac{1}{\sqrt{x^2 - a^2}}\,dx = \cosh^{-1}\left(\frac{x}{a}\right)$$
$$\text{or } \ln\left(x + \sqrt{x^2 - a^2}\right), (x > a).$$

If $y = \tanh^{-1}\left(\dfrac{x}{a}\right)$, then $\tanh y = \dfrac{x}{a}$.

Differentiating $\tanh y = \dfrac{x}{a}$, you obtain

$$\text{sech}^2\, y\, \frac{dy}{dx} = \frac{1}{a}$$

$$\Rightarrow \quad \frac{dy}{dx} = \frac{1}{a\,\text{sech}^2\, y} = \frac{1}{a(1-\tanh^2 y)}$$

which gives

$$\frac{dy}{dx} = \frac{1}{a\left[1-\left(\dfrac{x}{a}\right)^2\right]} = \frac{a}{a^2-x^2}$$

Therefore,

$$\frac{d}{dx}\tanh^{-1}\left(\frac{x}{a}\right) = \frac{a}{a^2-x^2} \text{ and it follows that, } \int \frac{1}{a^2-x^2}\,dx = \frac{1}{a}\tanh^{-1}\left(\frac{x}{a}\right)$$

$$(|x| < a)$$

Note that you can integrate $\dfrac{1}{a^2-x^2}$ using partial fractions:

$$\int \frac{1}{a^2-x^2}\,dx = \frac{1}{2a}\int\left(\frac{1}{a+x} + \frac{1}{a-x}\right)dx = \frac{1}{2a}\ln\left(\frac{a+x}{a-x}\right) + c, \text{ which gives the other}$$

form shown in the *Formulae and Statistical Tables* booklet.

This result is the logarithmic form of $\tanh^{-1}\left(\dfrac{x}{a}\right)$. Hence, it is unusual to use a function in $\tanh^{-1} x$ in differentiation or integration.

Example 11

Question

Differentiate

a **i** $\sinh^{-1}\left(\dfrac{x}{3}\right)$ **ii** $\sinh^{-1} 4x$ **b** $\cosh^{-1}\left(\dfrac{x}{5}\right)$

Answer

a **i** $\dfrac{d}{dx}\sinh^{-1}\left(\dfrac{x}{3}\right) = \dfrac{1}{\sqrt{9+x^2}}$

ii $\dfrac{d}{dx}\sinh^{-1} 4x = \dfrac{d}{dx}\sinh^{-1}\left(\dfrac{x}{\frac{1}{4}}\right) = \dfrac{1}{\sqrt{\frac{1}{16}+x^2}}$

$\dfrac{d}{dx}\sinh^{-1} 4x = \dfrac{4}{\sqrt{1+16x^2}}$

b $\dfrac{d}{dx}\cosh^{-1}\left(\dfrac{x}{5}\right) = \dfrac{1}{\sqrt{x^2-25}}$

Before using the appropriate integration formulae, make sure that the coefficient of x^2 is 1 as you did with inverse trigonometric functions.

Example 12

Question

Find **a** $\displaystyle\int_0^2 \frac{1}{\sqrt{4+x^2}}\,dx$ **b** $\displaystyle\int_0^1 \frac{1}{\sqrt{4+3x^2}}\,dx$

Answer

a $\displaystyle\int_0^2 \frac{1}{\sqrt{4+x^2}}\,dx = \left[\sinh^{-1}\frac{x}{2}\right]_0^2$

$= \sinh^{-1}1 - \sinh^{-1}0$

$= \sinh^{-1}1 = \ln\left(1+\sqrt{2}\right)$

> **Note**
> The denominator of the term is '$\sqrt{4+x^2}$' so use
> $\displaystyle\int \frac{1}{\sqrt{a^2+x^2}}\,dx = \sinh^{-1}\left(\frac{x}{a}\right)$, with $a=2$.

b $\displaystyle\int_0^1 \frac{1}{\sqrt{4+3x^2}}\,dx = \frac{1}{\sqrt{3}}\int_0^1 \frac{1}{\sqrt{\frac{4}{3}+x^2}}\,dx$

$= \frac{1}{\sqrt{3}}\int_0^1 \frac{1}{\sqrt{\left(\frac{2}{\sqrt{3}}\right)^2+x^2}}\,dx$

> **Note**
> Before integrating, you must reduce the coefficient of x^2 to unity.

> **Note**
> The denominator is in the form $\sqrt{a^2+x^2}$ so use
> $\displaystyle\int \frac{1}{\sqrt{a^2+x^2}}\,dx = \sinh^{-1}\left(\frac{x}{a}\right)$, with $a=\frac{2}{\sqrt{3}}$.

$\Rightarrow \qquad = \frac{1}{\sqrt{3}}\left[\sinh^{-1}\left(\frac{\sqrt{3}x}{2}\right)\right]_0^1$

$= \frac{1}{\sqrt{3}}\left[\sinh^{-1}\left(\frac{\sqrt{3}}{2}\right) - \sinh^{-1}0\right]$

$= \frac{1}{\sqrt{3}}\left[\ln\left(\frac{\sqrt{3}}{2}+\sqrt{\frac{3}{4}+1}\right)\right]$

$= \frac{1}{\sqrt{3}}\ln\left(\frac{\sqrt{3}+\sqrt{7}}{2}\right)$

Therefore,

$\displaystyle\int_0^1 \frac{1}{\sqrt{4+3x^2}}\,dx = \frac{1}{\sqrt{3}}\ln\left(\frac{\sqrt{3}+\sqrt{7}}{2}\right)$

Example 13

Question

Find $\displaystyle\int_3^6 \frac{1}{\sqrt{x^2-9}}\,dx$.

Answer

Using the appropriate integral formula,

$\displaystyle\int_3^6 \frac{1}{\sqrt{x^2-9}}\,dx = \left[\cosh^{-1}\left(\frac{x}{3}\right)\right]_3^6$

$= \cosh^{-1}2 - \cosh^{-1}1 = \ln(2+\sqrt{3}) - 0$

Therefore,

$\displaystyle\int_3^6 \frac{1}{\sqrt{x^2-9}}\,dx = \ln(2+\sqrt{3})$

> **Note**
> The denominator is in the form $\sqrt{x^2-a^2}$ so use
> $\displaystyle\int \frac{1}{\sqrt{x^2-a^2}}\,dx = \cosh^{-1}\left(\frac{x}{a}\right)$, with $a=3$.

You saw in Chapter *22 The Calculus of Inverse of Trigonometrical Functions* that the standard integral of inverse trigonometric functions can be used to integrate functions in the form $\dfrac{1}{ax^2+bx+c}$. The standard integrals of inverse hyperbolic functions can also be used to integrate functions of this form. If the roots are not real, you can complete the square and use the appropriate standard integral for an inverse hyperbolic function.

Example 14

Find $\displaystyle\int \dfrac{1}{\sqrt{4x^2-8x-16}}\,dx.$

Note

Start by completing the square to factorise the quadratic in the denominator, and reduce the coefficient of x^2 to unity.

$$\sqrt{4x^2-8x-16}=\sqrt{4}\sqrt{x^2-2x-4}$$
$$=2\sqrt{(x-1)^2-5}$$

Therefore,

$$\int \dfrac{1}{\sqrt{4x^2-8x-16}}\,dx=\dfrac{1}{2}\int \dfrac{dx}{\sqrt{(x-1)^2-5}}$$
$$=\dfrac{1}{2}\cosh^{-1}\left(\dfrac{x-1}{\sqrt{5}}\right)+c$$

Therefore,

$$\int \dfrac{1}{\sqrt{4x^2-8x-16}}\,dx=\dfrac{1}{2}\ln\left(\sqrt{(x-1)^2-5}+x-1\right)-\dfrac{1}{2}\ln\sqrt{5}+c$$
$$=\dfrac{1}{2}\ln\left((x-1)\sqrt{5}+\sqrt{(x-1)^{\frac{2}{5-1}}}\right)+c$$
$$=\dfrac{1}{2}\ln\left(\sqrt{x^2-2x-4}+x-1\right)+c'$$

Exercise 3

1 Differentiate each of the following with respect to x.

  a $\sinh^{-1}5x$ **b** $\sinh^{-1}\sqrt{2}x$ **c** $\cosh^{-1}\dfrac{3}{4}x$

  d $\sinh^{-1}x^2$ **e** $\text{sech}^{-1}x$ **f** $\coth^{-1}x$

2 Find each of the following integrals.

  a $\displaystyle\int \dfrac{1}{\sqrt{x^2-4}}$ **b** $\displaystyle\int \dfrac{1}{\sqrt{4x^2-25}}$ **c** $\displaystyle\int \dfrac{1}{\sqrt{9+x^2}}$ **d** $\displaystyle\int \dfrac{1}{\sqrt{25+16x^2}}$

3 Evaluate each of the following definite integrals, giving the exact value of your answer.

  a $\displaystyle\int_0^2 \dfrac{1}{\sqrt{4+x^2}}\,dx$ **b** $\displaystyle\int_4^8 \dfrac{1}{\sqrt{x^2-16}}\,dx$ **c** $\displaystyle\int_0^2 \dfrac{1}{\sqrt{4+3x^2}}\,dx$

4 Evaluate each of the following integrals, giving your answer in terms of logarithms.

  a $\displaystyle\int_1^2 \dfrac{1}{\sqrt{25x^2-4}}\,dx$ **b** $\displaystyle\int_1^2 \dfrac{1}{\sqrt{4+9x^2}}\,dx$ **c** $\displaystyle\int_3^4 \dfrac{1}{\sqrt{(x-1)^2-3}}\,dx$

  d $\displaystyle\int_0^1 \dfrac{1}{\sqrt{4(x+1)^2+5}}\,dx$ **e** $\displaystyle\int_0^2 \dfrac{1}{\sqrt{4+8x+x^2}}\,dx$ **f** $\displaystyle\int_0^1 \dfrac{1}{\sqrt{16x^2+20x+35}}\,dx$

Summary

▶ There are six hyperbolic functions, with definitions as follows:

$$\sinh x = \frac{1}{2}(e^x - e^{-x}) \qquad \cosh x = \frac{1}{2}(e^x + e^{-x}) \qquad \tanh x = \frac{\sinh x}{\cosh x} = \frac{e^x - e^{-x}}{e^x + e^{-x}}$$

$$\operatorname{cosech} x = \frac{1}{\sinh x} \qquad \operatorname{sech} x = \frac{1}{\cosh x} \qquad \coth x = \frac{1}{\tanh x}$$

▶ You obtain the graph of $y = \cosh x$ by finding the mean values of a few corresponding pairs of values of $y = e^x$ and $y = e^{-x}$, and then plotting these mean values.

▶ To produce the graph of $y = \sinh x$, you find half the difference between a few corresponding pairs of values of $y = e^x$ and $y = e^{-x}$, and then plot these values.

▶ The tanh curve is obtained by dividing the values of $\sinh x$ by the values of $\cosh x$; the graph of $y = \tanh x$ lies between the asymptotes $y = 1$ and $y = -1$.

▶ There are a number of standard hyperbolic identities that you should know and be able to prove.

  • $\cosh^2 x - \sinh^2 x = 1$

  • $\coth^2 x - 1 = \operatorname{cosech}^2 x$

  • $1 - \tanh^2 x = \operatorname{sech}^2 x$

  • $\cosh(A + B) = \cosh A \cosh B + \sinh A \sinh B$

  • $\sinh(A - B) = \sinh A \cosh B - \cosh A \sinh B$

  • $\sinh A + \sinh B \equiv 2\sinh\left(\frac{A+B}{2}\right)\cosh\left(\frac{A-B}{2}\right)$

▶ You can solve equations such as $a\sinh x + b\cosh x = c$ using the definition of each function and then solving a quadratic equation in e^x. Or you can use the standard methods used to solve trigonometric equations.

▶ The derivatives for the hyperbolic functions are as follows; you need to be able to prove these results:

  • $\frac{d}{dx}\sinh x = \cosh x; \quad \frac{d}{dx}\sinh ax = a\cosh ax$

  • $\frac{d}{dx}\cosh x = \sinh x; \quad \frac{d}{dx}\cosh ax = a\sinh ax$

  • $\frac{d}{dx}\tanh x = \operatorname{sech}^2 x; \quad \frac{d}{dx}\tanh ax = a\operatorname{sech}^2 ax$

  • To differentiate functions such as $\cosh^2 x$, you express it as $(\cosh x)^2$ and then apply the chain rule.

▶ The results of integrating the hyperbolic functions are as follows:

  • $\int \cosh ax\, dx = \frac{1}{a}\sinh ax + c$

  • $\int \sinh ax\, dx = \frac{1}{a}\cosh ax + c$

  • $\int \tanh x\ ax\, dx = \int \frac{\sinh ax}{\cosh ax}\, dx = \frac{1}{a}\ln\cosh ax + c$

  • $\int \operatorname{sech}^2 ax\, dx = \frac{1}{a}\tanh ax + c$

- The inverse of sinh is expressed as $\sinh^{-1}$, likewise for the other hyperbolic functions. If $y = \sinh^{-1}x$, then $\sinh y = x$. Likewise for $\cosh^{-1}x$, $\tanh^{-1}x$, $\operatorname{cosech}^{-1}x$, $\operatorname{sech}^{-1}x$ and $\coth^{-1}x$.

- You need to be familiar with, and recognise, the graphs of $\sinh^{-1}x$, $\cosh^{-1}x$ and $\tanh^{-1}x$. You can sketch these inverse hyperbolic functions by reflecting the curve of the relevant hyperbolic function in the line $y = x$. In each case, you must find the gradient of the hyperbolic curve at the origin.

- It is possible to express inverse functions in logarithmic form using the following formulae

  - $\sinh^{-1}x = \ln(x + \sqrt{x^2 + 1})$

  - $\cosh^{-1}x = \ln(x \pm \sqrt{x^2 - 1})\ x \geq 1$ (the positive sign gives the **value**)

  - $\tanh^{-1}x = \dfrac{1}{2}\ln\left(\dfrac{1+x}{1-x}\right)$, where $|x| < 1$

- You can use the definitions of inverse hyperbolic functions to prove differentiation formulae such as

  - $\dfrac{d}{dx}\sinh^{-1}x = \dfrac{1}{\sqrt{1+x^2}}$ • $\dfrac{d}{dx}\sinh^{-1}\left(\dfrac{x}{a}\right) = \dfrac{1}{\sqrt{a^2+x^2}}$

  - $\dfrac{d}{dx}\cosh^{-1}x = \dfrac{1}{\sqrt{x^2-1}}$ • $\dfrac{d}{dx}\cosh^{-1}\left(\dfrac{x}{a}\right) = \dfrac{1}{\sqrt{x^2-a^2}}$

  - $\dfrac{d}{dx}\tanh^{-1}x = \dfrac{1}{1-x^2}$ • $\dfrac{d}{dx}\tanh^{-1}\left(\dfrac{x}{a}\right) = \dfrac{a}{a^2-x^2}$

- The results of integrating inverse hyperbolic functions are as follows:

  - $\displaystyle\int \dfrac{1}{\sqrt{1+x^2}}\,dx = \sinh^{-1}x$

  - $\displaystyle\int \dfrac{1}{\sqrt{a^2+x^2}}\,dx = \sinh^{-1}\left(\dfrac{x}{a}\right) \text{ or } \ln\left(x + \sqrt{x^2+a^2}\right)$

  - $\displaystyle\int \dfrac{1}{\sqrt{x^2-1}}\,dx = \cosh^{-1}x$

  - $\displaystyle\int \dfrac{1}{\sqrt{x^2-a^2}}\,dx = \cosh^{-1}\left(\dfrac{x}{a}\right) \text{ or } \ln(x + \sqrt{x^2-a^2}), (x > a)$

  - $\displaystyle\int \dfrac{1}{1-x^2}\,dx = \tanh^{-1}x$

  - $\displaystyle\int \dfrac{1}{a^2-x^2}\,dx = \dfrac{1}{2a}\ln\left|\dfrac{a+x}{a-x}\right| = \dfrac{1}{a}\tanh^{-1}\left(\dfrac{x}{a}\right)\ (|x|<a)$

Review exercises

1. Solve $3\operatorname{sech}x - 2 = 5\tanh x$, correct to two significant figures.

2. Find the exact solutions of $5\cosh x + 4\sinh x = 9$.

3. Differentiate

   a $4\cosh^4 7x$ b $2\sinh^3 6x$

4 a Find the integral $\displaystyle\int_{0}^{1.5} \frac{dx}{4x^2+8x+29}$.

b Find $\displaystyle\int \frac{dx}{\sqrt{4x^2+8x+2}}$.

5 Find $\displaystyle\int \frac{dx}{\sqrt{4x^2+8x+53}}$.

Practice examination questions

1 a Show that $\dfrac{1}{5\cosh x - 3\sinh x} = \dfrac{e^x}{m+e^{2x}}$, where m is an integer. **(3 marks)**

b Use the substitution $u = e^x$ to show that

$$\int_{0}^{\ln 2} \frac{1}{5\cosh x - 3\sinh x}\, dx = \frac{\pi}{8} - \frac{1}{2}\tan^{-1}\frac{1}{2} \qquad \text{(5 marks)}$$

AQA MFP2 June 2013

2 a Show that $12\cosh x - 4\sinh x = 4e^x + 8e^{-x}$. **(2 marks)**

b Solve the equation $12\cosh x - 4\sinh x = 33$ giving your answers in the form $k\ln 2$. **(5 marks)**

AQA MFP2 January 2013

3 a Sketch the graph of $y = \tanh x$. **(2 marks)**

b Given that $u = \tanh x$, use the definitions of $\sinh x$ and $\cosh x$ in terms of e^x and e^{-x} to show that $x = \dfrac{1}{2}\ln\left(\dfrac{1+u}{1-u}\right)$. **(6 marks)**

c i Show that the equation $3\operatorname{sech}^2 x + 7\tanh x = 5$ can be written as $3\tanh^2 x - 7\tanh x + 2 = 0$. **(2 marks)**

ii Show that the equation $\tanh^2 x - 7\tanh x + 2 = 0$ has only one solution for x.

Find this solution in the form $\dfrac{1}{2}\ln a$, where a is an integer. **(5 marks)**

AQA MFP2 June 2009

4 a Using the identities

$$\cosh^2 t - \sinh^2 t = 1, \quad \tanh t = \frac{\sinh t}{\cosh t} \quad \text{and} \quad \operatorname{sech} t = \frac{1}{\cosh t}$$

show that:

i $\tanh^2 t + \operatorname{sech}^2 t = 1$ **(2 marks)**

ii $\dfrac{d}{dt}(\tanh t) = \operatorname{sech}^2 t$ **(3 marks)**

iii $\dfrac{d}{dt}(\operatorname{sech} t) = -\operatorname{sech} t \tanh t$ **(3 marks)**

b A curve C is given parametrically by

$$x = \operatorname{sech} t, \quad y = 4 - \tanh t$$

i Show that the arc length, s, of C between the points where $t = 0$ and $t = \dfrac{1}{2}\ln 3$ is given by

$$s = \int_{0}^{\frac{1}{2}\ln 3} \operatorname{sech} t\, dt \qquad \text{(4 marks)}$$

ii Using the substitution $u = e^t$, find the exact value of s. **(6 marks)**

AQA MFP2 June 2010

25 Differential Equations of First and Second Order

Introduction

Whenever you solve an indefinite integral, you are in fact solving a differential equation. Differential equations are vital to science and engineering. When scientists model radioactive decay, they are using a differential equation. When you look at a cup of coffee cooling, Newton's law of cooling leads us to a differential equation. In mechanics, when modelling the movement of a damped spring, you are forced to consider an underlying differential equation. More complex (partial) differential equations can be used in applications such as the air flow around a wing, in heat flow through an iron bar and in multiple other fields.

Recap

You will need to remember...
- Integration skills.
- How to solve separable first-order differential equations.

Objectives

By the end of this chapter, you should know how to:

- Recognise and use general and particular solutions of first order linear differential equations of the form $\frac{dy}{dx} + Py = Q$, provided that P and Q are functions only of x.

- Recognise and use general and particular solutions of second order equations of the form $a\frac{d^2y}{dx^2} + b\frac{dy}{dx} + c = 0$, for real a, b and c.

- Find solutions to equations of the form $a\frac{d^2y}{dx^2} + b\frac{dy}{dx} + c = f(x)$, where $f(x) = e^{kx}$, $\cos kx$, $\sin kx$, or a polynomial.

25.1 First-order linear equations

First-order linear differential equations are of the form $\frac{dy}{dx} + Py = Q$, where P and Q are functions of x.

Solving using an integrating factor

You can solve such an equation by first multiplying both sides by the **integrating factor** $e^{\int P dx}$.

Multiplying $\frac{dy}{dx} + Py = Q$ by $e^{\int P dx}$, you get

$$e^{\int P dx}\frac{dy}{dx} + Pe^{\int P dx}y = Qe^{\int P dx}$$

Since the left-hand side is the differential of $ye^{\int P dx}$, you therefore have

$$\frac{d}{dx}\left(ye^{\int P dx}\right) = Qe^{\int P dx}$$

which gives

$$ye^{\int P dx} = \int Qe^{\int P dx}dx$$

The right-hand side is often integrated by parts.

Example 1

Question

If $\dfrac{dy}{dx}+3y=x$, find y.

Answer

The integrating factor is $e^{\int 3dx}=e^{3x}$.

Multiplying both sides by e^{3x},

$$e^{3x}\dfrac{dy}{dx}+e^{3x}3y=xe^{3x}$$

$$\Rightarrow\quad \dfrac{d}{dx}(ye^{3x})=xe^{3x}$$

Integrating by parts,

$$ye^{3x}=\int xe^{3x}\,dx$$

$$=\dfrac{1}{3}e^{3x}\times x-\int\dfrac{1}{3}e^{3x}\,dx$$

Therefore

$$ye^{3x}=\dfrac{1}{3}xe^{3x}-\dfrac{1}{9}e^{3x}+c$$

Multiplying both sides by e^{-3x},

$$y=\dfrac{1}{3}x-\dfrac{1}{9}+ce^{-3x}$$

Note

You must also multiply c by e^{-3x}

Note

The constant term, c, has now become a function of x.

Example 2

Question

Solve the differential equation $x\dfrac{dy}{dx}-2y=x^4$.

Answer

$$\dfrac{dy}{dx}-\dfrac{2y}{x}=x^3$$

The integrating factor is

$$e^{\int-\left(\frac{2}{x}\right)dx}=e^{-2\ln x}=e^{\ln x^{-2}}$$

Since $e^{\ln u}=u$, you have $e^{\ln x^{-2}}=\dfrac{1}{x^2}$.

$$\dfrac{1}{x^2}\dfrac{dy}{dx}-\dfrac{2}{x^3}y=x$$

Simplifying,

$$\dfrac{d}{dx}\left(\dfrac{1}{x^2}y\right)=x$$

$$\Rightarrow\quad \dfrac{1}{x^2}y=\int x\,dx$$

$$\Rightarrow\quad \dfrac{1}{x^2}y=\dfrac{x^2}{2}+c$$

Multiplying both sides by x^2, you obtain the **general solution**.

$$y=\dfrac{1}{2}x^4+cx^2$$

Tip

Divide both sides by x to make the first term $\dfrac{dy}{dx}$.

Note

Multiply the differential equation by the integrating factor, $\dfrac{1}{x^2}$.

Note

To obtain a **particular solution,** you need to be given a specific point which lies on the curve. Hence, you can find the value of c. This extra fact is called a **boundary condition.** *Example* 3 illustrates such a situation.

Example 3

Solve the differential equation $\dfrac{dy}{dx} + \dfrac{1}{x}y = x^2$, given that $y = 3$ when $x = 2$.

The integrating factor is $e^{\int \left(\frac{1}{x}\right)dx} = e^{\ln x} = x$.

Multiplying the differential equation by the integrating factor, x, you have

$$x\dfrac{dy}{dx} + y = x^3$$

which you express as

$$\dfrac{d}{dx}(xy) = x^3$$

$$\Rightarrow \qquad xy = \dfrac{1}{4}x^4 + c$$

When $x = 2$, $y = 3$, which gives

$$6 = 4 + c \quad \Rightarrow \quad c = 2$$

Therefore, the solution is

$$xy = \dfrac{1}{4}x^4 + 2 \quad \text{or} \quad y = \dfrac{1}{4}x^3 + \dfrac{2}{x}$$

Exercise 1

1 Simplify each of the following.

    a $e^{\ln x^2}$
        b $e^{\frac{1}{2}\ln(x^2+1)}$
        c $e^{-3\ln x}$

    d $e^{\int \tan x\, dx}$
      e $e^{\int \frac{x}{(x^2-1)}dx}$
      f $e^{3x\ln 2}$

In each of questions **2** to **7**, find the general solution.

2 $\dfrac{dy}{dx} + 3y = x$
          3 $\dfrac{dy}{dx} - 5y = e^{2x}$

4 $x\dfrac{dy}{dx} + y = x^2$
        5 $x\dfrac{dy}{dx} - 2y = x^3$

6 $\dfrac{dy}{dx} - \dfrac{4y}{x-1} = 5(x-1)^3$
   7 $\tan x\dfrac{dy}{dx} + y = e^{2x}\tan x$

25.2 Second-order differential equations

An equation is termed **second order** when it contains the second derivative, $\dfrac{d^2y}{dx^2}$, usually along with terms in $\dfrac{dy}{dx}$, y and x.

Solving using an auxiliary equation

This approach is used to solve the equations in the form $a\dfrac{d^2y}{dx^2} + b\dfrac{dy}{dx} + cy = 0$.
This method relies on you making the substitution $y = Ae^{nx}$.

Hence, you have

$$\dfrac{dy}{dx} = nAe^{nx} \quad \text{and} \quad \dfrac{d^2y}{dx^2} = n^2Ae^{nx}$$

which give

$$an^2Ae^{nx} + bnAe^{nx} + cAe^{nx} = 0$$

That is,

$$an^2 + bn + c = 0$$

This quadratic equation is called the **auxiliary equation**.

The solution of a second-order differential equation using this approach depends on the type of solution that satisfies its auxiliary equation. There are three types of solution of a quadratic equation:

1   Two real and different roots
2   Two real and equal roots
3   Two complex roots

Type 1 solution – two real and different roots

The auxiliary equation has two **real, different roots**, n_1 and n_2. So, the solution of $a\dfrac{d^2y}{dx^2} + b\dfrac{dy}{dx} + cy = 0$ is

$$y = Ae^{n_1 x} + Be^{n_2 x}$$

where A and B are arbitrary constants.

To verify that this is the full solution, you need to confirm that the following two conditions are true:

▶   There are two arbitrary constants, as it is a second-order differential equation.
▶   The solution does satisfy the equation $a\dfrac{d^2y}{dx^2} + b\dfrac{dy}{dx} + cy = 0$.

Notice that there are indeed the two required arbitrary constants.

To prove that the solution, $y = Ae^{n_1 x} + Be^{n_2 x}$, satisfies the differential equation, you substitute it and its derivatives in the LHS of

$$a\frac{d^2y}{dx^2} + b\frac{dy}{dx} + cy = 0$$

Which gives

$$a\frac{d^2y}{dx^2} + b\frac{dy}{dx} + cy = a(n_1^2 Ae^{n_1 x} + n_2^2 Be^{n_2 x}) + b(n_1 Ae^{n_1 x} + n_2 Be^{n_2 x}) + c(Ae^{n_1 x} + Be^{n_2 x})$$

$$= Ae^{n_1 x}(an_1^2 + bn_1 + c) + Be^{n_2 x}(an_2^2 + bn_2 + c)$$

$$= 0$$

since n_1 and n_2 are roots of the equation $an^2 + bn + c = 0$.

To find the values of A and B, you need two **boundary conditions**. Usually, these are either

▶   the values of y at two different values of x, or
▶   the value of y and that of $\dfrac{dy}{dx}$ for one value of x.

Example 4

Find y when $2\dfrac{d^2y}{dx^2} - \dfrac{dy}{dx} - 3y = 0$, given that $x = 0$ when $y = 2$ and y is finite as x tends to infinity.

Substituting $y = Ae^{nx}$ and its derivatives in $2\dfrac{d^2y}{dx^2} - \dfrac{dy}{dx} - 3y = 0$ you get

$$2n^2 - n - 3 = 0$$

$$\Rightarrow \quad (2n-3)(n+1) = 0$$

$$\Rightarrow \qquad\qquad n = \frac{3}{2} \text{ and } -1$$

Therefore, you have

$$y = Ae^{\frac{3}{2}x} + Be^{-x}$$

When $x = 0$, $y = 2$, which gives

$$2 = A + B$$

You know that as x tends to infinity, y is finite. Therefore, $A = 0$ because the limit of $e^{\frac{3}{2}x}$ as x tends to infinity is not finite.

Hence, $B = 2$, which gives $y = 2e^{-x}$.

Type 2 solution – two real and equal roots

The auxiliary equation has two **real, equal roots,** n. In this case, you cannot, as in Type 1, use just $y = Ae^{nx} + Be^{nx}$, since this simplifies to $y = (A + B)e^{nx}$ or $y = Ce^{nx}$, which has only **one** arbitrary constant. The solution is, therefore,

$$y = (A + Bx)e^{nx}$$

To prove this is the solution, you must show that it satisfies the equation

$$a\frac{d^2y}{dx^2} + b\frac{dy}{dx} + cy = 0$$

Differentiating $y = (A + Bx)e^{nx}$ twice, you get

$$\frac{dy}{dx} = Be^{nx} + ne^{nx}(A + Bx)$$

$$\frac{d^2y}{dx^2} = Bne^{nx} + n^2e^{nx}(A + Bx) + ne^{nx}B$$

$$= n^2(A + Bx)e^{nx} + 2nBe^{nx}$$

Substituting these in the LHS of $a\dfrac{d^2y}{dx^2} + b\dfrac{dy}{dx} + cy = 0$, you have

$$a\frac{d^2y}{dx^2} + b\frac{dy}{dx} + cy = a\left[n^2(A + Bx)e^{nx} + 2nBe^{nx}\right] + b\left[Be^{nx} + ne^{nx}(A + Bx)\right] + c(A + Bx)e^{nx}$$

$$= (A + Bx)e^{nx}(an^2 + bn + c) + (2na + b)Be^{nx}$$

Since n is a root of $an^2 + bn + c = 0$, the first term is zero.

Consider now the quadratic formula, $n = \dfrac{-b \pm \sqrt{b^2 - 4ac}}{2a}$. When its roots are coincident, $b^2 - 4ac = 0$. Therefore, you have

$$n = -\frac{b}{2a} \qquad \Rightarrow \qquad 2na + b = 0$$

So the second term is also zero.

Hence, $a\dfrac{d^2y}{dx^2}+b\dfrac{dy}{dx}+cy$ does equal zero, and $y=(A+Bx)e^{nx}$ is indeed the required solution.

Example 5

Solve $\dfrac{d^2y}{dx^2}+6\dfrac{dy}{dx}+9y=0$.

Substituting $y=Ae^{nx}$ and its derivatives in $\dfrac{d^2y}{dx^2}+6\dfrac{dy}{dx}+9y=0$, you get

$$n^2+6n+9=0$$

$$\Rightarrow \quad (n+3)(n+3)=0$$

$$\Rightarrow \qquad\qquad n=-3$$

Therefore, the general solution is

$$y=(A+Bx)e^{-3x}$$

Type 3 solution – two complex roots

The auxiliary equation has two **complex roots**, $n_1\pm in_2$.

Therefore, the solution of $a\dfrac{d^2y}{dx^2}+b\dfrac{dy}{dx}+cy=0$ is

$$y=Ae^{(n_1+in_2)x}+Be^{(n_1-in_2)x}$$

$$=e^{n_1x}(Ae^{in_2x}+Be^{-in_2x})$$

$$=e^{n_1x}\left[A\cos n_2x+iA\sin n_2x+B\cos(-n_2x)+iB\sin(-n_2x)\right]$$

$$=e^{n_1x}(A\cos n_2x+iA\sin n_2x+B\cos n_2x-iB\sin n_2x)$$

$$=e^{n_1x}\left[(A+B)\cos n_2x+i(A-B)\sin n_2x\right]$$

Since A and B are arbitrary constants, you can combine $(A+B)$ to give an arbitrary constant C, and you can combine $i(A-B)$ to give an arbitrary constant D. So, you have $y=e^{n_1x}(C\cos n_2x+D\sin n_2x)$.

Example 6

Solve $\dfrac{d^2y}{dx^2}-2\dfrac{dy}{dx}+3y=0$, given that $y=0$ and $\dfrac{dy}{dx}=6$, when $x=0$.

Substituting $y=Ae^{nx}$ and its derivatives in $\dfrac{d^2y}{dx^2}-2\dfrac{dy}{dx}+3y=0$, you get

$$n^2-2n+3=0$$

$$\Rightarrow \quad n=\dfrac{2\pm\sqrt{4-12}}{2}=1\pm\sqrt{2}i$$

Therefore, the general solution is

$$y=e^x(C\cos\sqrt{2}x+D\sin\sqrt{2}x)$$

To find C and D, you use the boundary conditions.

When $x=0$, $y=0$, which gives

$$0=C\cos 0+D\sin 0\Rightarrow C=0$$

(continued)

(continued)

Hence, you have

$$y = De^x \sin\sqrt{2}x$$

As one boundary condition is given in terms of $\dfrac{dy}{dx}$, you differentiate the above:

$$\frac{dy}{dx} = De^x \sin\sqrt{2}x + \sqrt{2}De^x \cos\sqrt{2}x$$

When $x = 0, \dfrac{dy}{dx} = 6$, which gives

$$6 = D\sin 0 + \sqrt{2}D\cos 0$$
$$\Rightarrow \qquad 6 = \sqrt{2}D \qquad \Rightarrow \qquad D = 3\sqrt{2}$$

Therefore, the solution is $y = 3\sqrt{2}e^x \sin\sqrt{2}x$.

> **Note**
>
> Sometimes it is more convenient to denote $\dfrac{dy}{dx}$ by y' or $f'(x)$, and $\dfrac{d^2y}{dx^2}$ by y'' or $f''(x)$, where $y = f(x)$.

Exercise 2

In questions **1** to **12**, find the general solution of each differential equation.

1 $\dfrac{d^2y}{dx^2} - 6\dfrac{dy}{dx} - 8y = 0$
  2 $\dfrac{d^2y}{dx^2} + 3\dfrac{dy}{dx} + 2y = 0$
  3 $2\dfrac{d^2y}{dx^2} - \dfrac{dy}{dx} - 6y = 0$

4 $3\dfrac{d^2y}{dx^2} + 4\dfrac{dy}{dx} - 7y = 0$
  5 $\dfrac{d^2x}{dt^2} - 7\dfrac{dx}{dt} - 8x = 0$
  6 $\dfrac{d^2x}{dt^2} - 11\dfrac{dx}{dt} + 28x = 0$

7 $\dfrac{d^2y}{dx^2} + 4\dfrac{dy}{dx} + 4y = 0$
  8 $\dfrac{d^2y}{dx^2} - 6\dfrac{dy}{dx} + 9y = 0$
  9 $\dfrac{d^2y}{dx^2} + \dfrac{dy}{dx} + y = 0$

10 $\dfrac{d^2y}{dx^2} + 4\dfrac{dy}{dx} + 8y = 0$
  11 $\dfrac{d^2x}{dt^2} - 6\dfrac{dx}{dt} + 7x = 0$
  12 $\dfrac{d^2x}{dt^2} + 2\dfrac{dx}{dt} + 13x = 0$

The complementary function and particular integral

This method is required for second-order differential equations in the form

$$a\frac{d^2y}{dx^2} + b\frac{dy}{dx} + cy = f(x).$$

If $y = g(x)$ is the solution of

$$a\frac{d^2y}{dx^2} + b\frac{dy}{dx} + cy = 0$$

and $y = h(x)$ is a solution of

$$a\frac{d^2y}{dx^2} + b\frac{dy}{dx} + cy = f(x)$$

then

$$y = h(x) + \lambda g(x)$$

is the general solution of

$$a\frac{d^2y}{dx^2} + b\frac{dy}{dx} + cy = f(x)$$

To prove this, substitute $y = h + \lambda g$ and its derivatives in the LHS of

$$a\frac{d^2y}{dx^2} + b\frac{dy}{dx} + cy = f(x)$$

you have

$$ay'' + by' + cy = a(h'' + \lambda g'') + b(h' + \lambda g') + c(h + \lambda g)$$

$$= ah'' + bh' + ch + \lambda(ag'' + bg' + cg)$$

$$= f(x)$$

since h is a solution of $ah'' + bh' + ch = f(x)$, and g is a solution of

$ag'' + bg' + cg = 0$

Therefore,

$$y = h(x) + \lambda g(x)$$

is the general solution of $ay'' + by' + cy = f(x)$.

$g(x)$ is called the **complementary function (CF)**, and $h(x)$ is called the **particular integral (PI)**.

The particular solution is obtained by inserting boundary conditions into the general solution.

Types of particular integral

The particular integral depends on the function $f(x)$.

You will consider three types of function $f(x)$:

▶ polynomial
▶ exponential
▶ trigonometric

When $f(x)$ is a **polynomial of degree n**, the particular integral will also be a polynomial of degree n.

> **Tip**
>
> You will only be expected to use this method for polynomials of degree 4 or less.

Example 7

By finding

a the complementary function

b the particular integral,

solve the equation

$$\frac{d^2x}{dt^2} + 3\frac{dx}{dt} - 4x = 8$$

a For the complementary function, you use

$$\frac{d^2x}{dt^2} + 3\frac{dx}{dt} - 4x = 0$$

Substituting $x = Ae^{nt}$ in the above equation, you get

$$n^2 + 3n - 4 = 0$$

$\Rightarrow \quad (n+4)(n-1) = 0$

$\Rightarrow \qquad\qquad n = 1 \quad \text{or} \quad -4$

So, the CF is $x = Ae^t + Be^{-4t}$.

b For the particular integral, $f(x)$ is a polynomial of degree 0. Hence, $x = c$ for the particular integral.

Substituting $x = c$ in $\frac{d^2x}{dt^2} + 3\frac{dx}{dt} - 4x = 8$,

$$-4c = 8 \quad \Rightarrow \quad c = -2$$

So, the PI is $x = -2$.

Therefore, the general solution is $x = Ae^t + Be^{-4t} - 2$.

Example 8

Find the solution of $\dfrac{d^2y}{dx^2} + 3\dfrac{dy}{dx} - 4y = 3 + 8x^2$, given that, when $x = 0, y = 0$ and $\dfrac{dy}{dx} = 1$.

To find the CF, you use

$$\dfrac{d^2y}{dx^2} + 3\dfrac{dy}{dx} - 4y = 0$$

Substituting $y = Ae^{nx}$ in the above equation,

$$n^2 + 3n - 4 = 0$$

$\Rightarrow \quad (n+4)(n-1) = 0$

$\Rightarrow \qquad\qquad n = 1 \quad \text{or} \quad -4$

So, the CF is $y = Ae^x + Be^{-4x}$.

To find the PI, you substitute $y = a + bx + cx^2$ and its derivatives in

$$\dfrac{d^2y}{dx^2} + 3\dfrac{dy}{dx} - 4y = 3 + 8x^2$$

which gives

$$2c + 3(b + 2cx) - 4(a + bx + cx^2) = 3 + 8x^2$$

Equating coefficients of x^2: $-4c = 8 \Rightarrow c = -2$

Equating coefficients of x: $6c - 4b = 0 \Rightarrow b = -3$

Letting $x = 0$ in the above equation, you get

$$2c + 3b - 4a = 3$$

$\Rightarrow \qquad\qquad a = -4$

So, the PI is $y = -4 - 3x - 2x^2$.

Therefore, the general solution is

$$y = Ae^x + Be^{-4x} - 4 - 3x - 2x^2$$

You now need to find values for A and B.

When $x = 0$, $y = 0$, which gives

$$0 = A + B - 4$$

$\Rightarrow \quad A + B = 4 \qquad [1]$

Differentiating $y = Ae^x + Be^{-4x} - 4 - 3x - 2x^2$, you have

$$\dfrac{dy}{dx} = Ae^x - 4Be^{-4x} - 3 - 4x$$

When $x = 0$, $\dfrac{dy}{dx} = 1$, which gives

$$1 = A - 4B - 3$$

$\Rightarrow \quad A - 4B = 4 \qquad [2]$

From [1] and [2], you get $A = 4$ and $B = 0$.

Therefore, the general solution is $y = 4e^x - 4 - 3x - 2x^2$.

When f(x) is an **exponential function** you proceed as follows:

Take, for example, the equation $\dfrac{d^2y}{dx^2} + 3\dfrac{dy}{dx} - 4y = 3e^{7x}$.

In this case, $f(x) = 3e^{7x}$. The particular integral will be of the same form: Ce^{7x}.

The CF is $y = Ae^x + Be^{-4x}$ see *Example 8*.

To find the PI, you substitute $y = Ce^{7x}$ and its derivatives in

$$\frac{d^2y}{dx^2} + 3\frac{dy}{dx} - 4y = 3e^{7x}$$

This gives

$$49Ce^{7x} + 21Ce^{7x} - 4Ce^{7x} = 3e^{7x}$$

$$\Rightarrow \qquad 66C = 3 \Rightarrow C = \frac{1}{22}$$

So, the PI is $y = \frac{1}{22}e^{7x}$.

Therefore, the general solution is $y = Ae^x + Be^{-4x} + \frac{e^{7x}}{22}$.

When **f(x)** is a **trigonometric function** of the form $a \sin nx$ you proceed as follows:

Take, for example, $f(x) = 4\sin 2x$. The particular integral will be of the form $C\sin 2x + D\cos 2x$.

Example 9

Solve $\frac{d^2y}{dx^2} + 3\frac{dy}{dx} - 4y = 4\sin 2x$.

The CF is $y = Ae^x + Be^{-4x}$ see *Example 8*.

Consider

$$y = C\sin 2x + D\cos 2x$$

Differentiating this, you have

$$y' = 2C\cos 2x - 2D\sin 2x$$
$$y'' = -4C\sin 2x - 4D\cos 2x$$

Substituting y' and y'' in $\frac{d^2y}{dx^2} + 3\frac{dy}{dx} - 4y = 4\sin 2x$, you get

$$-4C\sin 2x - 4D\cos 2x + 6C\cos 2x - 6D\sin 2x - 4C\sin 2x - 4D\cos 2x = 4\sin 2x$$

Equating coefficients of $\sin 2x$: $-8C - 6D = 4$

$$\Rightarrow \qquad -4C - 3D = 2 \qquad [1]$$

Equating coefficients of $\cos 2x$: $-8D + 6C = 0$

$$\Rightarrow \qquad -4D + 3C = 0 \qquad [2]$$

Solving the simultaneous equations [1] and [2], you get

$$C = -\frac{8}{25} \qquad \text{and} \qquad D = -\frac{6}{25}$$

Therefore, the PI is

$$y = -\frac{8}{25}\sin 2x - \frac{6}{25}\cos 2x$$

Hence, the general solution is

$$y = Ae^x + Be^{-4x} - \frac{8}{25}\sin 2x - \frac{6}{25}\cos 2x$$

Tip

Suppose you were simply to consider $y = C\sin 2x$ as the PI. Because there is only a $\sin 2x$ term on the right-hand side, you would obtain $\frac{dy}{dx} = 2C\cos 2x$ and $\frac{d^2y}{dx^2} = -4C\sin 2x$.

Substituting these in $\frac{d^2y}{dx^2} + 3\frac{dy}{dx} - 4y = 4\sin 2x$, you would obtain

$$-4C\sin 2x + 3 \times 2C\cos 2x - 4C\sin 2x = 4\sin 2x,$$

which includes only one term in $\cos 2x$ (from $\frac{dy}{dx}$). This means that this equation could not be solved. Hence, the PI used **must** contain **both** $\sin 2x$ **and** $\cos 2x$ terms.

Example 10

Question

Solve $\dfrac{d^2y}{dx^2} - \dfrac{dy}{dx} - 2y = 3e^{2x}$, given that $y = 0$ and $\dfrac{dy}{dx} = 11$ when $x = 0$.

Answer

To find the CF, use
$$\frac{d^2y}{dx^2} - \frac{dy}{dx} - 2y = 0$$
Substituting $y = Ae^{nx}$, you obtain
$$n^2 - n - 2 = 0$$
$$\Rightarrow \quad (n - 2)(n + 1) = 0$$
$$\Rightarrow \qquad\qquad n = 2 \quad \text{or} \quad -1$$
So, the CF is $y = Ae^{2x} + Be^{-x}$.

To find the PI, let $y = Cxe^{2x}$.

Differentiating $y = Cxe^{2x}$,
$$\frac{dy}{dx} = Ce^{2x} + 2Cxe^{2x}$$
$$\frac{d^2y}{dx^2} = 2Ce^{2x} + 2Ce^{2x} + 4Cxe^{2x}$$

Substituting these in $\dfrac{d^2y}{dx^2} - \dfrac{dy}{dx} - 2y = 3e^{2x}$, you get
$$4Ce^{2x} + 4Cxe^{2x} - Ce^{2x} - 2Cxe^{2x} - 2Cxe^{2x} = 3e^{2x}$$
$$3Ce^{2x} = 3e^{2x} \quad \Rightarrow \quad C = 1$$

Therefore, the PI is $y = xe^{2x}$.

Hence, the general solution is $y = Ae^{2x} + Be^{-x} + xe^{2x}$.

The boundary conditions tell us that:
$$y = 0 \text{ when } x = 0 \quad \Rightarrow \quad 0 = A + B$$
$$\frac{dy}{dx} = 2Ae^{2x} - Be^{-x} + e^{2x} + 2xe^{2x}$$
$$\frac{dy}{dx} = 11 \text{ when } x = 0 \quad \Rightarrow \quad 11 = 2A - B + 1 \quad \Rightarrow \quad 10 = 2A - B$$

Since $0 = A + B$, you have
$$A = \frac{10}{3} \quad \text{and} \quad B = -\frac{10}{3}$$

The solution is, therefore,
$$y = \left(\frac{10}{3} + x\right)e^{2x} - \frac{10}{3}e^{-x}$$

> **Note**
>
> xe^{2x} is used here because e^{2x} already forms part of the CF.

> **Note**
>
> The x-terms should cancel at this stage.

Example 11

Question

Solve $y'' - 4y' + 4y = 3e^{2x}$.

Answer

To find the CF, you substitute $y = Ae^{nx}$ and its derivatives in
$y'' - 4y' + 4y = 0$, which gives
$$n^2 - 4n + 4 = 0$$

(continued)

(continued)

$\Rightarrow \quad (n-2)(n-2)=0$

$\Rightarrow \qquad\qquad n=2 \qquad$ (repeated root)

Therefore, the CF is $y=(A+Bx)e^{2x}$.

To find the PI, you need to use a term in x^2e^{2x}, since both e^{2x} and xe^{2x} already form terms in the CF. Therefore, you let $y=Cx^2e^{2x}$, so

$y'=2Cx^2e^{2x}+2Cxe^{2x}$

$y''=4Cx^2e^{2x}+4Cxe^{2x}+2Ce^{2x}+4Cxe^{2x}$

$\quad=4Cx^2e^{2x}+8Cxe^{2x}+2Ce^{2x}$

Substituting these in $y''-4y'+4y=3e^{2x}$,

$4Cx^2e^{2x}+8Cxe^{2x}+2Ce^{2x}-4(2Cx^2e^{2x}+2Cxe^{2x})+4Cx^2e^{2x}=3e^{2x}$

$2Ce^{2x}=3e^{2x} \quad\Rightarrow\quad C=\dfrac{3}{2}$

Therefore, the PI is $y=\dfrac{3}{2}x^2e^{2x}$.

Therefore the general solution is $y=\left(A+Bx+\dfrac{3}{2}x^2\right)e^{2x}$.

> **Note**
>
> The terms in x^2 and x should cancel at this stage.

Example 12

Solve $y''+16y=2\cos 4x$.

To find the CF, you substitute $y=Ae^{nx}$ and its second derivative in

$y''+16y=0$, which gives

$n^2+16=0 \quad\Rightarrow\quad n=\pm 4i$

Therefore the CF is $y=A\cos 4x+B\sin 4x$.

The PI is given by

$\quad y=Cx\cos 4x+Dx\sin 4x$

$\quad y'=C\cos 4x-4Cx\sin 4x+D\sin 4x+4Dx\cos 4x$

$\quad y''=-4C\sin 4x-4C\sin 4x-16Cx\cos 4x+4D\cos 4x+4D\cos 4x-16Dx\sin 4x$

Substituting the above in $y''+16y=2\cos 4x$,

$-8C\sin 4x-16Cx\cos 4x+8D\cos 4x-16Dx\sin 4x+16Cx\cos 4x+$
$16Dx\sin 4x=2\cos 4x$

Simplifying, equating sin and cos terms, and remembering that the terms in x should cancel, you find

$C=0 \quad$ and $\quad D=\dfrac{1}{4}$

Therefore, the PI is $y=\dfrac{1}{4}x\sin 4x$.

Hence, the solution is

$y=A\sin 4x+B\cos 4x+\dfrac{x}{4}\sin 4x$

> **Note**
>
> For the PI you need to use terms in $x\cos 4x$ and $x\sin 4x$, since the CF already contains the terms $\cos 4x$ and $\sin 4x$. Therefore, the PI is given by $y=Cx\cos 4x+Dx\sin 4x$.

25.3 Using a complementary function and particular integral to solve a first order equation

Although you can solve many first-order differential equations without using complementary functions and particular integrals, you may be asked to solve a first-order differential equation using this method.

Example 13

Find the complementary function and a particular integral for the differential equation below, and hence write down the general solution.

$$\frac{dy}{dx} - 5y = 4x - e^{3x}$$

The complementary function is the solution of $\dfrac{dy}{dx} - 5y = 0$, which is $y = Ae^{5x}$.

The particular integral is of the form $y = bx + c + de^{3x}$. Substituting gives:

$$b + 3de^{3x} - 5bx - 5c - 5de^{3x} = 4x - e^{3x}$$

Equating terms, $b = -\dfrac{4}{5}$, $c = -\dfrac{4}{25}$ and $d = \dfrac{1}{2}$.

Therefore the general solution is $y = Ae^{5x} - \dfrac{4}{5}x - \dfrac{4}{25} + \dfrac{1}{2}e^{3x}$.

> **Tip**
>
> While there are exceptions, the use of this method to solve first order equations is somewhat time-consuming, and should be avoided unless you are instructed to use it.

Exercise 3

In questions **1** to **9**, find the general solution of each differential equation.

1 $\dfrac{d^2y}{dx^2} + 7\dfrac{dy}{dx} - 8y = 16x$

2 $\dfrac{d^2y}{dx^2} + 4\dfrac{dy}{dx} + 3y = 4e^{-2x}$

3 $2\dfrac{d^2y}{dx^2} - 3\dfrac{dy}{dx} - 5y = 10x^2 + 1$

4 $3\dfrac{d^2y}{dx^2} + 2\dfrac{dy}{dx} - y = 4\sin 5x$

5 $\dfrac{d^2x}{dt^2} - 4\dfrac{dx}{dt} - 5x = 3e^{3t}$

6 $\dfrac{d^2s}{dt^2} - 8\dfrac{ds}{dt} + 15s = 5\cos 2t$

7 $\dfrac{d^2y}{dx^2} + 5\dfrac{dy}{dx} + 4y = 2e^{-x}$

8 $\dfrac{d^2y}{dx^2} - 2\dfrac{dy}{dx} + 3y = 22e^{4x}$

9 $\dfrac{d^2y}{dx^2} + 6\dfrac{dy}{dx} + 10y = 3e^{-4x}$

Summary

You have learnt to solve:

▶ equations of the form $\dfrac{dy}{dx} + Py = Q$, provided that P and Q are functions only of x, by using the integrating factor $e^{\int P(x)\,dx}$

▶ second-order equations of the form $a\dfrac{d^2y}{dx^2} + b\dfrac{dy}{dx} + c = 0$, for real a, b and c, starting with $y = e^{nx}$

▶ solutions to equations of the form $a\dfrac{d^2y}{dx^2} + b\dfrac{dy}{dx} + c = f(x)$, using a particular integral.

Review exercises

1 Find the solution of $\dfrac{dy}{dx} - 7y = e^{3x}$, given that $y = 0$ when $x = 0$.

2 Solve $\dfrac{d^2y}{dx^2} - 6\dfrac{dy}{dx} + 9y = 5e^{3x}$, given that when $x = 0$, $y = 0$ and $\dfrac{dy}{dx} = 0$.

3 Solve $\dfrac{d^2x}{dt^2} - 2\dfrac{dx}{dt} + x = 4e^t$, given that when $t = 0$, $x = 1$, and when $t = 1$, $x = 4e$.

4 Solve $\dfrac{d^2x}{dt^2} + 16x = 3\cos 4t$, given that when $t = 0$, $x = 1$, and when $t = \dfrac{\pi}{8}$, $x = 0$.

Practice examination questions

1 a Find the values of the constants p and q for which $p\sin x + q\cos x$ is a particular integral of the differential equation

$$\dfrac{dy}{dx} + 5y = 13\cos x \qquad\qquad \text{(3 marks)}$$

  b Hence find the general solution of this differential equation. (3 marks)

  AQA MFP3 January 2011

2 a Show that $\dfrac{1}{x^2}$ is an integrating factor for the first-order differential equation

$$\dfrac{dy}{dx} - \dfrac{2}{x}y = x \qquad\qquad \text{(3 marks)}$$

  b Hence find the general solution of this differential equation, giving your answer in the form $y = f(x)$. (4 marks)

  AQA MFP3 January 2009

3 a Find the values of the constants a, b and c for which $a + b\sin 2x + c\cos 2x$ is a particular integral of the differential equation

$$\dfrac{dy}{dx} + 4y = 20 - 20\cos 2x \qquad\qquad \text{(4 marks)}$$

  b Hence find the solution of this differential equation, given that $y = 4$ when $x = 0$. (4 marks)

  AQA MFP3 June 2014

4 Solve the differential equation

$$\dfrac{d^2y}{dx^2} - 2\dfrac{dy}{dx} - 3y = 2e^{-x}$$

  given that $y \to 0$ as $x \to \infty$ and that $\dfrac{dy}{dx} = -3$ when $x = 0$. (10 marks)

  AQA MFP3 June 2014

5 a Find the values of the constants a, b and c for which $a + bx + cxe^{-3x}$ is a particular integral of the differential equation

$$\dfrac{d^2y}{dx^2} + 2\dfrac{dy}{dx} - 3y = 3x - 8e^{-3x} \qquad\qquad \text{(5 marks)}$$

  b Hence find the general solution of this differential equation. (3 marks)

  c Hence express y in terms of x, given that $y = 1$ when $x = 0$ and that $\dfrac{dy}{dx} \to -1$ as $x \to \infty$. (4 marks)

  AQA MFP3 June 2013

26 Vectors and Three-Dimensional Coordinate Geometry

Introduction

In this chapter, you discover how to use vectors in three dimensions. Vectors can be used with matrices to describe linear transformations. You will learn about the vector product, which can be used to find the volume of a parallelepiped (a prism with parallelogram faces). Both the vector product and the scalar product have applications in mechanics. For example, the vector product can be used to calculate torque.

Recap

You will need to remember...

▶ That the vectors **i**, **j** and **k** represent the three-dimensional vectors
$$\begin{pmatrix} 1 \\ 0 \\ 0 \end{pmatrix}, \begin{pmatrix} 0 \\ 1 \\ 0 \end{pmatrix} \text{ and } \begin{pmatrix} 0 \\ 0 \\ 1 \end{pmatrix} \text{ respectively.}$$

▶ How to find the magnitude of a given vector in $\mathbf{R}^2$ (xy plane) or $\mathbf{R}^3$ (xyz space).

▶ How to convert the vector equation of a line $\mathbf{r} = \mathbf{a} + t\mathbf{b}$ into Cartesian form $y = mx + c$, and vice versa.

▶ How to find the scalar product of two vectors, $\mathbf{a}.\mathbf{b} = |\mathbf{a}||\mathbf{b}|\cos\theta$, where θ is the angle between **a** and **b**, and use this to find the angle between two vectors.

▶ How to find the distance of a point from a line.

▶ How to calculate the determinant of a 3×3 matrix.

See Chapter 7 *Matrices and Transformations* if you need a reminder of determinants.

Objectives

By the end of this chapter, you should know how to:

▶ Calculate the vector product and know its properties.

▶ Use the vector product to calculate the area of a triangle or parallelogram.

▶ Recognise direction ratios and direction cosines.

▶ Apply vectors to geometric problems involving points, lines and planes.

▶ Use Cartesian coordinates to make calculations about points and lines, and to find the line of intersection of two non-parallel planes.

▶ Calculate the scalar triple product and use this to find the volume of a parallelepiped.

▶ Use the scalar triple product to identify coplanar vectors.

26.1 Vectors

You should know from previous studies that vectors have magnitude and direction, and a single vector is expressed as **a** or $\overrightarrow{AB}$, for example. Vectors can also be written in component form, $OP = x\mathbf{i} + y\mathbf{j}$ where **i** and **j** are unit vectors in the x and y directions respectively.

If you know the magnitude, r, and the direction, θ, of a vector, you can resolve the vector into components as $\overrightarrow{OP} = r\cos\theta\mathbf{i} + r\sin\theta\mathbf{j}$. If you know the components x and y, you can find the magnitude and direction as $r = \sqrt{x^2 + y^2}$ and $\tan\theta = \dfrac{y}{x}$.

In previous studies, you will also have learned how to multiply a vector by a scalar quantity and to find the scalar product of two vectors. In this chapter, you will learn about vector products and the scalar triple product. First, there is some background knowledge to cover so that it is possible to understand these new concepts.

Direction ratios

You know from previous studies that when one vector is a scalar multiple of another vector, the two vectors are parallel. For example, vector $\mathbf{a} = \begin{pmatrix} 3 \\ 4 \\ -5 \end{pmatrix}$ is

parallel to vector $\mathbf{b} = \begin{pmatrix} -15 \\ -20 \\ 25 \end{pmatrix}$ since $\mathbf{b} = -5\mathbf{a}$.

The direction of a vector can be specified by the ratios of the components in the **i, j** and **k** directions. These are called the **direction ratios** of the vector and are normally expressed as integers. For example, the direction ratios of the vector $28\mathbf{i} - 21\mathbf{j} - 14\mathbf{k}$ are $4 : -3 : -2$.

Usually, these would be changed to a format that uses the fewest number of negative terms, so $-4 : 3 : 2$.

The direction ratios specify the direction of a vector in three-dimensional space.

Note that two lines that do not intersect and are not parallel are said to be **skew**.

Direction cosines

The **direction cosines** are cosines of the angles between each of the coordinate axes (x, y, z) and a vector.

For instance, take the vector $\mathbf{a}$, which in three-dimensional space has the components (a_1, a_2, a_3). The angle that vector $\mathbf{a}$ makes with the **i**-axis can be

given by $\cos^{-1}\left(\dfrac{a_1}{|\mathbf{a}|}\right)$, where $|\mathbf{a}|$ is the magnitude of vector $\mathbf{a}$ and a_1 is the

component of $\mathbf{a}$ in the **i**-direction.

If θ_x is the angle that vector $\mathbf{a}$ makes with the **i**-axis, you have $\cos\theta_x = \dfrac{a_1}{|\mathbf{a}|}$.

Likewise for θ_y and θ_z, you have $\cos\theta_y = \dfrac{a_2}{|\mathbf{a}|}$ and $\cos\theta_z = \dfrac{a_3}{|\mathbf{a}|}$.

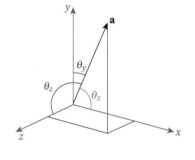

So, the three values $\dfrac{a_1}{|\mathbf{a}|}, \dfrac{a_2}{|\mathbf{a}|}$ and $\dfrac{a_3}{|\mathbf{a}|}$ are known as the direction cosines of vector $\mathbf{a}$.

They represent another way of specifying the vector's direction.

> If the direction cosines are l, m and n, then you will notice that
> $l^2 + m^2 + n^2 = 1$.

The relationship given above means that if you know two of the direction cosines of a vector, you can find the third.

You can use the direction cosine to calculate the angle between the vector and a specified axis.

Example 1

a   Find the direction cosines of the vector $3\mathbf{i} - 4\mathbf{j} + 5\mathbf{k}$.

b   Find the angle that the vector makes with the z-axis.

a $a_1 = 3$, $a_2 = -4$, $a_3 = 5$

$|\mathbf{a}| = \sqrt{3^2 + (-4)^2 + 5^2} = \sqrt{50} = 5\sqrt{2}$

Hence, the direction cosines are

respectively, $\dfrac{3}{5\sqrt{2}}, -\dfrac{4}{5\sqrt{2}}, \dfrac{5}{5\sqrt{2}}$.

b $\cos\theta_z = \dfrac{a_3}{|\mathbf{a}|}$

$\cos\theta_z = \dfrac{5}{5\sqrt{2}} = \dfrac{1}{\sqrt{2}}$

Therefore, the angle that the vector makes with the z-axis is $\dfrac{\pi}{4}$.

> **Note**
>
> The direction cosines are given by $\dfrac{a_1}{|\mathbf{a}|}, \dfrac{a_2}{|\mathbf{a}|}$ and $\dfrac{a_3}{|\mathbf{a}|}$, so write down the values of each component.

> **Note**
>
> Find the magnitude of $\begin{pmatrix} 3 \\ -4 \\ 5 \end{pmatrix}$.

> **Note**
>
> If θ is the angle that the vector makes with the z-axis.

Example 2

If two of the direction cosines are $\dfrac{1}{3}$, find the possible values of the third direction cosine.

$\dfrac{1}{9} + \dfrac{1}{9} + n^2 = 1$

$n = \pm\sqrt{1 - \dfrac{1}{9} - \dfrac{1}{9}} = \pm\dfrac{\sqrt{7}}{3}$

Therefore, the third direction cosine is $\pm\dfrac{\sqrt{7}}{3}$.

> **Note**
>
> The sum of the squares of the three direction cosines is 1.

Determinants

You saw in Chapter *7 Matrices and Transformations* how to calculate the determinant of a 2×2 matrix; recall that only a square matrix has a determinant. The 2×2 determinant of matrix $\begin{pmatrix} a & b \\ c & d \end{pmatrix}$ is written $\begin{vmatrix} a & b \\ c & d \end{vmatrix}$ and is calculated as $ad - bc$.

Given the matrix $\begin{pmatrix} a & b & c \\ d & e & f \\ g & h & i \end{pmatrix}$, the 3×3 determinant $\begin{vmatrix} a & b & c \\ d & e & f \\ g & h & i \end{vmatrix}$

$= a\begin{vmatrix} e & f \\ h & i \end{vmatrix} - b\begin{vmatrix} d & f \\ g & i \end{vmatrix} + c\begin{vmatrix} d & e \\ g & h \end{vmatrix}$ which is, $a(ei - fh) - b(di - fg) + c(dh - eg)$.

Note it is much easier to learn the method for evaluating a determinant than to remember its formula. The determinant of a 3×3 matrix is found by expanding the matrix along its first row. In turn, you take each element in the first row, cover up its column and the first row, and find the determinant of the 2×2 matrix that is left. You then combine the three results: $a(ei - fh)$. Notice the minus sign for the b-term, which relates to the fact that b is an odd number of places from the first element, a.

Example 3

Evaluate $\begin{vmatrix} 3 & 7 & 8 \\ 4 & 2 & 5 \\ 1 & 9 & 15 \end{vmatrix}$.

Note

Expand the matrix along its first row, remembering that b is negative, then for each component in the first row, cover up its column and the first row to create a 2×2 matrix.

Note

Find the determinant of each 2×2 matrix.

Note

You could also have identified the values of a, ei, fh etc. directly from the matrix and used $a(ei - fh) - b(di - fg) + c(dh - eg)$.

$$\begin{vmatrix} 3 & 7 & 8 \\ 4 & 2 & 5 \\ 1 & 9 & 15 \end{vmatrix} = 3\begin{vmatrix} 2 & 5 \\ 9 & 15 \end{vmatrix} - 7\begin{vmatrix} 4 & 5 \\ 1 & 15 \end{vmatrix} + 8\begin{vmatrix} 4 & 2 \\ 1 & 9 \end{vmatrix}$$

$$= 3(30 - 45) - 7(60 - 5) + 8(36 - 2)$$

$$= -45 - 385 + 272$$

$$= -158$$

Determinants, unlike matrices, always consist of a square array of elements.

Since determinants are always square, the expansion method just described can be applied to determinants of any size. Therefore, to evaluate the determinant of a 4×4 matrix, you first expand it along its top row to get an expression involving four 3×3 matrices, remembering to alternate the plus and minus signs. For example,

$$\begin{vmatrix} 1 & 3 & 4 & 2 \\ 5 & -1 & -3 & -4 \\ 2 & -3 & 4 & 7 \\ 1 & 8 & 5 & 6 \end{vmatrix} = 1\begin{vmatrix} -1 & -3 & -4 \\ -3 & 4 & 7 \\ 8 & 5 & 6 \end{vmatrix} - 3\begin{vmatrix} 5 & -3 & -4 \\ 2 & 4 & 7 \\ 1 & 5 & 6 \end{vmatrix} + 4\begin{vmatrix} 5 & -1 & -4 \\ 2 & -3 & 7 \\ 1 & 8 & 6 \end{vmatrix}$$

$$-2\begin{vmatrix} 5 & -1 & -3 \\ 2 & -3 & 4 \\ 1 & 8 & 5 \end{vmatrix}$$

You then proceed to evaluate each 3×3 matrix as before.

Recall that the determinant of the square matrix A is denoted either by $|A|$ or by $\det A$, and whenever two matrices A and B have the same order (size), $\det AB = \det A \times \det B$.

You first saw this in Chapter 7 Matrices and Transformations.

Exercise 1

1 For each vector, state

   i the direction ratios   **ii** the direction cosines.

   a $6\mathbf{i} + 12\mathbf{j} - 12\mathbf{k}$ **b** $3\mathbf{i} - 4\mathbf{j} - 5\mathbf{k}$

   c $12\mathbf{i} + 8\mathbf{j} - 20\mathbf{k}$ **d** $9\mathbf{i} - 18\mathbf{j} - 27\mathbf{k}$

2 Calculate the value of each of these determinants.

   a $\begin{vmatrix} 3 & 5 \\ -2 & 6 \end{vmatrix}$ **b** $\begin{vmatrix} 4 & -7 \\ 3 & 2 \end{vmatrix}$ **c** $\begin{vmatrix} 3 & 8 & 5 \\ 9 & 2 & -2 \\ 2 & 5 & 1 \end{vmatrix}$ **d** $\begin{vmatrix} 3 & 3 & 3 \\ 1 & -4 & 1 \\ 6 & -7 & 5 \end{vmatrix}$

   e $\begin{vmatrix} a & 5 & 1 \\ 6 & 3 & 3 \\ 8 & -2 & 4 \end{vmatrix}$ **f** $\begin{vmatrix} 4 & 3 & 1 \\ 1 & -5 & 2 \\ -5 & -1 & k \end{vmatrix}$

26.2 Vector product

The product of two vectors can be formed in two distinct ways: the scalar product and the **vector product** (sometimes called the **cross product**).

> The vector product of two vectors a and b is denoted by $\mathbf{a} \times \mathbf{b}$, and is defined as
>
> $$\mathbf{a} \times \mathbf{b} = |\mathbf{a}||\mathbf{b}|\sin\theta\,\hat{\mathbf{n}}$$
>
> where θ is the angle measured in the anticlockwise sense between a and b, and $\hat{\mathbf{n}}$ is a unit vector, such that a, b and $\hat{\mathbf{n}}$ (in that order) form a right-handed set.

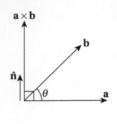

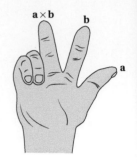

The vector product of **a** and **b** produces a vector perpendicular to both **a** and **b**.

Some important properties of the vector product

There are some important properties of the vector product that you must remember in order to use it properly.

The vector product is not commutative
Since $\mathbf{a} \times \mathbf{b} = |\mathbf{a}||\mathbf{b}|\sin\theta\,\hat{\mathbf{n}}$, it follows that $\mathbf{b} \times \mathbf{a} = -|\mathbf{a}||\mathbf{b}|\sin(-\theta)\,\hat{\mathbf{n}}$, as the three vectors must form a right-handed set.

Therefore, you have $\mathbf{a} \times \mathbf{b} = -\mathbf{b} \times \mathbf{a}$, which is known as the **anticommutative rule**.

The vector product of parallel vectors is zero
The angle, θ, between two parallel vectors, **a** and **b**, is 0° or 180°. Therefore, $\sin\theta = 0$, which gives the vector product $\mathbf{a} \times \mathbf{b} = \mathbf{0}$.

0 is called the **zero vector**. It is usually represented by an ordinary zero, 0.

Likewise, $\mathbf{a} \times \mathbf{a} = 0$, since the angle between **a** and **a** is zero. Hence, you have the following important result:

$$\mathbf{i} \times \mathbf{i} = \mathbf{j} \times \mathbf{j} = \mathbf{k} \times \mathbf{k} = \mathbf{0}$$

Note, in comparison, the scalar product $\mathbf{a} \cdot \mathbf{a} = a^2$.

The vector product of perpendicular vectors
Considering the unit vectors **i** and **j**, which are perpendicular, $\theta = 90°$ and you have the vector product $\mathbf{i} \times \mathbf{j} = 1 \times 1 \sin 90°\,\hat{\mathbf{n}} = \hat{\mathbf{n}}$.

Therefore, **i, j, $\hat{\mathbf{n}}$** form a right-handed set. But, by definition, **i, j, k** form a right-handed set. Therefore, $\hat{\mathbf{n}} = \mathbf{k}$.

Hence, you have

$$\mathbf{i} \times \mathbf{j} = \mathbf{k} \qquad\qquad \mathbf{j} \times \mathbf{i} = -\mathbf{k}$$

Similarly, you have

$$\mathbf{j} \times \mathbf{k} = \mathbf{i} \qquad\qquad \mathbf{k} \times \mathbf{j} = -\mathbf{i}$$
$$\mathbf{k} \times \mathbf{i} = \mathbf{j} \qquad\qquad \mathbf{i} \times \mathbf{k} = -\mathbf{j}$$

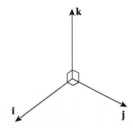

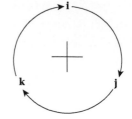

Notice from the diagram, that these vector products are **positive** when the alphabetical order in which **i**, **j** and **k** are taken is **clockwise**, but **negative** when this order is **anticlockwise**.

Note, in comparison, for perpendicular vectors **a** and **b**, the scalar product **a . b** = 0.

So, in summary,

> **For two vectors a and b:**
>
> ▶ $\mathbf{a} \times \mathbf{b} = -\mathbf{b} \times \mathbf{a}$
>
> ▶ $\mathbf{a} \times \mathbf{b} = 0$ when a and b are parallel
>
> $\mathbf{i} \times \mathbf{j} = \mathbf{k}$ $\mathbf{j} \times \mathbf{i} = -\mathbf{k}$
>
> $\mathbf{j} \times \mathbf{k} = \mathbf{i}$ $\mathbf{k} \times \mathbf{j} = -\mathbf{i}$
>
> $\mathbf{k} \times \mathbf{i} = \mathbf{j}$ $\mathbf{i} \times \mathbf{k} = -\mathbf{j}$

The vector product in component form

Expressing **a** and **b** in their component form, you have $\mathbf{a} = a_1\mathbf{i} + a_2\mathbf{j} + a_3\mathbf{k}$ and $\mathbf{b} = b_1\mathbf{i} + b_2\mathbf{j} + b_3\mathbf{k}$,

If $\mathbf{a} = a_1\mathbf{i} + a_2\mathbf{j} + a_3\mathbf{k}$ and $\mathbf{b} = b_1\mathbf{i} + b_2\mathbf{j} + b_3\mathbf{k}$, then

$$\mathbf{a} \times \mathbf{b} = (a_1\mathbf{i} + a_2\mathbf{j} + a_3\mathbf{k}) \times (b_1\mathbf{i} + b_2\mathbf{j} + b_3\mathbf{k})$$

$$= a_1\mathbf{i} \times b_1\mathbf{i} + a_2\mathbf{j} \times b_1\mathbf{i} + a_3\mathbf{k} \times b_1\mathbf{i} + a_1\mathbf{i} \times b_2\mathbf{j} + a_2\mathbf{j} \times b_2\mathbf{j} + a_3\mathbf{k} \times b_2\mathbf{j} + a_1\mathbf{i} \times b_3\mathbf{k}$$
$$+ a_2\mathbf{j} \times b_3\mathbf{k} + a_3\mathbf{k} \times b_3\mathbf{k}$$

$$= a_1 b_2\mathbf{k} - a_2 b_1\mathbf{k} + a_3 b_1\mathbf{j} - a_1 b_3\mathbf{j} + a_2 b_3\mathbf{i} - a_3 b_2\mathbf{i}$$

$$\mathbf{a} \times \mathbf{b} = (a_2 b_3 - a_3 b_2)\mathbf{i} - (a_1 b_3 - b_1 a_3)\mathbf{j} + (a_1 b_2 - a_2 b_1)\mathbf{k} \qquad [1]$$

From the definition of a 3×3 determinant, you obtain

$$\begin{vmatrix} \mathbf{i} & \mathbf{j} & \mathbf{k} \\ a_1 & a_2 & a_3 \\ b_1 & b_2 & b_3 \end{vmatrix} = \begin{vmatrix} a_2 & a_3 \\ b_2 & b_3 \end{vmatrix}\mathbf{i} - \begin{vmatrix} a_1 & a_3 \\ b_1 & b_3 \end{vmatrix}\mathbf{j} + \begin{vmatrix} a_1 & a_2 \\ b_1 & b_2 \end{vmatrix}\mathbf{k}$$

$$= (a_2 b_3 - a_3 b_2)\mathbf{i} - (a_1 b_3 - b_1 a_3)\mathbf{j} + (a_1 b_2 - a_2 b_1)\mathbf{k} \qquad [2]$$

Note that the RHS of [1] and [2] are identical. Therefore, you have

$$\mathbf{a} \times \mathbf{b} = \begin{vmatrix} \mathbf{i} & \mathbf{j} & \mathbf{k} \\ a_1 & a_2 & a_3 \\ b_1 & b_2 & b_3 \end{vmatrix}$$

Therefore,

> **You can calculate a vector product using the formula:**
>
> $$\mathbf{a} \times \mathbf{b} = \begin{vmatrix} \mathbf{i} & \mathbf{j} & \mathbf{k} \\ a_1 & a_2 & a_3 \\ b_1 & b_2 & b_3 \end{vmatrix}$$

Using this formula is the easiest way to calculate a vector product.

An alternative form of this determinant is using the transpose of this formula, shown below (this is the version you will see in the formulae booklet). Since the determinant of a matrix is the same as its transpose, the formulae are equivalent.

Vector product: $\mathbf{a} \times \mathbf{b} = |\mathbf{a}||\mathbf{b}|\sin\theta\hat{\mathbf{n}} = \begin{vmatrix} \mathbf{i} & a_1 & b_1 \\ \mathbf{j} & a_2 & b_2 \\ \mathbf{k} & a_3 & b_3 \end{vmatrix} = \begin{bmatrix} a_2 b_3 - a_3 b_2 \\ a_3 b_1 - a_1 b_3 \\ a_1 b_2 - a_2 b_1 \end{bmatrix}$

Example 4

Question

Evaluate $(2\mathbf{i} + 3\mathbf{j} - \mathbf{k}) \times (7\mathbf{i} + 4\mathbf{j} + 2\mathbf{k})$.

Answer

$\begin{pmatrix} 2 \\ 3 \\ -1 \end{pmatrix} \times \begin{pmatrix} 7 \\ 4 \\ 2 \end{pmatrix} = \begin{vmatrix} \mathbf{i} & \mathbf{j} & \mathbf{k} \\ 2 & 3 & -1 \\ 7 & 4 & 2 \end{vmatrix} = \mathbf{i}\begin{vmatrix} 3 & -1 \\ 4 & 2 \end{vmatrix} - \mathbf{j}\begin{vmatrix} 2 & -1 \\ 7 & 2 \end{vmatrix} + \mathbf{k}\begin{vmatrix} 2 & 3 \\ 7 & 4 \end{vmatrix}$

Therefore,

$\begin{pmatrix} 2 \\ 3 \\ -1 \end{pmatrix} \times \begin{pmatrix} 7 \\ 4 \\ 2 \end{pmatrix} = 10\mathbf{i} - 11\mathbf{j} - 13\mathbf{k}$

> **Note**
>
> Using
>
> $\begin{vmatrix} \mathbf{i} & \mathbf{j} & \mathbf{k} \\ a_1 & a_2 & a_3 \\ b_1 & b_2 & b_3 \end{vmatrix} = \begin{vmatrix} a_2 & a_3 \\ b_2 & b_3 \end{vmatrix}\mathbf{i} - \begin{vmatrix} a_1 & a_3 \\ b_1 & b_3 \end{vmatrix}\mathbf{j} + \begin{vmatrix} a_1 & a_2 \\ b_1 & b_2 \end{vmatrix}\mathbf{k}$

Example 5

Question

a   Evaluate $|\overrightarrow{AB} \times \overrightarrow{CD}|$, where A is $(6, -3, 0)$, B is $(3, -7, 1)$, C $(3, 7, -1)$ and D is $(4, 5, -3)$.

b   Hence find the shortest distance between AB and CD.

Answer

a

$\overrightarrow{AB} = \mathbf{b} - \mathbf{a} = \begin{pmatrix} 3 \\ -7 \\ 1 \end{pmatrix} - \begin{pmatrix} 6 \\ -3 \\ 0 \end{pmatrix} = \begin{pmatrix} -3 \\ -4 \\ 1 \end{pmatrix}$

$\overrightarrow{CD} = \mathbf{d} - \mathbf{c} = \begin{pmatrix} 4 \\ 5 \\ -3 \end{pmatrix} - \begin{pmatrix} 3 \\ 7 \\ -1 \end{pmatrix} = \begin{pmatrix} 1 \\ -2 \\ -2 \end{pmatrix}$

$\overrightarrow{AB} \times \overrightarrow{CD} = \begin{pmatrix} -3 \\ -4 \\ 1 \end{pmatrix} \times \begin{pmatrix} 1 \\ -2 \\ -2 \end{pmatrix} = \begin{vmatrix} \mathbf{i} & \mathbf{j} & \mathbf{k} \\ -3 & -4 & 1 \\ 1 & -2 & -2 \end{vmatrix}$

$\overrightarrow{AB} \times \overrightarrow{CD} = 10\mathbf{i} - 5\mathbf{j} + 10\mathbf{k}$

Therefore,

$|\overrightarrow{AB} \times \overrightarrow{CD}| = \sqrt{10^2 + (-5)^2 + 10^2} = 15$

> **Note**
>
> First find $\overrightarrow{AB}$ and $\overrightarrow{CD}$, and then find their vector product.

> **Note**
>
> Using $\mathbf{a} \times \mathbf{b} = \begin{vmatrix} \mathbf{i} & \mathbf{j} & \mathbf{k} \\ a_1 & a_2 & a_3 \\ b_1 & b_2 & b_3 \end{vmatrix}$.

> **Note**
>
> Using
> $(a_2 b_3 - a_3 b_2)\mathbf{i} - (a_1 b_3 - b_1 a_3)\mathbf{j} + (a_1 b_2 - a_2 b_1)\mathbf{k}$

(continued)

(continued)

b

P is the point with position vector $\begin{pmatrix} 6 \\ -3 \\ 0 \end{pmatrix} + t \begin{pmatrix} -3 \\ -4 \\ 1 \end{pmatrix}$

Q is the point with position vector $\begin{pmatrix} 4 \\ 5 \\ -3 \end{pmatrix} + s \begin{pmatrix} 1 \\ -2 \\ -2 \end{pmatrix}$

$\overrightarrow{QP} = \begin{pmatrix} 6 \\ -3 \\ 0 \end{pmatrix} + t \begin{pmatrix} -3 \\ -4 \\ 1 \end{pmatrix} - \left[\begin{pmatrix} 4 \\ 5 \\ -3 \end{pmatrix} + s \begin{pmatrix} 1 \\ -2 \\ -2 \end{pmatrix} \right] = k \begin{pmatrix} 10 \\ -5 \\ 10 \end{pmatrix}$

Therefore,

$2 - 3t - s = 10k$

$-8 - 4t + 2s = -5k$

$3 + t + 2s = 10k$

Solving these simultaneous equations: $k = 0.4$, $s = 1$ and $t = -1$.

The modulus of $\begin{pmatrix} 10 \\ -5 \\ 10 \end{pmatrix}$ is 15.

Therefore, the shortest distance between AB and CD is

$\left|k(\overrightarrow{AB} \times \overrightarrow{CD})\right| = 0.4 \times 15 = 6$

Area of a triangle and parallelogram

One of the applications of the vector product, is to calculate the area of a triangle and a parallelogram.

Consider the triangle ABC whose sides are $\mathbf{a}$, $\mathbf{b}$ and $\mathbf{c}$, as shown in the diagram. From the definition of the vector product, you have

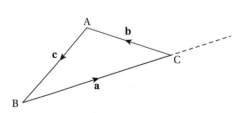

$$|\mathbf{a} \times \mathbf{b}| = |ab \sin \theta \, \hat{\mathbf{n}}|$$

where θ is the angle between $\mathbf{a}$ and $\mathbf{b}$.

However, the angle between $\mathbf{a}$ and $\mathbf{b}$ is $180° - C$, and $\sin(180° - C) = \sin C$.

Therefore, you obtain

$|\mathbf{a} \times \mathbf{b}| = |ab \sin(180° - C) \hat{\mathbf{n}}| = |ab \sin C \hat{\mathbf{n}}|$

Since $\hat{\mathbf{n}}$ is a unit vector, $|\mathbf{a} \times \mathbf{b}| = ab \sin C$. Hence, you have

Area of triangle $ABC = \dfrac{1}{2} ab \sin C = \dfrac{1}{2} |\mathbf{a} \times \mathbf{b}|$

Similarly, you can show that the area of triangle ABC is given by

$$\frac{1}{2}bc\sin A = \frac{1}{2}|\mathbf{b}\times\mathbf{c}| \text{ and } \frac{1}{2}ac\sin B = \frac{1}{2}|\mathbf{a}\times\mathbf{c}|$$

In general, to find the area of a triangle whose sides are a, b and c, you can use any of the formulae:

Area of a triangle $= \frac{1}{2}|\mathbf{a}\times\mathbf{b}|$ or $\frac{1}{2}|\mathbf{b}\times\mathbf{c}|$ or $\frac{1}{2}|\mathbf{a}\times\mathbf{c}|$

It is also possible to calculate the area of a parallelogram using vectors. Any two vectors with the angle θ between them form two sides of a parallelogram in three-dimensional space. The area of the parallelogram can be shown to be equal to the length of the vector product of the two vectors.

Area of a parallelogram $= |a\times b|$.

Example 6

Question

Find the area of triangle PQR where P is $(4, 2, 5)$, Q is $(3, -1, 6)$ and R is $(1, 4, 2)$.

Answer

$$\overrightarrow{PR} = \mathbf{r} - \mathbf{p} = \begin{pmatrix} 1 \\ 4 \\ 2 \end{pmatrix} - \begin{pmatrix} 4 \\ 2 \\ 5 \end{pmatrix} = \begin{pmatrix} -3 \\ 2 \\ -3 \end{pmatrix}$$

$$\overrightarrow{PQ} = \mathbf{q} - \mathbf{p} = \begin{pmatrix} 3 \\ -1 \\ 6 \end{pmatrix} - \begin{pmatrix} 4 \\ 2 \\ 5 \end{pmatrix} = \begin{pmatrix} -1 \\ -3 \\ 1 \end{pmatrix}$$

$$\overrightarrow{PR} \times \overrightarrow{PQ} = \begin{vmatrix} \mathbf{i} & \mathbf{j} & \mathbf{k} \\ -3 & 2 & -3 \\ -1 & -3 & 1 \end{vmatrix}$$

$$\overrightarrow{PR} \times \overrightarrow{PQ} = -7\mathbf{i} + 6\mathbf{j} + 11\mathbf{k}$$

$$|\overrightarrow{PR} \times \overrightarrow{PQ}| = \sqrt{49 + 36 + 121} = \sqrt{206}$$

Therefore, the area of triangle $PQR = \frac{1}{2}\sqrt{206}$

> **Note**
> First, find any two sides.

> **Note**
> Calculate the magnitude of the vector product.

> **Note**
> Using any one of the area formulae, for example, $\frac{1}{2}|\mathbf{a}\times\mathbf{b}|$.

> **Note**
> Using any one of the area formulae, you need to calculate the vector product; here the component form of the vector product has been used, $\mathbf{a}\times\mathbf{b} = \begin{vmatrix} \mathbf{i} & \mathbf{j} & \mathbf{k} \\ a_1 & a_2 & a_3 \\ b_1 & b_2 & b_3 \end{vmatrix}$.

Exercise 2

1. Find $\mathbf{a} \times \mathbf{b}$.

   a $\mathbf{a} = \begin{pmatrix} 1 \\ -4 \\ 3 \end{pmatrix}$ $\mathbf{b} = \begin{pmatrix} 2 \\ 3 \\ -1 \end{pmatrix}$

   b $\mathbf{a} = \begin{pmatrix} -3 \\ 4 \\ 5 \end{pmatrix}$ $\mathbf{b} = \begin{pmatrix} 2 \\ -3 \\ 4 \end{pmatrix}$

   c $\mathbf{a} = \begin{pmatrix} 4 \\ -4 \\ 2 \end{pmatrix}$ $\mathbf{b} = \begin{pmatrix} 1 \\ -5 \\ -3 \end{pmatrix}$

   d $\mathbf{a} = \begin{pmatrix} 1 \\ 4 \\ 6 \end{pmatrix}$ $\mathbf{b} = \begin{pmatrix} 3 \\ 2 \\ -5 \end{pmatrix}$

2. Find the area of a triangle ABC where A is $(1, 3, 2)$, B is $(3, 1, -2)$ and C is $(-1, 3, 5)$.

3 Find the area of a triangle PQR where P is $(1,-5,-3)$, Q is $(2,-1,6)$ and R is $(4,-1,-2)$.

4 The vectors **a** and **b** are given by $\mathbf{a}=\begin{pmatrix}2\\-1\\5\end{pmatrix}$ and $\mathbf{b}=\begin{pmatrix}4\\0\\2\end{pmatrix}$.

 a Find $\mathbf{a}\times\mathbf{b}$.

 b You are given the equations of two lines, $\mathbf{r}=\begin{pmatrix}1\\0\\1\end{pmatrix}+s\mathbf{a}$ and,

 $\mathbf{r}=\begin{pmatrix}1\\0\\3\end{pmatrix}+t\mathbf{b}$. What is the shortest distance between the two lines?

26.3 Applications of vectors to coordinate geometry

Vectors can be used to find the intersection of a line and a plane, to find the angle between two planes or between a line and a plane in two- and three-dimensional geometry. Before you look at these applications, you will first revisit the vector equation of a line and be introduced to the vector equation of a plane.

26.4 Equation of a line

You saw during your A-level mathematics studies that a line is defined by the direction of the line and a point on the line, resulting in the equation of the line as $\mathbf{r}=\mathbf{a}+\lambda\mathbf{b}$ where **a** is the position vector of a point on the line and **b** is a vector in the direction of the line.

Using the components of **r**, **a** and **b**, this becomes
$x\mathbf{i}+y\mathbf{j}+z\mathbf{k}=a_1\mathbf{i}+a_2\mathbf{j}+a_3\mathbf{k}+\lambda[b_1\mathbf{i}+b_2\mathbf{j}+b_3\mathbf{k}]$, which leads to the alternative form:

 The Cartesian equations of the straight line that passes through point A, with position vector $\mathbf{a}=a_1\mathbf{i}+a_2\mathbf{j}+a_3\mathbf{k}$, and has a direction given by vector $\mathbf{b}=b_1\mathbf{i}+b_2\mathbf{j}+b_3\mathbf{k}$, is $\dfrac{x-a_1}{b_1}=\dfrac{y-a_2}{b_2}-\dfrac{z-a_3}{b_3}=\lambda$.

If **r** is the position vector of a general point on a line, **a** is the position vector of a point on the line and **b** is a vector in the direction of the line, then the two vectors $(\mathbf{r}-\mathbf{a})$ and **b** are two vectors in the same direction. From the definition of a vector product, using $\mathbf{a}\times\mathbf{a}=0$ for parallel vectors, $(\mathbf{r}-\mathbf{a})\times\mathbf{b}=0$.
Therefore,

 The equation of a line can also be given in a vector product form, $(\mathbf{r}-\mathbf{a})\times\mathbf{b}=0$ where a is the position vector of a point on the line and b is a vector in the direction of the line.

26.5 Equation of a plane

A **plane** is a two-dimensional cross section (or slice) of three-dimensional space.

Equation in the form r = a + tb + sc

The position vector of *any* point on a plane can be expressed in terms of:

▶ **a**, the position vector of a point on the plane, and

▶ **b** and **c**, which are two *non-parallel* vectors *in* the plane.

From the diagram, you can see that the position vector of a point P on the plane is given by

$$\overrightarrow{OP} = \overrightarrow{OA} + \overrightarrow{AR} + \overrightarrow{RP}$$

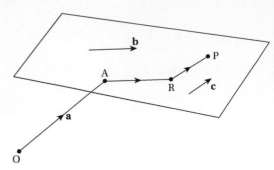

where $\overrightarrow{AR}$ is parallel to vector **b**, and $\overrightarrow{RP}$ is parallel to vector **c**.

Hence, $\overrightarrow{AR} = t\mathbf{b}$ and $\overrightarrow{RP} = s\mathbf{c}$, for some parameters t and s.

Therefore,

> The vector equation of a plane through point A is r = a + tb + sc, where b and c are non-parallel vectors in the plane, and t and s are parameters.

Equation in the form r = a + λ(b − a) + μ(c − a) = (1 − λ − μ)a + λb + μc

The plane can also be defined by using three non-collinear points on the plane. The plane through points A, B and C, has a point on the plane, A, and the two vectors $\overrightarrow{AB}$ (or **b** − **a**) and $\overrightarrow{AC}$ (or **c** − **a**) in the plane. Hence,

> The plane through non-collinear points A, B and C has vector equation,
> r = a + λ(b − a) + μ(c − a) = (1 − λ − μ)a + λb + μc.

Equation in the form r.n = d

Sometimes it is useful to have an equation of the plane that does not rely on parameters. Given **n** is a vector perpendicular to the plane, using the equation of the plane, $\mathbf{r} = \mathbf{a} + t\mathbf{b} + s\mathbf{c}$, you have

r.n = (a + tb + sc).n

$\qquad$ = **a.n** + t**b.n** + s**c.n**

Since **b** and **c** are perpendicular to **n**, **b.n = c.n = 0**.

Hence, you have **r.n = a.n**, where **r.n** is the same for all points **r** on the plane.

Therefore,

> The vector equation of the plane is r.n = d, where d is a constant that determines the position of the plane.

▶ If **n** is a unit vector, then d is the perpendicular distance of the plane from the origin.

▶ When d has the *same sign* for two planes, these planes are on the *same side* of the origin.

▶ When d has *opposite signs* for two planes, these planes are on *opposite sides* of the origin.

Cartesian form

You can find the Cartesian equation of a plane in the form $\mathbf{r.n} = d$. In a similar way to finding the Cartesian equation of a line, you take $\mathbf{r.n} = d$ and replace $\mathbf{r}$ by $x\mathbf{i} + y\mathbf{j} + z\mathbf{k}$, which gives the equation of a plane as

$$\begin{pmatrix} x \\ y \\ z \end{pmatrix}.\mathbf{n} = d$$

Let $\mathbf{n} = a\mathbf{i} + b\mathbf{j} + c\mathbf{k}$, therefore, using the scalar product,

The Cartesian equation of a plane is $\begin{pmatrix} x \\ y \\ z \end{pmatrix} \cdot \begin{pmatrix} a \\ b \\ c \end{pmatrix} = d$ or $ax + by + cz = d$,

where a, b, c and d are constants and the vector $\begin{pmatrix} a \\ b \\ c \end{pmatrix}$ is perpendicular to the plane.

In the *Formulae and Statistical Tables* booklet, you will see this as:

The plane through the point A with normal vector $\mathbf{n} = n_1\mathbf{i} + n_2\mathbf{j} + n_3\mathbf{k}$ has Cartesian equation $n_1 x + n_2 y + n_3 z = d$ where $d = \mathbf{a.n}$.

In summary,

A plane is identified by

▶ a vector perpendicular to the plane, and

▶ a point on the plane.

Example 7

Find the equation of the plane through $(3, 2, 7)$ which is perpendicular to the vector $\begin{pmatrix} 1 \\ -5 \\ 8 \end{pmatrix}$, giving its equation

a in vector form b in Cartesian form.

a $\mathbf{r} \cdot \begin{pmatrix} 1 \\ -5 \\ 8 \end{pmatrix} = \begin{pmatrix} 3 \\ 2 \\ 7 \end{pmatrix} \cdot \begin{pmatrix} 1 \\ -5 \\ 8 \end{pmatrix}$

> **Note**
> Using $\mathbf{r.n} = \mathbf{a.n}$

> **Note**
> Using the scalar product to calculate 49.

Therefore, the equation of the plane is $\mathbf{r} \cdot \begin{pmatrix} 1 \\ -5 \\ 8 \end{pmatrix} = 49$.

b $\begin{pmatrix} x \\ y \\ z \end{pmatrix} \cdot \begin{pmatrix} 1 \\ -5 \\ 8 \end{pmatrix} = 49$

> **Note**
> Using answer from part **a** and replacing $\mathbf{r}$ by $x\mathbf{i} + y\mathbf{j} + z\mathbf{k}$.

Therefore, the Cartesian equation is $x - 5y + 8z = 49$.

You can use the Cartesian equation of a plane to find a vector perpendicular to that plane; it is also possible to find a unit vector that is perpendicular to the plane.

Example 8

Find the unit vector perpendicular to the plane $2x + 3y - 7z = 11$.

A vector perpendicular to the given plane is $\begin{pmatrix} 2 \\ 3 \\ -7 \end{pmatrix}$.

The magnitude of this vector is $\sqrt{2^2 + 3^2 + (-7)^2} = \sqrt{62}$.

Therefore, the unit vector perpendicular to the given plane is $\begin{pmatrix} \dfrac{2}{\sqrt{62}} \\ \dfrac{3}{\sqrt{62}} \\ -\dfrac{7}{\sqrt{62}} \end{pmatrix}$.

> **Note**
>
> The vector $\begin{pmatrix} a \\ b \\ c \end{pmatrix}$ is perpendicular to the plane $ax + by + cz = d$.

> **Note**
>
> The unit vector perpendicular to the given plane must have magnitude one.

Given three points on a plane, you can find two vectors that join them, and use this to find the equation of a plane in any required format.

Example 9

Find the equation of a plane through $A(1, 4, 6)$, $B(2, 7, 5)$ and $C(-3, 8, 7)$, giving your answer in these forms:

a $\mathbf{r} = \mathbf{a} + t\mathbf{b} + s\mathbf{c}$ **b** $\mathbf{r}.\mathbf{n} = d$ **c** $ax + by + cz = d$

a $\vec{AB} = \mathbf{b} - \mathbf{a} = \begin{pmatrix} 2 \\ 7 \\ 5 \end{pmatrix} - \begin{pmatrix} 1 \\ 4 \\ 6 \end{pmatrix} = \begin{pmatrix} 1 \\ 3 \\ -1 \end{pmatrix}$

$\vec{AC} = \mathbf{c} - \mathbf{a} = \begin{pmatrix} -3 \\ 8 \\ 7 \end{pmatrix} - \begin{pmatrix} 1 \\ 4 \\ 6 \end{pmatrix} = \begin{pmatrix} -4 \\ 4 \\ 1 \end{pmatrix}$

Therefore, the equation is: $r = \begin{pmatrix} 1 \\ 4 \\ 6 \end{pmatrix} + t\begin{pmatrix} 1 \\ 3 \\ -1 \end{pmatrix} + s\begin{pmatrix} -4 \\ 4 \\ 1 \end{pmatrix}$

b $\vec{AB} \times \vec{AC} = \begin{pmatrix} 1 \\ 3 \\ -1 \end{pmatrix} \times \begin{pmatrix} -4 \\ 4 \\ 1 \end{pmatrix} = \begin{vmatrix} \mathbf{i} & \mathbf{j} & \mathbf{k} \\ 1 & 3 & -1 \\ -4 & 4 & 1 \end{vmatrix}$

$= 7\mathbf{i} + 3\mathbf{j} + 16\mathbf{k}$

> **Note**
>
> You can use any two vectors on the plane ABC.

> **Note**
>
> To find the equation of the plane in the form $\mathbf{r} = \mathbf{a} + t\mathbf{b} + s\mathbf{c}$, you need to identify **one** point on the plane. Here, the point chosen is $A(1, 4, 6)$.

> **Note**
>
> You and your classmates might not find solutions that appear identical, but you might both be right. For example, instead of choosing A, you could have chosen $B(2, 7, 5)$ or $C(-3, 8, 7)$. You could also have used the vectors $\vec{BC}$, $\vec{CA}$ and so on.

> **Note**
>
> To find the equation of a plane in the form $\mathbf{r}.\mathbf{n} = d$, you need to find a vector perpendicular to the plane ABC. By definition, the vector product of any two non-parallel vectors on the plane will be perpendicular to the plane, so use the two vectors from part **a** to make calculations easier.

(continued)

(continued)

Hence, the vector equation of the plane ABC is

$$\mathbf{r} \cdot \begin{pmatrix} 7 \\ 3 \\ 16 \end{pmatrix} = \begin{pmatrix} 2 \\ 7 \\ 5 \end{pmatrix} \cdot \begin{pmatrix} 7 \\ 3 \\ 16 \end{pmatrix} = 14 + 21 + 80$$

$$\Rightarrow \mathbf{r} \cdot \begin{pmatrix} 7 \\ 3 \\ 16 \end{pmatrix} = 115$$

c $7x + 3y + 16z = 115$

Note

Here you used the position vector of B, but you could have used A or C to obtain the same result.

Instead of determining a plane using three points, you can also define a plane using two lines it contains.

Example 10

Find the Cartesian equation of the plane containing the two lines

$$\mathbf{r} = \begin{pmatrix} 3 \\ 1 \\ 2 \end{pmatrix} + t\begin{pmatrix} -1 \\ 3 \\ -4 \end{pmatrix} \text{ and } \mathbf{r} = \begin{pmatrix} -2 \\ -3 \\ 7 \end{pmatrix} + s\begin{pmatrix} 2 \\ -1 \\ 5 \end{pmatrix}.$$

The vector perpendicular to this plane is

$$\begin{pmatrix} -1 \\ 3 \\ -4 \end{pmatrix} \times \begin{pmatrix} 2 \\ -1 \\ 5 \end{pmatrix} = \begin{vmatrix} \mathbf{i} & \mathbf{j} & \mathbf{k} \\ -1 & 3 & -4 \\ 2 & -1 & 5 \end{vmatrix} = 11\mathbf{i} - 3\mathbf{j} - 5\mathbf{k}$$

The equation of the plane is $11x - 3y - 5z = d$.

But the point $(3, 1, 2)$ is in the plane. So, $d = 11 \times 3 - 3 \times 1 - 5 \times 2 = 20$.

Therefore, the equation of the plane containing the two lines is

$11x - 3y - 5z = 20$.

Note

You know that two vectors are in the plane you are trying to find. These are the direction vectors of the lines,

$\begin{pmatrix} -1 \\ 3 \\ -4 \end{pmatrix}$ and $\begin{pmatrix} 2 \\ -1 \\ 5 \end{pmatrix}$.

Applying the equation of a plane in the r.n = d form

As mentioned at the start of the section, you can use vectors in two- and three-dimensional geometry. You can use the equation of a plane in the form $\mathbf{r.n} = d$ to calculate many different things such as the angle between two planes, the angle between a plane and a line and to find the distance of a plane from the origin.

The angle between two planes is the angle between the vectors perpendicular to the planes.

Example 11

Find the angle between the planes $3x + 4y + 5z = 7$ and $x + 2y - 2z = 11$.

Vectors perpendicular to the planes are $\begin{pmatrix} 3 \\ 4 \\ 5 \end{pmatrix}$ and $\begin{pmatrix} 1 \\ 2 \\ -2 \end{pmatrix}$.

Using $\cos\theta = \dfrac{\mathbf{a.b}}{|\mathbf{a}||\mathbf{b}|}$, where θ is the required angle,

$\cos\theta = \dfrac{3+8-10}{5\sqrt{2}\times 3}$

$\Rightarrow \theta = \cos^{-1}\left(\dfrac{1}{15\sqrt{2}}\right) = 87.3°\ (1\text{ dp})$

> **Note**
>
> Using the scalar product, $\mathbf{a.b} = |\mathbf{a}||\mathbf{b}|\cos\theta$, to find the angle between two vectors.

> **Note**
>
> Using, the vector $\begin{pmatrix} a \\ b \\ c \end{pmatrix}$ is perpendicular to the plane $ax + by + cz = d$.

In order to find the point where a line perpendicular to a plane meets that plane, you need to find the vector equation of the line and also use the equation of a plane in the form $\mathbf{r.n} = d$ as this involves a point on the plane and a vector perpendicular to the plane.

Example 12

A line from point $A(2, 7, 4)$ is perpendicular to the plane Π, which has the equation $3x - 5y + 2z + 2 = 0$. Find where the line from A meets Π.

Let T be the point where the line from $A(2, 7, 4)$ meets Π.

The equation of AT is

$\mathbf{r} = \begin{pmatrix} 2 \\ 7 \\ 4 \end{pmatrix} + t\begin{pmatrix} 3 \\ -5 \\ 2 \end{pmatrix}.$

Substituting $x = (2 + 3t)$, $y = (7 - 5t)$ and $z = (4 + 2t)$ into the equation of the plane Π,

$3(2 + 3t) - 5(7 - 5t) + 2(4 + 2t) + 2 = 0$

$\Rightarrow 38t = 19 \Rightarrow t = \dfrac{1}{2}$

Therefore, the point T is $\left(3\dfrac{1}{2}, 4\dfrac{1}{2}, 5\right)$.

> **Note**
>
> Using $ax + by + cz = d$, where a, b, c and d are constants and the vector $\begin{pmatrix} a \\ b \\ c \end{pmatrix}$ is perpendicular to the plane.

> **Note**
>
> Using the vector equation of a line, $\mathbf{r} = \mathbf{a} + \lambda\mathbf{b}$;
>
> $\mathbf{b} = \begin{pmatrix} 3 \\ -5 \\ 2 \end{pmatrix}$ since the line AT is perpendicular to the plane, and using the $\mathbf{r.n} = d$ form of the equation of a plane, $\begin{pmatrix} 3 \\ -5 \\ 2 \end{pmatrix}$ is a vector perpendicular to the plane.

To find the angle between a line and a plane you use the fact that the required angle is $90° - \theta$, where θ is the angle between the line and the vector perpendicular to the plane.

Example 13

Find the angle between the plane $3x + 4y - 5z = 6$ and the line $\mathbf{r} = \begin{pmatrix} 2 \\ 4 \\ 8 \end{pmatrix} + t\begin{pmatrix} 1 \\ 5 \\ -3 \end{pmatrix}.$

Angle perpendicular to the plane $= \begin{pmatrix} 3 \\ 4 \\ -5 \end{pmatrix}$

> **Note**
>
> Using the $\mathbf{r.n} = d$ Cartesian form of the equation of a plane.

(continued)

(continued)

The required angle $= 90° -$ angle between $\begin{pmatrix} 3 \\ 4 \\ -5 \end{pmatrix}$ and $\begin{pmatrix} 1 \\ 5 \\ -3 \end{pmatrix}$

Using $\cos\theta = \dfrac{\mathbf{a.b}}{|\mathbf{a}||\mathbf{b}|}$, $\cos\theta = \dfrac{3+20+15}{5\sqrt{2}\times\sqrt{35}}$

> **Note**
>
> Using the scalar product, $\mathbf{a.b} = |\mathbf{a}||\mathbf{b}|\cos\theta$, to find the angle between two vectors.

> **Note**
>
> $\begin{pmatrix} 1 \\ 5 \\ -3 \end{pmatrix}$ is the direction of the line **r**.

The required angle

$= 90° - \cos^{-1}\left(\dfrac{38}{\sqrt{35}\times5\sqrt{2}}\right) = 65.3°$ (1 dp)

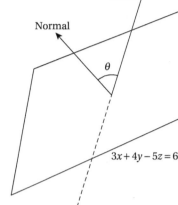

Line **r**

Normal

θ

$3x+4y-5z=6$

Two non-parallel planes always meet in a line, therefore you can use the equation of a plane to find the equation of a common line of the two planes, which is also known as the line of intersection. To obtain the equation of the common line, you need to find a point on it and its direction. To find the direction, you take the vector product of the vectors that are perpendicular to each plane respectively; to find the point, you solve simultaneous equations for the two planes to find a point where they meet.

> If the vectors n_1 and n_2 are the two directions perpendicular to the planes, then the direction of the line of intersection is $n_1 \times n_2$.

Example 14

Find the equation of the common line (line of intersection) of the two planes $\Pi_1\ 3x - y - 5z = 7$ and $\Pi_2\ 2x + 3y - 4z = -2$.

The vectors $\begin{pmatrix} 3 \\ -1 \\ -5 \end{pmatrix}$ and $\begin{pmatrix} 2 \\ 3 \\ -4 \end{pmatrix}$ are perpendicular to Π_1 and Π_2 respectively.

Therefore, $\begin{pmatrix} 3 \\ -1 \\ -5 \end{pmatrix} \times \begin{pmatrix} 2 \\ 3 \\ -4 \end{pmatrix}$ is perpendicular to both of these perpendiculars, and hence is in the direction of the common line.

Therefore, the direction of the common line is

$\begin{vmatrix} \mathbf{i} & \mathbf{j} & \mathbf{k} \\ 3 & -1 & -5 \\ 2 & 3 & -4 \end{vmatrix} = 19\mathbf{i}+2\mathbf{j}+11\mathbf{k}$

> **Note**
>
> Using the **r.n** = *d* form of the equation of a plane.

> **Note**
>
> Using the fact if the vectors n_1 and n_2 are the two directions perpendicular to the planes then the direction of the line of intersection is $n_1 \times n_2$.

> **Note**
>
> Using the component form of the vector product.

> **Note**
>
> To find a point on the common line, solving Π_1 and Π_2 will give only two equations to solve for three unknowns. So, you let $x = 0$ and solve the equations for the remaining two unknowns. However, if letting $x = 0$ causes problems because of the particular equations given, you may let either $y = 0$ or $z = 0$.

(continued)

Answer

(continued)

To find a point on the line, let $x = 0$.

Π_1 is $3x - y - 5z = 7$ When $x = 0$, Π_1 gives $-y - 5z = 7$

Π_2 is $2x + 3y - 4z = -2$ When $x = 0$, Π_2 gives $3y - 4z = -2$

Solving these simultaneous equations, $z = -1$, $y = -2$.

Therefore, the point $(0, -2, -1)$ lies on the common line, so its equation is

$$\mathbf{r} = \begin{pmatrix} 0 \\ -2 \\ -1 \end{pmatrix} + t \begin{pmatrix} 19 \\ 2 \\ 11 \end{pmatrix}.$$

> **Note**
>
> Using the vector equation of a line, $\mathbf{r} = \mathbf{a} + \lambda\mathbf{b}$.

Distance of a plane from the origin

Consider the equation of a plane in the form $\mathbf{r.n} = d$. If $\mathbf{n}$ is a unit vector ($\hat{\mathbf{n}}$), then d is the perpendicular distance of the plane from the origin. If $\mathbf{n}$ is *not* a unit vector and you would like to determine the distance from the origin, then divide each side of the equation by the magnitude of $\mathbf{n}$ to obtain the equation $\mathbf{r}.\,\dfrac{\mathbf{n}}{|\mathbf{n}|} = \dfrac{d}{|\mathbf{n}|}$. The distance of the plane from the origin is then $\dfrac{d}{|\mathbf{n}|}$ since $\dfrac{\mathbf{n}}{|\mathbf{n}|}$ is a unit vector.

Example 15

Question

Find the distance to the plane $3x + 4y - 5z = 21$ from the origin.

> **Note**
>
> Using the $\mathbf{r.n} = d$ form of the equation.

Answer

The equation of the plane is $\mathbf{r}.\begin{pmatrix} 3 \\ 4 \\ -5 \end{pmatrix} = 21.$

Magnitude of $\mathbf{n} = \sqrt{3^2 + 4^2 + (-5)^2} = 5\sqrt{2}$

$$\mathbf{r}.\begin{pmatrix} \dfrac{3}{5\sqrt{2}} \\[2mm] \dfrac{4}{5\sqrt{2}} \\[2mm] -\dfrac{5}{5\sqrt{2}} \end{pmatrix} = \dfrac{21}{5\sqrt{2}}$$

> **Note**
>
> Changing this to the form $\mathbf{r}.\hat{\mathbf{n}} = d$, where $\hat{\mathbf{n}}$ is a unit vector.

Therefore, the distance from the origin is $\dfrac{21}{5\sqrt{2}}$.

Distance of a plane from a point

To find the distance of a plane from a point, you first need to find the equation of the plane through the given point parallel to the given plane. Then you find the distance of each plane from the origin. The difference between these distances is equal to the distance of the given plane from the given point.

Example 16

Question

Find the distance from the point $(3, -2, 6)$ to the plane $3x + 4y - 5z = 21$.

Answer

Equation of plane parallel to $3x + 4y - 5z = 21$ through $(3, -2, 6)$:

$3x + 4y - 5z = (3 \times 3) - (2 \times 4) + (6 \times -5)$

$\Rightarrow 3x + 4y - 5z = 29$

$$\Rightarrow \mathbf{r} \cdot \begin{pmatrix} \dfrac{3}{5\sqrt{2}} \\[2mm] \dfrac{4}{5\sqrt{2}} \\[2mm] -\dfrac{5}{5\sqrt{2}} \end{pmatrix} = -\dfrac{29}{5\sqrt{2}}$$

Therefore, distance from point $(3, -22, 6)$ to the plane is

$\dfrac{29}{5\sqrt{2}} + \dfrac{21}{5\sqrt{2}} = \dfrac{50}{5\sqrt{2}} = 5\sqrt{2}$

> **Note**
>
> Adding the distance of the plane from the origin (from *Example 15*).

You can also calculate the distance of a plane from a point by calculating the equation of the line perpendicular to the plane using the vector equation of a line.

Example 17

Question

Find the distance from the point $(3, -2, 6)$ to the plane $3x + 4y - 5z = 21$.

Answer

Equation of line perpendicular to $3x + 4y - 5z = 21$ through $(3, -2, 6)$:

$$\mathbf{r} = \begin{pmatrix} 3 \\ -2 \\ 6 \end{pmatrix} + t \begin{pmatrix} 3 \\ 4 \\ -5 \end{pmatrix}$$

> **Note**
>
> Using the vector equation of a line, $\mathbf{r} = \mathbf{a} + t\mathbf{b}$.

The line meets the plane $3x + 4y - 5z = 21$ when

$3(3 + 3t) + 4(-2 + 4t) - 5(6 - 5t) = 21$

$\Rightarrow 50t = 50 \Rightarrow t = 1$

Using $\mathbf{r} = \mathbf{a} + t\mathbf{b}$, the line meets the plane at $(6, 2, 1)$.

Distance between the points $(3, -2, 6)$ and $(6, 2, 1)$ is $\sqrt{3^2 + 4^2 + 5^2} = 5\sqrt{2}$.

Therefore, distance from the point $(3, -2, 6)$ to the plane is $5\sqrt{2}$.

Exercise 3

1 Find the equation of a line through the points given, in the form

    i $\mathbf{r} = \mathbf{a} + t\mathbf{b}$ **ii** $(\mathbf{r} - \mathbf{a}) \times \mathbf{b} = 0$

    a $A(2, 4, -7)$ and $B(4, -1, -6)$ **b** $P(2, -5, 4)$ and $Q(4, -2, -3)$

2 In the form $\mathbf{a}.\mathbf{n} = d$, find the equation of the plane through

    a $A(4, 1, -5)$, $B(2, -1, -6)$, $C(-2, 3, 2)$

    b $P(2, 5, 3)$, $Q(4, 1, -2)$, $R(4, 3, 5)$

    c $D(4, 1, -3)$, $E(2, 3, 2)$, $F(-1, -3, 1)$

3 Find, in Cartesian form, the equation of the plane.

a $\mathbf{r} \cdot \begin{pmatrix} 3 \\ 1 \\ 7 \end{pmatrix} = 4$ **b** $\mathbf{r} \cdot \begin{pmatrix} 2 \\ 4 \\ 3 \end{pmatrix} = 8$ **c** $\mathbf{r} \cdot \begin{pmatrix} -1 \\ 5 \\ 3 \end{pmatrix} + 7 = 0$

4 Find the angle between each pair of planes.

   a $3x - y - 4z = 7,\ 2x + 3y - z = 11$ **b** $5x - 3y + 2z = 10,\ 2x - y - z = 8$

   c $7x + 4y - 2z = 5,\ 6x + 7y + z = 4$ **d** $x - 2y - 9z = 1,\ x + 3y + 2z = 0$

5 Find the angle between the line $\mathbf{r} = \begin{pmatrix} 1 \\ 4 \\ 5 \end{pmatrix} + t \begin{pmatrix} 2 \\ -3 \\ 4 \end{pmatrix}$ and the plane $2x + 4y - z = 7$.

6 Find the angle between the line $\mathbf{r} = \begin{pmatrix} 2 \\ -3 \\ 1 \end{pmatrix} + t \begin{pmatrix} 4 \\ 2 \\ -5 \end{pmatrix}$ and the plane $3x - y + 2z = 11$.

7 Write the equation of the plane $3x + 4y - 5z = 20$ in the form $\mathbf{r}.\hat{\mathbf{n}} = d$, where $\hat{\mathbf{n}}$ is a unit vector. Hence write down the distance from the plane to the origin.

8 Two planes have equations $3x - 3y + z = 7$ and $x + 2y - 3z = 6$. They meet in a line. Find the Cartesian equations of the line and its direction cosines.

9 The line l has equation $\mathbf{r} = (1 + 6t)\mathbf{i} + (3 - 5t)\mathbf{j} + (5 + 7t)\mathbf{k}$.

   a Write down a direction vector for l.

   b **i** Find direction cosines for l.

      ii Explain the geometrical significance of the direction cosines in relation to l.

   c Write down a vector equation for l in the form $(\mathbf{r} - \mathbf{a}) \times \mathbf{b} = 0$.

10 The line L has equation $\mathbf{r} = \begin{pmatrix} 4 \\ 2 \\ 5 \end{pmatrix} + t \begin{pmatrix} 3 \\ -1 \\ -2 \end{pmatrix}$.

   a Show that $P(16, -2, -3)$ lies on L.

   b Find the direction cosines of L and the acute angle between L and the y-axis.

   c The plane Π has Cartesian equation $2x - 5y + 3z = 38$. Write down a normal vector $\mathbf{n}$ to Π.

   d Find the acute angle between L and the normal vector $\mathbf{n}$.

   e Find the position vector of the point Q where L meets Π.

   f Determine the shortest distance from P to Π.

26.6 Scalar triple product and its applications

The **scalar triple product** of vectors $\mathbf{a}$, $\mathbf{b}$ and $\mathbf{c}$ is defined as $\mathbf{a}.\mathbf{b} \times \mathbf{c}$.

You must calculate $\mathbf{a}.\mathbf{b} \times \mathbf{c}$ as $\mathbf{a}.(\mathbf{b} \times \mathbf{c})$.

If you tried to calculate it as $(\mathbf{a}.\mathbf{b}) \times \mathbf{c}$, you would have the vector product of a scalar and a vector (since $\mathbf{a}.\mathbf{b}$ is a scalar product), which, by definition of a vector product, cannot exist.

The scalar triple product is found to have useful applications, including calculating the volume of a parallelepiped, which you will see later in the chapter.

Example 18

Calculate $\begin{pmatrix} 3 \\ 4 \\ 7 \end{pmatrix} \cdot \begin{pmatrix} 2 \\ 3 \\ -1 \end{pmatrix} \times \begin{pmatrix} 7 \\ 4 \\ 2 \end{pmatrix}$.

$\begin{pmatrix} 3 \\ 4 \\ 7 \end{pmatrix} \cdot \begin{pmatrix} 2 \\ 3 \\ -1 \end{pmatrix} \times \begin{pmatrix} 7 \\ 4 \\ 2 \end{pmatrix} = \begin{pmatrix} 3 \\ 4 \\ 7 \end{pmatrix} \cdot \left(\begin{pmatrix} 2 \\ 3 \\ -1 \end{pmatrix} \times \begin{pmatrix} 7 \\ 4 \\ 2 \end{pmatrix} \right)$

$= \begin{pmatrix} 3 \\ 4 \\ 7 \end{pmatrix} \cdot \begin{pmatrix} 10 \\ -11 \\ -13 \end{pmatrix} = 30 - 44 - 91 = -105$

> **Note**
>
> Using, $\mathbf{a}.\mathbf{b} \times \mathbf{c}$; you must calculate the vector
>
> product first; using $\mathbf{b} \times \mathbf{c} = \begin{vmatrix} \mathbf{i} & \mathbf{j} & \mathbf{k} \\ b_1 & b_2 & b_3 \\ c_1 & c_2 & c_3 \end{vmatrix}$ and
>
> $(b_2 c_3 - b_3 c_2)\mathbf{i} - (b_1 c_3 - b_3 c_1)\mathbf{j} + (b_1 c_2 - b_2 c_1)\mathbf{k}$,
> you get $10\mathbf{i} - 11\mathbf{j} - 13\mathbf{k}$.

Therefore $\begin{pmatrix} 3 \\ 4 \\ 7 \end{pmatrix} \cdot \begin{pmatrix} 2 \\ 3 \\ -1 \end{pmatrix} \times \begin{pmatrix} 7 \\ 4 \\ 2 \end{pmatrix} = -105$.

> **Note**
>
> Calculate the dot product.

A quicker way to find $\mathbf{a}.\mathbf{b} \times \mathbf{c}$ is as follows.

The vector product $\mathbf{b} \times \mathbf{c}$ is given by $\mathbf{b} \times \mathbf{c} = \begin{vmatrix} \mathbf{i} & \mathbf{j} & \mathbf{k} \\ b_1 & b_2 & b_3 \\ c_1 & c_2 & c_3 \end{vmatrix}$

$= \mathbf{i}(b_2 c_3 - b_3 c_2) - \mathbf{j}(b_1 c_3 - b_3 c_1) + \mathbf{k}(b_1 c_2 - b_2 c_1)$

Therefore, the scalar triple product $\mathbf{a}.\mathbf{b} \times \mathbf{c}$ is given by

$\mathbf{a}.\mathbf{b} \times \mathbf{c} = a_1(b_2 c_3 - b_3 c_2) - a_2(b1 c_3 - b_3 c_1) + a_3(b_1 c_2 - b_2 c_1)$

$\mathbf{a}.\mathbf{b} \times \mathbf{c} = \begin{vmatrix} a_1 & a_2 & a_3 \\ b_1 & b_2 & b_3 \\ c_1 & c_2 & c_3 \end{vmatrix}$

Applying this result to *Example 16*, you would have a 3×3 matrix, the determinant of which gives you the same result.

$\begin{pmatrix} 3 \\ 4 \\ 7 \end{pmatrix} \cdot \begin{pmatrix} 2 \\ 3 \\ -1 \end{pmatrix} \times \begin{pmatrix} 7 \\ 4 \\ 2 \end{pmatrix} = \begin{vmatrix} 3 & 4 & 7 \\ 2 & 3 & -1 \\ 7 & 4 & 2 \end{vmatrix}$

$= 3 \times 10 - 4 \times 11 + 7 \times -13$

$= -105$

Volume of a parallelepiped

A parallelepiped is a polyhedron with six faces, each of which is a parallelogram. Since the shape is a three-dimensional one, in order to calculate its volume, you need to use three vectors. The scalar triple product is a useful tool for doing this since at each vertex of a parallelepiped there are three adjacent edges that can be expressed as non-parallel vectors.

Consider the parallelepiped *OBDCAQRS*, which has adjacent edges $\overrightarrow{OA} = \mathbf{a}$, $\overrightarrow{OB} = \mathbf{b}$ and $\overrightarrow{OC} = \mathbf{c}$.

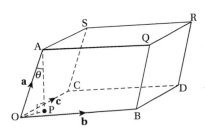

The volume of a parallelepiped is given by

Volume = Area of base × Perpendicular height

Therefore, the volume, V, of $OBDCAQRS$ is

$V = |\mathbf{b} \times \mathbf{c}| \times$ Perpendicular height

The perpendicular height, AP, is $|\mathbf{a}| \cos \theta$. Therefore, you have

$V = |\mathbf{b} \times \mathbf{c}| \times |\mathbf{a}| |\cos \theta|$

$\quad = |\mathbf{a}| |\mathbf{b} \times \mathbf{c}| |\cos \theta|$

Note that this is identical to the scalar product $\mathbf{x.y} = |\mathbf{x}| |\mathbf{y}| \cos \theta$, with $|\mathbf{x}| = |\mathbf{a}|$, $|\mathbf{y}| = |\mathbf{b} \times \mathbf{c}|$ and $\mathbf{b} \times \mathbf{c}$ having the same sense as PA. Therefore,

> The volume of a parallelepiped is given by $V = |\mathbf{a \cdot b} \times \mathbf{c}|$, where the vectors a, b and c represent three adjacent edges of the parallelepiped.

Note that a cuboid is a special case of a parallelepiped, so this equation can also be used to find the volume of a cuboid.

When calculating the volume of a parallelepiped, you can use any vectors that are measured between different pairs of points, but it is important to make sure that no two vectors are parallel. The best approach to determine three vectors that are not parallel, is to use one chosen vertex and find the vector from this vertex to each of three other points.

Example 19

Find the volume of parallelepiped $ABCDPQRS$, where A is $(3, 1, 7)$, B is $(2, 0, 4)$, D is $(7, 2, -1)$ and P is $(8, 3, 11)$.

> **Note**
>
> Using $V = |\mathbf{a.b} \times \mathbf{c}|$.

The volume, V, of parallelepiped $ABCDPQRS$ is given by $V = \overrightarrow{AP} \cdot \overrightarrow{AB} \times \overrightarrow{AD}$.

> **Note**
>
> Here, A is the chosen vertex.

$\overrightarrow{AP} = \mathbf{p} - \mathbf{a} = \begin{pmatrix} 8 \\ 3 \\ 11 \end{pmatrix} - \begin{pmatrix} 3 \\ 1 \\ 7 \end{pmatrix} = \begin{pmatrix} 5 \\ 2 \\ 4 \end{pmatrix}$

$\overrightarrow{AB} = \mathbf{b} - \mathbf{a} = \begin{pmatrix} 2 \\ 0 \\ 4 \end{pmatrix} - \begin{pmatrix} 3 \\ 1 \\ 7 \end{pmatrix} = \begin{pmatrix} -1 \\ -1 \\ -3 \end{pmatrix}$

> **Note**
>
> Find the vector from A to B.

$\overrightarrow{AD} = \mathbf{d} - \mathbf{a} = \begin{pmatrix} 7 \\ 2 \\ -1 \end{pmatrix} - \begin{pmatrix} 3 \\ 1 \\ 7 \end{pmatrix} = \begin{pmatrix} 4 \\ 1 \\ -8 \end{pmatrix}$

> **Note**
>
> Find the vector from A to D.

$\overrightarrow{AB} \times \overrightarrow{AD} = \begin{pmatrix} 11 \\ -20 \\ 3 \end{pmatrix}$

$V = \overrightarrow{AP} \cdot \overrightarrow{AB} \times \overrightarrow{AD}$

$\quad = \begin{pmatrix} 5 \\ 2 \\ 4 \end{pmatrix} \cdot \begin{pmatrix} 11 \\ -20 \\ 3 \end{pmatrix} = 55 - 40 + 12 = 27$

> **Note**
>
> You could also use the formula for the scalar triple product,
>
> $\mathbf{a.b} \times \mathbf{c} = \begin{vmatrix} a_1 & a_2 & a_3 \\ b_1 & b_2 & b_3 \\ c_1 & c_2 & c_3 \end{vmatrix}$,
>
> to find $V = \overrightarrow{AP} \cdot \overrightarrow{AB} \times \overrightarrow{AD}$
>
> $= \begin{pmatrix} 5 \\ 2 \\ 4 \end{pmatrix} \cdot \begin{pmatrix} -1 \\ -1 \\ -3 \end{pmatrix} \times \begin{pmatrix} 4 \\ 1 \\ -8 \end{pmatrix}$
>
> $V = \begin{vmatrix} 5 & 2 & 4 \\ -1 & -1 & -3 \\ 4 & 1 & -8 \end{vmatrix}$
>
> $= 5(8+3) - 2(8+12) + 4(-1+4)$
>
> $= 27$

> **Note**
>
> Using $\mathbf{a} \times \mathbf{b} = \begin{vmatrix} \mathbf{i} & \mathbf{j} & \mathbf{k} \\ a_1 & a_2 & a_3 \\ b_1 & b_2 & b_3 \end{vmatrix}$
>
> and $(a_2 b_3 - a_3 b_2)\mathbf{i} - (a_1 b_3 - a_3 b_1)\mathbf{j} + (a_1 b_2 - a_2 b_1)\mathbf{k}$.

Coplanar vectors

Vectors are coplanar if they all are contained in the same plane. You can use the scalar triple product to determine if three vectors are coplanar since,

> Three vectors are coplanar if their scalar triple product is zero.

Suppose that $\mathbf{a}.\mathbf{b} \times \mathbf{c} = \mathbf{a}.(|\mathbf{b}||\mathbf{c}| \sin \theta \, \hat{\mathbf{n}})$

$$= |\mathbf{a}||\mathbf{b}||\mathbf{c}| \sin \theta \cos \phi,$$

where θ is the angle between $\mathbf{b}$ and $\mathbf{c}$, and ϕ is the angle between $\mathbf{a}$ and $\hat{\mathbf{n}}$, which is perpendicular to the plane containing $\mathbf{b}$ and $\mathbf{c}$. Therefore, you get

$$\mathbf{a}.\mathbf{b} \times \mathbf{c} = |\mathbf{a}||\mathbf{b}||\mathbf{c}| \sin \theta \sin \psi,$$

where $\psi = (90° - \phi)$ is the angle between $\mathbf{a}$ and the plane containing $\mathbf{b}$ and $\mathbf{c}$.

Hence, when $\mathbf{a}$, $\mathbf{b}$ and $\mathbf{c}$ are **coplanar** ($\mathbf{a}$, $\mathbf{b}$ and $\mathbf{c}$ lie in the same plane), you have

$$\mathbf{a}.\mathbf{b} \times \mathbf{c} = 0$$

Instead of saying that $\mathbf{a}$, $\mathbf{b}$ and $\mathbf{c}$ are coplanar, you can say that the three vectors are **linearly dependent**, which means that one vector is a linear combination of the other two vectors. For example, by calculating $\mathbf{a} \cdot \mathbf{b} \times \mathbf{c}$, you can show that the

three vectors $\mathbf{a} = \begin{pmatrix} 2 \\ 8 \\ 12 \end{pmatrix}$, $\mathbf{b} = \begin{pmatrix} 2 \\ 5 \\ 9 \end{pmatrix}$, $\mathbf{c} = \begin{pmatrix} 1 \\ 1 \\ 3 \end{pmatrix}$ are coplanar. Alternatively, you could prove they are linearly dependent by observing they are related by the equation $\mathbf{b} - \frac{1}{2}\mathbf{a} = \mathbf{c}$.

You can also consider coplanar vectors in terms of the volume of a parallelepiped. If vectors are coplanar then it follows that the volume of the parallelepiped that they generate would be zero. It therefore follows that the scalar triple product would be zero for three coplanar vectors, since the volume of a parallelepiped is $V = |\mathbf{a} \cdot \mathbf{b} \times \mathbf{c}|$, where $\mathbf{a} \cdot \mathbf{b} \times \mathbf{c}$ is the scalar triple product.

Exercise 4

1. Are the vectors $\begin{pmatrix} 2 \\ 4 \\ -1 \end{pmatrix}$, $\begin{pmatrix} 7 \\ 2 \\ 3 \end{pmatrix}$ and $\begin{pmatrix} -1 \\ -3 \\ 5 \end{pmatrix}$ coplanar?

2. Find the value of $\begin{pmatrix} 3 \\ -2 \\ 1 \end{pmatrix} \cdot \begin{pmatrix} 2 \\ 1 \\ 4 \end{pmatrix} \times \begin{pmatrix} 1 \\ 5 \\ -2 \end{pmatrix}$.

3. Find the value of $\begin{pmatrix} 2 \\ 4 \\ -5 \end{pmatrix} \cdot \begin{pmatrix} 3 \\ 8 \\ 2 \end{pmatrix} \times \begin{pmatrix} 2 \\ -3 \\ 6 \end{pmatrix}$.

4. Find the value of $\begin{pmatrix} -1 \\ 2 \\ 5 \end{pmatrix} \cdot \begin{pmatrix} 2 \\ 3 \\ 1 \end{pmatrix} \times \begin{pmatrix} 3 \\ 8 \\ 4 \end{pmatrix}$.

5 Find the volume of a parallelepiped *ABCDEFGH*.

$$\overrightarrow{AB} = \begin{pmatrix} 1 \\ 3 \\ 2 \end{pmatrix} \quad \overrightarrow{AD} = \begin{pmatrix} -2 \\ 1 \\ -3 \end{pmatrix} \quad \overrightarrow{AE} = \begin{pmatrix} 5 \\ 2 \\ 7 \end{pmatrix}$$

6 Find the volume of a parallelepiped *ABCDEFGH*, where *A* is the point $(3, 5, 7)$, *B* is $(4, 2, 7)$, *C* is $(1, 3, 7)$ and *E* is $(1, 5, 8)$.

Summary

▶ A two-dimensional vector **a** can be considered as being composed of two parts relative to another vector **b**: one in the direction of a vector **b** (parallel to the *x*-axis), and the other perpendicular to the direction of vector **b**. The resolved part of vector **a** in the direction of vector **b** is $\dfrac{\mathbf{a}.\mathbf{b}}{|\mathbf{b}|}$.

▶ The direction of a vector can be specified by the ratios of the components in the **i**, **j** and **k** directions. These are called the direction ratios of the vector and are normally expressed as integers. For example, the direction ratios of the vector $28\mathbf{i} - 21\mathbf{j} - 14\mathbf{k}$ are $4 : -3 : -2$.

▶ If θ_x is the angle that vector **a** makes with the **i**-axis, you have $\cos\theta_x = \dfrac{a_1}{|\mathbf{a}|}$. Likewise for θ_y and θ_z, you have $\cos\theta_y = \dfrac{a_2}{|\mathbf{a}|}$ and $\cos\theta_z = \dfrac{a_3}{|\mathbf{a}|}$; these are the direction cosines of vector **a**, and they represent another way of specifying a vector's direction.

▶ If the direction cosines are *l*, *m* and *n*, then you will notice that $l^2 + m^2 + n^2 = 1$.

▶ Given the matrix $\begin{pmatrix} a & b & c \\ d & e & f \\ g & h & i \end{pmatrix}$, the 3×3 determinant $\begin{vmatrix} a & b & c \\ d & e & f \\ g & h & i \end{vmatrix}$ represents the expression $a\begin{vmatrix} e & f \\ h & i \end{vmatrix} - b\begin{vmatrix} d & f \\ g & i \end{vmatrix} + c\begin{vmatrix} d & e \\ g & h \end{vmatrix}$ which is $a(ei - fh) - b(di - fg) + c(dh - eg)$.

▶ The vector product of two vectors **a** and **b** is denoted by $\mathbf{a} \times \mathbf{b}$, and is defined as $\mathbf{a} \times \mathbf{b} = |\mathbf{a}||\mathbf{b}|\sin\theta\,\hat{\mathbf{n}}$, where θ is the angle measured in the anticlockwise sense between **a** and **b**, and $\hat{\mathbf{n}}$ is a unit vector, such that **a**, **b** and $\hat{\mathbf{n}}$ (in that order) form a right-handed set.

▶ The vector product is anticommutative: $\mathbf{a} \times \mathbf{b} = -\mathbf{b} \times \mathbf{a}$.

▶ The vector product of parallel vectors is zero: hence, $\mathbf{i} \times \mathbf{i} = \mathbf{j} \times \mathbf{j} = \mathbf{k} \times \mathbf{k} = \mathbf{0}$.

▶ The vector product of perpendicular vectors, **a** and **b**, leads to the following results:

$\mathbf{i} \times \mathbf{j} = \mathbf{k}$ $\mathbf{j} \times \mathbf{i} = -\mathbf{k}$

$\mathbf{j} \times \mathbf{k} = \mathbf{i}$ $\mathbf{k} \times \mathbf{j} = -\mathbf{i}$

$\mathbf{k} \times \mathbf{i} = \mathbf{j}$ $\mathbf{i} \times \mathbf{k} = -\mathbf{j}$

▶ Expressing vectors in their component form, and considering the calculation of the determinant of a 3×3 matrix, leads to the component form of the vector product

$$\mathbf{a} \times \mathbf{b} = \begin{vmatrix} \mathbf{i} & \mathbf{j} & \mathbf{k} \\ a_1 & a_2 & a_3 \\ b_1 & b_2 & b_3 \end{vmatrix} = (a_2 b_3 - a_3 b_2)\mathbf{i} - (a_1 b_3 - a_3 b_1)\mathbf{j} + (a_1 b_2 - a_2 b_1)\mathbf{k}$$

▶ In general, to find the area of a triangle whose sides are $\mathbf{a}$, $\mathbf{b}$ and $\mathbf{c}$, you can use any of the formulae

$$\frac{1}{2}|\mathbf{a} \times \mathbf{b}| \quad \text{or} \quad \frac{1}{2}|\mathbf{b} \times \mathbf{c}| \quad \text{or} \quad \frac{1}{2}|\mathbf{a} \times \mathbf{c}|$$

▶ The area of a parallelogram with two sides formed by the vectors $\mathbf{a}$ and $\mathbf{b}$ is $\mathbf{a} \times \mathbf{b}$.

▶ The equation of a line through a point with position vector $\mathbf{a}$ in the direction of vector $\mathbf{b}$, can be found using the formula $(\mathbf{r} - \mathbf{a}) \times \mathbf{b} = 0$.

▶ The vector equation of a plane can be written in the form
  - $\mathbf{r} = \mathbf{a} + t\mathbf{b} + s\mathbf{c}$, where $\mathbf{a}$ is a point on the plane and $\mathbf{b}$ and $\mathbf{c}$ are two non-parallel lines within the plane
  - $\mathbf{r} = \mathbf{a} + \lambda(\mathbf{b} - \mathbf{a}) + \mu(\mathbf{c} - \mathbf{a}) = (1 - \lambda - \mu)\mathbf{a} + \lambda\mathbf{b} + \mu\mathbf{c}$, where the plane passes through three non-collinear points A, B and C. The plane through points A, B and C, has a point on the plane, A, and the two vectors $\overrightarrow{AB}$ (or $\mathbf{b} - \mathbf{a}$) and $\overrightarrow{AC}$ (or $\mathbf{c} - \mathbf{a}$)
  - $\mathbf{r}.\mathbf{n} = d$, where d is a constant that determines the position of the plane.

▶ If $\mathbf{n}$ is a unit vector, then d is the perpendicular distance of the plane from the origin.

▶ When d has the same sign for two planes, these planes are on the same side of the origin.

▶ When d has opposite signs for two planes, these planes are on opposite sides of the origin.

▶ The Cartesian equation of a plane is $\begin{pmatrix} x \\ y \\ z \end{pmatrix}.\begin{pmatrix} a \\ b \\ c \end{pmatrix} = d$ or $ax + by + cz = d$, where

a, b, c and d are constants and the vector $\begin{pmatrix} a \\ b \\ c \end{pmatrix}$ is perpendicular to the plane.

▶ You can use the equation of a plane in the form $\mathbf{r}.\mathbf{n} = d$ to calculate the angle between two planes, the angle between a plane and a line, to find the distance of a plane from the origin or a point, and to find the line of intersection of planes that are not parallel.

▶ If the vectors $\mathbf{n}_1$ and $\mathbf{n}_2$ are the two directions perpendicular to the planes then the direction of the line of intersection is $\mathbf{n}_1 \times \mathbf{n}_2$.

▶ The scalar triple product of vectors $\mathbf{a}$, $\mathbf{b}$ and $\mathbf{c}$ is defined as $\mathbf{a}.\mathbf{b} \times \mathbf{c}$. You must calculate $\mathbf{a}.\mathbf{b} \times \mathbf{c}$ as $\mathbf{a}.(\mathbf{b} \times \mathbf{c})$.

▶ When the vectors are expressed in component for m, the scalar triple product can be calculated using $\mathbf{a}.\mathbf{b} \times \mathbf{c} = \begin{vmatrix} a_1 & a_2 & a_3 \\ b_1 & b_2 & b_3 \\ c_1 & c_2 & c_3 \end{vmatrix}$.

- The volume of a parallelepiped is given by $V = |\mathbf{a}.\mathbf{b} \times \mathbf{c}|$, where the vectors $\mathbf{a}$, $\mathbf{b}$ and $\mathbf{c}$ represent three adjacent edges of the parallelepiped.
- The scalar triple product can be used to determine that three vectors, $\mathbf{a}$, $\mathbf{b}$, $\mathbf{c}$, are coplanar precisely when $\mathbf{a}.\mathbf{b} \times \mathbf{c} = 0$.

Review exercises

1 Find the resolved part of $\mathbf{i} - 5\mathbf{j} - 2\mathbf{k}$ in the direction $\mathbf{j} - 7\mathbf{k}$.

2 Find the direction cosines of $\mathbf{i} + 2\mathbf{j} - 6\mathbf{k}$.

3 Find, in vector form, the equation of the plane through A(4, 0, 2), B(2, −1, −1), and C(1, 6, 2).

4 Find the angle between the line $\mathbf{r} = \begin{pmatrix} 1 \\ 2 \\ 1 \end{pmatrix} + t\begin{pmatrix} -2 \\ 3 \\ 1 \end{pmatrix}$ and the plane $3x - y + z = 1$.

5 Write the equation of the plane $3x - 4y + z = 8$ in the form $\mathbf{r}.\mathbf{n} = d$, where $\mathbf{n}$ is a unit vector, and hence write down the distance of the plane to the origin.

6 Find the volume of a parallelepiped *ABCDEFGH*, given that

$$\overrightarrow{AB} = \begin{pmatrix} 1 \\ 4 \\ -3 \end{pmatrix}, \overrightarrow{AD} = \begin{pmatrix} 2 \\ 5 \\ -2 \end{pmatrix} \text{ and } \overrightarrow{AE} = \begin{pmatrix} 4 \\ 1 \\ -1 \end{pmatrix}.$$

Practice examination questions

1 A line has vector equation $\left(\mathbf{r} - \begin{bmatrix} 3 \\ -2 \\ 6 \end{bmatrix}\right) \times \begin{pmatrix} 4 \\ 7 \\ -4 \end{pmatrix} = 0$.

  a Determine the direction cosines of this line. (3 marks)

  b Explain the geometrical significance of the direction cosines in relation to the line. (1 mark)

AQA MFP4 June 2012

2 The line *l* has equation $\mathbf{r} = (1 + 4t)\mathbf{i} + (-2 + 12t)\mathbf{j} + (1 - 3t)\mathbf{k}$.

  a Write down a direction vector for *l*. (1 mark)

  b **i** Find direction cosines for *l*. (2 marks)

     ii Explain the geometrical significance of the direction cosines in relation to *l*. (1 mark)

  c Write down a vector equation for *l* in the form $(\mathbf{r} - \mathbf{a}) \times \mathbf{b} = 0$. (2 marks)

AQA MFP4 January 2009

3 The non-zero vectors $\mathbf{a}$ and $\mathbf{b}$ have magnitudes a and b respectively.

  Let $c = |\mathbf{a} \times \mathbf{b}|$ and $d = |\mathbf{a}.\mathbf{b}|$.

  By considering the definitions of vector and scalar products, or otherwise, show that $c^2 + d^2 = a^2 b^2$. (3 marks)

AQA MFP4 January 2011

4 The fixed points A and B and the variable point C have position vectors

$$a = \begin{pmatrix} 3 \\ -4 \\ 1 \end{pmatrix}, b = \begin{pmatrix} 2 \\ 1 \\ -3 \end{pmatrix} \text{ and } c = \begin{pmatrix} 2-t \\ t \\ 5 \end{pmatrix}$$

respectively, relative to the origin O, where t is a scalar parameter.

a Find the equation of the line AB in the form $(\mathbf{r} - \mathbf{u}) \times \mathbf{v} = 0$. (3 marks)

b Determine $\mathbf{b} \times \mathbf{c}$ in terms of t. (4 marks)

c **i** Show that $\mathbf{a}.(\mathbf{b} \times \mathbf{c})$ is constant for all values of t, and state the value of this constant. (2 marks)

   ii Write down a geometrical conclusion that can be deduced from the answer to part **ci**. (1 mark)

AQA MFP4 June 2010

5 The points X, Y and Z have position vectors $\mathbf{x} = \begin{pmatrix} 2 \\ 3 \\ 2 \end{pmatrix}$, $\mathbf{y} = \begin{pmatrix} 5 \\ 7 \\ 4 \end{pmatrix}$ and $\mathbf{z} = \begin{pmatrix} -8 \\ 1 \\ a \end{pmatrix}$ respectively, relative to the origin O.

a Find

   i $\mathbf{x} \times \mathbf{y}$; (2 marks)

   ii $(\mathbf{x} \times \mathbf{y}) \cdot \mathbf{z}$ (2 marks)

b Using these results, or otherwise, find

   i the area of triangle OXY (2 marks)

   ii the value of a for which $\mathbf{x}$, $\mathbf{y}$ and $\mathbf{z}$ are linearly dependent. (2 marks)

AQA MFP4 January 2009

6 The points A, B, C and D have position vectors $\mathbf{a}$, $\mathbf{b}$, $\mathbf{c}$ and $\mathbf{d}$ respectively relative to the origin O, where

$$\mathbf{a} = \begin{bmatrix} 1 \\ 2 \\ -1 \end{bmatrix}, \mathbf{b} = \begin{bmatrix} 3 \\ 4 \\ 2 \end{bmatrix}, \mathbf{c} = \begin{bmatrix} -1 \\ 0 \\ 4 \end{bmatrix} \text{ and } \mathbf{d} = \begin{bmatrix} 4 \\ 1 \\ -2 \end{bmatrix}$$

a Find $\overrightarrow{AB} \times \overrightarrow{AC}$. (3 marks)

b The points A, B and C lie in the plane Π. Find a Cartesian equation for Π. (2 marks)

c Find the volume of the parallelepiped defined by $\overrightarrow{AB}$, $\overrightarrow{AC}$ and $\overrightarrow{AD}$. (3 marks)

AQA MFP4 June 2013

7 The planes Π_1 and Π_2 have Cartesian equations
$$2x + y - z = 3$$
$$3x - 2y + z = 5$$
respectively.

Find, in the form $\mathbf{r} = \mathbf{a} + \lambda\mathbf{d}$, a vector equation for the line L, which is the intersection of Π_1 and Π_2. (5 marks)

AQA MFP4 June 2012

Introduction

You already know how to use two simultaneous linear equations in x and y to solve real world questions. When solving such equations, there are three possible outcomes related to how the lines appear when drawn in $\mathbf{R}^2$ (xy plane): if there is no solution for x and y, the lines are parallel but distinct; if there is a unique solution for x and y, then the lines intersect in one place; or, the two equations are the same, and therefore the lines are identical and there are infinitely many solutions. In this chapter, you will look at a similar situation but in $\mathbf{R}^3$ (xyz space), hence you will be dealing with three planes rather than two lines.

Objectives

By the end of this chapter, you should know how to:
► Solve up to three linear equations in three unknowns.
► Explain the geometric interpretation of the solution of three equations in three unknowns.

Recap

You will need to remember...
► How to solve two simultaneous linear equations.
► How to store data as a matrix.
► How to find the determinant of a 3×3 matrix.
► How to create and use equations of a plane.

See Chapter *7 Matrices and Transformations* if you need to.

See Chapter *26 Vectors and Three-Dimensional Coordinate Geometry*.

27.1 Simultaneous linear equations

In earlier studies of mathematics, you have solved two equations with two unknowns. It is also possible to solve three linear equations in up to three unknowns, for example

$$a_1x + b_1y + c_1z + d_1 = 0$$
$$a_2x + b_2y + c_2z + d_2 = 0$$
$$a_3x + b_3y + c_3z + d_3 = 0$$

There are several ways to solve three linear equations and it is acceptable for you to use any method you like. For example, you might decide to use the elimination method you used for solving two simultaneous equations, and start by eliminating z and then y.

Example 1

A system of three equations is given. Find the solution of the three equations using elimination.

$2x + 5y + z = 15$	[1]
$3x - 4y - 5z = 1$	[2]
$5x + 3y + 2z = 10$	[3]
$5[1] + [2]$ gives $13x + 21y = 76$	[4]
$2[1] - [3]$ gives $-x + 7y = 20$	[5]

$[4] + 13[5]$ gives $112y = 336 \rightarrow y = 3$

> **Note**
>
> Eliminating z.

Substituting into [5] and then [1] gives $x = 1$, $z = -2$.

The solution of the three equations is $x = 1$, $y = 3$, $z = -2$.

However, in this chapter, you will see how to use **determinants** to produce a **general solution** that can be used to help solve three linear equations with three unknowns, but first you should understand the geometric interpretation of the solutions.

Geometric interpretation of three equations in three unknowns

Each of the three equations $a_i x + b_i y + c_i z + d_i = 0$ $(i = 1, 2, 3)$ can be considered as the equation of a plane in three-dimensional space.

You learned about the equation of a plane in Chapter *26 Vectors and Three-Dimensional Coordinate Geometry*.

With three planes, there are seven possible configurations:

▶ The three planes intersect in a single point. In this case, the three equations have a **unique solution**.

▶ The equations have no solutions and are said to be **inconsistent**. There are four ways this can happen:
  - The three planes form a triangular prism. In this case, there is no point where all three planes intersect so the equations are inconsistent.
  - Two of the planes are parallel and separate, and are intersected by the third plane. Again, there is no point where all three planes intersect, so the equations are inconsistent.
  - All three planes are parallel and separate; the planes have no common point and so the equations are inconsistent.

- Two of the planes are coincident and the third plane is parallel but separate; the planes have no common point and so the equations are inconsistent.

► The two remaining configurations correspond to the three equations having **infinitely many solutions**.

- The three planes have a common line, giving an infinite number of points (x, y, z) that satisfy all three equations. In this case, the equations are said to be **linearly dependent**, and the configuration is called a **sheaf of planes** or a **pencil of planes**.

- All three planes coincide, giving an infinite number of points that satisfy all three equations.

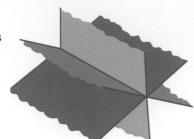

When you solve three linear equations in three unknowns algebraically, you can use your result to determine the geometrical representation of these equations if they were planes. So, if you were to find that there was a unique solution, you would expect the planes to intersect at a single point, as is the case in *Example 1*.

Example 2

Question

A system of equations is given by

$x + 2y + 2z = 7$

$3x - 9y + z = 11$

$2x - 11y - z = 4$

a Find the solution of this system of equations, showing all your working.

b interpret this solution geometrically

Answer

a

Determinant $\begin{vmatrix} 1 & 2 & 2 \\ 3 & -9 & 1 \\ 2 & -11 & -1 \end{vmatrix} = 9 + 11 \quad -2(-3-2) + 2(-33+18) \quad = 0$

hence the equations do not have a unique solution

$\quad$ let $y = t$,

$\quad\quad x + 2z = 7 - 2t \quad\quad$ [1]

$\quad\quad 3x + z = 11 + 9t \quad\quad$ [2]

$\quad\quad 2x - z = 4 + 11t \quad\quad$ [3]

$2[2] - [1]\ 5x = 15 + 20t$

$\quad\quad\quad x = 3 + 4t$

$2[1] - [3]\ 5z = 10 - 15t$

$\quad\quad\quad z = 2 - 3t$

solution is $x = 3 + 4t, y = t, z = 2 - 3t$

b None of the equations are a multiple of any other; hence the equations represent three planes which meet in a line – they form a sheaf.

Exercise 1

In questions **1** to **3**, use elimination or substitution.

a Find the solution of each of the systems of three equations.

b State the geometrical implications of your results.

1 $2x + 5y + z = 15$

$3x - 4y - 5z = 1$

$5x + 3y + 2z = 10$

2 $-3x + 2y + 5z = 1$

$4x - y - 3z = 6$

$2x - 3y - z = -8$

3 $4x + 5y + 4z = -3$

$x - 6y - 7z = 13$

$2x + 4y + z = -7$

27.2 Using determinants to find the number of solutions of three simultaneous equations

It is possible to use determinants to discover where three simultaneous equations have a unique solution.

If the equations' coefficients include an unknown, λ say, then the determinant will involve λ. There may be values of λ for which the determinant is zero, and for these values the equations do not have a unique solution. By substituting these values back into the equations you can decide whether the equations are then inconsistent (no solution) or consistent (an infinite number of solutions). All other values of λ will correspond to equations with a unique solution.

Example 3

Question

How many solutions are there to the simultaneous equations below?

$4x - \lambda y + 6z = 2 \qquad 2y + \lambda z = 1 \qquad x - 2y + 4z = 0$

Answer

$$\begin{vmatrix} a_1 & b_1 & c_1 \\ a_2 & b_2 & c_2 \\ a_3 & b_3 & c_3 \end{vmatrix} = \begin{vmatrix} 4 & -\lambda & 6 \\ 0 & 2 & \lambda \\ 1 & -2 & 4 \end{vmatrix}$$

$$= 4\begin{vmatrix} 2 & \lambda \\ -2 & 4 \end{vmatrix} + \lambda\begin{vmatrix} 0 & \lambda \\ 1 & 4 \end{vmatrix} + 6\begin{vmatrix} 0 & 2 \\ 1 & -2 \end{vmatrix}$$

$$= 4(8 + 2\lambda) + \lambda(-\lambda) + 6(-2)$$

$$= -\lambda^2 + 8\lambda + 20$$

Therefore, there is a unique solution unless $-\lambda^2 + 8\lambda + 20 = 0$,

$\Rightarrow \qquad \lambda^2 - 8\lambda - 20 = 0$

$\Rightarrow \qquad (\lambda - 10)(\lambda + 2) = 0$

> **Note**
>
> First, find the determinant of the system to see if there is a unique solution.

> **Note**
>
> Using the expansion of a 3×3 determinant.

(continued)

(continued)

There is a unique solution unless $\lambda = 10$ or $\lambda = -2$.

If $\lambda = 10$, the equations are

$4x - 10y + 6z = 2$ [1]

$2y + 10z = 1$ [2]

$x - 2y + 4z = 0$ [3]

$[1] - 4[3]$, gives

     $-2y - 10z = 2$

$\Rightarrow$ $2y + 10z = -2$

This contradicts equation [2], so the equations have no solution; they are inconsistent when $\lambda = 10$.

If $\lambda = -2$, the equations are

$4x + 2y + 6z = 2$ [4]

$2y - 2z = 1$ [5]

$x - 2y + 4z = 0$ [6]

$4[6] - [4]$

     $-10y + 10z = -2$

$\Rightarrow$ $2y - 2z = \dfrac{2}{5}$

This contradicts equation [5], so the equations have no solution. They are inconsistent when $\lambda = -2$.

Therefore, the determinant does not equal zero and there is a unique solution.

When there is not a unique solution, you cannot use the general formula for the solution of three equations; you must then proceed using some other method, such as elimination.

Example 4

Solve the equations.

$2x - 3y + 4z = 1$ [1] $3x - y = 2$ [2] $x + 2y - 4z = 1$ [3]

$$\begin{vmatrix} a_1 & b_1 & c_1 \\ a_2 & b_2 & c_2 \\ a_3 & b_3 & c_3 \end{vmatrix} = \begin{vmatrix} 2 & -3 & 4 \\ 3 & -1 & 0 \\ 1 & 2 & -4 \end{vmatrix} = 0$$

Therefore, there is not a unique solution.

Adding equations [1] and [3]

$3x - y = 2$

This is equation [2], therefore, equations are linearly dependent.

Let $x = t$.

(continued)

(continued)

Using equation [2], $y = 3t - 2$. Substituting into equation [3],

$$4z = t + 2(3t - 2) - 1$$

$$\Rightarrow \quad z = \frac{7t - 5}{4}$$

So, the solution is $\left(t, 3t - 2, \dfrac{7t - 5}{4} \right)$.

> **Note**
>
> Each value of the parameter t gives a different point. Since there is only one parameter, this solution represents a line.

> **Note**
>
> You cannot find a unique solution for two equations in three unknowns, so set one variable (in this case, x) to be a parameter and use this to calculate the others. Since x is no longer an unknown, you have only two unknowns in these two equations, and so you can solve them.

Exercise 2

1 Find the value of the determinant of the coefficients and hence determine whether the system of equations has a unique solution.

  a $5x - 5y + 2z = 7$ $x - 3y + 2z = 4$ $x + y + 4z = 8$

  b $4x + 3y + 7z = 21$ $-2x + 4y + 6z = 3$ $x + 2y + 5z = 18$

2 **a**   Show that the equations

       $x + 2y + 5z = 7$ $2x + 3y - 4z = 11$ $4x + 7y + pz = q$

       have a unique solution unless $p = 6$.

  b   If $p = 6$, find the value of q that ensures the equations are consistent.

3 A set of three planes is given by the system of equations

    $\lambda x + 2y - z = 4$ [1]

    $2x + \lambda y + z = 6 + \lambda$ [2]

    $2x + 3y + z = 0$ [3]

  a   Show that $\begin{vmatrix} \lambda & 2 & -1 \\ 2 & \lambda & 1 \\ 2 & 3 & 1 \end{vmatrix} = \lambda^2 - \lambda - 6$.

  b   Determine the number of solutions for the system of equations in the cases in which

       i $\lambda = 7$ **ii** $\lambda = -2$

Summary

▶ Three linear equations with three unknowns can be solved using elimination, or using determinants.

▶ There are three possible outcomes when solving three linear equations with up to three unknowns.

   •   There is a **unique solution** – geometrically, this means that the three planes intersect at one point.

   •   There is no solution and the equations are **inconsistent** – geometrically, this means that.

       –   the three planes form a triangular prism, or

       –   two of the planes are separate and parallel, and are intersected by the third plane, or

       –   all three planes are parallel and separate, or

       –   two of the planes are coincident and the third plane is parallel but separate.

- There are **infinitely many solutions** – geometrically, this means that
  - the three planes have a common line, giving an infinite number of points that satisfy all three equations; or
  - all three planes coincide, giving an infinite number of points that satisfy all three equations.

There is a unique solution exactly when $\begin{vmatrix} a_1 & b_1 & c_1 \\ a_2 & b_2 & c_2 \\ a_3 & b_3 & c_3 \end{vmatrix} \neq 0.$

Review exercises

1 Find the solution to the three equations.

$3x - 2y - 2z = 2$

$4x - y - z = 2$

$5x - 5y - 6z = -5$

2 Find the two values of a for which the system of equations

$3x - 2y + z = a$

$5x + 2y + az = 3$

$ax + 3y + z = b$

does not have a unique solution.

3 Find the two values of k for which the system of equations

$kx + 2y + z = 5$

$x + (k+1)y - 2z = 3$

$2x - ky + 3z = b$

does not have a unique solution. When $k = 4$, find the value of b that would make the system consistent.

Practice examination questions

1 A system of equations is given by

$x + 3y + 5z = -2$

$3x - 4y + 2z = 7$

$ax + 11y + 13z = b$

where a and b are constants.

a Find the unique solution of the system in the case when
$a = 3$ and $b = 2$. (5 marks)

b i Determine the value of a for which the system does
not have a unique solution. (3 marks)

ii For this value of a, find the value of b such that the system of
equations is consistent. (4 marks)

AQA MFP4 January 2008

2 The system of equations S is given in terms of the real parameters a and b by

$2x+y+3z=a+1$

$5x-2y+(a+1)z=3$

$ax+2y+4z=b$

a Find the two values of a for which S does not have a unique solution. (4 marks)

b In the case when $a=2$, determine the value of b for which S has infinitely many solutions. (4 marks)

AQA MFP4 June 2011

3 The system of equations

$2x-y-z=3$

$x+2y-3z=4$

$2x+y+az=b$

does not have a unique solution.

a Show that $a=-3$. (3 marks)

b Given further that the equations are inconsistent, find the possible values of b. (2 marks)

c State, with a reason, whether the vectors $\begin{pmatrix} 2 \\ 1 \\ 2 \end{pmatrix}$, $\begin{pmatrix} -1 \\ 2 \\ 1 \end{pmatrix}$ and $\begin{pmatrix} -1 \\ -3 \\ -3 \end{pmatrix}$

are linearly dependent or linearly independent. (1 mark)

AQA MFP4 June 2013

4 A set of three planes is given by the system of equations

$x+3y-z=10$

$2x+ky+z=-4$

$3x+5y+(k-2)z=k+4$

where k is a constant.

a Show that $\begin{vmatrix} 1 & 3 & -1 \\ 2 & k & 1 \\ 3 & 5 & k-2 \end{vmatrix} = k^2-5k+6$.

b In each of the following cases, determine the *number* of solutions of the given system of equations.

i $k=1$

ii $k=2$

iii $k=3$

c Give a geometrical interpretation of the significance of each of the three results in part **b** in relation to the three planes.

AQA MFP4 June 2006

28 Matrix Algebra

Introduction

Matrices can be used to understand transformations in $\mathbf{R}^3$ (xyz space), as they can in $\mathbf{R}^2$ (xy plane). You will discover that, in three dimensional space, rotations are about a line rather than about a point; and reflections are in a plane, rather than in a line. You will also learn about invariant points and invariant lines and eigenvectors. Eigenvalues and related eigenvectors have applications in sciences as diverse as geology and vibration analysis, and can even be used as part of a program to rank pages for a search engine.

Objectives

By the end of this chapter, you should know how to:
- Find the inverse of a 3×3 matrix.
- Find the 3×3 matrices that represent rotations about the coordinate axes.
- Find the 3×3 matrices that represent reflections in the planes $x = 0$, $y = 0$, $z = 0$, $x = y$, $x = z$ and $y = z$.
- Find the eigenvalues and eigenvectors of 2×2 and 3×3 matrices.
- Find the characteristic equation of a 3×3 matrix.

Recap

You will need to remember...
- How to multiply compatible matrices

See Chapter *7 Matrices and Transformations* for a reminder of the first four bullet points.

- That if $A = \begin{pmatrix} a & b \\ c & d \end{pmatrix}$ then the transpose, $A^T = \begin{pmatrix} a & c \\ b & d \end{pmatrix}$
- How to find the inverse of a 2×2 matrix: $\frac{1}{\det A} \begin{pmatrix} d & -b \\ -c & a \end{pmatrix}$
- That matrices can represent transformations of $\mathbf{R}^2$.
- About invariant points and lines of transformations in $\mathbf{R}^2$.

See Chapter *26 Vectors and Three-Dimensional Coordinate Geometry*.

- How to calculate the determinant of a 3×3 matrix.
- How to solve three linear equations in three unknowns.

See Chapter *27 Solutions of Linear Equations*.

28.1 Inverse matrices

You learned about the inverse of a 2×2 matrix in Chapter *7 Matrices and Transformations*.

The inverse of a 2×2 matrix $A = \begin{pmatrix} a & b \\ c & d \end{pmatrix}$ is $\frac{1}{\det A} \begin{pmatrix} d & -b \\ -c & a \end{pmatrix}$. You swap the position of a and d and then make b and c negative; then you divide by the determinant of A.

To find the inverse of a 3×3 matrix, you need to find the determinant of the matrix and the **minor determinant** of each element in the matrix.

The minor determinant

The minor determinant of an element of a matrix is the determinant of the matrix formed by deleting the row and column containing that element. For example, the minor determinant of the middle element, 2, of the matrix

$\begin{pmatrix} 5 & 6 & 9 \\ 7 & 2 & 1 \\ 3 & 4 & 8 \end{pmatrix}$ is the determinant of the matrix $\begin{pmatrix} 5 & 9 \\ 3 & 8 \end{pmatrix}$, which is $\begin{vmatrix} 5 & 9 \\ 3 & 8 \end{vmatrix} = 13$.

Finding the inverse of a 3 × 3 matrix

You proceed in the following order:

1 Find the value of the determinant, Δ, of the matrix.

2 Find the value of the minor determinant of each of the elements.

3 Form a new matrix from the minor values, inserting them in the positions corresponding to the elements from which they were derived. Insert a minus sign at each odd-numbered place, counting on from the top left entry of the matrix. These minor values with their associated signs (+ or −) are called the **cofactors** of the elements of the original matrix.

4 Find the transpose of the result by reflecting its elements along the leading diagonal.

You learned how to do this in Chapter *26 Vectors and Three-Dimensional Coordinate Geometry.*

You learned how to transpose matrices in Chapter *7 Matrices and Transformations.*

Hence, you have

The inverse of the 3 × 3 matrix, *A* is

$\begin{pmatrix} a & b & c \\ d & e & f \\ g & h & i \end{pmatrix}^{-1} = \frac{1}{\Delta} \begin{pmatrix} A & -B & C \\ -D & E & -F \\ G & -H & I \end{pmatrix}^{T}$,

where *A, B, C,* . . . are the minor determinants of the elements *a, b, c,* . . . respectively. Δ is the determinant of the original matrix.

Example 1

Find the inverse of *M*, where $M = \begin{pmatrix} 1 & 2 & 5 \\ 2 & 3 & 4 \\ 1 & 1 & 2 \end{pmatrix}$.

> **Note**
> Step 1, find the value of the determinant, Δ, of the matrix.

det *M* = 1(6 − 4) − 2(4 − 4) + 5(2 − 3) = −3

The minor determinants are:

$\begin{pmatrix} 2 & 0 & -1 \\ -1 & -3 & -1 \\ -7 & -6 & -1 \end{pmatrix}$

$\begin{pmatrix} 2 & -0 & -1 \\ +1 & -3 & +1 \\ -7 & +6 & -1 \end{pmatrix}$

> **Note**
> Step 2, find the minor determinants. For example, the minor determinant of the top right entry is det $\begin{pmatrix} 2 & 3 \\ 1 & 1 \end{pmatrix} = 2 - 3 = -1$.

> **Note**
> Step 3, form a new matrix using the minor determinants and then insert the appropriate signs (+ or −), counting on from the top left corner of the matrix.

(continued)

(continued)

Therefore,

$$M^{-1} = \frac{1}{-3}\begin{pmatrix} 2 & -0 & -1 \\ +1 & -3 & +1 \\ -7 & +6 & -1 \end{pmatrix}^{\mathrm{T}}$$

Note

Multiply by $\dfrac{1}{\det M}$, to form the matrix to be transposed.

Therefore,

$$M^{-1} = \frac{1}{-3}\begin{pmatrix} 2 & 1 & -7 \\ 0 & -3 & 6 \\ -1 & 1 & -1 \end{pmatrix} = \begin{pmatrix} -\dfrac{2}{3} & -\dfrac{1}{3} & \dfrac{7}{3} \\ 0 & 1 & -2 \\ \dfrac{1}{3} & -\dfrac{1}{3} & \dfrac{1}{3} \end{pmatrix}$$

Note

You obtain the transpose by reflecting the matrix in its leading diagonal.

Exercise 1

1 Evaluate PQ and QP, where

$$P = \begin{pmatrix} 6 & 4 \\ 2 & 3 \end{pmatrix} \quad \text{and} \quad Q = \begin{pmatrix} 1 & -2 \\ 2 & 3 \end{pmatrix}$$

What do you conclude from your results, and why has it happened?

You learned how to multiply two matrices together in Chapter *7 Matrices and Transformations*.

2 Find the inverse of $\begin{pmatrix} 3 & 4 \\ 4 & 5 \end{pmatrix}$.

3 Find the inverse of $\begin{pmatrix} 2 & 7 \\ 1 & 4 \end{pmatrix}$.

4 Find the inverse of $\begin{pmatrix} 1 & -2 & 1 \\ 3 & -1 & 5 \\ -1 & 4 & 0 \end{pmatrix}$.

5 Find the inverse of $\begin{pmatrix} 4 & 11 & 5 \\ 1 & 4 & 2 \\ 1 & 2 & 1 \end{pmatrix}$.

6 Find the inverse of $\begin{pmatrix} 3 & 4 & -2 \\ 2 & -1 & 5 \\ -3 & 4 & 1 \end{pmatrix}$.

28.2 Transformations

You saw in Chapter *7 Matrices and Transformations* that linear transformations in two-dimensional space could be represented using matrices.

It is also possible to represent linear transformations of a three-dimensional space onto a three-dimensional space, $\mathbf{R}^3$, by a matrix M, where

$$M\begin{pmatrix} x \\ y \\ z \end{pmatrix} = \begin{pmatrix} x_1 \\ y_1 \\ z_1 \end{pmatrix}$$

means that the image of (x, y, z) under the transformation, T, is (x_1, y_1, z_1).

For example, in three dimensions, you might represent T by the matrix

$$M = \begin{pmatrix} a & b & c \\ d & e & f \\ g & h & i \end{pmatrix}$$

Hence, to find, under T, the image of the point with position vector **i**, you calculate

$$\begin{pmatrix} a & b & c \\ d & e & f \\ g & h & i \end{pmatrix} \begin{pmatrix} 1 \\ 0 \\ 0 \end{pmatrix} = \begin{pmatrix} a \\ d \\ g \end{pmatrix}$$

So, under T, the image of the point $(1, 0, 0)$ is (a, d, g), which you can see is the first column of M.

To find which type of transformation is represented by a matrix, you can find the images of the vectors $(1, 0, 0)$ and $(0, 1, 0)$ and $(0, 0, 1)$. Common linear transformations are rotations about the origin, reflections in lines through the origin, stretches and shears.

In three dimensions, you find the images of the points $(1, 0, 0)$, $(0, 1, 0)$ and $(0, 0, 1)$, which are the vertices of the **unit cube**. In vector form, these are the images of the vectors **i**, **j** and **k**. These become the columns of the matrix representing the transformation, where the first column represents the image of **i**, the second is the image of **j** and the third is the image of **k**.

So, you calculate where a point is following the transformation and use this to find the matrix of a transformation. For example, a reflection in the plane $x = 0$ will result in the point $(1, 0, 0)$ being transformed to $(-1, 0, 0)$; meanwhile the points $(0, 1, 0)$ and $(0, 0, 1)$ are fixed. Therefore, the transformation matrix is

$$\begin{pmatrix} -1 & 0 & 0 \\ 0 & 1 & 0 \\ 0 & 0 & 1 \end{pmatrix}.$$

There are some common transformations that you should know:

▶ rotations of $\theta°$ about the

x-axis: $\begin{bmatrix} 1 & 0 & 0 \\ 0 & \cos\theta & -\sin\theta \\ 0 & \sin\theta & \cos\theta \end{bmatrix}$

y-axis: $\begin{bmatrix} \cos\theta & 0 & \sin\theta \\ 0 & 1 & 0 \\ -\sin\theta & 0 & \cos\theta \end{bmatrix}$

z-axis: $\begin{bmatrix} \cos\theta & -\sin\theta & 0 \\ \sin\theta & \cos\theta & 0 \\ 0 & 0 & 1 \end{bmatrix}$

▶ enlargement, scale factor λ: $\begin{bmatrix} \lambda & 0 & 0 \\ 0 & \lambda & 0 \\ 0 & 0 & \lambda \end{bmatrix}$

> **Tip**
>
> It is a good idea to know how to create matrices to represent transformations by considering the image of the vertices of the unit cube, rather than just relying on learning the formulae.

- reflections in the planes,

$$x = 0: \begin{bmatrix} -1 & 0 & 0 \\ 0 & 1 & 0 \\ 0 & 0 & 1 \end{bmatrix} \qquad\qquad x = y: \begin{bmatrix} 0 & 1 & 0 \\ 1 & 0 & 0 \\ 0 & 0 & 1 \end{bmatrix}$$

$$y = 0: \begin{bmatrix} 1 & 0 & 0 \\ 0 & -1 & 0 \\ 0 & 0 & 1 \end{bmatrix} \qquad\qquad y = z: \begin{bmatrix} 1 & 0 & 0 \\ 0 & 0 & 1 \\ 0 & 1 & 0 \end{bmatrix}$$

$$z = 0: \begin{bmatrix} 1 & 0 & 0 \\ 0 & 1 & 0 \\ 0 & 0 & -1 \end{bmatrix} \qquad\qquad x = z: \begin{bmatrix} 0 & 0 & 1 \\ 0 & 1 & 0 \\ 1 & 0 & 0 \end{bmatrix}$$

Here, the direction of rotation is taken to be as follows: if you look down from the positive z-direction, the rotation about the z-axis is anticlockwise, whereas from the positive y-direction, the rotation about the y-axis is clockwise.

- The 3×3 identity matrix is $\mathbf{I} = \begin{pmatrix} 1 & 0 & 0 \\ 0 & 1 & 0 \\ 0 & 0 & 1 \end{pmatrix}$. (Recall from Chapter 7 *Matrices and Transformations* that the identity of a matrix is any square matrix where all the elements in the leading diagonal are 1 and all other elements are 0.)

You can use the list above to help you recognise a type of transformation. Notice that the matrix used for the rotation always has 1 0 0 (in the top row for the x-axis rotation), 0 1 0 (in the middle row for the y-axis rotation) and 0 0 1 (in the lowest row for the z-axis rotation), which should make it simple to identify when a matrix represents such a rotation.

Determinants and areas

Recall from Chapter 7 *Matrices and Transformations* that after a linear transformation, the area of the unit square (and all other areas) might change. The new area of a unit square is given by $|\det \mathbf{M}|$ multiplied by the original area, where $\mathbf{M}$ is the matrix corresponding to the transformation. In other words, the determinant is the area scale factor of the transformation. If the determinant of the matrix is negative, this indicates that the transformation includes a reflection.

> When working in three-dimensions, $|\det \mathbf{M}|$ is the volume scale factor of the transformation.

You should know from earlier studies that if a shape is enlarged by a scale factor of k, then the volume of the shape becomes multiplied by k^3. So, k^3 is the value of the determinant and is the volume scale factor of the enlargement. For example, if a shape is enlarged by a scale factor of 3, then the volume of the shape becomes multiplied by 27, which is the value of the determinant of this enlargement.

> After transformation under T, the volume of the unit cube becomes $|\det \mathbf{M}|$.

Example 2

a Find the matrix M representing an enlargement, scale factor 2, with the origin as the centre of enlargement.

b What is the volume scale factor of the enlargement?

a The images of the vertices of the unit cube are:

$(1, 0, 0) \rightarrow (2, 0, 0)$

$(0, 1, 0) \rightarrow (0, 2, 0)$

$(0, 0, 1) \rightarrow (0, 0, 2)$

Hence,

$$M = \begin{pmatrix} 2 & 0 & 0 \\ 0 & 2 & 0 \\ 0 & 0 & 2 \end{pmatrix}$$

> **Note**
>
> Find the images of the points $(1, 0, 0)$, $(0, 1, 0)$ and $(0, 0, 1)$; here you multiply each element by the scale factor, 2.

> **Note**
>
> Alternatively, you could have
>
> used $\begin{pmatrix} \lambda & 0 & 0 \\ 0 & \lambda & 0 \\ 0 & 0 & \lambda \end{pmatrix}$
>
> with $\lambda = 2$.

> **Note**
>
> Using $\begin{vmatrix} a & b & c \\ d & e & f \\ g & h & i \end{vmatrix}$.

b Volume of unit cube = $|\det M|$

The determinant of M is 8,

so the volume scale factor of the transformation is 8.

You will remember that if a shape is enlarged by a scale factor of k, then the volume of the shape becomes multiplied by k^3. Thus if the shape is enlarged by a scale factor of 2, then the volume of the shape becomes multiplied by 8, which is the value of the determinant of this enlargement.

Example 3

Find the matrix M representing a reflection in the line $y = x$ in the xy plane.

The images of the vertices of the unit cube are:

$(1, 0, 0) \rightarrow (0, 1, 0)$

$(0, 1, 0) \rightarrow (1, 0, 0)$

$(0, 0, 1) \rightarrow (0, 0, 1)$

Hence, you have

$$M = \begin{bmatrix} 0 & 1 & 0 \\ 1 & 0 & 0 \\ 0 & 0 & 1 \end{bmatrix}$$

> **Note**
>
> Find the images of the points $(1, 0, 0)$, $(0, 1, 0)$ and $(0, 0, 1)$ under the transformation.

Recall from Chapter *7 Matrices and Transformations* that it is possible to combine two or more simple transformations to produce a more complicated transformation.

> If M represents a transformation, T, and N represents a transformation S, then the matrix NM represents the effect of performing the transformation T and then performing transformation S.

Since transformation T is applied first, to find the image of **x** you should find **Mx** and then premultiply by **N** to find **NMx**. Therefore the matrix **NM** represents the transformation of T followed by the transformation S.

Example 4

The transformation T is the composite transformation of

  S: an enlargement, scale factor 3

  R: a rotation through an angle of $45°$ about the y-axis

a Find the matrix **N** representing the composite transformation.

b Give the volume of the unit cube under the transformation T.

a S: Let the matrix be **A**, $\mathbf{A} = \begin{pmatrix} 3 & 0 & 0 \\ 0 & 3 & 0 \\ 0 & 0 & 3 \end{pmatrix}$

  R: Let the matrix be **B**, $\mathbf{B} = \begin{pmatrix} \dfrac{1}{\sqrt{2}} & 0 & \dfrac{1}{\sqrt{2}} \\ 0 & 1 & 0 \\ -\dfrac{1}{\sqrt{2}} & 0 & \dfrac{1}{\sqrt{2}} \end{pmatrix}$

$\mathbf{BA} = \begin{pmatrix} \dfrac{1}{\sqrt{2}} & 0 & \dfrac{1}{\sqrt{2}} \\ 0 & 1 & 0 \\ -\dfrac{1}{\sqrt{2}} & 0 & \dfrac{1}{\sqrt{2}} \end{pmatrix} \times \begin{pmatrix} 3 & 0 & 0 \\ 0 & 3 & 0 \\ 0 & 0 & 3 \end{pmatrix}$

$= \begin{pmatrix} \dfrac{3}{\sqrt{2}} & 0 & \dfrac{3}{\sqrt{2}} \\ 0 & 3 & 0 \\ -\dfrac{3}{\sqrt{2}} & 0 & \dfrac{3}{\sqrt{2}} \end{pmatrix}$

$\mathbf{N} = \begin{pmatrix} \dfrac{3}{\sqrt{2}} & 0 & \dfrac{3}{\sqrt{2}} \\ 0 & 3 & 0 \\ -\dfrac{3}{\sqrt{2}} & 0 & \dfrac{3}{\sqrt{2}} \end{pmatrix}$

b Volume of unit cube $= \det\mathbf{N}$

$= 27$

Note

Using $\begin{pmatrix} \lambda & 0 & 0 \\ 0 & \lambda & 0 \\ 0 & 0 & \lambda \end{pmatrix}$ with $\lambda = 3$.

Note

Find the matrices that represent the transformations S and R.

Note

Apply S first and premultiply by R to find BA.

Note

Using $\begin{pmatrix} \cos\theta & 0 & \sin\theta \\ 0 & 1 & 0 \\ -\sin\theta & 0 & \cos\theta \end{pmatrix}$, where $\theta = 45°$.

Note

Using $\begin{vmatrix} a & b & c \\ d & e & f \\ g & h & i \end{vmatrix}$.

Exercise 2

1 Find the matrix in $\mathbf{R}^3$ representing a reflection in the plane $z = 0$.

2 What transformation of $\mathbf{R}^3$ is represented by the matrix $\begin{pmatrix} \dfrac{12}{13} & 0 & \dfrac{5}{13} \\ 0 & 1 & 0 \\ -\dfrac{5}{13} & 0 & \dfrac{12}{13} \end{pmatrix}$?

3 Find the matrix that represents an enlargement by scale factor 2, and the matrix that represents a reflection in the plane $x = 0$. Calculate the matrix that represents the enlargement followed by the reflection.

4 a Find the matrix that represents a reflection in the plane $x = 0$, and the matrix that represents a reflection in the plane $y = x$.

   b Find the matrix that represents a reflection in the plane $x = 0$ followed by another reflection in the plane $y = x$.

   c Describe the single transformation given by combining the two reflections in part **b**.

5 Find the matrix representing the following transformations in $\mathbf{R}^3$:

   a An enlargement of scale factor 5.

   b A reflection in the plane $y = z$.

28.3 Invariant points and lines

Recall from Chapter *7 Matrices and Transformations* that an **invariant point** of the transformation T is a point that is unchanged by that transformation. That is, T$x = x$. **Note** that 0 is always an invariant point in any linear transformation.

For example, the only points that are unchanged by reflection in the plane $y = x$ are the points on the plane $y = x$ itself. Therefore, the only invariant points in this transformation are on the plane $y = x$.

Reflection in the plane $y = x$ also maps a line onto itself, so this is an **invariant line** of the reflection in the plane $y = x$. With the exception of the origin, each point on L does change: for example, $(3, -3, 0)$ becomes $(-3, 3, 0)$. However, each point on L is reflected to another point on L.

Every line in the plane is also an invariant line, since all of its points remain unchanged under the reflection.

In general, it is harder to find invariant lines than invariant points, so you start with the general method to find invariant points.

> To find the invariant points of a transformation represented by a matrix M, solve $Mx = x$ for the vector x. This determines the points that were not changed by M.

Example 5

Find the invariant points of the transformation whose matrix is $\begin{pmatrix} 2 & 1 & 3 \\ 1 & 1 & 4 \\ 1 & 5 & 0 \end{pmatrix}$.

$$Mx = \begin{pmatrix} 2 & 1 & 3 \\ 1 & 1 & 4 \\ 1 & 5 & 0 \end{pmatrix} \begin{pmatrix} x \\ y \\ z \end{pmatrix}$$

$$2x + y + 3z = x$$

$$x + y + 4z = y$$

$$x + 5y = z$$

$$x = -4\mu, \ y = \mu, \ z = \mu$$

So, $(-4\mu, \mu, \mu)$

So the line of points $x = -4y = -4z$ is invariant under the transformation.

> **Note**
>
> You have three simultaneous equations to solve.

> **Note**
>
> Solve the simultaneous equations using your preferred method; you learned how to solve three linear equations in three unknowns in Chapter 27 *Solutions of Linear Equations*.

> **Note**
>
> The invariant points obey the equation
> $$\begin{pmatrix} 2 & 1 & 3 \\ 1 & 1 & 4 \\ 1 & 5 & 0 \end{pmatrix} \begin{pmatrix} x \\ y \\ z \end{pmatrix} = \begin{pmatrix} x \\ y \\ z \end{pmatrix}, \text{ so start by}$$
> calculating $\begin{pmatrix} 2 & 1 & 3 \\ 1 & 1 & 4 \\ 1 & 5 & 0 \end{pmatrix} \begin{pmatrix} x \\ y \\ z \end{pmatrix} = \begin{pmatrix} x \\ y \\ z \end{pmatrix}.$

Eigenvectors and eigenvalues

An **eigenvector** of a linear transformation is a vector pointing in the direction of an invariant line under that transformation.

For example, let T be a reflection of $\mathbf{R}^2$ in the line $y = x$. The line $y = -x$ is invariant under T. Therefore, $\begin{pmatrix} 1 \\ -1 \end{pmatrix}$ is an eigenvector of T, since it is on the line $y = -x$.

Each point on $y = -x$ is reflected to a different point, other than the origin. $(2, -2)$ is on the invariant line $y = -x$, but it maps onto $(-2, 2)$.

You say that the **eigenvalue** for the eigenvector $\begin{pmatrix} 2 \\ -2 \end{pmatrix}$ under T is -1, since all the points on the line $y = -x$ map onto points whose coordinates are -1 times the original coordinates.

> **Note**
>
> Any multiple of $\begin{pmatrix} 1 \\ -1 \end{pmatrix}$ can be used, for example, $\begin{pmatrix} 2 \\ -2 \end{pmatrix}$.

If M is the matrix that represents T, then you can express the above as

$$M\begin{pmatrix} 2 \\ -2 \end{pmatrix} = -1 \times \begin{pmatrix} 2 \\ -2 \end{pmatrix}.$$

In general, if M is the matrix for a transformation T, then

$$M\begin{pmatrix} x \\ y \end{pmatrix} = \lambda \begin{pmatrix} x \\ y \end{pmatrix} \text{ means that } \begin{pmatrix} x \\ y \end{pmatrix} \text{ is an eigenvector of T, and } \lambda \text{ is the}$$

eigenvalue of T associated with $\begin{pmatrix} x \\ y \end{pmatrix}$.

In three-dimensional space, $M\begin{pmatrix} x \\ y \\ z \end{pmatrix} = \lambda \begin{pmatrix} x \\ y \\ z \end{pmatrix}$ means that $\begin{pmatrix} x \\ y \\ z \end{pmatrix}$ is an

eigenvector of T, and that λ is the eigenvalue of T associated with $\begin{pmatrix} x \\ y \\ z \end{pmatrix}$.

Once you have found the eigenvalues and corresponding eigenvectors, you can easily find the invariant lines through the origin.

If an eigenvector **v** has eigenvalue 1, then all points on the line $r = t\mathbf{v}$ are invariant points.

For any eigenvector **v**, the line $r = t\mathbf{v}$ is an invariant line.

The methods of Chapter 26 can help you to find the corresponding Cartesian equation.

Finding eigenvectors and eigenvalues

To find the eigenvalues of a transformation whose matrix is $M = \begin{pmatrix} a & b & c \\ d & e & f \\ g & h & i \end{pmatrix}$, then you must solve:

$$\begin{pmatrix} a & b & c \\ d & e & f \\ g & h & i \end{pmatrix} \begin{pmatrix} x \\ y \\ z \end{pmatrix} = \lambda \begin{pmatrix} x \\ y \\ z \end{pmatrix}$$

This gives

$ax + by + c = \lambda x$

$dx + ey + fz = \lambda y$

$gx + hy + iz = \lambda z$

from which you obtain

$(a - \lambda)x + by + cz = 0$

$dx + (e - \lambda)y + fz = 0$

$gx + hy + (i - \lambda)z = 0$

For the eigenvectors to be non-zero, these three equations must have non-unique solutions, as per the geometric interpretation.

Hence, you have

$$\begin{vmatrix} a-\lambda & b & c \\ d & e-\lambda & f \\ g & h & i-\lambda \end{vmatrix} = 0$$

That is, $\det(M - \lambda I) = 0$. This equation is known as the **characteristic equation** of the matrix M.

To find the eigenvalues and eigenvectors of a 3×3 matrix M, first find the eigenvalues, by solving the equation $\det(M - \lambda I) = 0$ for λI. Then find the eigenvectors by solving $Mx = \lambda x$ for each value of λ, where $x = \begin{pmatrix} x \\ y \\ z \end{pmatrix}$.

Note that you could use the same method for 2×2 matrices: to find eigenvalues, you again use the equation $\det(M - \lambda I) = 0$; and to find eigenvectors you solve $M \begin{pmatrix} x \\ y \end{pmatrix} = \lambda \begin{pmatrix} x \\ y \end{pmatrix}$.

See Chapter 27 Solutions of Linear Equations for a reminder of geometrical interpretations of the solution of three linear equations if you need to.

Example 6

The matrix $M = \begin{pmatrix} -5 & 6 \\ -6 & 7 \end{pmatrix}$ represents the plane transformation T.

Determine **a** the eigenvalue of M **b** a corresponding eigenvector of M.

a $\begin{vmatrix} -5-\lambda & 6 \\ -6 & 7-\lambda \end{vmatrix} = 0$

Note

Using det $(\boldsymbol{M} - \lambda \boldsymbol{I}) = 0$.

$\Rightarrow (-5-\lambda)(7-\lambda) + 36 = 0$

$\Rightarrow \quad \lambda^2 - 2\lambda - 35 + 36 = 0$

$\Rightarrow \quad \lambda^2 - 2\lambda + 1 = 0$

$\quad (\lambda - 1)(\lambda - 1) = 0$

$\Rightarrow \quad \lambda = 1$

Therefore, the eigenvalue is 1.

Note

Factorising the LHS.

b $\begin{pmatrix} -5 & 6 \\ -6 & 7 \end{pmatrix} \begin{pmatrix} x \\ y \end{pmatrix} = \begin{pmatrix} x \\ y \end{pmatrix}$

Note

The eigenvector for the eigenvalue 1 is given by the solution to this equation.

$\Rightarrow$ simultaneous equations:

$-5x + 6y = x$

$-6x + 7y = y$

Both of which give $6x = 6y$

$\Rightarrow x = y$

Therefore, the direction of the eigenvector is $\begin{pmatrix} 1 \\ 1 \end{pmatrix}$.

Hence, $\begin{pmatrix} 1 \\ 1 \end{pmatrix}$ is an eigenvector for the eigenvalue 1.

Note

Note that there is only one different equation from which to solve for two unknowns. Therefore, you cannot obtain a unique solution to such a set of equations. Hence you obtain $x = y$.

Example 7

Question

Given the matrix $\begin{pmatrix} 1 & 1 & 2 \\ 0 & 2 & 2 \\ -1 & 1 & 3 \end{pmatrix}$ find **a** the eigenvalues **b** the eigenvectors

Answer

a $\begin{vmatrix} 1-\lambda & 1 & 2 \\ 0 & 2-\lambda & 2 \\ -1 & 1 & 3-\lambda \end{vmatrix} = 0$

Note

Using det $(\boldsymbol{M} - \lambda \boldsymbol{I}) = 0$.

$(1-\lambda)[(2-\lambda)(3-\lambda) - 2] - 1(2) + 2(2-\lambda) = 0$

$\Rightarrow \quad (1-\lambda)(\lambda^2 - 5\lambda + 4) - 2 + 4 - 2\lambda = 0$

$\Rightarrow \quad \lambda^3 - 6\lambda^2 + 11\lambda - 6 = 0$

$\quad (\lambda - 1)(\lambda - 2)(\lambda - 3) = 0$

$\Rightarrow \quad \lambda = 1, 2, 3$

Therefore, the eigenvalues are 1, 2 and 3.

Note

Factorising the LHS.

(continued)

(continued)

b The eigenvector for the eigenvalue 1 is given by a solution to the equation

$$\begin{pmatrix} 1 & 1 & 2 \\ 0 & 2 & 2 \\ -1 & 1 & 3 \end{pmatrix} \begin{pmatrix} x \\ y \\ z \end{pmatrix} = \begin{pmatrix} x \\ y \\ z \end{pmatrix} \Rightarrow \begin{pmatrix} x+y+2z \\ 2y+2z \\ -x+y+3z \end{pmatrix} = \begin{pmatrix} x \\ y \\ z \end{pmatrix}$$

$\Rightarrow$ simultaneous equations:

$$x+y+2z=x \quad [1]$$
$$2y+2z=y \quad [2]$$
$$-x+y+3z=z \quad [3]$$

Let $z=t$ and solve the simultaneous equations for y:

$$x+y+2t=x \quad [4]$$
$$2y+2t=y \quad [5]$$
$$-x+y+3t=t \quad [6]$$

From, [4], you obtain $y=-2t$.

Therefore, the direction of the eigenvector is $\begin{pmatrix} 0 \\ -2t \\ t \end{pmatrix}$.

Hence, $\begin{pmatrix} 0 \\ -2 \\ 1 \end{pmatrix}$ is an eigenvector for the eigenvalue 1.

The eigenvector for the eigenvalue 2 is given by a solution to the equation

$$\begin{pmatrix} 1 & 1 & 2 \\ 0 & 2 & 2 \\ -1 & 1 & 3 \end{pmatrix} \begin{pmatrix} x \\ y \\ z \end{pmatrix} = 2 \begin{pmatrix} x \\ y \\ z \end{pmatrix}$$

$\Rightarrow$ simultaneous equations:

$$x+y-2z=2x \quad [7]$$
$$2y+2z=2y \quad [8]$$
$$-x+y+3z=2z \quad [9]$$

Let $y=t$. Then from [7], you obtain $x=t$.

Therefore, the direction of the eigenvector is $\begin{pmatrix} t \\ t \\ 0 \end{pmatrix}$.

Hence, $\begin{pmatrix} 1 \\ 1 \\ 0 \end{pmatrix}$ is an eigenvector for the eigenvalue 2.

The eigenvector for the eigenvalue 3 is given by a solution to the equation

$$\begin{pmatrix} 1 & 1 & 2 \\ 0 & 2 & 2 \\ -1 & 1 & 3 \end{pmatrix} \begin{pmatrix} x \\ y \\ z \end{pmatrix} = 3 \begin{pmatrix} x \\ y \\ z \end{pmatrix}$$

$\Rightarrow$ simultaneous equations:

$$x+y+2z=3x$$
$$2y+2z=3y$$
$$-x+y+3z=3z$$

> **Note**
>
> Note that there are only two different equations from which to solve for three unknowns. Therefore, you cannot obtain a unique solution to such a set of equations. Hence, let one of the unknowns be t. Also note that subtracting [3] from [1] gives $x=0$.

See Chapter *27 Solving Linear Equations* if you need to.

> **Note**
>
> This time, you do not let $z=t$, since [8] immediately gives $z=0$.

(continued)

which give

$$-2x + y + 2z = 0 \quad [10]$$
$$2z = y \quad [11]$$
$$-x + y = 0 \quad [12]$$

Let $x = t$. Then from [12] and [11], you have $y = t$ and $z = \dfrac{t}{2}$.

Therefore, the direction of the eigenvector is $\begin{pmatrix} t \\ t \\ \frac{1}{2}t \end{pmatrix}$.

Hence, $\begin{pmatrix} 1 \\ 1 \\ \frac{1}{2} \end{pmatrix}$ is an eigenvector for the eigenvalue 3.

Note

Since any scalar multiple of an eigenvector is also an eigenvector, you can write the eigenvector for 3 as $\begin{pmatrix} 2 \\ 2 \\ 1 \end{pmatrix}$.

Example 8

a Show that $\begin{pmatrix} 1 \\ -1 \\ -2 \end{pmatrix}$ is an eigenvector of the matrix A, where $A = \begin{pmatrix} 1 & 0 & -1 \\ 1 & 2 & 1 \\ 2 & 0 & 4 \end{pmatrix}$.

b Find the associated eigenvalue.

a, b If $\begin{pmatrix} 1 \\ -1 \\ -2 \end{pmatrix}$ is an eigenvector of A, then $A\begin{pmatrix} 1 \\ -1 \\ -2 \end{pmatrix} = \lambda\begin{pmatrix} 1 \\ -1 \\ -2 \end{pmatrix}$ where λ is the eigenvalue associated with $\begin{pmatrix} 1 \\ -1 \\ -2 \end{pmatrix}$.

Hence,

$$A\begin{pmatrix} 1 \\ -1 \\ -2 \end{pmatrix} = \begin{pmatrix} 1 & 0 & -1 \\ 1 & 2 & 1 \\ 2 & 0 & 4 \end{pmatrix}\begin{pmatrix} 1 \\ -1 \\ -2 \end{pmatrix} = \begin{pmatrix} 3 \\ -3 \\ -6 \end{pmatrix} \Rightarrow A\begin{pmatrix} 1 \\ -1 \\ -2 \end{pmatrix} = 3\begin{pmatrix} 1 \\ -1 \\ -2 \end{pmatrix}$$

$A\begin{pmatrix} 1 \\ -1 \\ -2 \end{pmatrix} = 3\begin{pmatrix} 1 \\ -1 \\ -2 \end{pmatrix}$ has the same form as $Ax = \lambda x$, therefore $\begin{pmatrix} 1 \\ -1 \\ -2 \end{pmatrix}$ is an eigenvector of A and its associated eigenvalue is 3.

Note

$M\begin{pmatrix} x \\ y \\ z \end{pmatrix} = \lambda\begin{pmatrix} x \\ y \\ z \end{pmatrix}$ means

that $\begin{pmatrix} x \\ y \\ z \end{pmatrix}$ is an eigenvector of T, and that λ is the eigenvalue of T associated with $\begin{pmatrix} x \\ y \\ z \end{pmatrix}$.

Exercise 3

① Find the characteristic equation of the matrix M where $M = \begin{pmatrix} 2 & 1 & 3 \\ -1 & 3 & -2 \\ 4 & 0 & 1 \end{pmatrix}$.

2 Find the characteristic equation of the matrix M where $M = \begin{pmatrix} 1 & 0 & 0 \\ -1 & 0 & -2 \\ 4 & 5 & 2 \end{pmatrix}$.

3 Find the invariant points of the transformation represented by

$$M = \begin{pmatrix} 0 & 1 & \dfrac{1}{\sqrt{3}} \\ 1 & 0 & -\dfrac{1}{\sqrt{3}} \\ -\dfrac{1}{\sqrt{3}} & \dfrac{1}{\sqrt{3}} & 0 \end{pmatrix}$$

4 The matrix $M = \begin{pmatrix} 5 & 12 \\ 2 & -5 \end{pmatrix}$.

    a Find the eigenvalues of M.   **b** Show that $\begin{pmatrix} 12 \\ 2 \end{pmatrix}$ is an eigenvector.

    c Find another eigenvector that is not a multiple of $\begin{pmatrix} 12 \\ 2 \end{pmatrix}$.

5 The matrix $M = \begin{pmatrix} 3 & 4 & -4 \\ 4 & 5 & 0 \\ -4 & 0 & 1 \end{pmatrix}$. Find its eigenvalues and corresponding eigenvectors.

Summary

You can now

▶ Find the inverse of the 3×3 matrix, A, denoted as A^{-1}

$$\begin{pmatrix} a & b & c \\ d & e & f \\ g & h & i \end{pmatrix}^{-1} = \frac{1}{\Delta} \begin{pmatrix} A & -B & C \\ -D & E & -F \\ G & -H & I \end{pmatrix}^{\mathrm{T}},$$

    where A, B, C,... are the minor determinants of the elements a, b, c,... respectively, and Δ is the determinant of A.

▶ Recognise that for two matrices A and B, $\det(AB) = \det A \det B$.

▶ Represent linear transformations in three-dimensions using matrices, such that

$$M \begin{pmatrix} x \\ y \\ z \end{pmatrix} = \begin{pmatrix} x_1 \\ y_1 \\ z_1 \end{pmatrix} \text{ means that the image of } (x, y, z) \text{ under the}$$

    transformation, T, is (x_1, y_1, z_1), where T is represented by the matrix M.

▶ Find which type of transformation is represented by a matrix and find the images of the points $(1, 0, 0)$, $(0, 1, 0)$ and $(0, 0, 1)$, which are the vertices of the unit cube. In vector form, these are the images of the vectors $\mathbf{i}$, $\mathbf{j}$ and $\mathbf{k}$ and they form the columns of the matrix. Alternatively, you can learn the formulae given in the chapter, for example, an enlargement of scale factor λ

is represented by the matrix $\begin{pmatrix} \lambda & 0 & 0 \\ 0 & \lambda & 0 \\ 0 & 0 & \lambda \end{pmatrix}$. However, it is advisable to know

how to create the matrices rather than learning the formulae.

▶ Understand that after a transformation represented by the 3×3 matrix M, the unit cube is transformed to a paralleliped of volume $|\det M|$, therefore, $|\det M|$ is the volume scale factor of the transformation.

▶ Understand that if M represents a transformation, T, and N represents a transformation S, then the matrix NM represents the effect of performing the transformation T and then performing transformation S.

▶ Find the invariant points of a transformation represented by the matrix M by solving the equation $Mx = x$ for the vector x or the equivalent equation $(M - I)\,x = 0$.

▶ Use eigenvectors to help you to find invariant lines of a transformation.

▶ Find the eigenvalues of matrices by using the characteristic equation of the matrix M, $\det(M - \lambda I) = 0$.

▶ Find the eigenvectors that correspond to each eigenvalue by solving the equation $Mx = \lambda x$, for each value of λ.

Review exercises

1. Find the inverse of $\begin{pmatrix} 3 & 1 \\ 2 & 6 \end{pmatrix}$.

2. Find the inverse of $\begin{pmatrix} 2 & 0 & 1 \\ 1 & 3 & 4 \\ -1 & 6 & 1 \end{pmatrix}$.

3. The matrix $M = \begin{pmatrix} 1 & -1 & 1 \\ 0 & -2 & 1 \\ 0 & 1 & 3 \end{pmatrix}$. Find its eigenvalues.

4. The matrix $M = \begin{pmatrix} 0 & 1 & 3 \\ -3 & 2 & 0 \\ 1 & 1 & 4 \end{pmatrix}$. Find its eigenvalues and the eigenvector

associated with one of the eigenvalues. Are there any invariant points associated with M?

5. Find the matrix representing the transformation of a reflection in the plane $x = z$.

Practice examination questions

1. The matrix $\begin{pmatrix} 1 & 0 & 0 \\ 0 & -0.6 & -0.8 \\ 0 & 0.8 & -0.6 \end{pmatrix}$ represents a rotation.

a State the axis of rotation. (1 mark)

b Find the angle of rotation, giving your answer to the nearest degree. (2 marks)

AQA MFP4 June 2014

2 a Write down the 3×3 matrices which represent the transformations A and B, where

    i A is a reflection in the plane $y = x$ (2 marks)

    ii B is a rotation about the z-axis through the angle θ, where $\theta = \frac{\pi}{2}$. (1 mark)

b i Find the matrix $\boldsymbol{R}$ which represents the composite transformation 'A followed by B'. (3 marks)

    ii Describe the single transformation represented by $\boldsymbol{R}$. (2 marks)

AQA MFP4 June 2009

3 The matrix $\boldsymbol{A} = \begin{pmatrix} k & 1 & 2 \\ 2 & k & 1 \\ 1 & 2 & k \end{pmatrix}$, where k is a real constant.

a i Show that there is a value of k for which $\boldsymbol{A}\boldsymbol{A}^{\mathrm{T}} = m\boldsymbol{I}$.

    where m is a rational number to be determined and $\boldsymbol{I}$ is the 3×3 identity matrix. (6 marks)

    ii Deduce the inverse matrix, $\boldsymbol{A}^{-1}$, of $\boldsymbol{A}$ for this value of k. (1 mark)

b i Find $\det \boldsymbol{A}$ in terms of k. (2 marks)

    ii In the case when $\boldsymbol{A}$ is singular, find the integer value of k and show that there are no other possible real values of k. (3 marks)

    iii Find the value of k for which $\lambda = 7$ is a real eigenvalue of $\boldsymbol{A}$. (2 marks)

AQA MFP4 June 2012

> You saw in Chapter 7 *Matrices and Transformations* that when $\det \boldsymbol{M} = 0$, $\boldsymbol{M}$ is said to be singular and does not have an inverse.

4 The matrix $\boldsymbol{M} = \begin{pmatrix} -11 & 9 \\ -16 & 13 \end{pmatrix}$ represents the plane transformation T.

a i Determine the eigenvalue, and a corresponding eigenvector, of $\boldsymbol{M}$. (4 marks)

    ii Hence write down the value of m for which $y = mx$ is the invariant line of T which passes through the origin, and explain why it is actually a line of invariant points. (2 marks)

    iii Show that, for this value of m, all lines with equations $y = mx + c$ are invariant lines of T. (3 marks)

AQA MFP4 June 2012

5 A plane transformation is represented by the 2×2 matrix $\boldsymbol{M}$. The eigenvalues of $\boldsymbol{M}$ are 1 and 2, with corresponding eigenvectors $\begin{pmatrix} 1 \\ 0 \end{pmatrix}$ and $\begin{pmatrix} 1 \\ 1 \end{pmatrix}$ respectively.

State the equations of the invariant lines of the transformation and explain which of these is also a line of invariant points. (3 marks)

AQA MFP4 June 2008

29 Moment Generating Functions

Introduction

In *Chapter 12 Probability Generating Functions*, you saw how to generate power series, including infinite series, where the coefficients were *probabilities* of the different values of the random variable. You will now see how to use functions to generate values of the mean and variance of a random variable.

Objectives

By the end of this chapter, you should know how to:

▶ Define and derive a moment generating function for discrete and continuous random variables and linear functions of random variables.

▶ Find the mean and variance of random variables using moment generating functions.

▶ Apply general results for moment generating functions to specific distributions including the Poisson, exponential and normal distributions and also to sums of independent random variables.

Recap

You will need to remember…

▶ The formulae for the mean and variance of a random variable:
$$\mu = \sum_{\forall x} xP\{X = x\}; \sigma^2 = \sum_{\forall x} x^2P\{X = x\} - \mu^2$$

▶ The expectation of functions of discrete random variables:
$$E[g(X)] = \sum_{\forall x} g(x)P\{X = x\}$$
where $g(X)$ is a function of the random variable X, and its continuous equivalent, $E[g(X)] = \int_{-\infty}^{\infty} g(x)f(x)dx$

▶ The formula for the sum of integers 1 to n: $\sum_{x=1}^{n} x = \frac{1}{2}n(n+1)$

▶ The formula for the sum of squares of integers 1 to n:
$$\sum_{x=1}^{n} x^2 = \frac{1}{6}n(n+1)(2n+1)$$

▶ The expansion of the exponential function $e^t = 1 + t + \frac{t^2}{2!} + \cdots$

29.1 Moment generating functions of random variables

The mean and variance of a random variable are collectively known as **moments**. The **moment generating function (mgf)** of a random variable can sometimes make it easier to find the mean and variance of that variable, compared to other methods. Before considering particular distributions, some general results relating to moment generating functions will be derived.

Although the concept of a moment generating function is identical for both discrete and continuous random variables, their mathematical treatment is different and so they will be dealt with separately.

Moment generating functions of discrete random variables

The moment generating function of a discrete random variable, X, with distribution function

$$P\{X = x_i\} = p_i; \ i = 1, 2, \ldots, \text{ is } M_X(t) = E[e^{tX}] = \sum_{\forall i} e^{tx_i} p_i$$

To see how this generates moments, differentiate the expression for $M_X(t)$ twice with respect to t:

$$M_X(t) = \sum_{\forall i} e^{tx_i} p_i$$

$$M_X'(t) = \sum_{\forall i} x_i e^{tx_i} p_i$$

$$M_X''(t) = \sum_{\forall i} x_i^2 e^{tx_i} p_i$$

Evaluated at $t = 0$, these become

$$M_X'(0) = \sum_{\forall i} x_i p_i \quad \text{and} \quad M_X''(0) = \sum_{\forall i} x_i^2 p_i$$

If μ and σ^2 are the mean and variance of X then, using the standard results for mean and variance, these expressions become

$$M_X'(0) = \sum_{\forall i} x_i p_i = E[X] = \mu$$

and $\quad M_X''(0) = \sum_{\forall i} x_i^2 p_i = E[X^2] = \sigma^2 + \mu^2$ since $\sigma^2 = E[X^2] - \mu^2$

$M_X'(0)$ is known as the first moment about the origin, also known as the mean of X, and $M_X''(0)$ the second moment about the origin.

Hence,

A random variable, X, with moment generating function $M_X(t)$ has mean and variance given by

$$\mu = M_X'(0) \text{ and } \sigma^2 = M_X''(0) - \mu^2$$

The use of moment generating functions is more general than suggested here. For example, they can be used to generate third and higher moments; however, these are not included in this course.

Example 1

Find the moment generating function for the random variable X, with distribution

$$P\{X = x\} = \frac{1}{a}; \ x = 1, 2, \ldots, a \text{ and use it to find } \mu \text{ and } \sigma^2, \text{ the mean and variance of } X.$$

$$M_X(t) = E[e^{tX}] = \sum_{x=1}^{a} \frac{1}{a} e^{tx} \qquad \text{Using } E[e^{tX}] = \sum_{\forall i} e^{tx_i} p_i$$

$$M_X'(t) = \frac{1}{a} \sum_{x=1}^{a} x e^{tx}$$

> **Note**
>
> Differentiate with respect to t twice. In differentiating these functions, as the differentiation is with respect to the variable t, you should think of x as a constant.

(continued)

$$M''{}_X(t) = \frac{1}{a}\sum_{x=1}^{a} x^2 e^{tx}$$

Using $\mu = M'{}_X(0)$ and $\displaystyle\sum_{x=1}^{n} x = \frac{1}{2}n(n+1)$

Hence, $\displaystyle \mu = M'{}_X(0) = \frac{1}{a}\sum_{x=1}^{a} x = \frac{a+1}{2}$

Using $\sigma^2 = M''{}_X(0) - \mu^2$

$$\sigma^2 = M''{}_X(0) - \mu^2 = \frac{1}{a}\sum_{x=1}^{a} x^2 - \left\{\frac{a+1}{2}\right\}^2$$

$$= \frac{1}{a}\left\{\frac{1}{6}a(a+1)(2a+1)\right\} - \left(\frac{a+1}{2}\right)^2$$

Using $\displaystyle\sum_{x=1}^{n} x^2 = \frac{1}{6}n(n+1)(2n+1)$

$$= \frac{a^2-1}{12}$$

Example 2

A random variable R has probability distribution $P\{R = r\} = ar^2$; $r = 1, 2, 3, 4, 5$.

a Show that $a = \dfrac{1}{55}$

b Find the moment generating function of R and use it to find the mean and variance of R.

a $a + 4a + 9a + 16a + 25a = 55a$

$55a = 1 \implies a = \dfrac{1}{55}$

b $M_R(t) = E[e^{tR}] = \dfrac{1}{55}\displaystyle\sum_{r=1}^{5} e^{tr} r^2$

Sum of all probabilities in sample space, in terms of a.

The total probability on sample space equals 1.

$M'{}_R(t) = \dfrac{1}{55}\displaystyle\sum_{r=1}^{5} e^{tr} r^3$

$M''{}_R(t) = \dfrac{1}{55}\displaystyle\sum_{r=1}^{5} e^{tr} r^4$

So

$\mu = M'{}_R(0) = \dfrac{1}{55}\displaystyle\sum_{r=1}^{5} r^3 = \dfrac{45}{11} = 4.09 \,(2\,\text{dp})$

$\sigma^2 = M''{}_R(0) - \mu^2 = \dfrac{1}{55}\displaystyle\sum_{r=1}^{5} r^4 - \left(\dfrac{45}{11}\right)^2 = \dfrac{644}{605} = 1.06\,(2\,\text{dp})$

> **Note**
>
> Using $M_X(t) = E[e^{tX}]$
> $$= \sum_{\forall i} e^{tx_i} p_i$$

Moment generating functions of continuous random variables

The moment generating functions of continuous variables follow easily from their discrete equivalents, by applying the formula for the expected value of a function of a continuous random variable

$$E[g(X)] = \int_{-\infty}^{\infty} g(x)\mathrm{f}(x)\,\mathrm{d}x$$

As before, the mgf of a random variable will be defined by $M_X(t) = E[e^{tX}]$

> The moment generating function of a continuous random variable, X, with density function $\mathrm{f}(x)$ is
> $$M_X(t) = E[e^{tX}] = \int_{-\infty}^{\infty} e^{tx}\mathrm{f}(x)\,\mathrm{d}x$$

As with the discrete case, the following results can be obtained for the mean and variance of a continuous random variable.

> A continuous random variable, X, with moment generating function $M_X(t)$ has mean and variance given by $\mu = M'_X(0)$ and $\sigma^2 = M''_X(0) - \mu^2$

Example 3

A random variable X has a density function given by $f(x) = \begin{cases} 1; & 0 < x < 1 \\ 0; & \text{otherwise} \end{cases}$

a Show that the moment generating function for X is $M_X(t) = \frac{e^t - 1}{t}$

b By expressing your answer to **part a** as a power series, find the mean and variance of X.

a $M_X(t) = \int_0^1 e^{tx} f(x) dx$

$= \int_0^1 e^{tx} dx$

$= \frac{e^{tx}}{t} \Big|_0^1 = \frac{e^t - 1}{t}$

> **Note**
>
> Using $M_X(t) = E[e^{tX}] = \int_{-\infty}^{\infty} e^{tx} f(x) dx$

b $M_X(t) = \frac{e^t - 1}{t} = \frac{1 + t + \frac{t^2}{2!} + \cdots - 1}{t} = 1 + \frac{t}{2!} + \frac{t^2}{3!} + \cdots$

> **Note**
>
> Using the expansion of e^t
> $e^t = 1 + t + \frac{t^2}{2!} + \cdots$

Therefore,

$M'_X(t) = \frac{1}{2!} + \frac{2t}{3!} + \frac{3t^2}{4!} + \cdots$ and $M''_X(t) = \frac{1}{3} + \frac{1}{4}t + \cdots$

$M'_X(0) = \frac{1}{2}$ and $M''_X(0) = \frac{1}{3}$

$\mu = M'_X(0) = \frac{1}{2}$ and $\sigma^2 = M''_X(0) - \mu^2 = \frac{1}{3} - \frac{1}{4} = \frac{1}{12}$

Moment generating functions of a linear function of X

It is possible, and very useful, to write down the moment generating function of a linear function of X, for example, $bX + a$, given the mgf for X itself.

$M_X(t) = E[e^{tX}] = \sum_{\forall i} e^{tX_i} p_i$

$M_{bX+a}(t) = E[e^{t(bX+a)}]$

$= \sum_{\forall i} e^{t(bx_i + a)} p_i$

$= \sum_{\forall i} e^{tbx_i} p_i e^{at}$

$= e^{at} \sum_{\forall i} e^{tbx_i} p_i = e^{at} M_X(bt)$

> **Note**
>
> An identical result can be obtained for continuous variables using
> $M_X(t) = E[e^{tx}]$
> $= \int_{-\infty}^{\infty} e^{tx} f(x)$

> If the random variable, X, has moment generating function $M_X(t)$ then the function $bX + a$ has moment generating function $M_{bX+a}(t) = e^{at} M_X(bt)$

Example 4

Question

Find the moment generating function, $M_X(t)$, for the random variable X, with distribution

$$P\{X = x\} = \frac{1}{10}; \ x = 1, 2, \ldots, 10 \text{ and use it to find the mgf of } Y, \text{ where } Y = 3X - 1$$

$$M_X(t) = E\left[e^{tX}\right] = \frac{1}{10} \sum_{x=1}^{10} e^{tx}$$

Then $M_Y(t) = M_{3X-1}(t) = e^{-t} M_X(3t) = \frac{1}{10} e^{-t} \sum_{x=1}^{10} e^{3tx}$

> **Note**
>
> Using $M_{bX+a}(t) = e^{at} M_X(bt)$

Example 5

a Show that the moment generating function for the random variable X, with distribution $P\{X = x\} = \frac{1}{4}; \ x = 1, 2, 3, 4$ can be written $M_X(t) = \frac{1}{4}\left(\frac{e^t(1-e^{4t})}{1-e^t}\right)$

b A fair tetrahedral dice is thrown once. You receive a point score of three times the number on the dice. You use 4 points to enter the game. Write down the mgf for Y, your overall point score.

a $M_X(t) = E\left[e^{tX}\right] = \frac{1}{4}\sum_{x=1}^{4} e^{tx} = \frac{1}{4}(e^t + e^{2t} + e^{3t} + e^{4t})$ which is a geometric series,

$a = e^t, r = e^t = \frac{1}{4}\left(\frac{e^t(1-e^{4t})}{1-e^t}\right)$

b $Y = 3X - 4$

$M_Y(t) = \frac{e^{-4t}}{4}\left(\frac{e^{3t}(1-e^{12t})}{1-e^{3t}}\right)$

Mgf for Y using $M_{(bX+a)}$
$(t) = e^{at} M_X(bt); \ a = -4 \text{ and } b = 3$

> **Note**
>
> Using the summation of a geometric progression:
>
> $S_n = \frac{a(1-r^n)}{1-r}$

Exercise 1

1 The random variable X has distribution given by

$P\{X = x\} = \frac{1}{2a}; \ x = 1, 2, \ldots, 2a$

  i Find the moment generating function for X.

  ii Use the moment generating function to find μ, the mean of X.

2 A random variable Y has probability distribution

$P\{Y = y\} = ky; \ y = 1, 2, 3, 4, 5, 6$

  i Show that $k = \frac{1}{21}$

  ii By finding its moment generating function, calculate the mean of Y.

3 A random variable X has moment generating function given by

$f(x) = a + bx + cx^2 + \ldots$

Show that the mean of X is b.

4 A random variable Y has moment generating function given by

$f(t) = 1 + 2t + 3t^2 + \ldots$

Show that the mean of Y is 2 and find the variance of Y.

5 Find the moment generating function, $M_X(t)$, for the random variable X, with distribution $P\{X = x\} = \frac{1}{6}; \ x = 1, 2, \ldots, 6$ and use it to find the moment generating function of Y, where $Y = 2X + 1$.

6 A continuous random variable R has a density function given by

$$f(r) = \begin{cases} \dfrac{1}{a}; & 0 < r < a \\ 0; & \text{otherwise} \end{cases}$$

a Show that the moment generating function for R is $M_R(t) = \dfrac{e^{at} - 1}{at}$

b By expressing your answer to **part a** as a power series, find the mean and variance of R in terms of a.

7 X is a continuous random variable with moment generating function given by

$$M_Y(t) = 1 + a_1 t + a_2 t^2 + \cdots$$

a Write down the mean of X.

b Find $E[X^2]$ and the variance of X.

29.2 Moment generating functions of some standard distributions

In this section, the moment generating functions for some standard distributions are derived by direct application of the formula $M_X(t) = E[e^{tX}]$. As shown earlier, the mgfs can be used to find expressions for the mean and variance of each distribution. The calculations of the means and variances will be done as part of the exercise questions.

Although most of the results are familiar, the mgf method produces some very elegant solutions.

Poisson distribution

The distribution function for a Poisson random variable, X, parameter μ, is given by the formula $P\{X = x\} = \dfrac{e^{-\mu}\mu^x}{x!}$; $x = 0, 1, 2, \ldots$

Its moment generating function is therefore,

$$M_X(t) = E[e^{tX}]$$

$$= \sum_{x=0}^{\infty} e^{tx} \frac{e^{-\mu}\mu^x}{x!}$$

$$= e^{-\mu} \sum_{x=0}^{\infty} \frac{(e^t \mu)^x}{x!}$$

(Compare these with the series expansion of e^x using $e^t = 1 + t + \dfrac{t^2}{2!} + \cdots$: $e^{-\mu}e^{e^t\mu}$ and $e^{\mu(e^t - 1)}$.)

> If X is a Poisson random variable with **parameter μ and distribution function** $P\{X = x\} = \dfrac{e^{-\mu}\mu^x}{x!}$; $x = 0, 1, 2, \ldots$ its moment generating function is $M_X(t) = e^{\mu(e^t - 1)}$

Exponential distribution

The density function for an exponential random variable, X, parameter μ, is given by the formula

$$f(x) = \begin{cases} ke^{-kx}, & x > 0 \\ 0, & x \leq 0 \end{cases}$$

Its moment generating function is, $M_X(t) = E[e^{tX}] = k \int_{-\infty}^{\infty} e^{tx} e^{-kx} \, dx$

$$= k \int_0^{\infty} e^{-x(k-t)} \, dx$$

$$= \left[\frac{e^{-x(k-t)}}{(t-k)} \right]_0^{\infty}$$

$$= k \left(0 - \frac{1}{t-k} \right)$$

$$= \frac{k}{k-t}.$$

Tip

This makes the (safe) assumption that t is small enough for $k - t > 0$ which is necessary to ensure that division by zero is avoided.

If X is an exponential random variable with density function

$$f(x) = \begin{cases} ke^{-kx}, & x > 0 \\ 0, & x \le 0 \end{cases}$$

its moment generating function is $M_X(t) = \dfrac{k}{k-t}$

Normal distribution

The standard normal distribution will be treated first; and the result for the general normal distribution, mean μ, and variance σ^2 will be derived using the result for linear functions of random variables (see page 403).

Standard normal distribution

The distribution function for a standard normal random variable, Z, is given by the formula $f(z) = \dfrac{1}{\sqrt{(2\pi)}} e^{-\frac{1}{2}z^2}; \; z \in \mathbb{R}$

Its moment generating function is therefore,

$$M_Z(t) = E[e^{tZ}] = \frac{1}{\sqrt{(2\pi)}} \int_{-\infty}^{\infty} e^{tz} e^{-\frac{1}{2}z^2} \, dz$$

$$= \frac{1}{\sqrt{(2\pi)}} \int_{-\infty}^{\infty} e^{tz - \frac{1}{2}z^2} \, dz$$

$$= \frac{1}{\sqrt{(2\pi)}} \int_{-\infty}^{\infty} e^{\frac{1}{2}(2tz - z^2)} \, dz$$

$$= \frac{1}{\sqrt{(2\pi)}} \int_{-\infty}^{\infty} e^{\frac{1}{2}(t^2 - (z-t)^2)} \, dz$$

$$= \frac{1}{\sqrt{(2\pi)}} e^{\frac{t^2}{2}} \int_{-\infty}^{\infty} e^{-\frac{1}{2}(z-t)^2} \, dz$$

Now, let $y = z - t$ (this change of variable allows the integral to be expressed in the required form)

$$M_X(t) = \frac{1}{\sqrt{(2\pi)}} e^{\frac{t^2}{2}} \int_{-\infty}^{\infty} e^{-\frac{1}{2}y^2} \, dy$$

Now,

$$\frac{1}{\sqrt{(2\pi)}} \int_{-\infty}^{\infty} e^{-\frac{1}{2}y^2} \, dy = 1$$

Tip

$f(z) = \dfrac{1}{\sqrt{(2\pi)}} e^{-\frac{1}{2}z^2}$ is the density function of Z and therefore the total area under its curve is 1.

Therefore, $M_z(t) = e^{\frac{t^2}{2}}$

If Z is a random variable with a standard normal distribution, that is,

with density function $f(z) = \frac{1}{\sqrt{(2\pi)}} e^{-\frac{1}{2}z^2}$ $z \in \mathbb{R}$

its moment generating function is $M_z(t) = e^{\frac{t^2}{2}}$

Normal distribution mean μ, variance σ^2

The distribution function for a normal random variable, X, mean μ, variance σ^2, is given by the formula

$$f(x) = \frac{1}{\sqrt{(2\pi\sigma^2)}} e^{-\frac{1}{2}\left(\frac{x-\mu}{\sigma}\right)^2}$$

Its moment generating function is therefore,

$$M_X(t) = E\left[e^{tX}\right] = \frac{1}{\sqrt{(2\pi\sigma^2)}} \int_{-\infty}^{\infty} e^{tx} e^{-\frac{1}{2}\left(\frac{x-\mu}{\sigma}\right)^2} dx$$

This is difficult to integrate. Instead, it is possible to use the above result for a standard normal random variable with the result for the moment generating function of a linear function of X, $bX + a$ (see page 403). This technique is used in Example 6 and gives the following result.

The moment generating function for a normal random variable, X, mean μ, variance σ^2, is given by

$$M_X(t) = e^{\mu t + \frac{1}{2}\sigma^2 t^2}$$

Note that this reduces to the result for a standard normal variable if $\mu = 0$ and $\sigma = 1$.

Example 6

$M_Z(t) = e^{\frac{t^2}{2}}$ is the moment generating function of Z, a standard normal random variable, and X is defined by the equation $Z = \frac{X - \mu}{\sigma}$

Find the moment generating function of X.

$X = \sigma Z + \mu$

The moment generating function for X is
$$M_X(t) = M_{\sigma Z + \mu}(t) = e^{\mu t} M_Z(\sigma t)$$

$$= e^{\mu t} e^{\frac{(\sigma t)^2}{2}}$$

$$= e^{\mu t + \frac{1}{2}\sigma^2 t^2} \qquad \text{Using } M_{bX+a}(t) = e^{at} M_X(bt)$$

So, $M_X(t) = e^{\mu t + \frac{1}{2}\sigma^2 t^2}$

Note

Since $M_Z(t) = e^{\frac{t^2}{2}}$ is the moment generating function of Z.

The techniques used to find the moment generating function of the Poisson, exponential and normal distributions can be used to find the mgfs of other distributions.

Example 7

A discrete uniform random variable U has distribution function given by

$$P\{U = u\} = \frac{1}{n}; \ u = 1, 2, 3, ..., n$$

Prove that the moment generating function of U is given by

$$M_U(t) = \frac{1}{n}\left(e^t\left(\frac{1 - e^{nt}}{1 - e^t}\right)\right)$$

The moment generating function of U is given by

$$M_U(t) = \mathrm{E}[e^{tU}] = \sum_{u=1}^{n} e^{tu}\frac{1}{n}$$

This gives you a geometric series.

$$= \frac{1}{n}\left(e^t + e^{2t} + ... + e^{nt}\right)$$

$$= \frac{1}{n}\left(e^t\left(\frac{1 - e^{nt}}{1 - e^t}\right)\right)$$

Using the sum of a geometric series.

Example 8

Find the moment generating function of a geometric random variable, X, with parameter p. Hence find its mean.

$$P\{X = x\} = (1 - p)^{x-1} p; \ x = 1, 2,...$$

Its moment generating function is

> **Note**
>
> Using $M_X(t) = \mathrm{E}\left[e^{tX}\right] = \sum_{\forall i} e^{tx_i} p_i$

Distribution function of a geometric random variable.

$$M_X(t) = \mathrm{E}\left[e^{tX}\right] = \sum_{x=1}^{\infty} e^{tx}(1 - p)^{x-1} p$$

$$M_X(t) = e^t p + e^{2t}(1 - p)p + e^{3t}(1 - p)^2 p + \cdots$$

This is a geometric progression and sums to infinity to give

Write out the first few terms of this summation.

$$M_X(t) = \frac{e^t p}{1 - (1 - p)e^t} = \frac{p}{e^{-t} - 1 + p}$$

$$M'_X(t) = \frac{e^{-t} p}{\left(e^{-t} - 1 + p\right)^2}$$

and so,

> **Note**
>
> Differentiate with respect to t using the quotient rule.

$$\mu = M'_X(0) = \frac{p}{p^2} = \frac{1}{p}$$

Exercise 2

1 X is an exponential random variable with density function

$$f(x) = \begin{cases} 2e^{-2x}, & x > 0 \\ 0, & x \le 0 \end{cases}$$

Find its moment generating function.

2 S is a Bernoulli random variable with parameter p. Find its moment generating function, and hence its mean and variance.

3 If X is a Poisson random variable with mean 4, show from first principles that its moment generating function is given by $M_X(t) = e^{4(e^t - 1)}$

4 The lifetime of an electrical component is modelled by the exponential distribution

$$f(x) = \begin{cases} \dfrac{1}{300} e^{-\frac{1}{300}x}, & x > 0 \\ \\ 0, & x \le 0 \end{cases}$$

Find the moment generating function for X and hence show that its mean is 300.

5 Given that $M_Z(t) = e^{\frac{t^2}{2}}$ is the moment generating function of Z, a standard normal random variable, and X is a normal random variable with mean μ, variance σ^2, show that the moment generating function of X is given by $M_X(t) = e^{\mu t + \frac{1}{2}\sigma^2 t^2}$

> **Tip**
>
> Use the result that if
> $X \sim N(\mu, \sigma^2)$ then $Z = \frac{X - \mu}{\sigma} \sim N(0, 1)$

6 The number of currants in a cake mix, N, follows a Poisson distribution, mean 28. Find the moment generating function for N and use it to find its variance.

7 14 cards are each numbered with integers 1–14. A card is drawn at random with replacement until a prime number is obtained. X is a random variable for the number of draws required up to and including the prime number. Find the moment generating function for X and hence find its mean.

29.3 Sums of independent random variables

Consider a set of n independent random variables X_i; $i = 1, 2, \ldots, n$

If X_T is the sum of these variables, then $X_T = \displaystyle\sum_{i=1}^{n} X_i$ and its moment generating function is

$$M_{X_T}(t) = E[e^{tX_T}] = E\left[e^t \sum_{i=1}^{n} X_i \right]$$

$$= E\left[\prod_{i=1}^{n} e^{tX_i} \right]$$

$$= \prod_{i=1}^{n} E[e^{tX_i}]$$

because the X_is are independent random variables. Therefore,

> Let X_T be the sum of a set of n independent random variables X_i;
> $i = 1, 2, \ldots, n$; that is $X_T = \displaystyle\sum_{i=1}^{n} X_i$ the moment generating function
> of X_T is given by $M_{X_T}(t) = \displaystyle\prod_{i=1}^{n} M_{X_i}(t)$

With the further assumption that the X_is are not only independent but also identically distributed, the following result follows,

Let X_T be the sum of a set of n independent and identically distributed random variables X_i; $i = 1, 2, \ldots, n$

The moment generating function of X_T is given by $M_{X_T}(t) = \left\{M_X(t)\right\}^n$
where $M_X(t)$ is the moment generating function for each X_i

It is worth noting that the X_is are independent and identically distributed (i.i.d.) random variables if X_i; $i = 1, 2, \ldots, n$ is an independent random sample from a large population (see, for example Exercise 3, Question 3).

Example 9

A fair coin is flipped until a head shows for the third time. Using moment generating functions, find the mean of the number of flips required. You are given that the moment generating function of a geometric random variable, parameter p is $\dfrac{p}{e^{-t} - 1 + p}$

Let X_1 be the number of throws required up to and including the 1st head, and X_2 and X_3 be the equivalent random variables up to and including the 2nd and 3rd heads, respectively.

$$M_{X_i}(t)\frac{\frac{1}{2}}{e^{-t} - \frac{1}{2}} = \frac{1}{2e^{-t} - 1}$$

This is the total number of throws required.

> **Note**
> The X_is have a geometric distribution, $p = \dfrac{1}{2}$

$$X_T = \sum_{i=1}^{3} X_i$$

X_T has moment generating function:

$$M_{X_T}(t) = \left\{M_{X_i}(t)\right\}^3 = \left\{\frac{1}{2e^{-t} - 1}\right\}^3$$

Using $M_{X_T}(t) = \left\{M_X(t)\right\}^n$

$$M'_{X_T}(t) = 3\left\{\frac{1}{2e^{-t} - 1}\right\}^2 \left\{\frac{2e^{-t}}{\left(2e^{-t} - 1\right)^2}\right\}$$

$$\mu_{X_T} = M'_{X_T}(0) = 3 \times 2 = 6$$

Example 10

If X_i; $i = 1, 2, \ldots, n$ is an independent random sample from a normal distribution, mean μ, variance σ^2, show that the random variable X_T given by $X_T = \sum_{i=1}^{n} X_i$ has a normal distribution, mean $n\mu$, variance $n\sigma^2$

$$X_T = \sum_{i=1}^{n} X_i$$

$$M_{X_i}(t) = e^{\mu t} e^{\frac{1}{2}\sigma^2 t^2}$$

This is the mgf of each of the X_is

$$M_{X_T}(t) = \left\{e^{\mu t} e^{\frac{1}{2}\sigma^2 t^2}\right\}^n$$

$$= e^{n\mu t} e^{\frac{1}{2}n\sigma^2 t^2}$$

$$M_{X_T}(t) = e^{n\mu t} e^{\frac{1}{2}n\sigma^2 t^2}$$ which is the mgf of a normal distribution mean $n\mu$, variance $n\sigma^2$. Therefore, X_T has a normal distribution mean $n\mu$, variance $n\sigma^2$.

The mgf of X_T

Exercise 3

(1) X_i; $i = 1, 2,..., n$ is an independent random sample from a Poisson distribution, parameter λ. Find the moment generating function for the random variable Y, where $Y = \sum_{i=1}^{4} X_i$

(2) X is a random variable for the number of random events in 10 seconds. The mean of X is 8. By using moment generating functions, show that the mean number of these events in 1 minute has a Poisson distribution with mean 48. You may assume that X follows a Poisson distribution.

(3) The density function for an exponential random variable, X, parameter μ, is given by

$$f(x) = \begin{cases} ke^{-kx}, & x > 0 \\ 0, & x \le 0 \end{cases}$$

A random sample of size 3 is taken and X_T is the total of the three values.

Find the moment generating function of X_T and, by differentiation, its mean value.

(4) **a** Find the moment generating function of a Bernoulli random variable, parameter p.

   X_i; $i = 1, 2,..., n$ is an independent random sample from a Bernoulli distribution, parameter, p.

   b Find the mgf of $\sum_{i=1}^{n} X_i$ and give its distribution.

(5) **a** Find the moment generating function of a binomial random variable, parameters n and p.

   Random variables X_1 and X_2 have the following distributions:

   $X_1 \sim$ Binomial (n_1, p), $X_2 \sim$ Binomial (n_2, p)

   b Use your answer to **part a** to find the mgf of $X_1 + X_2$

   c Hence, write down the distribution of $X_1 + X_2$

(6) A spinner can land on the scores −2, 0, 2 and 4 with equal probability.

   i Find the moment generating function for the score on a randomly chosen spin.

   Y is a random variable for the sum of the scores on two spins.

   ii Find the mgf of Y.

(7) Prove that, if X_i; $i = 1, 2,..., n$ is an independent random sample from a normal distribution, mean μ, variance σ^2, then the random variable $\overline{X}$, given by $\overline{X} = \frac{1}{n}\sum_{i=1}^{n} X_i$ has a normal distribution, mean μ, variance $\frac{\sigma^2}{n}$

Summary

▶ Moment generating functions can be used to find the mean and variance of discrete and continuous random variables.

▶ The moment generating function of a discrete random variable, X, with distribution function

$$P\{X = x_i\} = p_i; i = 1, 2,..., \text{ is } M_X(t) = E[e^{tX}] = \sum_{\forall i} e^{tx_i} p_i$$

- The moment generating function of a continuous random variable X with density function $f(x)$ is $M_X(t) = E[e^{tX}] = \int_{-\infty}^{\infty} e^{tx} f(x) dx$

- A random variable X with moment generating function $M_X(t)$ has mean and variance given by $\mu = M'_X(0)$ and $\sigma^2 = M''_X(0) - \mu^2$

- If the random variable X has moment generating function $M_X(t)$ then the function $bX + a$ has moment generating function $M_{bX+a}(t) = e^{at} M_X(bt)$

- If X is a Poisson random variable with parameter μ and distribution function
$$P\{X = x\} = \frac{e^{-\mu}\mu^x}{x!}; \; x = 0, 1, 2, \ldots \text{ its moment generating function is}$$
$$M_X(t) = e^{\mu(e^t - 1)}$$

- If X is an exponential random variable with density function
$$f(x) = \begin{cases} ke^{-kx}, & x > 0 \\ 0, & x \le 0 \end{cases} \text{ its mgf is } M_X(t) = \frac{k}{k - t}$$

- If Z is a random variable with a standard normal distribution, that is, with density function $f(z) = \frac{1}{\sqrt{(2\pi)}} e^{-\frac{1}{2}z^2}$ its mgf is $M_z(t) = e^{\frac{t^2}{2}}$

- The moment generating function for a normal random variable, mean μ, variance σ^2, is given by $M_X(t) = e^{\mu t + \frac{1}{2}\sigma^2 t^2}$

- If $X_T = \sum_{i=1}^{n} X_i$ is the sum of a set of n independent random variables X_i; $i = 1, 2, \ldots, n$, the moment generating function of X_T is given by
$$M_{X_T}(t) = \prod_{i=1}^{n} M_{X_i}(t)$$

- If $X_T = \sum_{i=1}^{n} X_i$ is the sum of a set of n independent and identically distributed random variables X_i; $i = 1, 2, \ldots, n$, the moment generating function of X_T is given by $M_{X_T}(t) = \{M_X(t)\}^n$

Review exercises

1 A random variable X has probability distribution
$P\{X = x\} = k(x + 1); x = 0, 1, 2, 3, 4$

   i Show that $k = \frac{1}{15}$

   ii By finding its moment generating function, calculate the mean of X.

2 A continuous random variable X has a density function given by
$$f(x) = \begin{cases} \dfrac{1}{b - a}; & a < x < b \\ 0; & \text{otherwise} \end{cases}$$

   a Show that the moment generating function for X is $M_X(t) = \dfrac{e^{bt} - e^{at}}{(b - a)t}$

   b Hence find the mgf of $2X + 4$.

3 X is an exponential random variable with density function

$$f(x) = \begin{cases} 3e^{-ax}, & x > 0 \\ 0, & x \leq 0 \end{cases}$$

  a Write down the value of a.

  b Find the moment generating function of X and hence its mean value.

4 Two independent random variables, X and Y, have distributions as follows:

  $X \sim N(\mu_x, \sigma_x^2); Y \sim N(\mu_y, \sigma_y^2)$

  Find the mgf of $W = X + Y$ in its simplest form and hence find the mean and variance of W.

  You are given that the mgf of a normal random variable with mean μ, variance σ^2 is $e^{\mu t + \frac{1}{2}\sigma^2 t^2}$

Practice examination questions

1 A random variable X has probability density function given by

$$f(x) = \begin{cases} ke^{-2x}; & x > 0 \\ 0; & \text{otherwise} \end{cases}$$

  a By integration, show that $k = 2$ (2 marks)

  b Find the moment generating function for X (3 marks)

  c Hence prove that the mean and variance of X are $\frac{1}{2}$ and $\frac{1}{4}$
  respectively.   (4 marks)

2 The random variable X has an exponential probability density function

$$f(x) = \begin{cases} 4e^{-4x}, & x > 0 \\ 0, & x \leq 0 \end{cases}$$

  a Derive the moment generating function, $M_X(t)$, of X.   (4 marks)

  b Deduce the moment generating function, $M_Y(t)$, of $Y = 3 + 4X$ (2 marks)

  c Hence find the mean and the variance of Y.   (5 marks)

3 The random variable X follows a Poisson distribution with mean 10.

  a Show that the moment generating function of X is given by
  $$M_X(t) = e^{10(e^t - 1)}$$ (4 marks)

  b Find the variance of X.   (3 marks)

  The random variable X_T is defined by $X_T = \sum_{i=1}^{n} X_i$, where X_i are independently

  distributed, each with moment generating function $M_X(t)$.

  c **i** Determine an expression for $M_{X_T}(t)$ (4 marks)

      ii Hence specify completely the distribution of X_T (2 marks)

4 X is a normal random variable with mean μ and variance σ^2. Z is a standard normal variable.

  a Prove that the moment generating function of Z is given
  by $M_Z(t) = e^{\frac{t^2}{2}}$ (5 marks)

  b Express X as a linear function of Z.   (1 mark)

  c Use the results in **parts a** and **b** to find the mgf of X.   (3 marks)

30 Estimators

Introduction

Inferential statistics deals with two main problems: estimating unknown population parameters, and testing hypotheses about those parameters. In this chapter, the focus will be on the theory relating to the desirable properties of estimators, giving the tools required to help solve practical problems in estimation.

Recap

You will need to remember...

▶ Results relating to A-level mathematics expectation algebra of $aX + b$:
  - $E[aX + b] = aE[X] + b$
  - $Var[aX + b] = a^2Var[X]$.
▶ How to find the mean and variance of a sum, or mean value of a set of independent random variables.
▶ How to calculate sample variance using $\dfrac{\sum(X - \bar{X})^2}{n}$.

▶ The sampling distribution of $\bar{X}$: if $X \sim N(\mu, \sigma^2)$ then $\bar{X} \sim N\left(\mu, \dfrac{\sigma^2}{n}\right)$

▶ The Central Limit Theorem: if X has a mean μ and variance σ^2, then the distribution of $\bar{X}$ tends to $N\left(\mu, \dfrac{\sigma^2}{n}\right)$ as $n \to \infty$.

Objectives

By the end of this chapter, you should know how to:
▶ Define the terms 'estimator' and 'estimate'.
▶ Define the three main desirable properties of estimators: unbiasedness, efficiency and consistency.
▶ Prove that a given estimator has one or more of these properties.
▶ Define the 'best' estimator among a set of unbiased estimators.
▶ Define, and calculate a value for, the relative efficiency of two estimators.

30.1 Review of key concepts

Estimators provide a means of obtaining estimates of population parameters. Before learning about the properties of estimators, it is necessary to consider the theory related to population parameters and statistical inference.

Population parameters and statistical inference

The probability distribution of a population is partly defined by giving its name (normal, Poisson, and so on). Having identified this, the distribution is then uniquely defined by giving the values of its **parameters**. These are quantities such as the mean and variance, which together uniquely define a normal distribution, or λ the quantity which defines a Poisson distribution. Often, some or all of these are unknown.

Statistical inference is where information about a population sample is used to inform knowledge about the population parameters.

For example, if the mean of a sample is known, it may be possible to say something about the mean of the population from which it is taken; it may provide an estimate for the population mean or suggest that a certain hypothesised value of the population mean is unlikely. In this sense, information about the population is being *inferred* from sample results.

Statistics and sampling distributions

In order to find out information about a population's parameters, it is common to take a random sample from that population then calculate a value for the quantity of interest.

For example, let $X_i; i = 1, 2, \ldots, n$ be an independent random sample from a population, P. It is possible to calculate the value of the sample mean, $\bar{X}$, where

$$\bar{X} = \frac{\sum_{i=1}^{n} X_i}{n}$$

$\bar{X}$ is a **statistic** because it is a random variable, whose value can be found from the sample values only.

> A statistic is a function of the random variables that make up the sample and which does not depend upon any unknown parameter.

A statistic must not depend upon unknown parameters because it must be possible to use the values which make up the sample to calculate a value for the statistic.

Since a statistic is a function of random variables, it is itself a random variable and, as such, has a probability distribution. This distribution is known as the **sampling distribution** of the statistic.

From your work on probability in A-level mathematics, you know that if $X_i; i = 1, 2, \ldots, n$ is an independent random sample from a normal population with mean μ, variance σ^2, then the distribution of the mean $\bar{X}$ is normal, with mean μ, variance $\frac{\sigma^2}{n}$. This is the sampling distribution of $\bar{X}$.

See *Chapter 29 Moment Generating Functions.*

Example 1

If the random variable X has mean μ, variance σ^2, and if $X_i; i = 1, 2, \ldots, n$ is an independent random sample from this distribution, find the mean and variance of

a $a\bar{X}$ where a is a constant

b $X_T = \sum_{i=1}^{n} X_i$

c $\dfrac{X_1 + 2X_2 + 3X_3}{6}$

a $E[a\bar{X}] = aE[\bar{X}] = a\mu \quad Var[a\bar{X}] = a^2 Var[\bar{X}] = \dfrac{a^2\sigma^2}{n}$

b $E[X_T] = \sum_{i=1}^{n} E[X_i] = n\mu \quad Var[X_T] = \sum_{i=1}^{n} Var[X_i] = n\sigma^2$

c $E\left[\dfrac{X_1 + 2X_2 + 3X_3}{6}\right] = \dfrac{1}{6}\{E[X_1] + 2E[X_2] + 3E[X_3]\} = \dfrac{1}{6}(\mu + 2\mu + 3\mu) = \dfrac{1}{6}(6\mu) = \mu$

$Var\left[\dfrac{X_1 + 2X_2 + 3X_3}{6}\right] = \dfrac{1}{36}\{Var[X_1] + 4Var[X_2] + 9Var[X_3]\} = \dfrac{1}{36}(\sigma^2 + 4\sigma^2 + 9\sigma^2) = \dfrac{7\sigma^2}{18}$

Note

$Var[X_T] = \sum_{i=1}^{n} Var[X_i]$

because the X_is are independent random variables.

Exercise 1

1 Identify each of the following as either a statistic, a random variable that is not a statistic, or a constant.

a $\dfrac{1}{4}\displaystyle\sum_{i=1}^{4} X_i$

b $\dfrac{1}{n}\displaystyle\sum_{i=1}^{n} X_i$, where n is the sample size

c $\overline{X} - \mu$

d σ^2

2 If X_i; $i = 1, 2, \ldots, n$ is an independent random sample from a distribution with mean μ, variance σ^2, find the mean and variance of

a $\overline{X} - \mu$ **b** $\dfrac{1}{6}\displaystyle\sum_{i=1}^{6} X_i$ **c** $\dfrac{2X_1 + 3X_2 + 5X_3}{10}$

3 Show that, if X_i; $i = 1, 2, \ldots, n$ is an independent random sample from a distribution with mean μ, variance σ^2, then $\dfrac{\overline{X} - \mu}{\frac{\sigma}{\sqrt{n}}}$ has mean 0, variance 1.

30.2 Estimators and estimates

The sample statistic $\overline{X}$ can be used to estimate the value of the unknown population parameter μ. Other population parameters, for example standard deviation or correlation coefficient, can also be estimated using sample statistics. In this context, statistics such as $\overline{X}$ are called **estimators**. A particular value of an estimator, $\bar{x}$ for example, is called an **estimate** of the parameter.

Estimators can have certain desirable properties, such as unbiasedness, efficiency and consistency.

Unbiased estimators

> An estimator U is unbiased for the population parameter u if the expected value of U equals u; that is $\mathrm{E}[U] = u$

This is clearly a good property for an estimator because it means that, in the long run, the average value of U will neither overestimate nor underestimate the value of u; it *is unbiased* for u.

Example 2

X_i; $i = 1, 2, 3, 4$ is a random sample of size 4 from a population with unknown mean μ. Which of the following estimators are unbiased for μ?

i $\dfrac{X_1 + X_2 + X_3 + X_4}{4}$ **ii** $\dfrac{2X_1 + X_2 + X_3 + 2X_4}{4}$

iii $\dfrac{2X_1 + X_2 + X_3 + 2X_4}{6}$

i $\quad E\left[\dfrac{X_1 + X_2 + X_3 + X_4}{4}\right] = \dfrac{1}{4}\{E[X_1] + E[X_2] + E[X_3] + E[X_4]\} = \dfrac{1}{4}4\mu = \mu$

ii $\quad E\left[\dfrac{2X_1 + X_2 + X_3 + 2X_4}{4}\right] = \dfrac{1}{4}\{2E[X_1] + E[X_2] + E[X_3] + 2E[X_4]\}$

$$= \dfrac{1}{4}6\mu = \dfrac{3}{2}\mu$$

iii $\quad E\left[\dfrac{2X_1 + X_2 + X_3 + 2X_4}{6}\right] = \dfrac{1}{6}\{2E[X_1] + E[X_2] + E[X_3] + 2E[X_4]\}$

$$= \dfrac{1}{6}6\mu = \mu$$

i and **iii** are therefore unbiased.

Population mean and variance are two important parameters and it is necessary to have the unbiased estimators of these.

> If X_i; $i = 1, 2, \ldots, n$ is an independent random sample from a population with mean μ and variance σ^2, then $\bar{X} = \dfrac{\sum x}{n}$ and $S^2 = \dfrac{\sum(x - \bar{x})^2}{n-1}$ are unbiased estimators of μ and σ^2, respectively.

These results are proved in **Example 3** and **Example 4**.

Example 3

Show that, for an independent random sample X_i; $i = 1, 2, \ldots, n$,

the sample statistic $\bar{X}$, given by $\bar{X} = \dfrac{\displaystyle\sum_{i=1}^{n} x_i}{n}$ is an unbiased estimator

of the population mean μ.

$$E[\bar{X}] = E\left[\dfrac{\displaystyle\sum_{i=1}^{n} x_i}{n}\right]$$

$$= \dfrac{1}{n}E\left[\sum_{i=1}^{n} X_i\right]$$

$$= \dfrac{1}{n}\left[\sum_{i=1}^{n} E[X_i]\right]$$

$$= \dfrac{1}{n}\left[\sum_{i=1}^{n} \mu\right]$$

$$= \dfrac{1}{n}n\mu = \mu$$

Therefore $\bar{X}$ is an unbiased estimator of μ.

Example 4

a   Show that the sample variance, given by $\dfrac{1}{n}\displaystyle\sum_{i=1}^{n} X_i^2 - \bar{X}^2$ is not unbiased for the population variance, σ^2.

b   Use your result to find an unbiased estimator for σ^2.

Tip

Throughout this course, S^2 will be used to denote the unbiased estimator of population variance. The sample variance will be denoted simply by its formulae

$\dfrac{\sum(X - \bar{X})^2}{n}$ or $\dfrac{1}{n}\sum X_i^2 - \bar{X}^2$

a $E\left[\dfrac{1}{n}\displaystyle\sum_{i=1}^{n}X_i^2 - \bar{X}^2\right]$

$= \dfrac{1}{n}E\left[\displaystyle\sum_{i=1}^{n}X_i^2\right] - E[\bar{X}^2]$

$= \dfrac{1}{n}\left[\displaystyle\sum_{i=1}^{n}E\left[X_i^2\right]\right] - E[\bar{X}^2]$

$= \dfrac{1}{n}\left[\displaystyle\sum_{i=1}^{n}(\sigma^2 + \mu^2)\right] - \left[\dfrac{\sigma^2}{n} + \mu^2\right]$

$= \dfrac{1}{n}(n\sigma^2 + n\mu^2) - \left[\dfrac{\sigma^2}{n} + \mu^2\right]$

$= \sigma^2 + \mu^2 - \dfrac{\sigma^2}{n} - \mu^2$

$= \sigma^2 - \dfrac{\sigma^2}{n} = \dfrac{n-1}{n}\sigma^2$

$\dfrac{1}{n}\displaystyle\sum_{i=1}^{n}X_i^2 - \bar{X}^2$ is therefore not unbiased for σ^2, since in the long run it will

slightly underestimate the population variance, since $\dfrac{n-1}{n} < 1$ and therefore

$\dfrac{n-1}{n}\sigma^2 < \sigma^2$

b $E\left[\dfrac{1}{n}\displaystyle\sum_{i=1}^{n}X_i^2 - \bar{X}^2\right] = \dfrac{n-1}{n}\sigma^2$

$E\left[\dfrac{n}{n-1}\left\{\dfrac{1}{n}\displaystyle\sum_{i=1}^{n}X_i^2 - \bar{X}^2\right\}\right] = \sigma^2$

Therefore, an unbiased estimator for σ^2 is

$S^2 = \dfrac{1}{n-1}\left[\displaystyle\sum_{i=1}^{n}X_i^2 - n\bar{X}^2\right]$

Since

$\sigma^2 = E\left[\displaystyle\sum_{i=1}^{n}X_i^2\right] - \mu^2$ and

$E[\bar{X}^2] - \mu_{\bar{X}}^2 = \mathrm{Var}[\bar{X}] = \dfrac{\sigma^2}{n}$

This is usually the best form for calculations. The *Formulae and Statistical Tables* booklet also gives the alternative forms:

$S^2 = \dfrac{1}{n-1}\displaystyle\sum_{i=1}^{n}\left(X_i - \bar{X}\right)^2$

$= \dfrac{1}{n-1}\left[\displaystyle\sum_{i=1}^{n}X_i^2 - \dfrac{\left(\displaystyle\sum_{i=1}^{n}X_i\right)^2}{n}\right]$

When you estimate a population parameter, it is common to combine the estimators from several samples from the population. It can be important to ensure that the pooled estimator is unbiased. The following examples show how this can be achieved.

Example 5

Samples of size n_1, n_2 and n_3 are taken from a population with mean μ. The sample means are $\bar{X}_1$, $\bar{X}_2$ and $\bar{X}_3$, respectively. Show that the pooled estimator $M = \dfrac{n_1\bar{X}_1 + n_2\bar{X}_2 + n_3\bar{X}_3}{n_1 + n_2 + n_3}$ is unbiased for the population mean.

$E[M] = E\left[\dfrac{n_1\bar{X}_1 + n_2\bar{X}_2 + n_3\bar{X}_3}{n_1 + n_2 + n_3}\right]$

$= \dfrac{n_1 E[\bar{X}_1] + n_2 E[\bar{X}_2] + n_3 E[\bar{X}_3]}{n_1 + n_2 + n_3}$

$= \dfrac{n_1\mu + n_2\mu + n_3\mu}{n_1 + n_2 + n_3} = \mu$

Example 6

Samples of size n_X and n_Y are taken from a population with variance σ^2. The unbiased estimates of σ^2 are S_X^2 and S_Y^2 respectively.

Find an expression for n so that the pooled estimator of σ^2, $S_P^2 = \frac{(n_X-1)S_X^2+(n_Y-1)S_Y^2}{n}$ is unbiased for the population variance.

$$E\left[S_P^2\right] = E\left[\frac{\left(n_X-1\right)S_X^2+(n_Y-1)S_Y^2}{n}\right]$$

$$= \frac{1}{n}\left[(n_X-1)E\left[S_X^2\right]+(n_Y-1)E\left[S_Y^2\right]\right]$$

$$= \frac{1}{n}\left[(n_X-1)\sigma^2+(n_Y-1)\sigma^2\right]$$

$$= \frac{\sigma^2}{n}\left[(n_X-1)+(n_Y-1)\right]$$

$$= \frac{\sigma^2}{n}(n_X+n_Y-2)$$

Therefore, if $n = n_X + n_Y - 2$, S_P^2 is unbiased for σ^2.

The results from **Examples 5** and **6** give the following result:

> Samples of size n_X and n_Y are taken from a population with mean μ, variance σ^2. The sample means are $\bar{X}$ and $\bar{Y}$ and the two unbiased estimators of σ^2 are S_X^2 and S_Y^2 respectively.
> If $\bar{X}_P$ and S_P^2 are unbiased estimators for μ and σ^2 based upon the two samples, then $\bar{X}_P = \frac{n_X\bar{X}+n_Y\bar{Y}}{n_X+n_Y}$ and $S_P^2 = \frac{(n_X-1)S_X^2+(n_Y-1)S_Y^2}{n_X+n_Y-2}$.

Consistent estimators

Estimators with small variance are valued because their estimates are relatively likely to be close to the value of the parameter they are estimating. **Consistency** is where the variance of an estimator tends to zero as the sample size increases, and is therefore a desirable property. In mathematical terms:

> An estimator, T, is a consistent estimator for a population parameter if the variance of T tends towards 0 as the sample size n tends to ∞

You might like to think of consistent estimators and unbiased estimators as follows: consistency suggests that the estimator successfully targets a particular estimate as the sample size increases. Unbiasedness says that it targets the *right* estimate.

Example 7

$\bar{X}$ is an estimator for the mean, μ, of a normal distribution with variance σ^2. Show that $\bar{X}$ is a consistent estimator of μ.

$$\mathrm{Var}\left[\bar{X}\right] = \frac{\sigma^2}{n}$$

As $n \to \infty$, $\frac{\sigma^2}{n} \to 0$. $\bar{X}$ is therefore a consistent estimator.

Tip

This result can be extended to more than two samples.

Formulae booklet

The formula for S_P^2 is given in your *Formulae and Statistical Tables* booklet.

Example 8

$T = \dfrac{2X_1 + X_2 + X_3 + \cdots + X_n}{n+1}$ is an estimator for population mean μ.

Is T **a**   an unbiased estimator for μ?   **b**   a consistent estimator for μ?

a $E[T] = E\left[\dfrac{2X_1 + X_2 + X_3 + \cdots + X_n}{n+1}\right]$

$\qquad = \dfrac{2E[X_1] + E[X_2] + E[X_3] + \cdots + E[X_n]}{n+1}$

$\qquad = \dfrac{(n+1)\mu}{n+1} = \mu$

Therefore T is unbiased for μ.

b $\text{Var}[T] = \text{Var}\left[\dfrac{2X_1 + X_2 + X_3 + \cdots + X_n}{n+1}\right]$

$\qquad = \dfrac{1}{(n+1)^2}\left[4\text{Var}[X_1] + \text{Var}[X_2] + \text{Var}[X_3] + \cdots + \text{Var}[X_n]\right]$

$\qquad = \dfrac{n+3}{(n+1)^2}\sigma^2$

$\qquad = \left[\dfrac{1}{n+1} + \dfrac{2}{(n+1)^2}\right]\sigma^2$

which tends to zero as $n \to \infty$

Therefore T is a consistent estimator for μ.

Efficient estimators

In **Example 2** (see page 394) you saw that it is possible to have more than one unbiased estimator for a particular population parameter. Choosing between them can be done on the basis of their variance.

Since small variance suggests a good estimator, it follows that the reciprocal of variance is a good measure of efficiency.

> Among a set of unbiased estimators of a population parameter, the most efficient (sometimes referred to the best) estimator is the one with the smallest variance.

The efficiency of two estimators can be compared using their **relative efficiency**.

> The relative efficiency of two estimators, A and B, is defined by the equation:
>
> $$\text{Efficiency of } A \text{ relative to } B = \dfrac{\dfrac{1}{\text{Var}[A]}}{\dfrac{1}{\text{Var}[B]}}$$

> **Formulae booklet**
>
> This formula is in the *Formulae and Statistical Tables* booklet. It can also be written as: Efficiency of A relative to $B = \dfrac{\text{Var}[B]}{\text{Var}[A]}$

Since the most efficient estimator is the one with the smallest variance, it follows from these formulae that: the efficiency of A relative to B is greater than 1 if A is more efficient than B, and less than 1 if A is less efficient than B.

Example 9

X_i; $i = 1, 2, 3$ is an independent random sample from a population with mean μ, variance σ^2.

a Show that, when estimating μ, the estimators $M = \dfrac{\displaystyle\sum_{i=1}^{3} X_i}{3}$ and $N = \dfrac{X_1 + 2X_2 + 3X_3}{6}$ are both unbiased for μ.

b Find the variances of M and N.

c Show that M is a more efficient estimator than N and find the efficiency of N relative to M.

a $E[M] = E\left[\dfrac{\displaystyle\sum_{i=1}^{3} X_i}{3}\right] = \dfrac{1}{3} E\left[\displaystyle\sum_{i=1}^{3} X_i\right] = \mu$

$E[N] = E\left[\dfrac{X_1 + 2X_2 + 3X_3}{6}\right]$

$= \dfrac{1}{6} E[X_1 + 2X_2 + 3X_3] = \dfrac{1}{6}(\mu + 2\mu + 3\mu) = \mu$

Therefore, both M and N are unbiased for μ.

Using $\text{Var}\left[\bar{X}\right] = \dfrac{\sigma^2}{n}$

b $\text{Var}[M] = \dfrac{\sigma^2}{3}$

$\text{Var}[N] = \dfrac{1}{36}\left\{\text{Var}\left[X_1\right] + 4\text{Var}\left[X_2\right] + 9\text{Var}\left[X_3\right]\right\} = \dfrac{7}{18}\sigma^2$

c $\dfrac{\sigma^2}{3} < \dfrac{7}{18}\sigma^2 \quad \text{Var}[M] < \text{Var}[N]$

Therefore, M is a more efficient estimator than N.

Efficiency of N relative to $M = \dfrac{\frac{1}{\text{Var}[N]}}{\frac{1}{\text{Var}[M]}} = \dfrac{6}{7}$

It is worth noting that for the estimator $\dfrac{a_1 X_1 + a_2 X_2 + \cdots + a_n X_n}{n}$ variance is minimised on the condition that $a_1 = a_2 = \cdots = a_n = 1$. Therefore, no linear combination of the sample values can give a better unbiased estimate of μ than $\bar{X}$.

Exercise 2

1 X_i; $i = 1, 2, 3$ is a random sample of size 3 from a population with unknown mean μ. By finding the expected value of each of the following estimators, state which are unbiased for μ.

i $T_1 = \dfrac{X_1 + X_2 + X_3}{3}$ **ii** $T_2 = \dfrac{2X_1 + X_2 + X_3}{4}$

iii $T_3 = \dfrac{2X_1 + X_2 + X_3}{6}$

2 Find a condition on a and b such that the estimator, T, given by $T = \dfrac{aX_1 + bX_2}{2}$ is unbiased for the population mean μ.

③ The haemoglobin levels of a sample of 12 adult female patients on discharge from hospital, in g/dl, are:

12.1 13.5 12.9 15.1 14.2 16.1 14.9 13.1 12.0 15.4 14.0 13.1

Find unbiased estimates for the population mean and variance.

④ $X_i; i = 1, 2, ..., n$ is an independent random sample of size n from a population with unknown mean μ. Show that the statistic $T = \frac{3X_1 + 2X_2 + X_3 + \cdots + X_n}{n+3}$ is an unbiased and consistent estimator for population mean μ.

⑤ a A random sample of size n is taken from a population with mean μ.

Show that $\overline{X}_1 = \frac{1}{n}\sum_{i=1}^{n} X_i$ is consistent for μ.

b An independent sample of size $n+1$ is taken from the same population and $\overline{X}_2 = \frac{1}{n+1}\sum_{i=1}^{n+1} X_i$

Show that $\overline{X}_2$ is also consistent but more efficient than $\overline{X}_1$.

⑥ The yields of 12 plants of a new strain of pepper are, in kg:

3.1 3.5 4.1 3.9 3.6 2.9 3.6 4.5 4.1 3.8 4.0 2.8

A further 15 plants have yields with sample mean 3.90 and variance 0.34.

By combining the results, find unbiased estimates of the population mean and variance based upon the 27 plants.

⑦ For large sample size, the unbiased estimate S^2 can be approximated by the sample variance given by $\frac{1}{n}\sum_{i=1}^{n}(X_i - \overline{X})^2$. If $S^2 = 12$, find the percentage error in using the sample variance if the sample size is a 10 b 100.

⑧ Samples of size n_X and n_Y are taken from a normal population with mean μ, variance σ^2. The statistic $\frac{n_X\overline{X} + n_Y\overline{Y}}{n_X + n_Y}$ is used as a pooled estimator of μ. Show that this statistic is unbiased for μ and find its variance.

⑨ Show that the estimator $\hat{\mu} = a_1 X_1 + a_2 X_2 + \cdots + a_n X_n$ is unbiased for the population mean μ if $\sum_{i=1}^{n} a_i = 1$.

Show also that, if $a_1 = a_2 = \cdots = a_n$, $\hat{\mu}$ is a consistent estimator.

⑩ X_1, X_2 is an independent random sample of size 2 from a population with mean μ. The statistic $T = a_1 X_1 + a_2 X_2$ is unbiased for μ. Find a condition on a_1 and a_2 so that T is the most efficient among all estimators of that form.

⑪ Samples of size 10 and 15 are taken from a normal population with mean μ, variance σ^2. Let the sample means be $\overline{X}_1$ and $\overline{X}_2$. The statistics T_1 and T_1 are defined by the equations

$$T_1 = \frac{\overline{X}_1 + \overline{X}_2}{2}; \quad T_2 = \frac{2\overline{X}_1 + 3\overline{X}_2}{5}$$

a Find the means and variances of the two estimators.

b Find the efficiency of T_1 relative to T_2. Which is the more efficient estimator? You should give a reason for your answer.

Summary

▶ Statistical inference is where information about a sample from a population is used to inform knowledge about the population parameters.

▶ A statistic is a function of the random variables that make up the sample and which does not depend upon any unknown parameter.

▶ An estimator, U, is unbiased for the population parameter u if the expected value of U equals u; that is $E[U] = u$

▶ If $X_i; i = 1, 2, ..., n$ is an independent random sample from a population with mean μ and variance σ^2, then $\bar{X} = \dfrac{\sum X}{n}$ and $S^2 = \dfrac{\sum(X - \bar{X})^2}{n-1}$ are unbiased estimators of μ and σ^2, respectively.

▶ Unbiased estimators for μ and σ^2 based upon two samples are,

$$\bar{X}_P = \frac{n_X \bar{X} + n_Y \bar{Y}}{n_X + n_Y} \text{ and } S_P^2 = \frac{(n_X - 1)S_X^2 + (n_Y - 1)S_Y^2}{n_X + n_Y - 2}$$

where n_X and n_Y are the sample sizes, the sample means are $\bar{X}$ and $\bar{Y}$ and the two unbiased estimators of σ^2 are S_X^2 and S_Y^2, respectively.

▶ An estimator, T, is a consistent estimator for a population parameter if the variance of T tends towards 0 as the sample size n tends to ∞.

▶ Among a set of unbiased estimators of a population parameter, the most efficient estimator is the one with the smallest variance.

▶ The relative efficiency of two estimators, A and B, is defined by the equation: Efficiency of A relative to $B = \dfrac{\frac{1}{\text{Var}[A]}}{\frac{1}{\text{Var}[B]}}$

Review exercises

1 **a** Define the term 'statistic' as used in the theory of estimation.

b $X_i; i = 1, 2, ..., n$ is an independent random sample from a population with mean μ and variance σ^2. Give the sampling distribution of the sample mean $\bar{X}$ and use this to explain why $\bar{X}$ is both unbiased and consistent for μ.

2 $X_i; i = 1, 2, 3$ is a random sample of size 3 from a population with unknown mean μ. Prove that the following estimators are unbiased.

a $T_1 = \dfrac{2X_1 + 2X_2 + X_3}{5}$

b $T_2 = \dfrac{2X_1 + X_2 + 3X_3}{6}$

3 **a** In estimating a population parameter p, two estimators T_1 and T_2 are used. Explain in terms of their means and variances, what is meant by saying that they are both unbiased for p but T_1 is more efficient than T_2.

b Prove for $n = 3$, that, among all unbiased estimators of the form $a_1 X_1 + a_2 X_2 + \cdots + a_n X_n$, the minimum variance estimator is the sample mean, $\bar{X}$. (Note: this result applies for all positive integer n.)

4 a Explain what is meant by the phrase 'T is an unbiased estimator of the population parameter t'.

b A random sample of n observations is taken from a population with mean μ, variance σ^2. The statistic is defined by the equation

$$S^2 = \frac{1}{n-1}\left[\sum_{i=1}^{n} X_i^2 - n\bar{X}^2\right]$$

Prove that S^2 is unbiased for σ^2.

5 X_i; $i = 1, 2, \ldots, n$ is an independent random sample of size n from a population with mean μ. V is to be used as an estimator of μ where $V = a_1 X_1 + a_2 X_2 + \ldots + a_n X_n$. Find a condition on the a_is so that V is a consistent estimator for μ.

6 X_i; $i = 1, 2$ is a random sample of size 2 from a population with unknown mean μ and variance σ^2. The following estimators have been identified to estimate μ:

$$T_1 = \frac{2X_1 + X_2}{3} \qquad T_2 = 3X_1 - X_2 - X_1 \qquad T_3 = X_1$$

a Show that all three estimators are unbiased and find the most efficient among them.

b Find the efficiency of T_3 relative to T_2.

Practice examination questions

1 The random variable X has a probability density function

$$f(x) = \begin{cases} \frac{1}{4}\sin\left(\frac{x}{2}\right) & 0 \leq x \leq 2\pi \\ 0 & \text{otherwise} \end{cases}$$

For this distribution, you are given that
$E(X) = \pi$ and $E(X^2) = 2\pi^2 - 8$

a Find, in terms of π, the variance of X. (2 marks)

b The mean of a random sample of n observations, $X_1, X_2, X_3, \ldots, X_n$, is denoted by $\bar{X}$.

 i Write down, in terms of π and n, expressions for the mean and the variance of $\bar{X}$. (2 marks)

 ii Explain why $\bar{X}$ is an unbiased and consistent estimator for π. (2 marks)

c i For a random sample of size 5, the median, M, is an unbiased estimator for π with variance $\pi^2 - \frac{2072}{225}$

 For such a sample, calculate the relative efficiency of M with respect to $\bar{X}$, giving your answer to three decimal places.

 Hence give a reason why $\bar{X}$ should be preferred to M as an estimator for π. (3 marks)

 ii A particular random sample of 5 observatioas of X gave the following values:

 0.80   1.99   3.12   3.89   6.20

A Given that $P(X \geq 2\pi) = 0$, use this sample to find an inequality for π.

B Obtain the values of $\overline{X}$ and m for this sample.

C Comment on your answers in part **(c)(ii)(B)** in the light of your answers to part **(c)(i)**. (5 marks)

AQA MS04 June 2010

2 **a** The statistic T is derived from a random sample taken from a population which has an unknown parameter θ. T is an unbiased estimator for θ.

What does the statement "T is an unbiased estimator for θ" imply? (1 mark)

b A random sample of size n is taken from each of two independent populations.

The first population has mean μ and variance σ^2, and $\overline{X}$ denotes the sample mean.

The second population has mean $\frac{\mu}{3}$ and variance $b\sigma^2$, where b is a positive constant, and $\overline{Y}$ denotes the sample mean.

Two unbiased estimators for μ are defined by

$$T_1 = 4\overline{X} - a\overline{Y} \quad \text{and} \quad T_2 = \frac{1}{9}(8\overline{X} + 3\overline{Y})$$

i Determine the value of a. (3 marks)

ii Show that $\text{Var}(T_1) = \frac{\sigma^2}{n}(16 + 81b)$ and find a simplified expression for $\text{Var}(T_2)$. (5 marks)

iii Calculate the relative efficiency of T_2 with respect to T_1 and decide, giving a reason, which of T_1 or T_2 is the more efficient estimator for μ. (4 marks)

AQA MS04 June 2011

3 **a** State why, in choosing between two unbiased estimators of a parameter, the one with the smaller variance is preferred. (2 marks)

b The random variable $\overline{X}_1$ denotes the mean of a random sample of size n_1 taken from a normal population with mean μ_1 and variance σ_1^2. The random variable $\overline{X}_2$ denotes the mean of a random sample of size n_2 taken from an independent normal population with mean μ_2 and variance σ_2^2

i Show that $\overline{X}_1$ and $\overline{X}_2$ is an unbiased estimator of $\mu_1 - \mu_2$ and write down the variance of this estimator. (3 marks)

ii It is given that $n_1 + n_2 = n$, where n is a fixed number, and that n_1 and n_2 are so large that they may be assumed to be continuous variables. Given further that the variance of $\overline{X}_1$ and $\overline{X}_2$ has a minimum value, show that this minimum value occurs when

$n_1 : n_2 = \sigma_1 : \sigma_2$ (4 marks)

iii Find the values of n_1 and n_2 which minimise the variance of $\overline{X}_1$ and $\overline{X}_2$ in the case when $\sigma_1^2 = 0.0025$, $\sigma_2^2 = 0.0081$ and $n = 280$. (2 marks)

AQA MS04 June 2012

4 The conductivity, γ, of metal wire is estimated by observing a related random variable R, which has probability density function

$$f(r) = \begin{cases} \dfrac{2r}{\gamma^2} & 0 \leq r \leq \gamma \\ 0 & \text{otherwise} \end{cases}$$

a **i** Show that $\dfrac{3}{2}R$ is an unbiased estimator of γ. (4 marks)

    ii Given that the variance of R is $\dfrac{1}{18}\gamma^2$ find in terms of γ the variance of $\dfrac{3}{2}R$. (2 marks)

b The conductivity can also be estimated by making observations of a random variable, S, which has mean $\dfrac{1}{4}\gamma$ and variance $\dfrac{1}{16}\gamma^2$

    The random variable T is defined by $T = S_1 + S_2 + S_3$ where S_1, S_2 and S_3 are three independent observations of S.

    i Find the value of the constant k such that kT is an unbiased estimator of γ. (2 marks)

    ii Hence find the relative efficiency of $\dfrac{3}{2}R$ with respect to kT. (4 marks)

    iii State, with justification, which of $\dfrac{3}{2}R$ and kT is a preferred unbiased estimator of γ.

AQA MS04 June 2013

5 Two independent random samples of observations of sizes n_1 and n_2 are made of a random variable X, which has mean μ and variance σ^2. The sample means are denoted by $\overline{X}_1$ and $\overline{X}_2$ respectively.

a Show that $T = k\overline{X}_1 + (1-k)\overline{X}_2$ is an unbiased estimator of μ. (2 marks)

b Show that V, the variance of T, is given by

$$V = k^2\frac{\sigma^2}{n_1} + (1-k)^2\frac{\sigma^2}{n_2}$$ (2 marks)

c Find the value of k for which $\dfrac{\mathrm{d}V}{\mathrm{d}k} = 0$. (3 marks)

d For the value of k found in **part (c)**:

    i find an expression for T (2 marks)

    ii interpret the expression found in **part (d)(i)** (1 mark)

    iii find $\dfrac{\mathrm{d}^2V}{\mathrm{d}k^2}$ and hence comment on what you can deduce about V. (2 marks)

AQA MS04 June 2014

31 Estimation

Introduction

In the previous chapter you learned how, when estimating a population parameter, to choose an estimator with good properties. These estimators give a single value as the estimate and are called **point estimators**. You will now learn how to construct *intervals* that are likely to contain the parameter being estimated. This material is of great practical importance since many real-life applications of inferential statistics relate to estimating population parameters.

Recap

You will need to remember...

▶ That a point estimate gives you a single value but does not indicate how close to the true value the estimate is.

See chapter 30 Estimators.

▶ How to use normal distribution tables to find probabilities and z-values.

▶ The sampling distribution of $\bar{X}$: if $X \sim N(\mu, \sigma^2)$ then $\bar{X} \sim N\left(\mu, \dfrac{\sigma^2}{n}\right)$

▶ The Central Limit Theorem: if X has a mean μ and variance σ^2, then the distribution of $\bar{X}$ tends to $N\left(\mu, \dfrac{\sigma^2}{n}\right)$ as $n \to \infty$

▶ The normal approximation to the Poisson distribution: if X has a Poisson distribution, mean λ, then, for large λ, X can be approximated by Y where $Y \sim N(\lambda, \lambda)$.

Objectives

By the end of this chapter, you should know how to:

▶ Define a confidence interval for an unknown population mean.

▶ Construct a confidence interval for the population mean when:
  - The population is normally distributed and variance is known
  - The normal distribution has been used to approximate large samples
  - The population is normally distributed with unknown variance and small sample size.

▶ Calculate the sample size required to produce a confidence interval of a given width.

▶ Make inferences about the population mean based upon its confidence interval.

See chapter *30 Estimators* for a reminder of how to choose estimators with good properties.

31.1 Confidence intervals for the mean of a distribution

A **confidence interval** for a population mean is an interval within which the mean is thought to lie with a given degree of confidence. The interval varies from one sample to the next of the same population. To calculate the interval, it is necessary to start with an estimator for the population mean.

As the statistic $\bar{X}$ has some good properties when estimating population mean, this will be used.

As shown in the diagram, different samples give different values of the sample mean and therefore different confidence intervals. In the case of 95% confidence intervals, on average 95% of these intervals would contain the true population mean. Increasing the degree of confidence with the same sample

size is possible but means widening the interval with the resultant loss of precision in the estimate.

The focus of the chapter will only be on confidence intervals that are symmetrical about the mean.

Normal distribution with known variance

The sample mean, $\bar{X}$, can be used to find a confidence interval for the mean of a normally distributed population with known variance, σ^2.

Since $X \sim N(\mu, \sigma^2)$, $\bar{X} \sim N\left(\mu, \frac{\sigma^2}{n}\right)$.

It is possible to construct an interval, centred on μ, that has a 0.95 probability of containing $\bar{X}$. Using normal distribution theory, if this interval is $\mu \pm a$, then

$$P\left\{\mu - a < \bar{X} < \mu + a\right\} = 0.95$$

$$P\left\{\frac{-a}{\frac{\sigma}{\sqrt{n}}} < \frac{\bar{X} - \mu}{\frac{\sigma}{\sqrt{n}}} < \frac{a}{\frac{\sigma}{\sqrt{n}}}\right\} = 0.95$$

$$P\left\{\frac{-a}{\frac{\sigma}{\sqrt{n}}} < Z < \frac{a}{\frac{\sigma}{\sqrt{n}}}\right\} = 0.95 \qquad \text{[E1]}$$

> **You should know this result from previous studies.**

> **Tip**
>
> Using the standard error of the mean, $\left(\frac{\sigma}{\sqrt{n}}\right)$, from the sampling distribution of $\bar{X}$

However, from normal distribution table, you get a critical value of 1.96 at the 95% confidence level, $P\left\{-1.96 < Z < 1.96\right\} = 0.95$ [E2]

Therefore, comparing [E1] and [E2], you get $\dfrac{a}{\frac{\sigma}{\sqrt{n}}} = 1.96 \;\Rightarrow\; a = 1.96\dfrac{\sigma}{\sqrt{n}}$

and so $P\left\{\mu - 1.96\dfrac{\sigma}{\sqrt{n}} < \bar{X} < \mu + 1.96\dfrac{\sigma}{\sqrt{n}}\right\} = 0.95$ [E3]

or, equivalently, $P\left\{\bar{X} - 1.96\dfrac{\sigma}{\sqrt{n}} < \mu < \bar{X} + 1.96\dfrac{\sigma}{\sqrt{n}}\right\} = 0.95$ [E4]

> **Tip**
>
> Recall that this means that 95% of the area under a normal curve is within 1.96 standard deviations of the mean.

Equation [E4] shows that the probability that μ lies within the random interval $\bar{X} \pm 1.96\frac{\sigma}{\sqrt{n}}$ is 0.95.

Remember to be careful about the way language is used here. The probability that a (random) interval, such as $\bar{X} \pm 1.96\frac{\sigma}{\sqrt{n}}$, contains μ could be 0.95. If a specific value of $\bar{X}$ is substituted into the interval, the corresponding (non-random) interval contains μ with probability either 1 or 0 depending upon whether or not the interval contains μ. This explains the use of the phrase 'confidence interval' rather than 'probability interval'.

Following the construction above, in general, you have

> If $\bar{x}$ is the mean of a sample of size n from a normally distributed population with known variance, σ^2, then $\bar{x} \pm 1.96\frac{\sigma}{\sqrt{n}}$ is a 95% confidence interval for μ.

> **Note**
>
> The structure of the confidence interval is always the same: the estimate of mean value ± an 'error' term. It is worth noting that in this context the standard deviation of $\bar{X}$ is known as the **standard error**.

(3) An independent random sample from a normal population with variance 12 gave a 92% confidence interval for the population mean of $(11.96, 16.24)$. Find the sample size used.

(4) A sample of size 60 was taken from a normal population and gave the following statistics:
$\sum x = 115, \sum x^2 = 249$

  a Find an approximate 95% confidence interval for the mean and use it to test the hypothesis that the population mean is 2.2. You should use a significance level of 5%.

  b Justify the use of your chosen distribution.

Practice examination questions

(1) The waiting time at a hospital's A&E department may be modelled by a normal distribution with mean μ and standard deviation $\frac{\mu}{2}$.

The department's manager wishes a 95% confidence interval for μ to be constructed such that it has a width of at most $0.2\,\mu$.

Calculate, to the nearest 10, an estimate of the minimum sample size necessary in order to achieve the manager's wish.   (5 marks)

AQA MS03 June 2011

(2) Vanya collected five samples of air and measured the carbon dioxide content of each sample, in parts per million by volume (ppmv). The results were as follows.

387   375   382   379   381

  a Assuming that these data form a random sample from a normal distribution with mean μ ppmv, construct a 90% confidence interval for μ.   (6 marks)

  b Vanya repeated her sampling procedure on each of 30 days and, for each day's results, a 90% confidence interval for μ was constructed. On how many of these 30 days would you expect μ to lie outside that day's confidence interval?   [1 mark]

AQA MS2B June 2014

(3) Dimitra is an athlete who competes in 400 m races. The times, in seconds, for her first six races of the 2012 season were 54.86 53.09 53.75 52.88 51.97 51.81

  a Assuming that these data form a random sample from a normal distribution, construct a 95% confidence interval for the mean time of Dimitra's races in the 2012 season, giving the limits to two decimal places.   (5 marks)

  b For the 2011 season, Dimitra's mean time for her races was 53.41 seconds. After her first six races of the 2012 season, her coach claimed that the data showed that she would be more successful in races during the 2012 season than during the 2011 season. Make two comments about the coach's claim.   (2 marks)

AQA MS2B January 2013

Further Hypothesis Testing 1 – Means and Variances

Introduction

This chapter builds on your knowledge of hypothesis testing (from A-level Mathematics) by investigating techniques to compare the means and variances of two populations. These techniques are at the heart of inferential statistics and allow powerful statements to be made when comparing real populations. You will also learn how to evaluate a test by calculating its power.

Objectives

By the end of this chapter, you should know how to:

► Define and calculate the power of a hypothesis test.

► Test for the difference between two population means in various circumstances.

► Test for a value of a population variance and test for the equality of two variances.

Recap

You will need to remember...

► The basic stages of a hypothesis test:
  • formation of a null and an alternative hypothesis
  • identification of a sampling distribution and its critical region
  • decision making based upon an observed value.

► The definition of a type I and a type II error
  • a type I error means rejecting the null hypothesis when it is true
  • a type II error means failing to reject the null hypothesis when it is false.

► The unbiased estimate of population variance:

$$s^2 = \frac{1}{n-1}\left\{\sum x^2 - \frac{(\sum x)^2}{n}\right\} = \frac{\sum(x-\bar{x})^2}{n-1}$$

► The sampling distribution of $\overline{X}$: if $X \sim N(\mu, \sigma^2)$ then $\overline{X} \sim N\left(\mu, \frac{\sigma^2}{n}\right)$

► The distribution of $\overline{X} \pm \overline{Y}$: if $X \sim N(\mu_X, \sigma_X^2)$ and $Y \sim N(\mu_Y, \sigma_Y^2)$

then $\overline{X} \pm \overline{Y} \sim N\left(\mu_X \pm \mu_Y, \frac{\sigma_X^2}{n_X} + \frac{\sigma_Y^2}{n_Y}\right)$ for independent X and Y.

► The normal approximation to the binomial distribution: if X has a binomial distribution, parameters n and p, then, for large n, $X \sim N(np, np(1-p))$ approximately.

► The normal approximation to the Poisson distribution: if X has a Poisson distribution, mean λ, then, for large λ, X can be approximated by Y where $Y \sim N(\lambda, \lambda)$.

► The central limit theorem: if X has a mean μ and variance σ^2, then the distribution of $\overline{X}$ tends to $N\left(\mu, \frac{\sigma^2}{n}\right)$ as $n \to \infty$.

32.1 Power of a hypothesis test

In your study of A-level Mathematics, you learned about type I and type II errors in hypothesis testing. In this section, these ideas are developed to include the calculation of the probability of a type II error and the important idea of the power of a test.

Type II errors and probability calculations

A **type I error** means rejecting the null hypothesis when it is true. The probability of this occurring is equal to the significance level of the test.

A **type II error** means failing to reject the null hypothesis when it is false. Calculating the probability of a type II error is more complicated, since it depends on how far the parameter values in the null hypothesis are from their true values. If the null hypothesis is close to being correct, failing to reject it is relatively likely and the probability of a type II error could be high, even if the test is quite good. However, in general, a good test is one in which, among other qualities, there is a relatively low probability of a type II error.

In order to develop a method for finding the probability of a type II error, consider the following problem.

In a canning factory, a machine fills cans with soda. The nominal volume of the liquid is 330 ml, and it is known that the standard deviation of the volume per can is 3 ml. The manufacturer claims that the machine dispenses a mean volume greater than 330 ml. To test this claim, 10 cans are chosen at random and the mean volume calculated.

Let X be a random variable for the volume of a randomly chosen can. It is necessary to assume a distribution for X and this type of variable is usually normally distributed.

$X \sim N(\mu, 3^2)$ where μ is the (unknown) population mean

Using the sampling distribution of $\overline{X}$ and a sample of size 10, $\overline{X} \sim N\left(\mu, \frac{3^2}{10}\right)$ where $\overline{X}$ is the sample mean, the random variable of interest

Form the null and alternative hypotheses:

$H_0: \mu = 330$

$H_1: \mu > 330$

Therefore, under H_0, $\overline{X} \sim N(330, 0.9)$

Standardising, you have

$Z = \dfrac{\overline{X} - 330}{\sqrt{0.9}}$ which follows a standard normal distribution

Using the values of z in the percentage points table for normal distribution, at a significance level of 5% and for a one-tailed test, the critical value of z is 1.645. This means that, under the null hypothesis, the probability of getting a value of $z > 1.645$ is 5% and a value of $z < 1.645$ is 95%.

So, in terms of z, the acceptance region is therefore $z < 1.645$. Expressing this region in terms of the random variable $\overline{X}$, you have

$\overline{x} < 330 + 1.645 \times \sqrt{0.9} = 331.561$

This is found using standard theory related to the normal distribution.

Now, suppose that the true population mean volume per can was actually 331.6 ml, so that the null hypothesis was false. It would be possible to calculate the probability of accepting the null hypothesis given this population mean; that is, the probability of a type II error for this particular value of μ.

Given that $\mu = 331.6$, $\overline{X} \sim N(331.6, 0.9)$

$P(\text{type II error}) = P\{\text{accepting } H_0 | H_0 \text{ false}\}$

$= P\{\overline{X} < 331.561 | \mu = 331.6\}$

$= P\left\{Z < \dfrac{331.561 - 331.6}{\sqrt{0.9}}\right\}$

$= P\{Z < -0.0411\} = 1 - P\{Z < 0.0411\}$

$= 1 - 0.516$

$= 0.484$

Using $P(z < -a) = 1 - P(Z < a)$

So, the probability that the test described results in a type II error, is 0.484.

This example suggests the following general procedure:

> **To find the probability of a type II error**
>
> **Stage A (calculations based on the value of the parameter in the null hypothesis)**
>
> 1. Identify the relevant statistic, P, and state the null and alternative hypotheses.
> 2. Find the sampling distribution of P under the null hypothesis.
> 3. Using the significance level, find the critical region and its complement, the acceptance region, of the distribution of P.
>
> **Stage B (calculations based on a value of the parameter included in the alternative hypothesis)**
>
> 4. Find the distribution of the statistic P, given the value of the population parameter in the alternative hypothesis
> 5. Find the probability of the random variable P being in the previously found acceptance region, given the distribution in part 4 above.

> **Tip**
>
> This may be presented as the 'true' value of the parameter in the question.

Example 1

Question

An independent random sample of size 12 is taken from a normally distributed population with standard deviation 5. It is suggested that the mean of the population is 14. A 5% significance test is performed, to check this suggestion against the alternative hypothesis that the mean is less than 14. Find the probability of a type II error if the mean subsequently turns out to be 13.1.

Answer

Let $\overline{X}$ be a random variable for the sample mean.

$H_0 : \mu = 14$

$H_1 : \mu < 14$ Step 1

$\overline{X} \sim N\left(\mu, \dfrac{25}{12}\right)$

You are told the population is normally distributed.

(continued)

(continued)

Under the null hypothesis, $\overline{X} \sim N\left(14, \dfrac{25}{12}\right)$ Step 2

Standardise using $Z = \dfrac{\overline{X} - \mu}{\frac{\sigma}{\sqrt{n}}}$

Critical region: $\overline{x} < 14 - 1.645 \times \dfrac{5}{\sqrt{12}} = 11.63 \,(2\,\text{dp})$

Acceptance region: $\overline{x} > 11.63$ Step 3

If the mean is 13.1, $\overline{X} \sim N\left(13.1, \dfrac{25}{12}\right)$ Step 4

P(type II error) $= \text{P}\{\text{accepting } H_0 | H_0 \text{ false}\}$ Step 5

$\qquad = \text{P}\{\overline{X} > 11.63 | \mu = 13.1\}$

$\qquad = \text{P}\left\{Z < \dfrac{(13.1 - 11.63)\sqrt{12}}{5}\right\}$

$\qquad = \text{P}\{Z < 1.018\} = 0.846$

Example 2

A Geiger counter indicates radioactive decay by a series of clicks heard through a loudspeaker, one click per decay. For a particular sample of radioactive material, the total number of decays over 5 minutes was observed. The source of the activity is thought to be a material with a mean number of decays of 9 per 10-second period.

A test is to be performed to check this suggestion against the hypothesis that the source has a greater mean.

Calculate the probability of making a type II error, if the mean number of decays for the material is actually 10.7. State any distributional assumptions made. The test should have a 0.05 probability of a type I error, and you may assume that the decays occur at random and the radioactivity of the material is constant over time (because of its long half-life).

Let X be a random variable for the total number of decays in a randomly chosen 5 minute interval.

$X \sim \text{Poisson}(\lambda)$

Assumption: Random events in time follow a Poisson distribution, mean λ

H_0: $\lambda = 270$

H_1: $\lambda > 270$

Under H_0, $X \sim \text{Poisson}(270)$

Hence X has the approximate distribution $N(270, 270)$ under H_0

Critical region: $x > 270 + 1.645 \times \sqrt{270}$
$= 297.03\,(2\,\text{dp})$

Acceptance region: $x < 297.03$

If $\lambda = 10.7$, $X \sim N(321, 321)$

Note
A mean of 9 per 10 seconds is equivalent to $9 \times 30 = 270$ per 5 minutes.

Note
Use the 95% point of a standard normal distribution because the significance level of the test equals the probability of a type I error, 0.05

Note
Since the mean of X is large, approximate to a normal distribution $X \sim N(\lambda, \lambda)$.

Note
H_0: $\lambda = 270$, H_1: $\lambda > 270$, so acceptance region is to the left of the mean.

$10.7 \times 30 = 321$ per 5 minutes

(continued)

(continued)

$P(\text{type II error}) = P\{\text{accepting } H_0 | H_0 \text{ false}\}$

$= P\{X < 297.03 | \lambda = 10.7\}$

$= P\left\{Z < \dfrac{297.03 - 321}{\sqrt{321}}\right\}$

$= 1 - P\left\{Z < \dfrac{(321 - 297.03)}{\sqrt{321}}\right\}$

$= 1 - P\{Z < 1.338\} = 0.0905$

If $\lambda = 10.7$, $X \sim N(321, 321)$ approximately

> **Note**
>
> Using $P(Z < -a) = 1 - P(Z < a)$, so $P(Z < -1.339) = 1 - P(Z < 1.339)$

Power

A type II error occurs when the null hypothesis is accepted when it is false. The complementary event, rejecting the null hypothesis when it *should* be rejected, is clearly a good characteristic of any test. This gives a measure of the **power** of a test.

> The power of a hypothesis test is given by
>
> $P = P\{H_0 \text{ is rejected} | H_0 \text{ is false}\} = 1 - P\{\text{type II error}\}$

> **Note**
>
> Power is usually given as a percentage.

Example 3

A six-sided dice, with faces numbered $1 - 6$, was rolled 240 times. A test, based on the number of 1s obtained, was performed to check whether the dice was fair. Calculate the power of the test if the probability of a 1 was actually 0.12. The significance level of the test should be approximately 0.05.

Let X be the number of 1s in 240 throws of the coin.

$X \sim B(240, p)$ where p is the probability of a 1 in any given throw

$X \sim N(240p, 240p(1 - p))$ approximately

This is a binomial distribution, $n = 240$.

$H_0: p = \dfrac{1}{6}$

$H_1: p \neq \dfrac{1}{6}$.

That is, $X \sim N\left(40, \dfrac{100}{3}\right)$ approximately

> **Note**
>
> Null hypothesis is that the dice is fair.

Critical region:

$x > 40 + 1.96 \times \sqrt{\dfrac{100}{3}}; x < 40 - 1.96 \times \sqrt{\dfrac{100}{3}}$

or $x > 51.32; x < 28.68$ (2 dp).

X takes integer values only

Acceptance region: $29 \leq x \leq 51$

If $p = 0.12$, $X \sim N(28.8, 25.344)$

$P(\text{type II error}) = P\{\text{accepting } H_0 | H_0 \text{ false}\}$

$= P\{28.5 < X < 51.5 | p = 0.12\}$

$= 0.524$ (3 dp)

> **Note**
>
> Using 1 − P(type II error):
> 1 − 0.524 = 0.476

> **Note**
>
> Using the normal approximation to the binomial distribution for large n, $X \sim N(np, np(1 - p))$

That is, $X \sim N\left(40, \dfrac{100}{3}\right)$ approximately

Power of test is 47.6%

Exercise 1

1. A random variable X has a normal distribution with mean μ and variance given. Tests of the null hypothesis against the alternative is performed at the significance levels given.

   Find the acceptance region for each test if the sample size n on which the test is based is as given and the estimator is $\overline{X}$.

P(type II error)

a $\sigma^2 = 25$; H_0: $\mu = 15$, H_1: $\mu \neq 15$; $n = 20$; significance level of 4%

b $\sigma^2 = 20$; H_0: $\mu = 11$, H_1: $\mu < 11$; $n = 14$; significance level of 5%

2 R is a random variable with unknown mean and a variance of 18. A random sample of size 32 is taken and gives a mean of 17. Test at a significance level of 5% whether the population mean is 20. Find the probability of making a type II error if the mean is actually 17.5. State any assumptions made.

3 X is a random variable with the distribution $N(\mu, 36)$. A random sample of size 30 is taken and gives a mean of 35. Test at a significance level of 5% whether the population mean is 32. Find the probability of making a type II error if the population mean is 34.

4 A random variable X has a normal distribution with mean μ and variance 24. A test of the hypothesis is performed that $\mu = 26$ against the alternative that $\mu > 26$. The significance level of the test is 5%.

 i Find the acceptance region for this test, if the sample size on which the test is based is 14 and the estimator $\overline{X}$ is used.

 ii Find the probability of a type II error if $\mu = 30$.

 iii Find the power of this test.

5 In order to check the settings of a packing machine, the weights of 12 bags of potatoes are measured and the mean calculated. The nominal weight of the bags is 1.5 kg. It is known that the standard deviation of the weights is 0.1 kg. A test, with significance level 0.04, is performed to find whether the population mean is greater than the nominal weight. Find the power of the test if the true population mean is 1.6 kg.

32.2 Tests for the difference between the means of two independent distributions

In this section, you will see how to test whether or not two samples, taken from separate populations, provide evidence that the populations have different means.

Difference of means – normal populations with known variance

The first problem to consider assumes that the populations are both normally distributed with known variances. Consider two populations with independent normal distributions with means μ_X and μ_Y and variances, σ_X^2 and σ_Y^2. Further, suppose that the two independent random samples taken from these populations are of sizes n_X and n_Y respectively, and denoted X_i; $i = 1, 2, ..., n_X$ and Y_i; $i = 1, 2, ..., n_Y$.

The null hypothesis to be tested is that the two population means are equal. Assuming that the alternative hypothesis is that they are not equal, the hypotheses can be written as:

H_0: $\mu_X - \mu_Y = 0$

H_1: $\mu_X - \mu_Y \neq 0$

In order to test for a value of $\mu_X - \mu_Y$ it is necessary to know the distribution of the corresponding statistic $\bar{X} - \bar{Y}$.

Since $X_i \sim N\left(\mu_X, \sigma_X^2\right)$; $i = 1, 2, \ldots, n_X$ and $Y_i \sim N\left(\mu_Y, \sigma_Y^2\right)$; $i = 1, 2, \ldots, n_Y$

the distributions of sample means are $\bar{X} \sim N\left(\mu_X, \frac{\sigma_X^2}{n_X}\right)$ and $\bar{Y} \sim N\left(\mu_Y, \frac{\sigma_Y^2}{n_Y}\right)$

where $\bar{X} = \dfrac{\sum\limits_{i=1}^{n_X} X_i}{n_X}$ and $\bar{Y} = \dfrac{\sum\limits_{i=1}^{n_Y} Y_i}{n_Y}$

Now, $E\left[\bar{X} - \bar{Y}\right] = E\left[\bar{X}\right] - E\left[\bar{Y}\right]$ and $\text{Var}\left[\bar{X} - \bar{Y}\right] = \text{Var}\left[\bar{X}\right] + \text{Var}\left[\bar{Y}\right]$ as $\bar{X}$ and $\bar{Y}$ are independent random variables

Therefore, the statistic $\bar{X} - \bar{Y}$ has a normal distribution given by

$$\bar{X} - \bar{Y} \sim N\left(\mu_X - \mu_Y, \frac{\sigma_X^2}{n_X} + \frac{\sigma_Y^2}{n_Y}\right)$$

Under the null hypothesis, therefore $\bar{X} - \bar{Y} \sim N\left(0, \frac{\sigma_X^2}{n_X} + \frac{\sigma_Y^2}{n_Y}\right)$

Normalising gives $\dfrac{\bar{X} - \bar{Y}}{\sqrt{\left\{\dfrac{\sigma_X^2}{n_X} + \dfrac{\sigma_Y^2}{n_Y}\right\}}} \sim N(0, 1)$

> **Tip**
>
> If $X \sim N(\mu, \sigma^2)$,
> then $\frac{X - \mu}{\sigma} \sim N(0, 1)$

This suggests the following test:

Let X_i; $i = 1, 2, \ldots, n_X$ and Y_i; $i = 1, 2, \ldots, n_Y$ be random samples from two populations with independent normal distributions, means μ_X and μ_Y and known variances σ_X^2 and σ_Y^2 respectively. Hypotheses:

H_0: $\mu_X - \mu_Y = 0$

H_1: $\mu_X - \mu_Y \neq 0$

Test statistic and distribution under the null hypothesis:

$$Z = \frac{(\bar{X} - \bar{Y})}{\sqrt{\dfrac{\sigma_X^2}{n_X} + \dfrac{\sigma_Y^2}{n_Y}}} \sim N(0, 1).$$

Type: two-tailed test

> **Tip**
>
> For the alternative hypotheses
> $\mu_X - \mu_Y > 0$ or $\mu_X - \mu_Y < 0$
> use a one-tailed test

The methods for finding critical regions in these problems are the same as those you used when studying hypothesis testing (in your A-level mathematics course).

Example 4

Two normally distributed populations P and R have variances 12 and 16, respectively. Samples of size $n_P = 50$ and $n_R = 65$ are taken from these populations and have means $\bar{x}_P = 8.4$ and $\bar{x}_R = 9.2$.

Test at 5% whether these populations have different means.

Let the populations have means μ_P and μ_R and variances σ_P^2 and σ_R^2

$$H_0: \mu_P - \mu_R = 0$$
$$H_1: \mu_P - \mu_R \neq 0$$

$$Z = \frac{\bar{X}_P - \bar{X}_R}{\sqrt{\dfrac{\sigma_P^2}{n_P} + \dfrac{\sigma_R^2}{n_R}}} = \frac{8.4 - 9.2}{\sqrt{\dfrac{12}{50} + \dfrac{16}{65}}} = -1.15 \; (2 \text{ dp})$$

This is the value of the test statistic.

Critical region is the set of z-values outside the interval $(-1.96, 1.96)$

-1.15 is *within* the interval $(-1.96, 1.96)$

Therefore no reason to reject H_0

The evidence suggests that there is no significant reason to believe that the means are different.

Example 5

Two independent, normally distributed random variable, X and Y have variances 12 and 17, respectively. Samples of size $n_X = 20$ and $n_Y = 25$ are taken and have means $\bar{x} = 31.0$ and $\bar{y} = 29.1$. Test at 10% whether these random variables have the same mean against the alternative hypotheses that:

a they have different means   **b** X has a larger mean.

Let the random variables have means μ_X and μ_Y and variances σ_X^2 and σ_Y^2

$$Z = \frac{\bar{X} - \bar{Y}}{\sqrt{\dfrac{\sigma_X^2}{n_X} + \dfrac{\sigma_Y^2}{n_Y}}} = \frac{31.0 - 29.1}{\sqrt{\dfrac{12}{20} + \dfrac{17}{25}}} = 1.68 \; (2 \text{ dp})$$

a $H_0: \mu_X - \mu_Y = 0$
     $H_1: \mu_X - \mu_Y \neq 0$

This is the value of the test statistic. The alternative hypothesis suggests a two-tailed test.

Critical region is the set of z-values outside the interval $(-1.645, 1.645)$.

1.68 is *outside* the interval $(-1.645, 1.645)$.

Therefore reject H_0. The evidence suggests that there is significant reason to believe that the means are different.

b $H_0: \mu_X - \mu_Y = 0$
     $H_1: \mu_X - \mu_Y > 0$

The alternative hypothesis suggests a one-tailed test.

The critical region is the set of values $z > 1.282$.

$1.68 > 1.282$.

Therefore H_0 is rejected. The evidence suggests that the mean of X is greater than the mean of Y.

Example 6

In a trial on the effectiveness of drugs on reducing cholesterol levels, 30 participants received statins and reduced their levels by an average of 1.1 mmol/L. A further 34 participants used diet-based methods to control their levels and achieved an average reduction of 0.96 mmol/L.

(continued)

(continued)

Assuming that the reductions follow a normal distribution and that the standard deviation of cholesterol reduction is 0.27 mmol/L for both populations, test at 5% significance whether the statin treatment is an improvement on dietary methods.

Let X and Y be random variables for the reduction in cholesterol for a patient using statin and dietary methods, respectively, and let μ_X and μ_Y, σ_X^2 and σ_Y^2 be their means and variances.

$H_0: \mu_X - \mu_Y = 0$

$H_1: \mu_X - \mu_Y > 0$

The wording in the question suggests a one-tailed test.

$$z = \frac{1.1 - 0.96}{\sqrt{\left\{\dfrac{0.27^2}{30} + \dfrac{0.27^2}{34}\right\}}} = 2.07 \ (2 \ \mathrm{dp})$$

Critical region $z > 1.645$

Reject H_0. The evidence suggests that statins are an improvement on dietary methods of cholesterol reduction.

> **Note**
>
> You are asked to test whether the statin treatment is an improvement.

Difference of means – using normal approximations for large samples

Testing population means when the underlying population is normally distributed with known variance is unusual; this amount of information is rarely known. In this section, two problems are considered: sampling from non-normal distributions with known variance, and sampling from *any* distribution with unknown variance. Both of these require the use of large samples.

Variances known

If the population variances are known but the population is not normally distributed, the method described in the previous section can be used without alteration, but the results are approximate. This is possible due to the **central limit theorem** which says that even if the underlying population is not normal, *if the sample size is large*, it is possible to assume that the sample mean $\bar{X}$ is *approximately* normal.

> **Tip**
>
> If X has a mean μ and variance σ^2 then the distribution of $\bar{X}$ tends to $N\left(\mu, \dfrac{\sigma^2}{n}\right)$ as $n \rightarrow \infty$

With the notation of the previous section, a one- or two-tailed test for the difference between means for two non-normal populations with known variances, can be performed using the statistic:

$$Z = \frac{(\bar{X} - \bar{Y})}{\sqrt{\dfrac{\sigma_X^2}{n_X} + \dfrac{\sigma_Y^2}{n_Y}}}$$

which under the null hypothesis, has the approximate standard normal distribution N(0, 1) for large sample size.

Example 7

Samples are taken from two populations with variances 14 and 21. The sample sizes are 32 and 30 and have means 9.1 and 9.7, respectively. Test at 5% whether these populations have the same mean.

Let the populations have means μ_1 and μ_2 and variances σ_1^2 and σ_2^2 and let the sample means be $\bar{X}_1$ and $\bar{X}_2$.

$H_0: \mu_1 - \mu_2 = 0$

$H_1: \mu_1 - \mu_2 \neq 0$

Test statistic:

$$Z = \frac{\bar{X}_1 - \bar{X}_2}{\sqrt{\dfrac{\sigma_1^2}{n_1} + \dfrac{\sigma_2^2}{n_2}}} = \frac{9.1 - 9.7}{\sqrt{\dfrac{14}{32} + \dfrac{21}{30}}} = -0.56 \ (2 \text{ dp})$$

The critical region is outside the interval $(-1.96, 1.96)$

$Z = -0.56$ which is within the acceptance region. The evidence suggests that there is no reason to believe the means are different.

> **Note**
>
> The sample sizes are large enough to assume an approximate normal distribution.

Variances unknown

As discussed in *Chapter 31 Estimators*, when the population variances are unknown and the sample size is large, it is possible to replace the variances in the test statistic with their sample equivalents, without significant further loss of accuracy. This is possible because when the samples are large, it can be assumed that the population and sample values are approximately equal.

Using the notation of the previous section, if the population variances are unknown and s_X^2 and s_Y^2 are their sample estimates, a one- or two-tailed test for the difference between means can be performed using the statistic:

$$Z = \frac{(\bar{X} - \bar{Y})}{\sqrt{\dfrac{s_X^2}{n_X} + \dfrac{s_Y^2}{n_Y}}}$$

which under the null hypothesis, has an approximate standard normal distribution $N(0, 1)$

The above statistic applies to normal populations, as well as other populations if the sample size is large.

Example 8

X and Y are two independent random variables. Samples of size $n_X = 30$ and $n_Y = 35$ are taken from the two populations. They have means $\bar{x} = 18.4$ and $\bar{y} = 19.1$ and sample variances $s_X^2 = 12.4$ and $s_Y^2 = 16.1$, respectively. Test at 5% whether these random variables have the same mean against the alternative hypotheses that Y has a larger mean.

Let the random variables have means μ_X and μ_Y

Test statistic:

$$Z = \frac{\bar{X} - \bar{Y}}{\sqrt{\dfrac{s_X^2}{n_X} + \dfrac{s_Y^2}{n_Y}}} = \frac{18.4 - 19.1}{\sqrt{\dfrac{12.4}{30} + \dfrac{16.1}{35}}} = -0.75 \, (2 \, \text{dp})$$

H_0: $\mu_X - \mu_Y = 0$

H_1: $\mu_X - \mu_Y < 0$

This is a one-tailed test.

The critical region is $Z < -1.645$

Therefore H_0 is not rejected. The evidence suggests that the mean of Y is not greater than the mean of X.

−1.645 is the 5% point of a standard normal variable.

Difference between means – small samples from normal populations

The final problem to consider is that of the difference between means when the variances are not known but, unlike the previous section, the samples are not considered large enough to assume that the two sample variances approximately equal their population equivalents. To make the mathematics manageable, it will be assumed that the variances of the two populations are equal.

When discussing confidence intervals in *Chapter 31 Estimators*, you learned about the statistic $T = \frac{\bar{X} - \mu}{\frac{S}{\sqrt{n}}}$ where S is the unbiased estimate of the population standard deviation.

If the underlying population is normally distributed, this statistic follows a t-distribution on $n - 1$ degrees of freedom. To identify a test statistic for the difference of two means, start from the test statistic for known variances

(see page 430) $Z = \dfrac{(\bar{X} - \bar{Y})}{\sqrt{\dfrac{\sigma_X^2}{n_X} + \dfrac{\sigma_Y^2}{n_Y}}}$

> **Cross-reference**
>
> See section 'Small samples from normal populations with unknown variance' in *Chapter 31 Estimators* for a reminder if you need to.

If the population variances are assumed to be equal (an assumption which will be made throughout this section), this statistic becomes $Z = \dfrac{(\bar{X} - \bar{Y})}{\sqrt{\sigma^2 \left(\dfrac{1}{n_X} + \dfrac{1}{n_Y} \right)}}$ where σ is the common population standard deviation.

σ will be replaced by a value obtained from the pooled samples. The replacement of the population variance by a sample estimate is more complicated in this case, but after some complex statistical theory, the following result is obtained:

> Let X_i; $i = 1, 2, \ldots, n_X$ and Y_i; $i = 1, 2, \ldots, n_X$ be independent random samples from two populations with normal distributions, means μ_X and μ_Y and equal but unknown variances σ^2.
> A two-tailed test for the difference between means is as follows
>
> Hypotheses:
>
> H_0: $\mu_X - \mu_Y = 0$
>
> H_1: $\mu_X - \mu_Y \neq 0$

Test statistic and distribution under the null hypothesis:

$$T = \frac{(\overline{X} - \overline{Y})}{\sqrt{\left\{S_p^{\,2}\left(\dfrac{1}{n_X} + \dfrac{1}{n_Y}\right)\right\}}} \sim t_{n_X + n_Y - 2}$$

where $S_p^{\,2}$ is a pooled estimate of the equal population variance σ^2 given by:

$$S_p^{\,2} = \frac{(n_X - 1)S_X^{\,2} + (n_Y - 1)S_Y^{\,2}}{n_X + n_Y - 2}$$

and $S_X^{\,2}$ and $S_Y^{\,2}$ are unbiased estimators of the common population variance σ^2.

Type: two-tailed test

> **Tip**
>
> For the alternative hypotheses $\mu_X - \mu_Y > 0$ or $\mu_X - \mu_Y < 0$, use a one-tailed test.

Example 9

Samples of size $n_P = 12$ and $n_R = 15$ are taken from two populations and have means $\overline{x}_P = 8.4$ and $\overline{x}_R = 9.2$. Estimates of the common population variance are $s_P^2 = 3.4$ and $s_R^2 = 2.8$. Test at 5% whether these populations have the same mean. You should assume that the populations are normally distributed and that their variances are equal.

Let the populations have means μ_P and μ_R

H_0: $\mu_P - \mu_R = 0$

H_1: $\mu_P - \mu_R \neq 0$

The pooled estimate of population variances is

$$S_p^{\,2} = \frac{11 \times 3.4 + 14 \times 2.8}{25} = 3.06 \text{ (2 dp)}$$

and $t = \dfrac{(8.4 - 9.2)}{\sqrt{\left\{3.06\left(\dfrac{1}{12} + \dfrac{1}{15}\right)\right\}}} = -1.18$ (2 dp)

> **Note**
>
> Substitute this value into the test statistic
> $$T = \frac{(\overline{X} - \overline{Y})}{\sqrt{\left\{S_p^{\,2}\left(\dfrac{1}{n_X} + \dfrac{1}{n_Y}\right)\right\}}} \sim t_{n_X + n_Y - 2}$$

From t-tables with 25 degrees of freedom, the critical region is outside the interval $(-2.060, 2.060)$.

Therefore, there is no reason to reject H_0. The evidence suggests that there is no reason to believe that the means are different.

> **Note**
>
> The small sample sizes and the assumption of equal population variances suggest a t-distribution solution.

> **Note**
>
> Use Table 5 in the *Formulae and Statistical Tables* booklet, with $v = 25$. See *Chapter 31 Estimators* if you need a reminder of how to use this table.

Example 10

The means of two independent random samples from different populations are 25.1 and 24.4. The two samples are of sizes 23 and 12, and give population variance estimates of 4.9 and 6.3, respectively. Both samples come from normally distributed populations with equal variance. Test at a significance level of 10%, the hypothesis that the means of the two populations are equal, against the alternative hypothesis that the second sample comes from a population with a smaller mean.

Let the two samples have means μ_X and μ_Y and equal (unknown) variances.

$H_0: \mu_X - \mu_Y = 0$

$H_1: \mu_X - \mu_Y > 0$

$S_p^2 = \dfrac{22 \times 4.9 + 11 \times 6.3}{33} = 5.37 \ (2 \text{ dp})$

and $t = \dfrac{(25.1 - 24.4)}{\sqrt{\left\{ 5.37 \left(\dfrac{1}{23} + \dfrac{1}{12} \right) \right\}}} = 0.85 \ (2 \text{ dp})$

This is a one-tailed test.

From t-tables with 33 degrees of freedom, the critical region is $t > 1.308$.

Therefore, there is no reason to reject H_0. The evidence suggests that there is no reason to believe that the second sample comes from a population with a smaller mean.

> **Note**
>
> Use the 90th percentage point in Table 5 because this is a one-tailed test with H_1: $\mu_P - \mu_R > 0$; $v = 33$ since $n_x + n_y - 2 = 33$.

Exercise 2

1. In each of the following cases, you are given sample sizes, sample means and variances for two samples from normal distributions. Using an exact test, test at a 5% significance level the null hypothesis that the two samples come from populations with the same mean:

   a $n_X = 10$, $\bar{x} = 9.8$, $\sigma_X^2 = 1.4$; $n_Y = 15$, $\bar{y} = 12.1$, $\sigma_Y^2 = 1.4$

   b $n_X = 30$, $\bar{x} = 2.3$, $\sigma_X^2 = 1.0$; $n_Y = 30$, $\bar{y} = 3.4$, $\sigma_Y^2 = 1.8$

   c $n_X = 16$, $\bar{x} = 27$, $\sigma_X^2 = 9$; $n_Y = 20$, $\bar{y} = 22$, $\sigma_Y^2 = 9$

2. Two random variables, X and Y, have independent, normal distributions with variances $\sigma_X^2 = 3.2$ and $\sigma_Y^2 = 5.4$. Samples are taken from the two populations and give the following statistics: $n_X = 18$, $\bar{x} = 9.4$ and $n_Y = 25$, $\bar{y} = 10.9$

   Test at a 5% significance level, the null hypothesis that X and Y have the same mean.

3. In each of the following cases, test at the given significance level, the null hypothesis that the two samples come from populations with the same mean against the given alternative hypothesis. You should assume that all samples come from normal populations.

   a $n_X = 20$, $\bar{x} = 19.8$, $\sigma_X^2 = 1.7$; $n_Y = 15$, $\bar{y} = 19.5$, $\sigma_Y^2 = 1.0$.   10%.
      $H_1: \mu_X > \mu_Y$.

   b $n_X = 15$, $\bar{x} = 9.7$, $\sigma_X^2 = 1.8$; $n_Y = 18$, $\bar{y} = 8.2$, $\sigma_Y^2 = 1.8$.   5%. $H_1: \mu_X > \mu_Y$.

   c $n_X = 8$, $\bar{x} = 24$, $\sigma_X^2 = 12$; $n_Y = 15$, $\bar{y} = 27$, $\sigma_Y^2 = 8$.   2%. $H_1: \mu_X < \mu_Y$.

4. The means of two independent random samples of sizes 23 and 32 are 14.1 and 16.3, respectively. Test at a significance level of 2%, the hypothesis that both samples come from populations with the same mean. You should assume that both samples come from populations that are normally distributed with variance 7.

5 In each of the following, the sample means, sample sizes and population standard deviations or variances are given for two populations. Test at the stated significance level, whether or not the data suggests equal population means.

a $\bar{x} = 26$; $\sigma_X = 4$, $n_X = 45$;
$\bar{y} = 29$; $\sigma_Y = 4$, $n_Y = 45$; 10%

b $\bar{x} = 17.9$; $\sigma_X = 2.4$, $n_X = 35$;
$\bar{y} = 16.4$; $\sigma_Y = 5.1$, $n_Y = 45$; 5%

c $\bar{x} = 38$; $\sigma_X^2 = 9$, $n_X = 50$;
$\bar{y} = 39$; $\sigma_X^2 = 5$, $n_Y = 60$; 2%

6 Two independent populations have variances 12.1 and 8.2. Samples of size 40 and 35 are taken from these populations and have means 25.1 and 25.8, respectively. Test at 5%, whether or not the populations have the same mean.

Is it possible to say whether your test is exact or approximate? Give reasons for your conclusion.

7 Two random variables X and Y have independent distributions with variances $\sigma_X^2 = 20$ and $\sigma_Y^2 = 30$. Samples are taken from the two populations and give the following results:

$n_X = 52$, $\bar{x} = 9.4$ and $n_Y = 41$, $\bar{y} = 10.5$

Using a normally distributed statistic, test at a 5% significance level the null hypothesis that X and Y have the same mean.

What assumption has been made in order to use a normally distributed statistic?

8 In each of the following cases, you are given two sample means and estimates of the corresponding population variances. Test at a 5% significance level, the null hypothesis that the two samples come from populations with the same mean. You should assume that the two populations are normally distributed with equal variances.

a $n_X = 12$, $\bar{x} = 10.8$, $s_X^2 = 1.9$; $n_Y = 18$, $\bar{y} = 12.1$, $s_Y^2 = 1.4$

b $n_X = 23$, $\bar{x} = 4.1$, $s_X^2 = 2.0$; $n_Y = 19$, $\bar{y} = 3.4$, $s_Y^2 = 1.6$

c $n_X = 16$, $\bar{x} = 29$, $s_X^2 = 10$; $n_Y = 23$, $\bar{y} = 27$, $s_Y^2 = 12$

9 Samples of size 25 and 18 are taken from two independent, normally distributed populations and have means of 23.1 and 25.3, respectively. Given that the sample variances are both equal to 1.8, test at 5% whether or not the populations have the same mean. You should assume that the population variances are equal.

10 Two random variables X and Y have independent, normal distributions with equal variances. Samples are taken from the two populations and give the following statistics:

$n_X = 15$, $\bar{x} = 10.3$ and $n_Y = 21$, $\bar{y} = 11.1$

Estimates of the population variances are $s_X^2 = 3.2$ and $s_Y^2 = 4.4$.

Test at a 10% significance level whether X and Y have the same mean.

32.3 Tests for the difference between the means of two non-independent distributions

So far you have learned how to test for the difference between two population means when the samples used come from independent populations. You will now consider a related problem where the two samples consist of paired values, where one value in the first sample corresponds to one, and only one, value in the second sample.

For example, a school uses a 'mathematics mastery' approach to teaching students. For each unit, students were given a test before the unit was taught, called the pre-test, and another test at the end of the unit, called the post-test. From a given year group, a random sample of students is chosen, these students take both tests and the two scores are recorded for each student. Each student's pre-test score is therefore paired with their post-test score.

It is now possible to investigate the quality of learning by testing the *differences* in scores for each student. This use of a **paired sample** is better than comparing the pre-test scores of one sample of students with the post-test scores of a different, independent sample because, by linking the two scores for a particular student, changes in the test results can be attributed to the teaching itself, rather than differences between students.

In this context, testing the hypothesis that the mean values of the pre- and post-scores are equal, becomes testing that the *difference* between the scores has mean zero. Suppose that 13 students were chosen at random from the year group.

Let x_i be the pre-test result and y_i the post-test result for the ith student $i = 1, 2, ..., 13$. Let the difference in the ith student's results be D_i. The table gives the marks for the 13 students with the corresponding D_i values:

Student	1	2	3	4	5	6	7	8	9	10	11	12	13
Pre-test (%) x_i	42	48	45	36	57	33	33	45	42	27	38	60	29
Post-test (%) y_i	69	45	53	58	44	38	47	51	62	54	46	60	47
$D_i = Y_i - X_i$	27	−3	8	22	−13	5	14	6	20	27	8	0	18

As long as it is assumed that the *differences are normally distributed*, it is possible to adapt the techniques described earlier in this chapter as follows:

First, assume that the D_i are normally distributed with mean μ_D and variance, σ_D^2

If the mean of the differences is $\overline{D_i}$, then

$$\overline{D_i} = \frac{\sum\limits_{i=1}^{n} D_i}{n} \sim N\left(\mu_D, \frac{\sigma_D^2}{n}\right) \text{ where } n \text{ is the sample size}$$

Therefore, $\dfrac{\overline{D_i} - \mu_D}{\dfrac{\sigma_D}{\sqrt{n}}} \sim N(0, 1)$

Replacing σ_D with its unbiased estimator S_D, you get the test statistic:

$$T = \frac{\overline{D_i} - \mu_D}{\frac{S_D}{\sqrt{n}}} \sim t_{n-1}$$

The population standard deviation is replaced with its unbiased estimate, resulting in a statistic with a t-distribution, because the sample is small and the variance is unknown.

With this test statistic and distribution, it is now possible to define the procedure for testing the difference of means for small, paired samples:

Let (X_i, Y_i); $i = 1, 2, 3, ..., n$ be a paired sample from two populations with (unknown) means μ_X and μ_Y. Further, assume that the difference, $D_i = Y_i - X_i$, follows a normal distribution, mean μ_D, variance σ_D^2. Let S_D^2 be the unbiased estimator of σ_D^2.

Hypotheses:

$H_0: \mu_D = 0$

$H_1: \mu_D \neq 0$

Test statistic and distribution under the null hypothesis:

$$T = \frac{\overline{D_i}}{\frac{S_D}{\sqrt{n}}}$$ which has a t-distribution with $n - 1$ degrees of freedom.

Type: Two-tailed test

> **Tip**
>
> For the alternative hypotheses $\mu_D > 0$ or $\mu_D < 0$, use a one-tailed test.

You can calculate S_D^2 using the formula given the *Formulae and Statistical Tables* booklet:

$$S_D^2 = \frac{1}{n_D - 1} \left\{ \sum_{i=1}^{n_D} D_i^2 - \frac{\left(\sum_{i=1}^{n_D} D_i \right)^2}{n_D} \right\}$$

It is now possible to return to the Mathematics Mastery problem. To test the hypothesis that there is no difference between the pre-test and post-test results against the alternative that there is a difference,

$H_0: \mu_D = 0$ versus $H_1: \mu_D \neq 0$

$$\overline{D_i} = \frac{\sum_{i=1}^{n} D_i}{n} = \frac{139}{13}$$

Mean of the differences.

$$S_D^2 = \frac{1}{n_D - 1} \left\{ \sum_{i=1}^{n_D} D_i^2 - \frac{\left(\sum_{i=1}^{n_D} D_i \right)^2}{n_D} \right\} = \frac{1}{12} \left\{ 3229 - \frac{(139)^2}{13} \right\} = 145.23$$

Substituting these two values into the expression for T gives

$$T = \frac{\overline{D_i}}{\frac{S_D}{\sqrt{n}}} = \frac{139}{13} \times \sqrt{\frac{13}{145.23}} = 3.20 \, (2 \, \text{dp})$$

Under the null hypothesis, $T = \dfrac{\overline{D_i}}{\dfrac{S_D}{\sqrt{n}}} \sim t_{12}$

Using the t-distribution percentage points table (Table 5) with 12 degrees of freedom ($n-1$), the critical region for a 5% significance test is outside the interval $(-2.179, 2.179)$.

The observed t-value is outside the critical region, and therefore H_0 is rejected. There is significant evidence to conclude that the means of the two samples are different.

Example 11

(X_i, Y_i); $i = 1, 2, \ldots, n$ is a paired sample from two populations with means μ_X and μ_Y. Assuming that the differences $D_i (= X_i - Y_i)$ are normally distributed, test the hypothesis that the two population means are equal against the alternative that they are different. Use a significance level of 10%.

x_i	23	18	20	16	19	12	9	15	24
y_i	25	21	26	21	17	10	9	18	27

Start by calculating $D_i (= X_i - Y_i)$

$D_i = X_i - Y_i$	2	3	6	5	-2	-2	0	3	3

$H_0: \mu_D = 0$ versus $H_1: \mu_D \neq 0$

$\overline{D_i} = 2$

$S_D^2 = \dfrac{1}{8}\left\{100 - \dfrac{(18)^2}{9}\right\} = 8$ Using $\overline{D_i} = \dfrac{\sum\limits_{i=1}^{n} D_i}{n}$

State the hypotheses.

$T = 2 \times \sqrt{\dfrac{9}{8}} = 2.121 \, (3 \, \text{dp})$

Substitute S_D and $\overline{D_i}$ into the expression for T.

Using the t-distribution percentage points table (Table 5) with 8 degrees of freedom, the critical region for a 10% significance test is outside the interval $(-1.860, 1.860)$.

The observed t-value is inside the critical region and therefore H_0 is rejected. There is significant evidence to conclude that the means of the two samples are different.

> **Note**
>
> Under the null hypothesis,
>
> $T = \dfrac{\overline{D_i}}{\dfrac{S_D}{\sqrt{n}}} \sim t_8$

Exercise 3

① For each of the following tables, (X_i, Y_i); $i = 1, 2, \ldots, n$ is a paired sample from two populations with means μ_X and μ_Y. Assume that the differences $D_i (= X_i - Y_i)$ are normally distributed. Test the given null hypothesis against the given alternative at the significance level indicated.

a $H_0: \mu_X = \mu_Y$; $H_1: \mu_X < \mu_Y$ 10% significance

x_i	23	18	20	16	19	12	9	15	24
y_i	25	21	26	21	17	10	9	18	27

b $H_0: \mu_X = \mu_Y$; $H_1: \mu_X \neq \mu_Y$ 5% significance

x_i	8.1	6.3	9.9	7.2	5.9	10.1
y_i	9.2	10.4	8.3	9.1	5.1	10.0

c $H_0: \mu_X = \mu_Y$; $H_1: \mu_X > \mu_Y$ 1% significance

x_i	164	186	201	174	192	156	183	161
y_i	151	181	210	175	183	153	186	142

2 Each of the following gives data on a paired sample. Assume that the differences, $D_i (= X_i - Y_i)$ are normally distributed. Test the given null hypothesis against the given alternative, at the significance level indicated:

a $H_0: \mu_X = \mu_Y$; $H_1: \mu_X \neq \mu_Y$ 10% significance

d_i	7	8	12	−1	−1	1	2	13

b $H_0: \mu_X = \mu_Y$; $H_1: \mu_X < \mu_Y$ 5% significance

d_i	1.4	3.1	−1.9	−1.4	−3.2	−1.0	2.1	−1.4

3 Each of the following gives data on a paired sample. Assume that the differences $D_i (= X_i - Y_i)$ are normally distributed. Test the given null hypothesis against the given alternative, at the significance level indicated:

a $\Sigma d_i = 15.2$, $\Sigma d_i^2 = 121.2$, $n = 10$. $H_0: \mu_X = \mu_Y$; $H_1: \mu_X < \mu_Y$ 5% significance

b $\Sigma d_i = -33$, $\Sigma d_i^2 = 375$, $n = 8$. $H_0: \mu_X = \mu_Y$; $H_1: \mu_X \neq \mu_Y$ 10% significance

c $\Sigma d_i = -34$, $\Sigma d_i^2 = 718$, $n = 12$. $H_0: \mu_X = \mu_Y$; $H_1: \mu_X < \mu_Y$ 1% significance

4 The heart rates, in beats per minute, are taken for 9 students at 08:00h and 21:00h. The results are shown below:

Heart rate at 08.00, x_i	60	72	69	80	55	58	61	49	52
Heart rate at 21.00, y_i	62	75	68	77	64	60	68	48	59

Using a paired sample *t*-test, test at 5% whether the population from which the sample is taken has different mean heart rates for the two times of day. State any distributional assumptions made.

5 The moisture content of five pieces of timber is measured using two different methods for each. The results, in percentages, is as follows:

Piece	1	2	3	4	5
Method A	12	21	32	18	30
Method B	14	25	31	24	32

a State a necessary distributional assumption in order to use a paired *t* test, to test the hypothesis that the mean moisture content is the same for the two methods.

b Given that all the necessary assumptions are valid, test the hypothesis in **part a** using the 2% level of significance.

32.4 Tests related to population variances

The tests described in this chapter have so far involved population mean values, and in some of these cases, the population variance was assumed to be known. These are somewhat contrived situations, since if the mean of a population is unknown, it is likely that the variance will also be unknown. This suggests that we also need tests on values of population variance, or equivalently on standard deviation.

The theory related to variance is generally more difficult than that of the mean, therefore, proofs will be sketched out rather than rigorously derived. Unless otherwise stated, these proofs do not need to be learned, but are given here to help with understanding and to increase problem-solving skills.

Test for a given value of population variance

Sample values can be used to test a hypothesis that the population variance equals some given value. In order to do this, it is first necessary to introduce a new distribution.

The chi-squared distribution
Consider a standard normal random variable Z. It is common in statistical theory to use the sum of the squares of a set of n independent standard normal variables, that is $\sum_{i=1}^{n} Z_i^2$

Cross-reference

This is the same parameter as was introduced in connection with the *t*-distribution in *Chapter 31 Estimators*.

This sum of squares statistic follows one of a family of distributions called **chi-squared (χ^2) distributions**, which are dependent on one parameter v the *number of degrees of freedom* (df).

So, $X^2 = \sum_{i=1}^{n} Z_i^2$.

The shape of the chi-squared distribution is shown here for two values of v,

As expected from its definition, it takes 0 and positive values only.

The parameter v usually takes positive integer values since, in this context, $v = n - 1$ and n takes positive integer values.

The χ^2 distribution with v degrees of freedom will be written χ^2_v.

Table 6 in the *Formulae and Statistical Tables* booklet gives percentage points for the chi-squared distribution.

Example 12

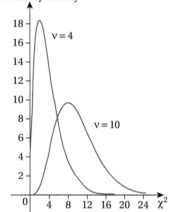

Probability density

Cross-reference

You use this table in a similar way to Table 5 (see page 435) for the *t*-distributions.

Question

Use tables to find the constants a, b, c and d:

a $\quad P\{X^2_9 > 3.325\} = a$ b $\quad P\{X^2_{13} < b\} = 0.99$

c $\quad P\{8.260 < X^2_{20} < 31.410\} = c$ d $\quad P\{X^2_{24} > d\} = 0.95$

Answer

a $\quad a = P\{X^2_9 > 3.325\} = 1 - P\{X^2_9 < 3.325\} = 1 - 0.05 = 0.95$

b $\quad P\{X^2_{13} < 27.688\} = 0.99$ from tables, therefore $b = 27.688$

c $\quad c = P\{8.260 < X^2_{20} < 31.410\} = P\{X^2_{20} < 31.410\} - P\{X^2_{20} < 8.260\} = 0.95 - 0.01 = 0.94$

d $\quad P\{X^2_{24} > d\} = 0.95$. Therefore $P\{X^2_{24} < d\} = 0.05$. From tables, $d = 13.848$

Using the chi-squared distribution to test for a given value of σ^2
To develop a test for σ^2, first consider an independent random sample of size n taken from a normal population with mean μ and variance σ^2. Let the random sample be denoted by X_i; $i = 1, 2, \ldots, n$ and consider the random variable $X_1 - \overline{X}$.

Using standard expectation algebra, it is possible to show that $X_1 - \overline{X} \sim N\left(0, \frac{n-1}{n}\sigma^2\right)$

Therefore $Z_1 = \dfrac{X_1 - \overline{X}}{\sqrt{\dfrac{n-1}{n}\sigma^2}} \sim N(0,1)$

Squaring, you get

$$Z_1^2 = \frac{n}{\sigma^2} \times \frac{(X_1 - \overline{X})^2}{(n-1)}$$

Then, summing over all similar variables in the whole sample gives the test statistic:

$$X^2 = \frac{n}{\sigma^2} \times S^2 \text{ since } \sum_{i=1}^{n} \frac{(X_i - \overline{X})^2}{(n-1)} = S^2$$

$\frac{n}{\sigma^2} \times S^2$ is therefore the sum of the squares of n standard normal variables

It follows from the theory of the χ^2 distribution given above, plus further considerations about independence, that:

The statistic X^2, given by $X^2 = \frac{n-1}{\sigma^2} \times S^2$ has a chi-squared

distribution with $n-1$ degrees of freedom; that is $\frac{n-1}{\sigma^2} \times S^2 \sim \chi^2_{n-1}$

It is now possible to define a test for a given value of σ^2:

Let X_i; $i = 1, 2, ..., n$ be a random sample from a population with a normal distribution, mean μ and variance σ^2. Let the sample estimate of population variance be S^2. A two-tailed test for a given value of the population variance is as follows:

$H_0: \sigma^2 = \sigma_0^2$
$H_1: \sigma^2 \neq \sigma_0^2$

Test statistic and distribution under the null hypothesis: $\frac{n-1}{\sigma_0^2} \times S^2$

which has a chi-squared distribution with $n-1$ degrees of freedom; that

is $\frac{n-1}{\sigma_0^2} \times S^2 \sim \chi^2_{n-1}$

Type: Two tailed test

> **Note**
> $n-1$ has taken the place of n because the summation described above is over $n-1$, not n, *independent* terms.

> **Note**
> For the alternative hypotheses, $H_1: \sigma^2 > \sigma_0^2$ or $H_1: \sigma^2 < \sigma_0^2$ use a one-tailed test.

Example 13

An independent random sample of size 30 is taken from a normally distributed population, and found to give an estimate of the population standard deviation of 1.8. Test at a significance level of 5%, whether or not the true population standard deviation is 2.5.

$H_0: \sigma = 2.5$
$H_1: \sigma \neq 2.5$
Test statistic: $X^2 = \frac{n-1}{\sigma_0^2} \times S^2 = \frac{29}{2.5^2} \times 1.8^2 = 15.03$

The X^2_{29} critical values for a 5% two-tailed test are 16.047 and 45.722.

Therefore, there is significant reason to reject the null hypothesis. The evidence suggests that the population standard deviation is not 2.5.

> **Note**
> Assuming an underlying normal distribution, under the null hypothesis, this value should be considered as a single observation from a X^2_{29} distribution.

Example 14

Question

A bus company claims that, as a result of changes to the route, journey times from Oxford to Witney have maintained their average time but become more consistent. Before the changes, journey times had a mean of 42 minutes and a standard deviation of 2.2 minutes. After the changes, a random sample of 12 journey times was as follows:

43 44 39 41 42 40 38 39 41 43 42 43

Test at a significance level of 5% the company's claim.

Answer

H_0: $\sigma = 2.2$ H_1: $\sigma < 2.2$, where σ is the population standard deviation in minutes.

$$S^2 = \frac{1}{11}\left\{20459 - \frac{495^2}{12}\right\} = \frac{161}{44} = 3.66 \text{ (2 dp)}$$

S is the unbiased estimate of population standard deviation.

$$\frac{n-1}{\sigma_0^{\,2}} \times S^2 = \frac{11}{2.2^2} \times \frac{161}{44} = 8.32 \text{ (2 dp)}$$

The χ^2_{11} critical value for a 5% one-tailed test is 4.575. Therefore, there is no reason to reject the null hypothesis. The evidence suggests that the bus company's claim is unjustified.

Reject H_0 if $X^2 < 4.575$

> **Note**
>
> The null hypothesis suggests that the variation of journey times remains constant following the changes. The alternative hypothesis suggests that there is less variation.

> **Note**
>
> Assuming an underlying normal distribution, under the null hypothesis, this value should be considered as a single observation from a X^2_{11} distribution.

Test for a comparison of two values of population variance

When deriving a test for the difference of two means based upon small sample data earlier in the chapter (see page 434), it was assumed that the variances of the two populations were equal. It would be useful to have a test to check this assumption. To develop the theory for this test, it is again necessary to introduce a new distribution, one which has many other statistical applications.

The *F* distribution

Consider two independent random variables, U and V, each with a χ^2-distribution with v_1 and v_2 degrees of freedom, respectively: $U \sim \chi^2_{v_1}$ and $V \sim \chi^2_{v_2}$

Using these distributions, you can form the following quotient, called the *F-statistic*

$$F = \frac{\dfrac{U}{v_1}}{\dfrac{V}{v_2}}$$

This statistic follows one of a family of distributions called *F* distributions, which are dependent on two parameters, the number of degrees of freedom of U and V. The shape of the *F* distribution for large v_1 and v_2 is shown.

As this distribution has two parameters, v_1 and v_2 the associated tables of values are more cumbersome than those of the other distributions considered. Table 7 in the *Formulae and Statistical Tables* booklet is divided into four sections, corresponding to the upper percentage points of 95%, 97.5%, 99% and 99.5%.

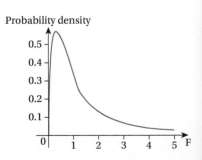

The lower percentage points, 5%, 2.5%, 1% and 0.5%, are found using the relation

$$P\left(F_{v_1, v_2} < c\right) = P\left(F_{v_2, v_1} > \frac{1}{c}\right)$$

This follows from the result:

$$\frac{\frac{U}{v_1}}{\frac{V}{v_2}} \sim F_{v_1, v_2} \quad \Leftrightarrow \quad \frac{\frac{V}{v_2}}{\frac{U}{v_1}} \sim F_{v_2, v_1}$$

Example 15

Use tables to find the constants a, b, c and d:

a $\quad P\{F_{9, 4} < a\} = 0.95$
b $\quad P\{F_{12, 2} < b\} = 0.99$
c $\quad P\{F_{20, 4} < c\} = 0.01$
d $\quad P\{F_{15, 10} > d\} = 0.05$

Note

Use the table for $p - 0.95$, then find the value that corresponds to the column $v_1 = 9$ and $v_2 = 4$.

a $\quad P\{F_{9, 4} < a\} = 0.95$. From tables, $P\{F_{9, 4} < 6.00\} = 0.95$. $a = 6.00$

b $\quad P\{F_{13, 2} < b\} = 0.99$. From tables, $P\{F_{12, 2} < 99.42\} = 0.99$. $b = 99.42$

c $\quad P\{F_{20, 4} < c\} = P\left\{F_{4, 20} > \frac{1}{c}\right\}$

$\qquad = 1 - P\left\{F_{4, 20} < \frac{1}{c}\right\} = 0.01 \qquad$ Using $P(F > a) = 1 - P(F < a)$

$P\left\{F_{4, 20} < \frac{1}{c}\right\} = 0.99$

$\frac{1}{c} = 4.431; c = 0.226$

d $\quad P\{F_{15, 10} > d\} = 1 - P\{F_{15, 10} < d\} = 0.05$

$\quad P\{F_{15, 10} < d\} = 0.95$. $d = 2.845$

Lower percentage point so use

$$P(F_{v_1, v_2} < c) = P\left(F_{v_2, v_1} > \frac{1}{c}\right)$$

Using the *F* distribution to compare population variances

It is now possible to derive a test to decide whether two independent random samples come from populations with equal variances.

First, consider two independent normal populations with variances σ_X^2 and σ_Y^2. Suppose that two independent random samples of sizes n_X and n_Y are taken from these populations, and let S_X^2 and S_Y^2 be unbiased estimators of those variances.

Recall from the theory given earlier in this chapter (see page 443) that the variables $(n_X - 1)\frac{S_X^2}{\sigma_X^2}$ and $(n_Y - 1)\frac{S_Y^2}{\sigma_Y^2}$ both have χ^2 distributions with $n_X - 1$ and $n_Y - 1$ degrees of freedom, respectively. You now also know that if U and V are both χ^2 distributed with v_1 and v_2 degrees of freedom, respectively, then

$$\frac{U}{v_1} \div \frac{V}{v_2} \sim F_{v_1, v_2}$$

It follows that $\dfrac{\frac{S_X^2}{\sigma_X^2}}{\frac{S_Y^2}{\sigma_Y^2}}$ has an F-distribution with $n_X - 1$ and $n_Y - 1$ degrees of freedom.

Under the null hypothesis that the two population variances are equal

$$\frac{S_X^2}{S_Y^2} \sim F_{n_X - 1, n_Y - 1}$$

Let $X_i; i = 1, 2, \ldots, n_X$ and $Y_i; i = 1, 2, \ldots, n_Y$ be two random samples from populations with independent normal distributions, means μ_X and μ_Y and variances σ_X^2 and σ_Y^2, respectively. A two-tailed test for the equality of population variances is as follows:

$H_0: \sigma_X^2 = \sigma_Y^2$
$H_1: \sigma_X^2 \neq \sigma_Y^2$

Test statistic and distribution under the null hypothesis: $\dfrac{S_X^2}{S_Y^2}$

which has an F distribution with $n_X - 1$ and $n_Y - 1$ degrees of freedom. That is

$$\frac{S_X^2}{S_Y^2} \sim F_{n_X - 1, \, n_Y - 1}$$

Example 16

Question

A sample is taken from each of two independent, normally distributed populations. The samples are of sizes 13 and 16 and give sample variances of 5932.3 and 2853.7, respectively. Test at a significance level of 5% the hypothesis that the population variances are equal.

Answer

$H_0: \sigma_X^2 = \sigma_Y^2$

$H_1: \sigma_X^2 \neq \sigma_Y^2$ State your hypotheses.

where X and Y are random variables for the two populations.

$\dfrac{s_X^2}{s_Y^2} = \dfrac{5932.3}{2853.7} = 2.079 \text{ (3 dp)}$

Upper critical value: $F_{12, 15} = 2.963$

No reason to reject the null hypothesis that the population means are the same.

This is the F-statistic.

Note

Using the 97.5% point of an $F_{12, 15}$ distribution, since this is a two-tail test.

Example 17

Question

30 plots of land are each planted with equal numbers of cabbage seedlings. Half of these are treated with fertilizer A and half with fertilizer B. On reaching maturity, the total weight of the cabbages, in kg, for each plot is recorded. X and Y are random variables for the total weight of cabbage per plot following treatment with A and B respectively. These weights give the following statistics:
$\sum x = 15.46, \sum x^2 = 16.18; \sum y = 15.65, \sum y^2 = 16.95$.

Test at 5% whether the fertilizer B produce crops with a larger variation in weight than fertilizer A.

Answer

$H_0: \sigma_X^2 = \sigma_Y^2$

$H_1: \sigma_X^2 < \sigma_Y^2$ This is a one-tailed test.

$s^2 = \dfrac{1}{n-1}\left\{ \sum x^2 - \dfrac{(\sum x)^2}{n} \right\} = \dfrac{\sum (x - \bar{x})^2}{n-1}$ The test statistic.

$s_X^2 = 0.018$ and $s_Y^2 = 0.044$

$\dfrac{s_X^2}{s_Y^2} = 0.41$

Critical F-value: 0.40 (linear interpolation)

No reason to reject the null hypothesis that the variances are equal.

Note

Further research would be advisable since the observed F-value is so close to the critical value.

Note

Using the formula for the unbiased estimate of population variance S^2.

Note

95% point of F(14, 14) is 2.49 (2 dp)

Exercise 4

1 Use tables to find the following probabilities:

  a $P\{X^2_9 < 2.088\}$
        b $P\{X^2_{17} < 30.191\}$

  c $P\{X^2_{20} > 31.410\}$
      d $P\{6.262 < X^2_{15} < 24.996\}$

  e $P\{X^2_1 \geq 5.024\}$
       f $P\{1.145 < X^2_5 < 11.070\}$

2 In the following probability statements, find the constants $a - f$:

  a $P\{X^2_{14} < a\} = 0.99$
      b $P\{X^2_{17} < b\} = 0.01$

  c $P\{X^2_{22} > c\} = 0.975$
      d $P\{X^2_{10} < d\} = 0.05$

  e $P\{X^2_e < 35.563\} = 0.9$
    f $P\{f < X^2_9 < 2.088\} = 0.005$

3 In the following probability statements, find the constants $a - d$:

  a $P\{F_{1,4} < a\} = 0.99$
      b $P\{F_{5,8} < b\} = 0.01$

  c $P\{F_{8,4} > c\} = 0.975$
      d $P\{F_{15,20} < d\} = 0.05$

4 For each of the following F distributions, find the given percentage point:

  a $v_1 = 5, v_2 = 9; 95\%$
      b $v_1 = 10, v_2 = 15; 99.5\%$

  c $v_1 = 2, v_2 = 8; 99\%$
      d $v_1 = 5, v_2 = 7; 5\%$

  e $v_1 = 20, v_2 = 30; 1\%$
     f $v_1 = 22, v_2 = 30; 97.5\%$

5 In each of the following, an independent random sample of size n is taken from a normally distributed population and found to give an estimate of the population standard deviation of s. Test at a significance level of $\alpha\%$ whether the true population standard deviation is σ. H_1 is the alternative hypothesis.

  a $n = 25, s = 2.3, H_0: \sigma = 3.1, \alpha = 5, H_1: \sigma < 3.1$

  b $n = 20, s = 3.4, H_0: \sigma = 3.0, \alpha = 10, H_1: \sigma \neq 3.0$

  c $n = 14, s^2 = 8.2, H_0: \sigma = 3, \alpha = 2.5, H_1: \sigma < 3$

  d $n = 9, s^2 = 8, H_0: \sigma^2 = 15, \alpha = 5, H_1: \sigma^2 \neq 15$

6 An independent random sample of size 35 is taken from a normally distributed population and found to give an estimate of the population standard deviation of 5.4. Test at a significance level of 2.5%, whether the true population standard deviation is 7.5 against the alternative hypothesis that it is less than 7.5.

7 A train company wishes to check its consistency of journey times with a competitor. A sample is taken from each of the two company's times on comparable journeys. The samples are of sizes 21 and 16 and give sample standard deviations of 27.1 and 19.8, respectively. Test at a significance level of 5%, the hypothesis that the consistency of times, as measured by the population variances, is equal for the two competitors. State any assumptions made.

8 Independent random samples of size 51 are taken from two independent, normally distributed populations, P and R. The standard deviations of these samples are 17 and 27, respectively. Test at a significance level of 2.5%, the hypothesis that the standard deviation of population P is less than that of R.

9 The values in an independent random sample of size 8 from a normally distributed population are as follows:

12.4   13.1   12.9   14.2   11.3   12.6   13.2   10.9

Test at a significance level of 10%, whether the true population standard deviation is 1.4 against the alternative hypothesis that it is less than 1.4.

10 The birth weights of 10 babies born at full term in a hospital on one day in March 2015 are, in kilogrammes, to the nearest 0.001 kg:

3.213   3.502   3.035   2.646   3.813   4.490   3.324   3.763   2.787   2.934

It is suggested that the population standard deviation for full-term births at this hospital is 0.6 kg. Does the data support this? Use a 5% test and state any assumptions being made.

Summary

▶ A type I error means rejecting the null hypothesis when it is true; the probability of this occurring is equal to the significance level of the test.

▶ A type II error means failing to reject the null hypothesis when it is false; the probability of this occurring is given by P{type II error} = P{ H_0 is accepted | H_0 is false}.

▶ The power of a hypothesis test is the probability that the null hypothesis is rejected when it *should* be rejected; it is is given by P = P{H_0 is rejected|H_0 is false} = 1 − P{type II error}.

▶ You can test whether or not two samples taken from independent populations provide evidence that the populations have different means, by using different test statistics according to the properties of the population. If X_i; $i = 1, 2, \ldots, n_X$ and Y_i; $i = 1, 2, \ldots, n_Y$ are random samples from two populations with independent normal distributions, means μ_X and μ_Y, then the difference of means for:

• normal distributions with known variances σ_X^2 and σ_Y^2 has test statistic and distribution under the null hypothesis of:

$$Z = \frac{\left(\bar{X} - \bar{Y}\right)}{\sqrt{\dfrac{\sigma_X^2}{n_X} + \dfrac{\sigma_Y^2}{n_Y}}} \sim N(0, 1)$$

• non-normal distributions with known variances σ_X^2 and σ_Y^2 and large sample has test statistic and distribution under the null hypothesis of,

$Z = \frac{(\bar{X} - \bar{Y})}{\sqrt{\frac{\sigma_X^2}{n_X} + \frac{\sigma_Y^2}{n_Y}}} \sim N(0, 1)$ approximately.

• any distribution with unknown variance and large sample size, where s_X^2 and s_Y^2 are their sample estimates, has test statistic and distribution under the null hypothesis of $Z = \frac{(\bar{X} - \bar{Y})}{\sqrt{\frac{s_X^2}{n_X} + \frac{s_Y^2}{n_Y}}} \sim N(0, 1)$ approximately

- a normal distribution with unknown variance and small sample size has test statistic and distribution under the null hypothesis of:

$$T = \frac{(\bar{X} - \bar{Y})}{\sqrt{\left\{S_p^2\left(\frac{1}{n_X} + \frac{1}{n_Y}\right)\right\}}} \sim t_{n_X+n_Y-2}$$ where S_p^2 is a pooled estimate of the equal

population variance σ^2 given by $S_p^2 = \frac{(n_X-1)S_X^2 + (n_Y-1)S_Y^2}{n_X+n_Y-2}$

▶ You can also test for the difference of means of two non-independent distributions. If (X_i, Y_i); $i = 2, 3, \ldots, n$ is a paired sample from two populations with (unknown) means μ_X and μ_Y where the difference $D_i = X_i - Y_i$ follows a normal distribution, mean μ_D, variance σ_D^2, and S_D^2 is the unbiased estimator of σ_D^2, the test statistic and distribution under the null hypothesis is:

$$T = \frac{\bar{D_i}}{\frac{S_D}{\sqrt{n}}}$$

which has a t-distribution on $n - 1$ degrees of freedom.

▶ You can do tests on values of population variance to see if it is likely to be equal to a given value. For a given value of the population variance, the test statistic and distribution under the null hypothesis is:

$$\frac{n-1}{\sigma_0^2} \times S^2 \sim \chi_{n-1}^2$$

▶ You can test if the variance of two populations are equal. If X_i; $i = 1, 2, \ldots, n_X$ and Y_i; $i = 1, 2, \ldots, n_Y$ be two random samples from populations with independent normal distributions, means μ_X and μ_Y and variances σ_X^2 and σ_Y^2, respectively. The test statistic and distribution under the null hypothesis is $\frac{S_X^2}{S_Y^2} \sim F_{n_X-1, n_Y-1}$ where S_X^2 and S_Y^2 are the unbiased estimators of σ_X^2 and σ_Y^2.

Review exercises

1 A random variable X has a normal distribution with mean μ and variance 30. A test of the hypothesis is performed that $\mu = 19$ against the alternative that $\mu < 19$, at a significance level of 5%. Find the acceptance region for this test, if the sample size on which the test is based is 18 and the estimator $\bar{X}$ is used.

2 X is a random variable with the distribution $N(\mu, 4)$. A random sample of size 32 is taken and gives a mean of 17.

a Test at a significance level of 5% whether the population mean is 16.

b Find the probability of making a type II error if the population mean is actually 14.5.

c Find the power of the test.

3 For two samples with the following statistics, test at a 5% significance level the null hypothesis that the samples come from normally distributed populations with the same mean: $n_X = 10$, $\bar{x} = 12.1$, $\sigma_X^2 = 1.9$; $n_Y = 17$, $\bar{y} = 13.3$, $\sigma_Y^2 = 1.5$

4 The table below shows a paired sample from two populations with means μ_X and μ_Y. Assume that the differences $D_i (= X_i - Y_i)$ are normally distributed. Test the null hypothesis against the alternative, at the significance level indicated:

$H_0: \mu_X = \mu_Y$; $H_1: \mu_X < \mu_Y$. 10% significance

x_i	51	50	38	42	48	45	39	47	42
y_i	56	49	45	46	52	44	45	50	47

5 An independent random sample of size 41 is taken from a normally distributed population and found to give an estimate of the population standard deviation of 2.2. Test at a significance level of 5%, whether the true population standard deviation is 1.9 against the alternative hypothesis that it is greater than 1.9.

Practice examination questions

1 Each household within an area has two types of bin: a black one for general trash and a green one for garden trash.

The mass, in kilograms, of trash emptied from a black bin can be modelled by the random variable $B \sim N(\mu_B, 0.5625)$.

The mass, in kilograms, of trash emptied from a green bin can be modelled by the random variable $G \sim N(\mu_G, 0.9025)$.

The mean mass of trash emptied from a random sample of 20 black bins was 21.35 kg. The mean mass of trash emptied from an independent random sample of 15 green bins was 21.90 kg.

Test, at the 5% level of significance, the hypothesis that $\mu_B = \mu_G$ (6 marks)

AQA, FM04 (Further Statistics Unit 2 Specimen (9665) 2018

2 The tensile strength of rope is measured in kilograms.

The standard deviation of the tensile strength of a particular design of 10 mm diameter rope is known to be 285 kilograms. A retail organisation, which buys such rope from two manufacturers, A and B, wishes to compare their ropes for mean tensile strength.

The mean tensile strength $\bar{x}$ of a random sample of 80 lengths from manufacturer A was 3770 kilograms.

The mean tensile strength $\bar{y}$ of a random sample of 120 lengths from manufacturer B was 3695 kilograms.

a i   Test, at the 5% level of significance, the hypothesis that there is no difference between the mean tensile strength of rope from manufacturer A and that of rope from manufacturer B.   (6 marks)

   ii Why was it **not** necessary to know the distributions of tensile strength in order for your test in **part a i** to be valid?   (1 mark)

b i   Deduce that, for your test in **part a i**, the critical values of $(\bar{x} - \bar{y})$ are ±80.63, correct to two decimal places.   (2 marks)

   ii In fact, the mean tensile strength of rope from manufacturer A exceeds that of rope from manufacturer B by 125 kilograms.

Determine the probability of a Type II error for a test of the hypothesis in **part a i** at the 5% level of significance, based upon a random sample of 80 lengths from manufacturer A and a random sample of 120 lengths from manufacturer B. (4 marks)

AQA MS03 June 2011

3 An ergonomist is investigating the effect of training on the speed with which workers in a factory can assemble a particular product. The ergonomist selects a random sample of 8 workers who have not received the training and a random sample of 6 workers who have received the training. The ergonomist records the time taken, in minutes, for each of these selected workers to assemble the product. The results are shown in the table below.

Untrained	10.4	8.9	10.1	9.0	9.4	9.6	10.0	10.2
Trained	9.0	8.3	9.5	8.0	9.2	8.2		

a State **two** necessary assumptions in order to test the hypothesis, that the mean time taken by the untrained workers is the same as the mean time taken by the trained workers. [2 marks]

b Given that all the necessary assumptions are valid, test the hypothesis in part **a** using the 2% level of significance. [10 marks]

AQA MS04 June 2014

4 A random sample of 8 male 400-metre runners was selected and their personal best times, in seconds, achieved at sea level and at altitude were recorded, as shown in the table.

Runner	A	B	C	D	E	F	G	H
Sea level time	48.4	47.6	48.4	45.9	47.1	50.4	48.2	50.8
Altitude time	47.9	47.1	47.7	45.7	46.8	50.3	47.9	50.3

A coach believes that the personal best times of male 400-metre runners at sea level will be more than 0.2 seconds slower than those at altitude.

Assuming that the differences between times at sea level and at altitude are normally distributed, investigate the coach's belief at the 1% level of significance. (10 marks)

AQA MS04 June 2010

5 A random sample of 12 batches of steel was taken from a production line. The carbon content, measured in grams per cubic metre, of each of these batches of steel is given below.

3.6   4.3   3.8   4.1   3.7   3.6   3.9   4.2   4.3   4.1   4.4   3.5

Assuming that this sample came from an underlying normal population, test, at the 1% level of significance, whether the standard deviation of the carbon content of steel from this production line is 0.7. (7 marks)

AQA MS04 June 2011

33 Further Hypothesis Testing 2 – Goodness of Fit Tests and Tests of No Association

Introduction

The tests in *Chapter 32 Further Hypothesis Testing 1 – Means and Variances* have all assumed that the random variable in question has a known distribution type (normal, binomial, ...), and that the only uncertainty is the value of the distribution's parameters. Sometimes it is the distribution type itself that is unknown and, in these cases, it is often possible to test hypotheses about the fit of data to a distribution. This problem, and testing for a possible association between two variables, will be dealt with in this chapter. This theory has considerable practical value; it may, for example, be helpful to know whether real data relating to the heights of a certain group of people is normally distributed.

Objectives

By the end of this chapter, you should know how to:

► Test the goodness of fit of given data to a particular probability distribution, such as the Poisson, exponential and normal distributions, when
  • The parameters of the distribution are known
  • The parameters of the distribution are unknown
  • The number of expected frequencies for an interval are not greater than 5.
► Test if two variables are independent of each other when the data is given in the form of a bivariate table.
► Use the Yates' correction in no association tests and know when it is appropriate to do so.

Recap

You will need to remember...

► Basic properties of probability measure, including probabilities relating to mutually exclusive and independent events, and conditional probability.
► The properties of probability distributions, such as the binomial distribution.
► The circumstances where given probability distributions (including the binomial, Poisson, exponential and normal distributions) provide a good mathematical model.
► The basic stages of a hypothesis test:
  • Formation of a null and an alternative hypothesis
  • Identification of a probability distribution and its critical region
  • Decision making based upon an observed value.
► The use of chi-squared tables and cumulative binomial and Poisson tables
► The definition of a cumulative probability distribution:

$$F\{t_1\} = P\{T < t_1\} = \int_{t=-\infty}^{t_1} f(t)\, dt$$

33.1 Goodness of fit tests

When the distribution of a set of data is unknown and you are testing for the fit of the data to a particular distribution, there are two cases to be studied depending on whether the parameter value to be tested is known, or has to be estimated from the data.

Known parameters

Consider the very simple random experiment where an ordinary dice is thrown 120 times with the following results:

Score [x_i]	1	2	3	4	5	6
Frequency	25	21	15	19	23	17

Do these data suggest that the dice is fair?

If X is a random variable for the score on a randomly chosen roll, this question is equivalent to asking: does X have a discrete uniform distribution on the integers $1-6$?

Null hypothesis:

H_0: $P\{X=i\} = \frac{1}{6}$, $i = 1, 2, ..., 6$

Denote the observed frequency of x_i by O_i and let E_i be the corresponding frequency of x_i expected under the null hypothesis.

If the null hypothesis is true, each value of O_i is likely to be close to the corresponding E_i and therefore each $O_i - E_i$ is likely to be small. However, 'close' and 'small' are relative terms and depend upon the values of the E_is. This suggests that the 'relative closeness' given by the values of $\frac{O_i - E_i}{E_i}$ could indicate how well the observed data fits the hypothesized distribution.

Since a single value obtained from the whole data set is required, it seems reasonable to consider $\sum_{\forall i} \frac{O_i - E_i}{E_i}$ as the required test statistic, with small values suggesting a good fit.

The terms of this statistic can be positive or negative, which means that the sum of the terms could be small, even for a poorly fitting distribution. Karl Pearson, a British mathematician (1857 – 1936), resolved this problem by showing that the test statistic $\sum_{\forall i} \frac{(O_i - E_i)^2}{E_i}$ has an approximate chi-squared distribution if the null hypothesis is true, and that the approximation gets better as the sample size increases.

This gives a procedure to test how well data fits a particular distribution, and is therefore a '**goodness of fit**' test:

> Let H_0 be a null hypothesis specifying completely a probability distribution. For k possible outcomes of a repeated experiment where O_i; $i = 1, 2, ..., k$ represents the observed frequencies, and E_i; $i = 1, 2, ...,$ k represents the expected frequencies from the distribution specified in H_0 then the statistic $X^2 = \sum_{i=1}^{k} \frac{(O_i - E_i)^2}{E_i}$ approximately follows a chi-squared distribution on $k - 1$ degrees of freedom (χ^2_{k-1}), if the null hypothesis is true.

The phrase '**specifying completely**...' means that there are no unknown parameters in the distribution.

It is now possible to solve the problem given at the start of this section.

Under the null hypothesis, H_0: $P\{X=i\} = \frac{1}{6}$, $i = 1, 2, ..., 6$, and after 120 rolls, the expected frequencies are each $20 \left(=120 \times \frac{1}{6}\right)$.

Cross-reference

See *Chapter 11 Uniform and Geometric Distributions* if you need to.

Cross-reference

You were introduced to chi-squared distributions in *Chapter 32 Further Hypothesis Testing 1 – Means and Variances*.

This gives the following table of observed and expected frequencies:

Score [x_i]	1	2	3	4	5	6
O_i	25	21	15	19	23	17
E_i	20	20	20	20	20	20

Therefore, $\chi^2 = \sum_{\forall i} \dfrac{(O_i - E_i)^2}{E_i} = 3.5$, from $\dfrac{(25-20)^2}{20} + \quad + \dfrac{(17-20)^2}{20}$

Since $k = 6$, the number of degrees of freedom is $6 - 1 = 5$ and, for a significance level of 2.5%, the critical region (the 97.5% point of the χ^2_5 distribution), is $\chi^2 > 12.833$. (These tests are all one-tailed since only large values of the distribution result in H_0 being rejected.)

Since the observed value of χ^2 is 3.5, there is no significant evidence to reject the hypothesis that the dice has a discrete uniform distribution on the integers $1-6$, and therefore that the dice is fair.

Example 1

Question

An experiment consists of flipping a coin 4 times. This experiment is repeated 300 times and the number of heads showing each time is recorded. The results are as follows:

Number of heads [x]	0	1	2	3	4
Frequency	46	95	102	44	13

Test at a significance level of 5%, the hypothesis that the coin is biased with the probability of a head equal to 0.4.

Answer

Let X be the number of heads in a randomly chosen set of 4 flips.

H_0: $X \sim B(4, 0.4)$

Under the null hypothesis, $P\{X = x\} = \begin{pmatrix} 4 \\ x \end{pmatrix} 0.4^x 0.6^{4-x}$; $x = 0, 1, 2, 3, 4$

Number of heads [x]	0	1	2	3	4
$P\{X = x\}$	0.1296	0.3456	0.3456	0.1536	0.0256
Expected frequencies	38.88	103.68	103.68	46.08	7.68

$X^2 = \sum_{\forall i} \dfrac{(O_i - E_i)^2}{E_i} = 5.84$ (2 dp)

Critical region: $\chi^2 > 9.488$

No reason to reject H_0; the evidence suggests that there is no reason to reject the hypothesis that the probability of a head is 0.4.

Note

Since X is the number of heads (successes) in 4 independent and identical Bernoulli trials.

Formulae booklet

The binomial distribution function is given in the *Formulae and Statistical Tables* booklet.

Note

This is the binomial probability distribution function, where $\begin{pmatrix} n \\ x \end{pmatrix} = \dfrac{n!}{(n-x)!\,x!}$ and x is the number of successes. Substitute x-values into the distribution function, or use tables of cumulative binomial probabilities (see Table 1 in the *Formulae and Statistical Tables* booklet) with $n = 4$, $p = 0.4$.

Note

Recall that $P(X = 1) = P(X \leq 1) - P(X \leq 0)$ and so on. Expected frequencies are found using $300 \times P\{X = x\}$.

Note

Using the goodness of fit test statistic

Note

The 95% point of a χ^2_4 distribution

When dealing with normally distributed populations, since the expected frequencies will be found from cumulative probability tables, it makes sense to focus on the upper class boundary of each interval.

Example 2

The following table gives the weight gain, over a four-month period, in grams, of 100 babies born in a hospital in England.

Weight gain (x g)	Frequency
$x < 2800$	7
$2800 \leq x < 3000$	12
$3000 \leq x < 3200$	29
$3200 \leq x < 3400$	36
$3400 \leq x < 3600$	10
$3600 \leq x < 3800$	6

Test at 10% whether this data comes from a normal population with mean 3200 g, standard deviation 300 g.

$H_0: X \sim N(3200, 300^2)$

Weight gain (x g)	Upper class boundary (u.c.b)	u.c.b. z-value $\left(= \dfrac{x - \mu}{\sigma} \right)$	$P\{Z < z\}$ under H_0	Probability under H_0	Expected frequency under H_0
$x < 2800$	2800	-1.3333	0.0912	0.0912	9.12
$2800 \leq x < 3000$	3000	-0.6667	0.2525	0.1613	16.13
$3000 \leq x < 3200$	3200	0	0.5	0.2475	24.75
$3200 \leq x < 3400$	3400	0.6667	0.7475	0.2475	24.75
$3400 \leq x < 3600$	3600	1.3333	0.9088	0.1613	16.13
$x \geq 3600$	∞	∞	1	0.0912	9.12

> **Note**
>
> The probabilities corresponding to each interval are found by subtraction of the relevant cumulative probabilities.

$$X^2 = \sum_{\forall i} \frac{(O_i - E_i)^2}{E_i} = 10.79 \text{ (2 dp)}$$

Critical region: $\chi^2 > 9.236$

Reject H_0; the evidence suggests that the data does not come from a normal population with mean 3200 g, standard deviation 300 g.

> **Note**
>
> The 90% point of a χ^2_5 distribution: since $k = 6$, the number of degrees of freedom is $6 - 1 = 5$

The above procedure has demonstrated the basic problem of testing whether or not a set of data comes from a population with a particular distribution. In circumstances where the parameters of the distribution are not known, or when the expected frequencies are small, the procedure has to be altered slightly.

Unknown parameters

Sometimes you will be asked to test whether data comes from a given distribution, but will not be given the value of any of the parameters of the distribution. In this case, these values have to be estimated.

In addition to having to estimate the population parameters, the only other change to the method described previously involves the number of degrees of freedom of the relevant chi-squared distibution.

You saw in *Chapter 32 Further Hypothesis Testing 1 – Means and Variances* that when using the chi-squared distribution to test for a given value of σ^2, the number of degrees of freedom for a problem is the number of *independent* terms in the chi-squared summation. When the parameters are known, the number of independent terms is $k - 1$, due to the constraint that the sum of the expected frequencies must equal the sum of the observed frequencies. When the parameters are unknown you still have this constraint, but there are also more constraints imposed by the fact that the parameters are estimated. For each estimated parameter, you lose a degree of freedom.

In general:

> A chi-squared test of goodness of fit, for a repeated experiment with k possible outcomes, has the number of degrees of freedom v given by
> $v = k -$ number of constraints on the expected frequencies
> $\quad = k - 1 -$ number of estimated parameters

It is often helpful to think of the number of degrees of freedom as the number of expected frequencies that can be chosen so that, to satisfy the constraints, the rest are set. This explains the origins of the phrase 'degrees of freedom'.

Example 3

It is suggested that the lifetime T, in hours, of a certain electrical component follows a exponential distribution. 200 components are randomly selected and tested. Their lifetimes are as follows:

Lifetime (t hours)	Frequency
$t < 50$	45
$50 \leq t < 100$	35
$100 \leq t < 150$	26
$150 \leq t < 200$	18
$200 \leq t < 300$	29
$300 \leq t < 400$	21
$400 \leq t < 600$	26

Test this suggestion at a significance level of 1%.

H_0: $T \sim$ Exponential (λ)

$\bar{t} = 188.75$

$\bar{t} = \dfrac{1}{\lambda}$

$\lambda = \dfrac{1}{\bar{t}} = 0.005298$

$f(t) = \lambda e^{-\lambda t};\ t \geq 0$

$F\{t_1\} = P\{T < t_1\} = \displaystyle\int_{t=0}^{t_1} \lambda e^{-\lambda t}\, dt$

$= e^{-\lambda t}\Big|_{t_1}^{0} = 1 - e^{-\lambda t_1}$

> **Note**
> Calculate the sample mean using the data in the table $\bar{t} = \dfrac{\Sigma tf}{\Sigma f}$

> **Note**
> Using $\bar{t}$ as an estimate of the population mean, where the mean of an exponential distribution is $\frac{1}{\lambda}$ (given in the *Formulae and Statistical Tables* booklet)

λ is not given and therefore has to be estimated.

> **Note**
> Using the exponential distribution $f(x) = \lambda e^{-\lambda x}$ given in the *Formulae and Statistical Tables* booklet

> **Note**
> This is the cumulative distribution function of T. With continuous random variables, it is necessary to find the cumulative distribution function in order to calculate probabilities.

Lifetime (t hours)	u.c.b. t_1	$P\{T < t_1\}$ under H_0	Probability under H_0	Expected frequency under H_0	Frequency
$t < 50$	50	0.2327	0.2327	46.54	45
$50 \leq t < 100$	100	0.4113	0.1786	35.72	35
$100 \leq t < 150$	150	0.5483	0.1370	27.40	26
$150 \leq t < 200$	200	0.6534	0.1051	21.02	18
$200 \leq t < 300$	300	0.7960	0.1426	28.52	29
$300 \leq t < 400$	400	0.8799	0.0839	16.78	21
$t > 400$	600	1	0.1201	24.02	26

$X^2 = \displaystyle\sum_{\forall i} \dfrac{(O_i - E_i)^2}{E_i} = 1.80$ (2 dp)

Degrees of freedom: 5

Critical region: $\chi^2 > 15.086$

There is no reason to reject H_0. There is no reason to believe that the data does not come from an exponential distribution.

Using the test statistic

> **Note**
> The number of constraints is 2: the totals of O_i and E_i are equal (1) and one parameter, the mean, was estimated (2); number of degrees of freedom = number of intervals (7) − number of constraints (2)

The 99% point of a χ^2_5 distribution

Small expected frequencies

In deriving the chi-squared goodness of fit test earlier in the chapter (on page 453), you saw that the test statistic $\sum_{\forall i} \dfrac{(O_i - E_i)^2}{E_i}$ has an approximate chi-squared distribution if the null hypothesis is true; and also that the approximation gets better as the number of experiments increases. It has been shown that the approximation is good, as long as the expected frequencies are all greater than 5.

So, in cases where the expected frequencies are not all greater than 5, you need to pool the cells until all expected frequencies are greater than 5. 'Pooling the cells' simply means adding together the expected frequencies in each of the intervals with $E < 5$, until you get a combined value of $E > 5$; you also need to add together the corresponding values of O. This process creates a new interval with $E > 5$. As a result, you will have fewer intervals after pooling has taken place.

Example 4

The number of cars passing a rural traffic survey station during a 4-minute interval, was recorded for 200 intervals. Test, at a 1% significance level, the hypothesis that the cars are arriving randomly.

Number of cars x	0	1	2	3	4	5	6	7	8
Frequency	24	49	52	34	19	11	7	3	1

$\bar{x} = 2.285$

Let X be the number of cars arriving in a randomly chosen 4-minute interval.

H_0: $X \sim$ Poisson(2.285)

Under the null hypothesis,

$P\{X = x\} = \dfrac{e^{-2.285}2.285^x}{x!}$ $x = 0, 1,$

Note

Substitute x-values into the distribution function. In this case it is not possible to use cumulative Poisson distribution tables because the probabilities are not given for this mean value. The expected frequencies are found using $200 \times P\{X = x\}$.

Calculate the sample mean from the table; this value will be used as an estimate of the population mean, λ.

Note

This is because the number of random events in a given interval of time follows a Poisson distribution, mean λ.

x	0	1	2	3	4	5	6	7	≥ 8
$P\{X=x\}$	0.1018	0.2326	0.2657	0.2024	0.1156	0.0528	0.0201	0.0066	0.0024
Expected frequencies	20.36	46.52	53.14	40.48	23.12	10.56	4.02	1.32	0.48
Observed frequency	24	49	52	34	19	11	7	3	1

The last three intervals all have expected frequencies < 5 so you need to pool them. Pooling gives the interval $x \geq 6$, with $E = 5.82$ and $O = 11$. This means that you now have 7 intervals. This pooling gives the following table:

Note

Using the Poisson probability distribution function, mean 2.285; $Po(\lambda) = e^{-\lambda}\dfrac{\lambda^x}{x!}$ is in the *Formulae and Statistical Tables* booklet

x	0	1	2	3	4	5	≥ 6
$P\{X=x\}$	0.1018	0.2326	0.2657	0.2024	0.1156	0.0528	0.0291
Expected frequencies	20.36	46.52	53.14	40.48	23.12	10.56	5.82
Observed frequency	24	49	52	34	19	11	11

Note

Number of constraints is 2 (O and E totals have to be equal and the population mean is an estimate). Number of degrees of freedom = number of intervals *after pooling* (7) − number of constraints (2)

$X^2 = \displaystyle\sum_{\forall i} \dfrac{(O_i - E_i)^2}{E_i} = 7.21$ (2 dp)

Degrees of freedom: 5

Critical region: $\chi^2 > 15.086$

There is no reason to reject H_0. There is no reason to reject the hypothesis that the cars are arriving randomly.

The test statistic

The 99% point of a χ^2_5 distribution

Exercise 1

1 A tetrahedral die is thrown 160 times with the following results:

Score [x_i]	1	2	3	4
Frequency	30	28	53	49

Does this data suggest that the die is fair? You should use a 5% significance test.

2 An experiment consists of flipping a coin until a head shows. The experiment is repeated 100 times and X, the number of throws up to and including the head, is recorded. The results are as follows:

Number of heads [x]	1	2	3	4	≥ 5
Frequency	58	34	6	1	1

Test at a significance level of 5% the hypothesis that the coin is fair. You should state clearly the probability distribution of X, assuming the null hypothesis to be true.

3 The following table gives the diastolic blood pressures, in mm of mercury, of 150 women aged 18 – 55 years from a health and fitness club:

Blood pressure (x mm)	Frequency
$x < 60$	6
$60 \leq x < 70$	19
$70 \leq x < 80$	39
$80 \leq x < 90$	66
$90 \leq x < 100$	14
$100 \leq x < 110$	6

It is suggested that the diastolic blood pressure for women of this age should be normally distributed with a mean of 82 mm and standard deviation of 10 mm.

a Complete the following table of expected frequencies based upon this assumption.

Blood pressure (x mm)	Expected frequency
$x < 60$	2.09
$60 \leq x < 70$	15.18
$70 \leq x < 80$	
$80 \leq x < 90$	
$90 \leq x < 100$	
$x \geq 100$	5.37

b Test at a 1% significance level whether this data comes from a normal population with mean 82 mm, standard deviation 10 mm.

4 The number of emails Dan receives each day over a period of 60 days is shown in the following table:

Number of emails	0	1	2	3	4	5	6
Frequency	2	7	12	13	11	8	7

He believes that the number of emails per day follows a Poisson distribution

a Explain why this variable may be Poisson distributed.

b Calculate the mean number of emails Dan receives per day, giving your answer to 3 dp.

c Copy and complete the following table of expected frequencies for a Poisson distribution with mean equal to that calculated in **part b**.

Number of emails	0	1	2	3	4	5	≥ 6
Expected Frequency	2.29	7.47	12.21				6.79

d Perform a chi-squared test at a 5% significance level to test Dan's belief, explaining why the first two cells should be pooled and giving the number of degrees of freedom for the test.

33.2 Tests of no association

The ideas of the previous section on testing the fit of data to a particular distribution can be extended to investigating the association between variables. Although the problem appears to be very different, the techniques involved in their solution rely on the same theory.

Contingency tables

To develop a method for testing no association between two quantities, you first need to know how to create a contingency table. Let's consider the following scenario.

A school is in the process of re-evaluating its sports provision, and you have been asked to act as statistics consultant to advise the school on possible changes. To this end, you ask 100 A-level students to name their favourite school sport. The results of this, broken down by gender, is as follows:

Sport Gender	Football	Cricket	Swimming	Athletics	Other	Total
Male	16	7	15	11	6	55
Female	7	6	16	9	7	45
Total	23	13	31	20	13	100

This is known as a **contingency table**, where the members of a sample are classified according to two characteristics and the results tabulated.

The totals 23, 13, … and 55 and 45 are known as the **marginal totals**. For example, 23 is the total number of people that prefer football; 45 is the total number of females in the sample of 100 students. The **grand total** is 100.

Testing an association between two quantities

Given the school sports example, you are now required to answer the questions 'Do females and males differ in their take up of these sports?'. Or rather, 'Is there an association between gender and choice of sport?'.

In order to test this, it is necessary to calculate the expected frequencies based upon a null hypothesis.

H_0: there is no association between gender and chosen sport

No association between gender and sport means that these methods of classifying the students are independent of each other. If a person is chosen at random, assuming the null hypothesis to be true:

P{the person choses football *and* is male} = P{the person choses football} × P{the person is male}

These probabilities are not known, but they can be estimated from their sample proportions found from the table. If a person is chosen at random:

P{the person choses football} $\approx \frac{23}{100}$; P{the person is male} $\approx \frac{55}{100}$

Therefore, *under the null hypothesis*:

P{the person choses football *and* is male} $\approx \frac{23}{100} \times \frac{55}{100}$

Therefore, the expected number of people in this category is approximately:

$\frac{23}{100} \times \frac{55}{100} \times 100 = \frac{23 \times 55}{100}$

This example gives a procedure for finding the expected frequencies, based on the null hypothesis, for a test of no association.

> **For any given cell in a contingency table, the expected frequency is found by multiplying together the corresponding marginal totals, then dividing the result by the grand total.**

So, the expected frequencies for the 10 cells of the contingency table in the school sports example can now be calculated and are shown in brackets.

Sport Gender	Football	Cricket	Swimming	Athletics	Other	Total
Male	16 (12.65)	7 (7.15)	15 (17.05)	11 (11)	6 (7.15)	55
Female	7 (10.35)	6 (5.85)	16 (13.95)	9 (9)	7 (5.85)	45
Total	23	13	31	20	13	100

Having obtained a set of observed and expected frequencies, these can be compared using the chi-squared distribution, as discussed in the tests for goodness of fit.

$$X^2 = \sum_{\forall i} \frac{(O_i - E_i)^2}{E_i} = 2.94 \text{ (2 dp)}$$

The number of degrees of freedom for the chi-squared distribution is the number of independent terms in the chi-squared summation. For this example, the number of independent terms is: the number of expected frequencies that must be calculated, so that all the others are fixed by the marginal total constraints. If the expected frequencies in the cells shaded blue are calculated, the remaining ones are fixed; for example, if there are 12.65 males expected to choose football, the equivalent number for females must be 10.35, because the total number who chose football is 23.

> In general, if the contingency table has r rows and c columns, then the total number of free cells, v, is given by
> $$v = (r-1)(c-1)$$

For the sports problem, $r = 2$, $c = 5$ and therefore $v = 1 \times 4 = 4$

The critical region for this test for a significance level of 5% is $\chi^2_4 = 9.488$

There is no reason to reject the hypothesis that females and males have the same preference for different types of sport.

Yates' continuity correction

The chi-squared test of no association is an approximate test, where the approximation is good apart from the case of a 2×2 contingency table, where the number of degrees of freedom equals 1. In this case, Yates' continuity correction is applied to improve the accuracy.

> For a 2×2 contingency table, the test statistic for a chi-squared test of no association is
> $$X^2 = \sum_{\forall i} \frac{\left(\left|O_i - E_i\right| - \frac{1}{2}\right)^2}{E_i}$$

$\left|O_i - E_i\right|$ is the modulus of $O_i - E_i$, which means that it is the value of $O_i - E_i$ *taken as positive*; for example $\left|-4\right| = 4$

Example 5

As part of a survey into patient satisfaction on discharge from hospital, a large hospital asks each patient to rate their general satisfaction with the service provided as they leave. In a preliminary analysis of the results, satisfaction level (satisfactory or not satisfactory) is recorded against time of discharge (weekend or weekday) for 54 patients.

Time of discharge Satisfaction level	Weekday	Weekend	Total
Satisfied	27	15	42
Not satisfied	4	8	12
Total	31	23	54

Test at a significance level of 5%, if the time of discharge affects patients' level of satisfaction with the hospital.

Expected frequencies to 2 decimal places are:

satisfied and weekday: $\frac{42 \times 31}{54} = 24.11$; satisfied and weekend: $\frac{42 \times 23}{54} = 17.89$

not satisfied and weekday: $\frac{12 \times 31}{54} = 6.89$; not satisfied and weekend: $\frac{12 \times 23}{54} = 5.11$

$$X^2 = \sum_{\forall i} \frac{\left(|O_i - E_i| - \frac{1}{2} \right)^2}{E_i} = 2.50 \text{ (2 dp)}$$

$r = 2$, $c = 2$ and therefore $v = 1 \times 1 = 1$

The critical region for this test is $\chi^2_1 > 3.841$

There is no reason to reject the hypothesis that patients have the same level of satisfaction on discharge, if they leave on a weekend or a weekday.

> **Note**
>
> By calculating the sum of
>
> $$\frac{\left(|27 - 24.11| - \frac{1}{2} \right)^2}{24.11} +$$

Exercise 2

1 A total of 1000 electrical components are produced by 3 operatives: A, B and C.

	Defective	Not defective	Total
A	25	356	381
B	35	265	300
C	31	288	319
Total	91	909	1000

Are the operators equally effective in producing working components? Your test should be at a 10% significance level.

2 120 medical students were asked their intended specialism on qualification. Their results, broken down by gender, were as follows:

Speciality	Medicine	Surgery	Anesthesiology	Other
Male	25	28	5	7
Female	25	12	10	8

a Test at 5% whether there is an association between gender and specialism.

b Comment on the strength of your conclusion, and also on the expected frequencies for the female medical students.

3 90 batches of milk, all from different farms, were each tested by one of two food inspectors and each graded as pass, fail, or requires a re-test. The results of the tests were as follows:

Result	Pass	Fail	Re-test	Total
Inspector 1	27	5	10	42
Inspector 2	22	9	17	48
Total	49	14	27	90

a Copy and complete the following table of frequencies expected, on the assumption that the two inspectors are consistent with each other in their decision making.

Result	Pass	Fail	Re-test	Total
Inspector 1	22.87	6.53		42
Inspector 2	26.13			48
Total	49	14	27	90

b Test at a significance level of 5%, the assumption made in **part a**.

c Subsequently, it was found that the two inspectors were using different standards to inform their decision making. Has a type I error, a type II error, or neither, been made in the conclusion to **part b**?

4 In a drugs trial, a group of 44 healthy individuals were each offered either a course of vitamin tablets or, for comparison purposes, a course of an identical tablet with no therapeutic value (a **placebo**). At the end of the trial, the participants were asked whether they felt an improvement in their general well-being. The results were as follows:

Outcome	Vitamin tablet	Placebo	Total
Improvement	10	9	19
No improvement	13	12	25
Total	23	21	44

a Test at a significance level of 10%, whether the vitamin tablet is different from the placebo in terms of improvement rates.

b By considering the overall chi-squared value, comment on your conclusion in **part a**.

5 Two classes of students sit the same examination and each student is graded either fail, pass or distinction. The results are as follows:

Result Class	Fail	Pass	Distinction	Total
A	6	17	7	30
B	12	17	4	33
Total	18	34	11	63

The teacher of class B believes that there is no significant difference in exam performance for her students, compared to the class A students.

a Copy and complete the following table of frequencies expected, on the assumption that she is correct.

Result Class	Fail	Pass	Distinction	Total
A	8.57	16.19		30
B	9.43			33
Total	18	34	11	63

b Test at a significance level of 5%, the teacher's belief.

c Comment on the expected frequencies for class B in relation to those expected under the hypothesis of no association.

Summary

▶ When the distribution of a set of data is unknown, you can propose a distribution to model the data, and then test for the fit of the data to that distribution. This is known as a goodness of fit test.

▶ Testing goodness of fit with known parameters:

the statistic $X^2 = \sum_{i=1}^{k} \frac{(O_i - E_i)^2}{E_i}$

follows a chi-squared distribution on $k - 1$ degrees of freedom (χ^2_{k-1}) if the null hypothesis is true.

▶ Testing goodness of fit with unknown parameters:

the statistic $X^2 = \sum_{i=1}^{k} \frac{(O_i - E_i)^2}{E_i}$

follows a chi-squared distribution on $k - 1 - a$ degrees of freedm ($\chi^2_{k-1} - a$) where a is the number of parameters estimated, if the null hypothesis is true.

▶ If any of the expected frequencies in a chi-squared goodness of fit test are less than or equal to 5, pool the cells until all expected frequencies are greater than 5.

▶ A chi-squared test of no association between two quantities can be used to determine if the quantities are independent. The test is performed on data in a contingency table using the test statistic:

$X^2 = \sum_{i=1}^{k} \frac{(O_i - E_i)^2}{E_i}$

▶ For any given cell in a contingency table, the expected frequency is found by multiplying together the corresponding marginal totals, then dividing the result by the grand total.

▶ If the contingency table has r rows and c columns, then the total number of degrees of freedom, v, is given by $v = (r - 1)(c - 1)$

▶ For a 2×2 contingency table, use the Yates' continuity correction test statistic:

$X^2 = \sum_{\forall i} \frac{\left(|O_i - E_i| - \frac{1}{2}\right)^2}{E_i}$

Review exercises

1 An experiment consists of flipping a coin three times. The experiment is repeated 80 times and X, the number of heads, is recorded. The results are as follows:

Number of heads [x]	0	1	2	3
Frequency	34	32	11	3

Test at a significance level of 10%, the hypothesis that the coin is biased with the probability of obtaining a head in any given throw of 0.3. State clearly the distribution of X, assuming the null hypothesis to be true and give the number of degrees of freedom of the test.

2 The following famous data gives information about the number of deaths from horse kicks for 10 units of the Prussian Army over 20 years (that is, 200 observations) in the last few decades of the nineteenth century.

Number of deaths [x]	0	1	2	3	4
Frequency	109	65	22	3	1

Test at 0.5%, whether this data follows a Poisson distribution.

3 A road safety officer wishes to find out if cars of certain colours are more likely to be involved in road traffic accidents at particular times of the day. She investigates whether dark coloured cars are more likely to be involved in an accident at night time compared to lighter coloured cars. She obtains data (time of accident and car colour) for 200 accidents in a large city over a 4-month period and produces the following table:

Colour	Dark	Light	White	
Time of day	Daytime	Twilight	Night time	Total
	21	59	18	98
	11	24	7	42
	48	6	6	60
	80	89	31	200

a Copy and complete the following table of expected frequencies.

Colour Time of day	Dark	Light	White	Total
Daytime	39.20	43.61	15.19	98
Twilight	16.80			42
Night time	24			60
Total	80	89	31	200

b Test at 5%, whether there is an association between the time of day an accident occurs and car colour.

Practice examination questions

1 The random variable X may be modelled by an exponential distribution with probability density function:

$$f(x) = \begin{cases} \lambda e^{-\lambda x} & x \geq 0 \\ 0 & \text{otherwise} \end{cases}$$

a Show that, for $x \geq 0$, the cumulative distribution function is given by $F(x) = 1 - e^{-\lambda x}$ (2 marks)

b For an exponential distribution with $\lambda = 0.5$, find values for the two probabilities missing from the following table, giving your answers to four decimal places.

Interval	0–1	1–2	2–3	3–4	4–5	≥5
Probability	0.3935	0.2387	0.1447	0.0878		

(2 marks)

c An engineering company manufactures mechanical components for power stations.

A random sample of 80 of these components was selected. The length of time (in years) to the failure of each component was recorded. These lifetimes are shown in the table.

Lifetime (years)	0–1	1–2	2–3	3–4	4–5	≥5
Number of components	34	20	9	6	2	9

Use a χ^2 test to determine, at the 10% level of significance, whether the exponential distribution with $\lambda = 0.5$ is a suitable model for the lifetime (in years) of these mechanical components. (8 marks)

AQA MS04 June 2010

2 A cosmologist claimed that the lifetime of a certain particle, measured in picoseconds, can be modelled by the random variable T, which has cumulative distribution function:

$$F(t) \begin{cases} 0 & t < 0 \\ \dfrac{1}{2}t^3 - \dfrac{3}{16}t^4 & 0 \le t \le 2 \\ 1 & t > 2 \end{cases}$$

To test this claim, the cosmologist first divided the interval [0, 2] into five equal intervals, and then recorded into which of these intervals the lifetimes of 50 randomly selected such particles fell.

The results are shown in the table.

Interval	0–0.4	0.4–0.8	08–1.2	1.2–1.6	1.6–2.0
Number of particles	2	9	12	22	5

a Assuming that the cosmologist's claim is correct:

  i evaluate $F(t)$ for $t = 0.4, 0.8, 1.2, 1.6$ and 2.0

  ii complete the table. (4 marks)

Interval	0–0.4	0.4–0.8	08–1.2	1.2–1.6	1.6–2.0
Probability	0.0272	0.1520			

b Hence, use a χ^2 test, at the 5% level of significance, to investigate the cosmologist's claim. (9 marks)

AQA MS04 June 2011

Further Hypothesis Testing 2 – Goodness of Fit Tests and Tests of No Association 457

3 A horticulturalist was considering the germination rate of a large batch of runner-bean seeds. He selected a random sample of 500 seeds and planted them in 100 rows of 5 seeds. He then recorded the number of seeds germinating in each of the 100 rows. The results that he obtained are shown in the table.

Number of seeds germinating (x)	0	1	2	3	4	5
Number of rows (f)	25	41	20	12	2	0

a Calculate the mean number of seeds germinating per row, and hence show that an estimated value for p, the probability that a seed germinates, is 0.25. *(2 marks)*

b The model B(5, 0.25) is suggested as suitable for the number of seeds germinating per row. Calculate the expected frequencies for this model. *(4 marks)*

c Use a χ^2 goodness of fit test, at the 5% level of significance, to investigate whether the model B(5, p) is suitable for the number of seeds germinating per row. *(7 marks)*

d Does your conclusion in **part (c)** support the view that the probability that a seed germinates is the same whichever row it is planted in? Explain your answer. *(2 marks)*

AQA MS04 June 2012

4 It is claimed that a new drug is effective in the prevention of sickness in holiday-makers. A sample of 100 holiday-makers was surveyed, with the following results.

	Sickness	No sickness	Total
Drug taken	24	56	80
No drug taken	11	9	20
Total	35	65	100

Assuming that the 100 holiday-makers are a random sample, use a χ^2 test, at the 5% level of significance, to investigate the claim. *(8 marks)*

AQA MS2B June 2010

5 Emily believed that the performances of 16-year-old students in their GCSEs are associated with the schools that they attend. To investigate her belief, Emily collected data on the GCSE results for 2010 from four schools in her area.

The table shows Emily's collected data, denoted by O_i, together with the corresponding expected frequencies E_i necessary for a χ^2 test.

	≥ 5 GCSEs		1 ≤ GCSEs ≤ 5		No GCSEs	
	O_i	E_i	O_i	E_i	O_i	E_i
Jolliffe College for the Arts	187	193.15	93	90.62	30	26.23
Volpe Science Academy	175	184.43	97	86.52	24	25.05
Radok Music School	183	183.31	78	86.23	34	24.96
Bailey Language School	265	248.61	112	116.63	22	33.76

Emily used these values to correctly conduct a χ^2 test at the 1% level of significance.

a State the null hypothesis that Emily used. (1 mark)

b Find the value of the test statistic χ^2, giving your answer to one decimal place. (3 marks)

c State, in context, the conclusion that Emily should reach based on the results of her χ^2 test. (3 marks)

d Make **one** comment on the GCSE performances of 16-year-old students attending Bailey Language School. (1 mark)

e Emily's friend, Joanna, used the same data to correctly conduct a χ^2 test using the 10% level of significance. State, with justification, the conclusion that Joanna should reach. (2 marks)

AQA MS2B June 2011

6 A large multinational company recruits employees from all four countries in the UK. For a sample of 250 recruits, the **percentages** of males and females from each of the countries are shown in **Table 1**.

Table 1

	England	Scotland	Wales	Northern Ireland
Male	22.8	17.6	10.8	6.8
Female	15.6	17.2	7.6	1.6

a Add the frequencies to the contingency table, **Table 2**, below. (2 marks)

b Carry out a χ^2 test at the 10% significance level to investigate whether there is an association between country and gender of recruits. (8 marks)

c By comparing observed and expected values, make *one* comment about the distribution of *female* recruits. (1 mark)

Table 2

	England	Scotland	Wales	Northern Ireland	Total
Male					145
Female					105
Total					250

AQA MS2B June 2014

Answers

1 Loci, Graphs and Algebra

Exercise 1

1 $(x-3)^2 = -14y + 21$ **2** $(x-2)^2 + (y+3)^2 = 16$ **3** $(y-3)^2 = -14x - 21$ **4** $(x-4)^2 + (y+4)^2 = 32$

Exercise 2

1 $x=-3$ and $y=\dfrac{1}{2}$ **2** $x=-2$ and $y=0$ **3** $x=3$ and $y=2$ **4** Student's own response

5 Student's own response

6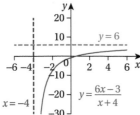

7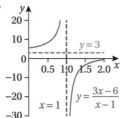

8 $x=\dfrac{5}{3}$ and $y=\dfrac{2}{3}$ 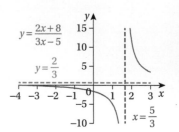

Exercise 3

1 $x=-4$ and 1 $y=0$ **2** $x=-2$ and 3 $y=0$ **3** $x=3$ and $y=1$ **4** $x=1$ and -7 $y=8$

5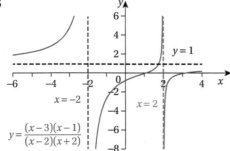

6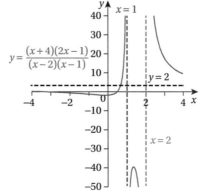

7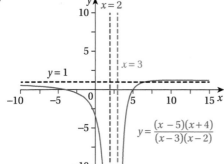

8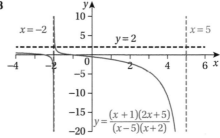

9 a $y>1, y \geq \dfrac{49}{25}$ **b** $y>1, y<\dfrac{5}{13}$ **10 a** $\max(0,1)$ $\min\left(-6, \dfrac{49}{25}\right)$ **b** $\max\left(-0.5, \dfrac{5}{13}\right)$

11 a $x>-1, x<-2$ **b** $x>3$

12 a $x<-2, -1<x<0$ **b** $x>8, 2<x<3$ **c** $5<x<\dfrac{1}{2}(3+\sqrt{65}), \dfrac{1}{2}(3-\sqrt{65})<x<-1$

13 a $1>x>-2$ **b** $x>1, x<-\dfrac{3}{2}$

Exercise 4

1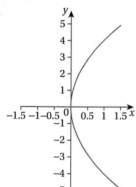

2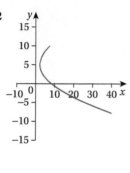

3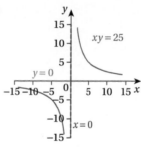

4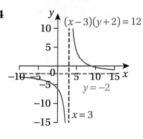

5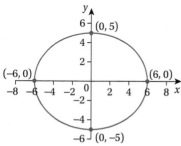

6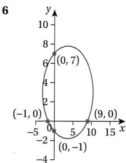

7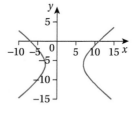

8 $(25 - 9m^2x^2) + 18m^2x - (9m^2 + 225) = 0$

9 $x = 5, y = -3$

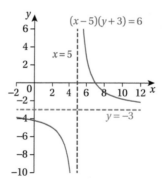

10 a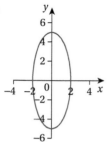

b Student's own response

c $a = -1, b = 3, c = 39$

Review exercise

1 $(y+1)^2 = 10x - 25$

2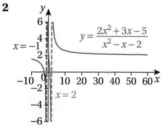

3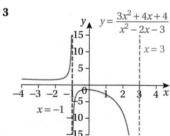

4 $-10 < x < -3$

5 $x < -2, [x \neq -1]$

6
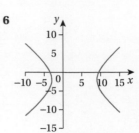

7 a Student's own response

b $10y^2 - 54 + 2ky + k^2 + 72 = 0$

Practice examination questions

1 a i $x = 3, y = 0$ **ii and iii**

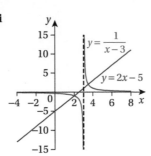

b i $2, 3.5$ **ii** $2 < x < 3, x > 3.5$

2 a i and ii

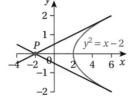

b i $m^2x^2 + 4(m^2 - 1)x + (4m^2 + x) = 0$ **ii** $16m^2 = 1$ **iii** $x = 6, y \pm 2$

3 a $a = 2, b = 4$ **b** $(m^2 - 4)x^2 - 2m^2x + (m^2 + 16) = 0$ **c** $3m^2 = 16$ **d** $y = \pm 4\sqrt{3}$

4 a $y = 1, x = -1, x = 3$ **b i** $(k - 1)x^2 - 2(k - 1)x - (3k + 1) = 0$ **ii** $k^2 - k >= 0$ **iii** $(1, 0)$

c

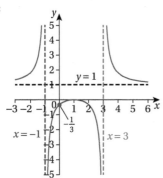

5 a $(2, 0), (6, 0)$ **b i** $(1 + 4m^2)x^2 - 8x + 12 = 0$ **ii** $\dfrac{1}{\sqrt{12}}$ **iii** $\left(3, \dfrac{3}{\sqrt{12}}\right)$

6 a $x = -1, x = 2, y = 0$ **b** $x = 1, -2$ **c**   **d** $-2 \leq x < -1, 1 \leq x < 2$

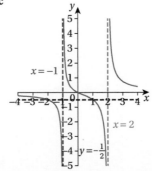

2 Complex Numbers

Exercise 1

1 **a** $-i$ **b** 1 **c** -1

2 **a** $3+2i$ **b** $6-3i$ **c** $-4+3i$ **d** $2i$

3 **a** $3-4i$ **b** $2+6i$ **c** $-4+3i$

4 **a** $-2\pm\sqrt{3}i$ **b** $-1\pm\sqrt{5}i$

5 **a** $-1\pm\sqrt{3}i$ **b** $\frac{3}{2}\pm\frac{\sqrt{15}}{2}i$ **c** $\frac{1}{4}(-1\pm\sqrt{7}i)$

Exercise 2

1 **a** $10-2i$ **b** $1-i$ **c** $-1+2i$

2 **a** $3+11i$ **b** $26+2i$ **c** $10+11i$

3 **a** $\frac{5}{17}+\frac{14}{17}i$ **b** $\frac{23}{26}+\frac{11}{26}i$

4 **a** $x=4, y=-2$ **b** $x=-11, y=22$ **c** $x=\frac{13}{5}, y=\frac{9}{5}$ **5** $\frac{28}{13}+\left(\frac{36}{13}\right)i$

6 **a** $z=3+6i$ **b** $z=-\frac{7}{2}+\frac{i}{4}$ **c** $z=1+\frac{3i}{5}$ **d** $z=-\frac{4}{3}+\frac{8i}{9}$

Exercise 3

1 **a**
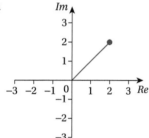

b
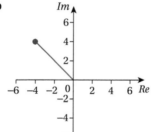

c
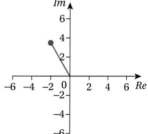

d
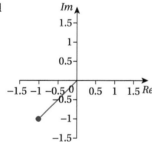

e
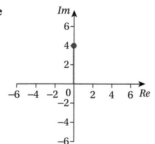

f
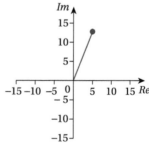

g
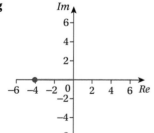

h
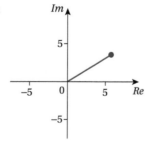

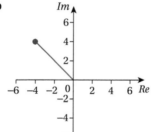

2 a $2\sqrt{2}; \dfrac{\pi}{4}$

b $3\sqrt{2}; \dfrac{3\pi}{4}$

c $4; \dfrac{2\pi}{3}$

d $\sqrt{2}; -\dfrac{3\pi}{4}$

e $4; \dfrac{\pi}{2}$

f $13; \tan^{-1}\dfrac{12}{5}$

g $4; \pi$

h $7; \tan^{-1}\dfrac{\sqrt{13}}{6}$

3 a i $-7+41i$

    ii $-117+44i$

b i 5

  ii 25

    iii 125

c i 0.9273

  ii 1.8546

    iii 2.7819

4 a $1+\sqrt{3}i$

b $2\sqrt{2}+2\sqrt{2}i$

5 $|z| = \sqrt{\dfrac{53}{5}};$ $\arg z = -\tan^{-1}\dfrac{22}{29}$

Exercise 4

1 a
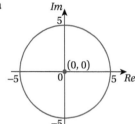

b
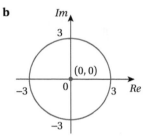

c
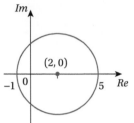

d
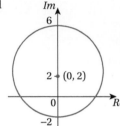

2 a
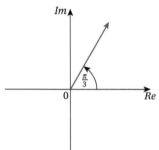

b
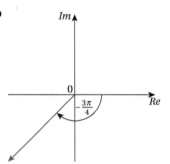

c
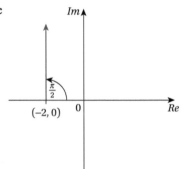

3 a
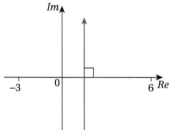

b
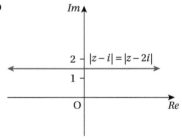

c
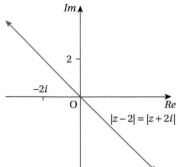

d
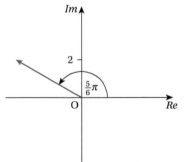

e
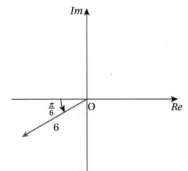

Review exercise

1 a $1+i$ **b** $26+2i$ **c** $10+11i$

2 a $1-2i$ **b** $\dfrac{4}{13}-\dfrac{19}{13}i$ **3** $3.3+0.9i$

4 a $-\dfrac{3}{2}\pm\dfrac{3i}{2}$ **b** $\dfrac{5}{2}\pm\dfrac{5\sqrt{3}i}{2}$

5 a 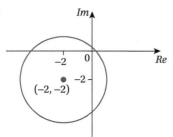   **b** 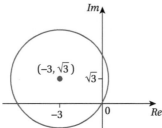   **c**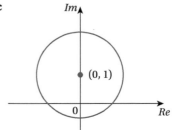

6 a 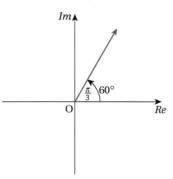   **b** 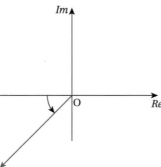   **c**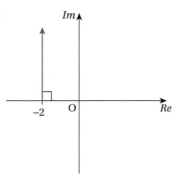

7 a 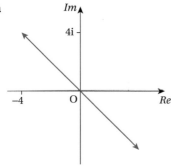   **b** 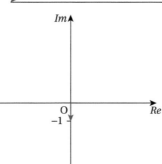   **c**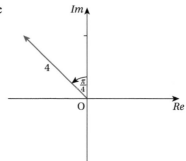

8 $1.077,\ -\pi+\tan^{-1}\dfrac{23}{14}$

Practice examination questions

1 $\dfrac{11}{2}+\dfrac{13}{2}i$ **2 a** $-3\pm5i$ **b i** $-3+i$ **ii** $m=-1\ n=2$

3 a $3x-y+i(x-3y+4)$ **b** $z=\dfrac{1}{2}+\dfrac{3i}{2}$ **4 a** $(x^2+y^2-1)-2ix$ **b** $z=4\pm3i$

5 a Student's own response   **b** 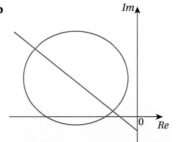   **c** $-10+9i$ **6**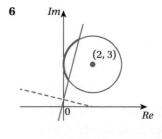

3 Roots and Coefficients of a Quadratic Equation

Exercise 1

1 a $-3, -7$ **b** $11, 5$ **c** $-5, -4$ **d** $-\dfrac{11}{3}, \dfrac{2}{3}$

2 a $x^2 - 7x + 15 = 0$ **b** $x^2 + 3x + 5 = 0$ **c** $x^2 + 2x - 4 = 0$ **d** $x^2 + 5x - 11 = 0$

3 a sum $= -2$, product $= -5$ **b** $-2, -\dfrac{7}{2}$ **c** sum $= 7$ product $= 4$

d sum $= \dfrac{2}{3}$ product $= 5$ **4** Student's own response

Exercise 2

1 $16x^2 - 89x + 25 = 0$ **2 a** $-\dfrac{7}{5}, -\dfrac{12}{5}$ **b** $\alpha^2 + \beta^2 = \dfrac{169}{25}$ **3** $6x^2 - 5x - 3 = 0$

4 a 5 **b** 19 **c** 80

5 a $\alpha + \beta = 4, \alpha\beta = 9$ **b** $81x^2 + 164x + 6368 = 0$ **6 a** $\alpha + \beta = \dfrac{9}{4}, \alpha\beta = 2$ **b** $16x^2 - 53x + 168 = 0$

Review exercise

1 a $-3, -1$ **b** $-\dfrac{2}{3}, -4$ **2 a** $-\dfrac{7}{2}, 4$ **b** Student's own response

3 $21x^2 + 12x - 1$ **4 a** $\dfrac{4}{3}, \dfrac{1}{2}$ **b** Student's own response **c** $108x^2 - 228x + 121 = 0$

Practice examination questions

1 a $-\dfrac{3}{2}, -3$ **b** $\alpha^3 + \beta^3 = \dfrac{135}{8}$ **c** $24x^2 + 81x - 146 = 0$

2 a $\dfrac{7}{5}, \dfrac{1}{5}$ **b** $\dfrac{\alpha}{\beta} + \dfrac{\beta}{\alpha} = \dfrac{39}{5}$ **c** $5x^2 - 42x + 65 = 0$

3 a $-\dfrac{7}{2}, 4$ **b** $\alpha^2 + \beta^2 = \dfrac{17}{4}$ **c** $64x^2 - 17x + 4 = 0$

4 a $-\dfrac{3}{2}, \dfrac{3}{4}$ **b** $\alpha^2 + \beta^2 = \dfrac{3}{4}$ **c** $4x^2 + 12x + 21 = 0$

5 a $6, 18$ **b** $x^2 + 324 = 0$ **c** $\pm 18i$

4 Series

Exercise 1

1 $\dfrac{2n}{3}(n+1)(n+2)$ **2** $\dfrac{n}{2}(n+1)(n^2+n+1)$ **3** $\dfrac{n}{3}(n^2-7)$ **4** 8235

Exercise 2

1 $\dfrac{n}{3}(n+1)(n+2)$ **2** $\dfrac{n}{4}(n+1)(n+2)(n+3)$ **3** $\dfrac{x(1-x^{2n})}{(1-x^2)^2} - \dfrac{nx^{2n+1}}{1-x^2}$ **4** $a = 8. \dfrac{n}{2}(n+1)$ **5** 3

Review exercise

1 21265 **2** $\dfrac{n}{6}(4n^2 - 27n - 1)$ **3** $(n+1)! - 1$

Practice examination questions

1 2906061

2 a sum from $r = 1$ to n, $r^2(4r - 3) = kn(n+1)(2n^2 - 1)$ **b** 2486190

3 a $\dfrac{n}{4}(n+1)(n^2+n+2)$ **b** $n = 10$

4 a $f(r) - f(r-1) = (2r-1)^3$ **b** sum of series $r = n+1$ to $2n$, $(2r-1)^3 = 3n^2(10n^2 - 1)$

5 a $f(r) - f(r-1) = r^3$ **b** sum of series $r = n$ to $2n$, $r^3 = \dfrac{3}{4}n^2(n+1)(5n+1)$

6 a $f(r+1) - f(r) = r(3r+1)$ **b** 867500

5 Trigonometry

Exercise 1

1 a i $n\pi+(-1)^n\dfrac{\pi}{4}$ **ii** $180n° + ((-1)^n 45°)$ **b** 45, 135

2 a i $2n\pi\pm\dfrac{2\pi}{3}$ **ii** $360n° \pm 120°$ **b** 120, 240

3 a i $\dfrac{n\pi}{2}+(-1)^n\dfrac{\pi}{12}$ **ii** $90n° + (-1)^n 15°$ **b** 15, 75, 195, 255

4 a i $\dfrac{n\pi}{3}+\dfrac{\pi}{12}$ **ii** $60n° + 15°$ **b** 15, 75, 135, 195, 255, 315

5 $\pi n+\dfrac{\pi}{6}$

Exercise 2

1 $\dfrac{\pi n}{2}+(-1)^n\dfrac{\pi}{4}-\dfrac{\pi}{8}$ **2** $(360°n\pm 135°)-30°$

3 a $\dfrac{2\pi n}{3}+\dfrac{\pi}{9}\pm\dfrac{1}{9\pi}$ **b** $x=0,\dfrac{2\pi}{9},\dfrac{2\pi}{3},\dfrac{8\pi}{9},\dfrac{4\pi}{3},\dfrac{14\pi}{9},2\pi,\dfrac{20\pi}{9},\dfrac{8\pi}{3},\dfrac{26\pi}{9},\dfrac{10\pi}{3},\dfrac{32\pi}{9},4\pi$

4 $n\pi+\dfrac{-\pi}{2}(-1)^n$

5 a i $n\dfrac{\pi}{2}+\dfrac{\pi}{24}$ **ii** $90n° + 7.5°$ **b i** $\dfrac{2n\pi}{3}+0.591+\dfrac{1}{3}$ **ii** $120n° \pm 33.85+\dfrac{1}{3}$

  c i $n\dfrac{\pi}{4}+\dfrac{\pi}{16}\quad\dfrac{\pi n}{4}+(-1)^n\dfrac{\pi}{16}$ **ii** $45n+(-1)^n 11.25$ **d i** $\dfrac{2n\pi}{3}+\dfrac{\pi}{3}$ **ii** $120n° \pm 60°$

  e i $n\pi+\dfrac{\pi}{2},2n\pi\pm\dfrac{\pi}{3}$ **ii** $180n° + 90°, 360n° \pm 60°$

6 $\dfrac{n\pi}{2}-\dfrac{\pi}{24}$

Review Exercises

1 $\dfrac{60n+(-1)^n 50}{3}-\dfrac{20}{3}$ **2** $\dfrac{\pi n}{4}+(-1)^n\dfrac{\pi}{24}+\dfrac{\pi}{24}$ **3** $\dfrac{\pi n}{2}\pm\dfrac{\pi}{12}+\dfrac{\pi}{6}$

4 $6\pi n\pm\dfrac{\pi}{2}-\dfrac{\pi}{2}$ **5** $\dfrac{\pi}{6}+\dfrac{n\pi}{3}$

Practice examination questions

1 a $n\pi+\dfrac{\pi}{24},n\pi+\dfrac{5\pi}{24}$ **b** $\dfrac{125\pi}{64}$ **2** $\dfrac{n\pi}{2}+\dfrac{\pi}{16}$

3 a $\dfrac{24n\pi+7\pi}{15},\dfrac{24n\pi+\pi}{15}$ **b** $\dfrac{3848\pi}{15}$ **4** $\dfrac{\pi}{8}+\dfrac{1}{2}n\pi$ or $\dfrac{-\pi}{24}+\dfrac{1}{2}n\pi$

5 a $2n\pi+\dfrac{5\pi}{6}$ **b** $2n\pi+\dfrac{5\pi}{6},2n\pi+\dfrac{\pi}{6}$

6 Calculus

Exercise 1

1 a h   **b** 0   **2** $6x^2+6xh+2h^2+6x-2$; 70
3 a $8+h$ **b** let h tend to zero; 8
4 a student's own response   **b** let h tend to zero, gradient is zero; hence P is a stationary point.

Exercise 2

1 8.03   **2** 2.99   **3** 450   **4** 3.001

5 $\dfrac{3}{16\pi}$ cm/s   **6** 2 cm²/s   **7** 2.4 cm²/s   **8** 5.75 cm³

Exercise 3

1 $\dfrac{3}{2}$ **2** Does not exist   **3** Does not exist   **4** Does not exist

Review exercise

1 a $11 + 2h$ **b** 11   **2** $2x + h$; 6   **3** $\dfrac{8}{3}$ cm²/s

4 $\dfrac{4}{12\pi}$ **5** student's own response

Practice examination questions

1 a $-7 + h$ **b** let h tend to zero; -7

2 a $125 + 75h + 15h^2 + h^3$ **b i** $65 + 14h + h^2$ **ii** 65

3 a Student's own response   **b** gradient is zero

4 $\dfrac{2}{5}$ **5 a** no finite value   **b** $\dfrac{3}{2}$

6 a $x^{\frac{-1}{2}}$ tends to infinity as x tends to zero   **b i** $\dfrac{1}{2}$ **ii** no finite value

7 a $V = \dfrac{3\pi r^2}{10^6}$ **b** $\dfrac{1000}{3\pi}$ metres per minute

7 Matrices and Transformations

Exercise 1

1 $\begin{pmatrix} 7 & -1 \\ 13 & 14 \end{pmatrix}$ **2 a** $\begin{pmatrix} 14 & 0 \\ 8 & 5 \end{pmatrix}, \begin{pmatrix} 2 & -2 \\ 18 & 17 \end{pmatrix}$ **b** PQ $\neq$ QP The product of matrices depends on which one comes first.

3 a $\begin{pmatrix} 1-k & -k \\ -1 & -k \end{pmatrix}$ $k = 0, 2$ **b** $\begin{pmatrix} 1-k & -k \\ -1 & -k \end{pmatrix}$ **4** $\begin{pmatrix} 7 & 2 \\ 3 & 6 \end{pmatrix}$ $n = 6$

5 a $\begin{pmatrix} 0 & -9 \\ 9 & 0 \end{pmatrix}$ **b** $\begin{pmatrix} 9 & 0 \\ 0 & 9 \end{pmatrix}$ **c** $\begin{pmatrix} -81 & 0 \\ 0 & -81 \end{pmatrix}$ is not equal to $\begin{pmatrix} 81 & 0 \\ 0 & 81 \end{pmatrix}$

6 a $\begin{pmatrix} 0 & -9 \\ 9 & 0 \end{pmatrix}$ **b** $\begin{pmatrix} 9 & 0 \\ 0 & 9 \end{pmatrix}$ **c** Student's own response

7 a $\begin{pmatrix} 4 & 0 \\ 0 & 4 \end{pmatrix}$ **b** $\begin{pmatrix} 16 & 0 \\ 0 & 16 \end{pmatrix}$

Exercise 2

1 $\begin{pmatrix} 4 & 0 \\ 0 & 4 \end{pmatrix}$ and $\begin{pmatrix} 16 & 0 \\ 0 & 16 \end{pmatrix}$ **2 a** $\dfrac{1}{15-4k}\begin{pmatrix} 5 & -k \\ -4 & 3 \end{pmatrix}$ **b** $\dfrac{1}{15-2l}\begin{pmatrix} 5 & -2 \\ -l & 3 \end{pmatrix}$

3 a $\begin{pmatrix} \dfrac{1}{2} & \dfrac{\sqrt{3}}{2} \\ -\dfrac{\sqrt{3}}{2} & \dfrac{1}{2} \end{pmatrix}$ **b** $\begin{pmatrix} 5 & 0 \\ 0 & 1 \end{pmatrix}$ **c** $\begin{pmatrix} 3 & 0 \\ 0 & 3 \end{pmatrix}$ **d** $\begin{pmatrix} \dfrac{5}{2} & \dfrac{5\sqrt{3}}{2} \\ -\dfrac{\sqrt{3}}{2} & \dfrac{1}{2} \end{pmatrix}$

4 a $p = 20$ **b** $\sqrt{20}$

5 a $p = 16$ **b** scale factor 4, $y = \tan 15°x$

1 $\begin{pmatrix} -\dfrac{3}{5} & -\dfrac{1}{5} \\ -\dfrac{7}{5} & -\dfrac{4}{5} \end{pmatrix}$

2 $\begin{pmatrix} 1 & 0 \\ 0 & 1 \end{pmatrix}, \begin{pmatrix} 1 & 0 \\ 0 & 1 \end{pmatrix}$

3 a $\begin{pmatrix} -\dfrac{1}{2} & -\dfrac{\sqrt{3}}{2} \\ \dfrac{\sqrt{3}}{2} & -\dfrac{1}{2} \end{pmatrix}$ **b** $\begin{pmatrix} 6 & 0 \\ 0 & 1 \end{pmatrix}$ **c** $\begin{pmatrix} 4 & 0 \\ 0 & 4 \end{pmatrix}$ **d** $\begin{pmatrix} -12 & -12(3^{0.5}) \\ 2(3^{0.5}) & -2 \end{pmatrix}$

4 a $(4, 0)$ and $(0, 4)$ **b** Student's own response **5 a** $(0, 0)$ **b** none

Practice examination questions

1 a i $\begin{pmatrix} p-3 & 1 \\ 2 & p-3 \end{pmatrix}$ **ii** $\begin{pmatrix} 3p+4 & p+6 \\ 12+2p & 4+3p \end{pmatrix}$ **b** $p=-7, k=-27$

2 a i $\begin{pmatrix} 0 & -1 \\ -1 & 0 \end{pmatrix}$ **ii** $\begin{pmatrix} 1 & 0 \\ 0 & 7 \end{pmatrix}$ **b** $\begin{pmatrix} 0 & -1 \\ -7 & 0 \end{pmatrix}$

   c i 6 **ii** $2\sqrt{3}, y = \tan 105° \, x$

3 a $\begin{pmatrix} 0 & -4 \\ 4 & 0 \end{pmatrix}$ **b** 4 **c** Student's own response

4 a i $\begin{pmatrix} 0 & 2k \\ 2k & 0 \end{pmatrix}$ **ii** $\begin{pmatrix} 2k^2 & 0 \\ 0 & 2k^2 \end{pmatrix}$ **b** $\begin{pmatrix} 4k^2 & 0 \\ 0 & 4k^2 \end{pmatrix}$

   c i $\dfrac{1}{15-4k}\begin{pmatrix} 5 & -k \\ -4 & 3 \end{pmatrix}$ **ii** scale factor $\sqrt{2}$ line is $y = x \tan 22.5°$

5 a i a $\begin{pmatrix} \dfrac{\sqrt{3}}{2} & -\dfrac{1}{2} \\ \dfrac{1}{2} & \dfrac{\sqrt{3}}{2} \end{pmatrix}$ **ii** $\begin{pmatrix} \dfrac{1}{2} & \dfrac{\sqrt{3}}{2} \\ \dfrac{\sqrt{3}}{2} & -\dfrac{1}{2} \end{pmatrix}$

   b scale factor 2 line $y = \dfrac{1}{\sqrt{3}} x$ **c** $\begin{pmatrix} 0 & 4 \\ 4 & 0 \end{pmatrix}$ scale factror 4 reflection in $y = x$

8 Linear Graphs

Exercise 1

1 a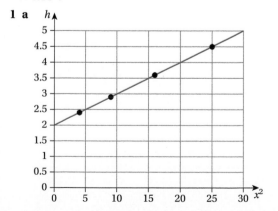

b $h = 10x^2 + 2$

2 a $a = 2$ $b = 5$ **b** $x = (10^{0.5})$

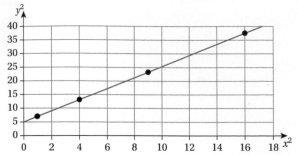

3 a Student's own response   **b** $a = 3$ $b = 2$; $Y = 3X + 2$
4 a Student's own response   **b** $a = 2.2$, $b = -0.86$, $Y = 2.2X - 0.86$

Exercise 2

1 b $= 2$ $n = 2$ $y = 2x^2$ **2** $k = 3$ $b = 1.8$ $y = 3 \times 1.8^x$
3 a Student's own response   **b** $a = 8.5$, $b = 1.7$ $y = 8.5 \times 1.7^x$
4 a Student's own response   **b** $a = 5$ $n = 3$ $y = 5x^3$

Review exercise

1 a $a = 100$ $b = 36$ **b** $a = 100$ $b = 36$ $y = 100x^2 + 36$

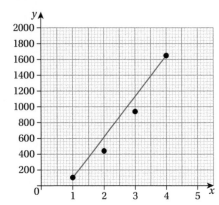

2 $a = 7$ $b = 3$ $y^2 = 7x^5 + 3x^3$

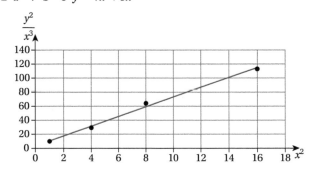

3 $a = 2$ $n = 4$ $y = 2x^4$

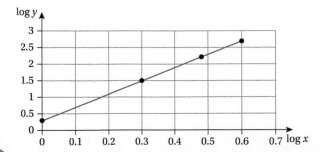

4 a

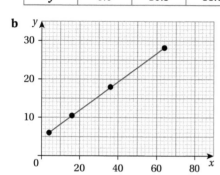

b $a = 4$ $b = 1.1$

Practice examination questions

1 a

x	2	4	6	8
X	4	16	36	64
y	6.0	10.5	18.0	28.2

b

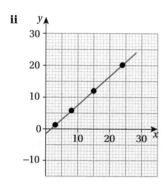

c i 5.3   **ii** $a = 0.37$ $b = 4.5$

2 a $Y = \log(a) + nX$ (which is a linear relationship between X and Y).

   b $a = -\dfrac{2}{3}$ $n = 10\,000$

3 a Student's own response   **b i** 12.6   **ii** 1.1

4 a Student's own response   **b i**

x	1	2	3	4
y	0.40	1.43	2.40	3.35
X	3	8	15	24
Y	1.20	5.72	12	20.1

   ii

   iii $a = 0.9$ $b = -1.5$

9 Numerical Methods

Exercise 1

1 1.12   **2** 0.87   **3** 2.46   **4** 84
5 1.41   **6** 6.54

Exercise 2

1 1.5   **2** 1.8   **3** 8.65   **4** 2.0465 radians

Review exercise

1 3.4899   **2** 4.4576   **3** 0.347296

Practice examination questions

1 2.3464   **2** 9.859   **3** 46.1
4 a Sign change   **b** 1.7   **5** 1.40396
6 a change of sign   **b** between 0.175 and 0.2   **c** 0.1934   **7** 4.04014

10 Bayes' Theorem

Exercise 1

1

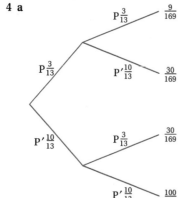

2

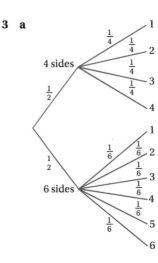

3 a   **b** $\dfrac{5}{24}$

4 a   **b** $\dfrac{60}{169}$

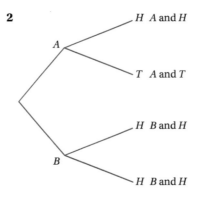

5

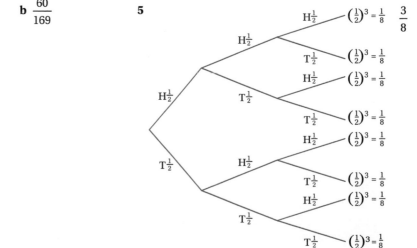

6

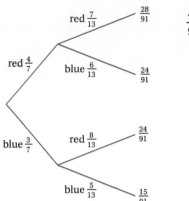

$\frac{48}{91}$

b i 0.257　　　**ii** 0.085　　　**iii** 0.056

7 a

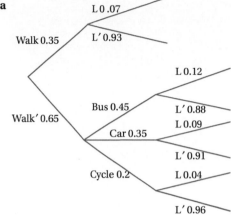

8

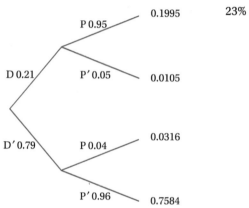

23%

Exercise 2

1 $\dfrac{3}{11}$　　　　　　**2 a** 0.7　　　　　　**b** 0.125

3 a 0.6　　　　**b** 0.29　　　　**c** 0.71　　　　　**d** 0.42

4 $\dfrac{42}{167}$　　　　　**5 a** $\dfrac{2}{3}$　　　　**b** $\dfrac{1}{2}$

6 Student's own response.　　　**7 a** $\dfrac{13}{27}$　　　**b** $\dfrac{4}{7}$

8 a 0.23　　　　**b** 17 (2 dp)　　　**9** 0.14 (2 dp)　　　**10** $\dfrac{4}{11}$

Review exercise

1 a

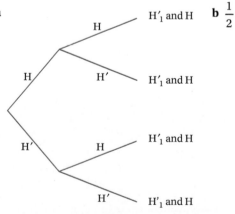

b $\dfrac{1}{2}$

2

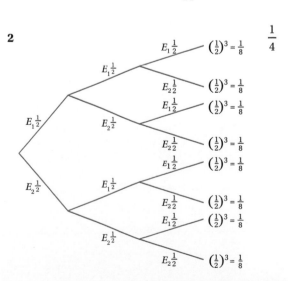

$\dfrac{1}{4}$

3

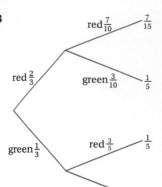

red $\frac{2}{3}$
- red $\frac{7}{10}$ → $\frac{7}{15}$
- green $\frac{3}{10}$ → $\frac{1}{5}$

green $\frac{1}{3}$
- red $\frac{3}{5}$ → $\frac{1}{5}$
- green $\frac{2}{5}$ → $\frac{2}{15}$

a $\dfrac{2}{15}$ **b** $\dfrac{1}{5}$ **c** $\dfrac{2}{5}$

4 $\dfrac{20}{41}$ **5 a** 0.6   **b** 0.44 (2 dp)   **c** 0.56 (2 dp)   **d** 0.54

Practice examination questions

1 a

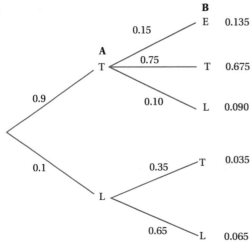

```
                                    B
                              E    0.135
                    0.15
              A
              T          0.75  T   0.675
    0.9
                         0.10  L   0.090

    0.1
                              T    0.035
                    0.35
              L
                         0.65  L   0.065
```

b i 0.85 (2 dp)   **ii** 0.95 (2 dp)   **iii** 0.04 (2 dp)
c 0.487 (3 dp)

2 a $\dfrac{191}{300}$

b $\dfrac{58}{191}$

3 a

```
                              A+ 0.09
                    0.90
              S       A− 0.002
                 0.02              + 0.00784
                 0.08            0.98
                      B
    0.10
                              0.02
                                   − 0.00016

    M

    0.90
                              A+  0.009
                    0.01
              NS      A− 0.72
                 0.80              + 0.00171
                 0.19            0.01
                      B
                              0.99
                                   − 0.16929
```

b i A 0.00216   **B** 0.01071
   ii 130
c i 0.901 to 0.902   **ii** 0.997 to 0.998

4 a

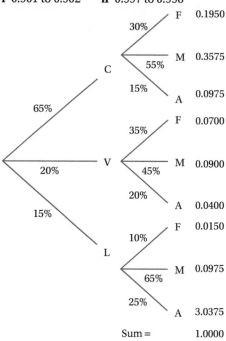

F	0.1950
M	0.3575
A	0.0975
F	0.0700
M	0.0900
A	0.0400
F	0.0150
M	0.0975
A	3.0375
Sum =	1.0000

b i 0.455   **ii** 0.214   **iii** 0.757
c 0.056

5 a

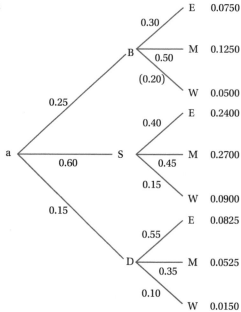

E	0.0750
M	0.1250
W	0.0500
E	0.2400
M	0.2700
W	0.0900
E	0.0825
M	0.0525
W	0.0150

b i $\dfrac{159}{400}$

   ii $\dfrac{11}{53}$

c 0.0955 to 0.0975

11 Discrete Uniform and Geometric Distributions

Exercise 1

1 Mean = 2.5, variance = 1.25   **2** Mean = 5, variance = 8

3 a discrete uniform on integers 2, 4,..., 20 each with probability $\dfrac{1}{10}$; mean = 11, and variance = 33

  b discrete uniform on integers 0 – 9 each with probability $\dfrac{1}{10}$; mean = $\dfrac{9}{2}$, variance = $\dfrac{33}{4}$

4 a Y **b** $p = \dfrac{1}{10}$

5 $\dfrac{13}{32}$ **6** $\dfrac{5}{32}$ **7** $\dfrac{151}{1440}$ **8** $\dfrac{3}{5}$

9 Student's own response.   **10 a** 10.5   **b** 10.74 (2 dp)

11 25 tokens   **12 a** $\dfrac{4n+1}{2}$ **b** Student's own response   **c** $\dfrac{9}{16}$

Exercise 2

1 a $\dfrac{1}{16}$ **b** $\dfrac{1}{8}$ **2 i** 0.09   **ii** 0.000009   **iii** 0.99   **iv** 0.1

3 Mean = 2, variance = 2   **4** Mean = 2, variance = 2   **5** 56   **6** Student's own response.

7 a $P\{Y = y\} = (1-p)^y p, \; y = 0, 1, 2, 3,.... \; \dfrac{1-p}{p}$ **b** Student's own response.

8 0.06 (2 dp)   **9** $P(S) = 1$

Review exercise

1 $\dfrac{5}{18}$ **2** $\dfrac{21}{50}$ **3** $\dfrac{1}{4}, \dfrac{3}{4}$ **4 a** 0.2   **b** 0.64

Practice examination questions

1 Student's own response.   **2 a** Student's own response   **b** Student's own response   **c** $\dfrac{1}{4}$

3 a Student's own response.   **b i** 0.36 (2 dp)   **ii** 0.07 (2 dp)

4 a i Student's own response   **ii** Student's own response   **b i** Student's own response   **ii** 17

5 £1.84   **6 a i** Student's own response   **ii** Student's own response

  b i $\dfrac{9}{16}$ **ii** $E(Y) = \dfrac{7}{3}$ $Var(Y) = \dfrac{4}{9}$

7 a Student's own response   **b i** 0.512   **ii** 42

12 Probability and Generating Functions

Exercise 1

1 $G_R(t) = 0.4t + 0.4\,t^2 + 0.2t^3$ **2** $k = \dfrac{1}{14}$; $G_X(t) = \dfrac{1}{14}t + \dfrac{2}{7}t^2 + \dfrac{9}{14}t^3$ **3** $G_X(t) = 0.5t + 0.3t^2 + 0.2t^4$

4 a $\dfrac{1}{30}$ **b** $\dfrac{1}{30}t + \dfrac{2}{15}t^2 + \dfrac{3}{10}t^3 + \dfrac{8}{15}t^4$ **c** mean = $\dfrac{10}{3}$, variance = $\dfrac{31}{45}$

5 $G_R(t) = \dfrac{1}{k}t + \dfrac{2}{k}t^2 + \dfrac{3}{k}t^3 + \dfrac{4}{k}t^4$. $\mu = 3$, $\sigma^2 = 1$ **6** $G_X(t) = \dfrac{81}{125} + \dfrac{44}{125}t^5$ $\dfrac{44}{25}$

7 a 2   **b** $G_R(t) = \dfrac{1}{4}t + \dfrac{1}{3}t^2 + \dfrac{5}{12}t^3$ **c** mean = $\dfrac{13}{6}$, variance = $\dfrac{23}{36}$

8 a $\dfrac{1}{10}$ **b** $P(R=1) = \dfrac{1}{5}$, $P(R=2) = \dfrac{3}{10}$, $P(R=4) = \dfrac{1}{2}$ **c** 2.8

9 a $a = 13$, $b = 9$ $\dfrac{66}{169}$

Exercise 2

1 a Discrete uniform on integers 1 - 10 with probabilities each $\frac{1}{10}$ **b** Bernoulli $(p = 0.6)$

  c Geometric $(p = 0.3)$ **d** Binomial $(n = 13\ p = 0.9)$

2 a $(0.6 + 0.4t)^7$ **b** $0.1t\left\{\frac{1}{1 - 0.9t}\right\}$ **c** $0.6 + 0.4t$

3 Student's own response.   **4** Student's own response.

5 $G_X(t) = \frac{t}{6} + \frac{t^3}{6} + \frac{t^5}{6} + \frac{t^7}{6} + \frac{t^9}{6} + \frac{t^{11}}{6}$ mean = 6, variance = $\frac{35}{3}$ **6** Student's own response.

7 a $\frac{2}{3} + \frac{t}{3}$ **b** mean = $\frac{1}{3}$, variance = $\frac{2}{9}$ **c** $\frac{1}{3} + \frac{2}{3}t$ mean = $\frac{2}{3}$, variance = $\frac{2}{9}$ **d** Equal because $X_R = 1 - X_{R'}$

8 a Student's own response.   **b** $E[R] = \frac{5}{2}$, $Var[R] = \frac{5}{4}$

Exercise 3

1 $\frac{0.64t^2}{(1 - 0.2t)^2}$ **2** $e^{-7(1-t)}$ Poisson, mean 7   **3** $\frac{t^2(1 - t^6)(1 - t^8)}{48(1 - t)^2}$ **4** $(0.375 + 0.5t + 0.125t^2)^n$

5 $G_T(t) = (0.6 + 0.4t)^3$; $\mu = 1.2$; $\sigma^2 = 0.72$

6 $\mu = np$; $\sigma^2 = np(1 - p)$. T has a binomial distribution, mean np, variance np(1–p), as expected since T is the number of successes in n Bernoulli trials.

7 Student's own response.   **8** $(0.375 + 0.5t + 0.125t^2)^3$; mean = 2.25

9 $G_X(t) = E[t^X] = \frac{1}{2}t + \frac{1}{4}t^2 + \frac{1}{4}t^4$; $E[T] = 6$

Review exercise

1 $0.4 + 0.2t^2 + 0.4t^5$. 2.4, 5.04   **a** Discrete uniform distribution, $p = \frac{1}{10}$ **b** $\mu = \frac{1}{10}\sum_{i=1}^{10} u_i, \sigma^2 = \frac{1}{10}\sum_{i=1}^{10} u_i^2 - \left\{\frac{1}{10}\sum_{i=1}^{10} u_i\right\}^2$

3 mean = $\frac{7}{2}$ variance = $\frac{35}{12}$ **4 a** Bernoulli $(p = 0.8)$ **b** mean = 0.8, variance = 0.16

5 a Geometric $(p = 0.2)$ **b** $G_X(x) = \frac{(0.2t)}{(1 - 0.8t)}$; $E[X] = 5$

6 a Binomial $\left(5, \frac{3}{5}\right)$ **b** mean = 3, variance = $\frac{6}{5}$ **7** $\frac{5}{16}$

Practice examination questions

1 a $G_B(t) = 1 - p + pt$ **b** a = 2; $P(S = 1) = \frac{1}{5}$, $P(S = 3) = \frac{2}{5}$, $P(S = 5) = \frac{2}{5}$

2 $G_X(t) = 0.3t + 0.3t^3 + 0.4t^4$ mean = 2.8, variance = 1.56   **3** $G_X(t) = \frac{t^1}{12} + \frac{t^2}{12} + \ldots + \frac{t^{12}}{12}$ mean = $\frac{13}{2}$, variance = $\frac{143}{12}$

4 $\frac{t}{3 - 2t}$ **5** Student's own response.

6 $\frac{0.21t^2}{1 - 1.05t + 0.26t^2}$

13 Linear Combinations of Discrete Random Variables

Exercise 1

1 2, 59   **2 a** 58.5   **b** 0.85   **c** 18.1, 5.1

3 Student's own response.   **4 a** −0.42   **b** −0.40

5 Student's own response.   **6 a** 23, 20   **b** 25, 5.8

7 18, 1   **8** 8a, 1. $p = 1$ because of linear relationship between X and Y.

9 a 0.29   **b** 0.42   **c** Bus. Cycling and lateness have positive correlation.

Exercise 2

1 a 155.4   **b i** 28.4, 20.8   **ii** −48.2, 138.2

2 a mean = 12, variance = 30   **b** Mean = 124, Variance = 165   **c** Mean = 104, Variance = 606

3 i 36, 18 **ii** 12, 2 **iii** 12, $\dfrac{7}{3}$ **4** Mean $= 12.875$ or $\dfrac{103}{8}$ Variance $= 0.458$ (3dp)

5 a Student's own response **b** Zero covariance or zero correlation coefficient don't, by themselves, imply independence.

1 −77.3, 1.29 **2 a** 62.3 **b** 0.6 **c** 29.1, 16.8. These are random variables.
3 a 37, 34.45 **b** Mean $= 4$ Variance $= 10.874$

Practice examination questions

1 Mean $= 1120$ Variance $= 3.34$ to 3.36 **2 i** 28, 23.4 **ii** −2, 52 **iii** 61, 43.4
3 i 141, 325 **ii** 93, 45
4 a i Distribution of X is symmetrical about 4. Student's own response **ii** −0.4, −0.54
 b i 7.7, 0.71 **ii** 0.3, 2.31

14 Constant Velocity in Two Dimensions

Exercise 1

1 7.62 **2 a** $(3\mathbf{i} + 4\mathbf{j})\,\mathrm{ms}^{-1}$ **b** $5\,\mathrm{ms}^{-1}$ **c** $(18\mathbf{i} + 18\mathbf{j})\,\mathrm{m}$
3 a i $\mathbf{p} = 3.63\mathbf{i} + 1.69\mathbf{j}$, $\mathbf{q} = 3.21\mathbf{i} + 3.83\mathbf{j}$ **ii** 8.79, 38.9° **b i** $\mathbf{p} = -1.03\mathbf{i} + 2.82\mathbf{j}$, $\mathbf{q} = 5.64\mathbf{i} - 2.05\mathbf{j}$ **ii** 4.67, 9.48°
4 7.02 km, 098.8° **5 a** $2.24\,\mathrm{ms}^{-1}$ **b** 45° **c** $(2t\mathbf{i} - t\mathbf{j})\,\mathrm{m}$ **d** 40.3 s
6 a $(4 - t)\mathbf{i} + (0.5t - 2)\mathbf{j}$ **b** Collide at $t = 4$, $13\mathbf{i} + 16\mathbf{j}$ **7 a** 991 km **b** 173°

Exercise 2

1 a $30.4\,\mathrm{ms}^{-1}$, 064.7° **b** $17.8\,\mathrm{kmh}^{-1}$, 099.1° **c** $19.4\,\mathrm{ms}^{-1}$, 249.2°
2 $11.7\,\mathrm{ms}^{-1}$ **3** $441.97\,\mathrm{kmh}^{-1}$, 014.6° **4 a** $6\,\mathrm{kmh}^{-1}$ **b** $10.4\,\mathrm{kmh}^{-1}$
5 a $5.39\,\mathrm{ms}^{-1}$, 80 m **b** 23.6° to AB, $4.58\,\mathrm{ms}^{-1}$ **6** $6.17\,\mathrm{ms}^{-1}$, 034.8°
7 a $408\,\mathrm{kmh}^{-1}$, 078.7° **b** 101.5°, $392\,\mathrm{kmh}^{-1}$ **8** Dead heat – both take 395 s
9 $\sqrt{v^2 - u^2} : v$ **10 a** $12\,\mathrm{ms}^{-1}$ **b** Student's own response

Exercise 3

1 a $7\mathbf{i} - 2\mathbf{j}$ **b** $-7\mathbf{i} + 2\mathbf{j}$ **2** $28.3\,\mathrm{kmh}^{-1}$, 122°
3 $20\,\mathrm{kmh}^{-1}$ from 053.1° **4 a** 3.79 km **b** 15.36
5 a 112.6° **b** 1 hour **6** Miss by 0.458 km **7** $42.5\,\mathrm{kmh}^{-1}$, 105.1°
8 a Student's own response **b** Student's own response **c** 1 s, 3 m
9 a The closest approach happens when V(A) is perpendicular to (A)V(B) **b** 4 km, 30°

Review exercise

1 Reduces by 7.56 km, so don't build **2** $16.1\,\mathrm{kmh}^{-1}$
3 a Student's own response **b** 1 hour 53 min
4 a 082.2° **b** 32.4 s **5** 0.571 km **6** 036.9°

Practice examination questions

1 a $218\,\mathrm{ms}^{-1}$ **b** 351°
2 a 193° **b i** 109°
 ii 665 s **iii** No cross wind, calm lake, instantaneous change of direction by the patrol boat
3 a $-10\mathbf{i} + 6\mathbf{j}$ **b** Student's own response **c** 0.103 **d** 1.89 km
4 a 035° **b i** 6.60 km **ii** 1236.5
5 a $-2\mathbf{i} + 5\mathbf{j}$ **b** $(100 - 2t)\mathbf{i} + (5t - 250)\mathbf{j}$ **c** Collide when $t = 50$

15 Dimensional Analysis

Exercise 1

1 a ML^2T^{-2} **b** ML^2T^{-3} **c** LT^{-1} **d** LT^{-1} **e** ML^2T^{-2}

2 a $ML^{-1}T^{-2}$ **b** ML^{-3} **c** T^{-1} **d** dimensionless
3 Yes ML^2T^{-2} **4** Yes MLT^{-1} **5** Student's own response **6 a** ML^2T^{-3} **b** Student's own response

7 T^{-1} **8 a** Student's own response   **b** Student's own response   **c** Student's own response

9 $ML^{-1}T^{-1}$, $kgm^{-1}s^{-1}$

Exercise 2

1 $\alpha = -\dfrac{1}{2}, \beta = \dfrac{3}{2}, A \equiv K\sqrt{\dfrac{c^3}{b}}$ **2** $\alpha = \dfrac{5}{3}, \beta = -\dfrac{5}{3}, \gamma = -\dfrac{1}{3}, A \equiv K\sqrt[3]{\dfrac{b^5}{c^5 d}}$ **3** $\alpha = 0, \beta = \dfrac{1}{2}, \gamma = -\dfrac{1}{2}, T = k\sqrt{\dfrac{l}{g}}$

4 $v = k\sqrt{\lambda g}$ **5** $F = \dfrac{kmv^2}{r}$ **6 a** Cannot find 4 variables from 3 equations   **b** $V = \dfrac{kr^4 p}{\eta l}$

Review exercise

1 a $[T] = T, [r] = L, [g] = LT^{-2}, [R] = L$ **b** Yes   **2** Student's own response

3 $h = \dfrac{kv^2}{g}$ **4** $v = k\sqrt{gr}$ **5** $f = \dfrac{k}{l^2}\sqrt{\dfrac{T}{\rho}}$

Practice examination questions

1 $\alpha = 1, \beta = -\dfrac{1}{2}, \gamma = \dfrac{1}{2}$ **2** $M^{-1}T$ **3 a** $M^{-1}L^3T^{-2}$ **b** $\alpha = -\dfrac{1}{2}, \beta = \dfrac{3}{2}, \gamma = -\dfrac{1}{2}$

4 $\alpha = 0, \beta = 2, \gamma = -1$ **5 a** Student's own response   **b** $T = k\sqrt{\dfrac{l}{g}}$

16 Collisions in One Dimension

Exercise 1

1 a A 12 Ns, B 30 Ns   **b** A 8 Ns, B −21 Ns   **c** A −8 Ns, B 30 Ns
2 a 84 Ns   **b** 10 Ns   **c** −35 Ns   **d** 6 N
3 $7\,ms^{-1}$ **4 a** 42 Ns   **b** 120 N   **5 a** 68 Ns   **b** $10.5\,ms^{-1}$
6 7.2 Ns   **7** 50 000 Ns, $16\,666\dfrac{2}{3}$ N   **8** $5\,ms^{-1}$ **9** 4 s

Exercise 2

1 a $7\,ms^{-1}$ **b** $1.8\,ms^{-1}$ **c** $-6\,ms^{-1}$
2 a $1.5\,ms^{-1}$ in opposite direction   **b** B would need to pass through A

3 a $3.2\,ms^{-1}$ **b** $0.8\,ms^{-1}$ **4 a** $4\dfrac{2}{3}\,ms^{-1}$ (direction reversed)   **b** Student's own response

5 $3.97\,ms^{-1}$ **6** 0.45 kg   **7** $4.5\,ms^{-1}$

Exercise 3

1 a 0.75   **b** 7.2   **2** 2.5 m

3 a 0.75   **b** 0.5   **c** $\dfrac{5}{9}$

4 a $2.51\,ms^{-1}$, $7.31\,ms^{-1}$ **b** $2.64\,ms^{-1}$, $4.24\,ms^{-1}$ **c** $0.91\,ms^{-1}$, $5.71\,ms^{-1}$ **d** $-2.71\,ms^{-1}$, $-0.91\,ms^{-1}$
5 $1\,ms^{-1}$, $2\,ms^{-1}$ **6 a** $5.6\,ms^{-1}$ **b** 0.2
7 0.6   **8** $-7.38\,ms^{-1}$, $4.62\,ms^{-1}$ **9** A has speed u, direction reversed, B is at rest
10 a Student's own response   **b** Yes

Exercise 4

1 $2.03\,ms^{-1}$, $2.34\,ms^{-1}$, $5.63\,ms^{-1}$ **2** Student's own response   **3** 0.538
4 a $2.24\,ms^{-1}$, 0.32 ms   **b** 0.925 m/s, 0.01 m/s in the direction away from the wall.

Review exercise

1 a 30 Ns   **b** 12 Ns   **c** $7\,ms^{-1}$ **d** $3\,ms^{-1}$
2 a 72 Ns   **b** 17.4 m/s
3 $0.2\,ms^{-1}$ **4** $\dfrac{5}{16}$ or 4.25   **5** $1.86\,ms^{-1}$, $5.76\,ms^{-1}$ towards wall

1 a $3 \, \text{kg}$ b $5 \, \text{kg}$ or $1.8 \, \text{kg}$ 2 a $10 \, \text{Ns}$ b 8230

3 a $\frac{1}{2}u(1+e)$ b 0.5

4 a $\frac{1}{2}u(5-3e), \frac{1}{2}u(e+5)$ b Student's own response   c $2.5 \, mu$

5 a $416\frac{2}{3}\text{Ns}$ b $1.94 \, \text{ms}^{-1}$

17 Roots and Polynomials

Exercise 1

1 a 0 b 11 c 0 d $-\frac{7}{3}$

2 a $-\frac{2}{3}$ b 7 c $\frac{7}{2}$

3 a $-\frac{7}{8}$ b $-\frac{1}{4}$ 4 a $-\frac{5}{3}$ b 4

5 a 0 b -10 c -9

6 $x^3+4x^2+4x-4=0$ 7 $q=-12$ or 12

Exercise 2

1 $i, -i, \frac{5}{2}\pm\sqrt{\frac{21}{2}}$ 2 $2i, -2i, 2, -\frac{5}{3}$ 3 a $3-4i, 2-11i$ b $k=-11$

4 3 5 $0<k<4$ 6 $1, -i, \frac{3}{2}\pm\frac{i\sqrt{11}}{2}$

7 a Student's own answer   b $3-i, -11$

Review exercise

1 $-7, -\frac{58}{7}$ 2 a Student's own response   b $p=0, q=2-3i$ c $24+\text{l}0i$

3 a $-7+2i$ b $\alpha=\frac{1}{50}(-47+21i)$ c $\frac{1}{50}(-203-2\text{l}i)$

4 a Student's own response   b i $2+3i, -1$ [twice]   ii $(z^2-4z+13)(z^2+2z+1)$

Practice examination questions

1 a $2+3i$ b i 13 ii 3 iii $p=-7 \, q=-39$

2 a i $-18+12i$ ii 0 b i -2 ii $9-6i$ iii $5-6i$ c $\beta=-3i$ $\gamma=2+3i$

3 a 6

    b i sum of squares <0, so roots are not all real; coefficients are real so you have a conjugate pair   ii 0

    c i $-1-3i$ and 2 ii $q=-20$

4 a $p=-4 \quad q=-2$ b 20

18 Proof by Induction and Finite Series

Exercise 1

1 Student's own response   2 Student's own response   3 Student's own response   4 Student's own response

5 Student's own response   6 Student's own response   7 Student's own response   8 Student's own response

Exercise 2

1 $\frac{3}{4}-\frac{1}{2n}-\frac{1}{2(n+1)}$ 2 $\frac{1}{2}-\frac{1}{n+2}$ 3 $\frac{9}{40}-\frac{1}{2(n+3)}-\frac{1}{2(n+4)}$ 4 $\frac{1}{4}-\frac{1}{2n}+\frac{1}{2(n+1)}$

Review exercise

1 Student's own response   2 Student's own response   3 $B=-\frac{1}{2}; \frac{1}{840}-\frac{1}{5304}=\frac{31}{30\,940}$

1 a Student's own response   b Student's own response   2 Student's own response

3 a Student's own response   b Student's own response

4 a $k = 7$ b Student's own response

5 a $k = 7$ b Student's own response

6 a $A = \dfrac{1}{2}$ $B = -\dfrac{1}{2}$ b $\dfrac{894}{1225}$

7 a Student's own response   b 316

8 a $A = \dfrac{1}{2};\ B = -\dfrac{1}{2}$ b Student's own response   c 250

19 Series and Limits

Exercise 1

1 a converge   b converge   c converge

2 a $1 - 2x + 10x^2 - 60x^3$ b $\dfrac{1}{2} - \dfrac{1}{16}x + \dfrac{5}{256}x^2 - \dfrac{15}{2048}x^3$

3 a $1 + \dfrac{21}{2}x + \dfrac{147}{2}x^2 + \dfrac{1715}{4}x^3$ b valid for $-\dfrac{2}{7} < x < \dfrac{2}{7}$ c $a = 64$ $b = 672$ $c = 4704$

4 a $1 + 3x + \dfrac{9}{2}x^2$ b $1 - \dfrac{1}{2}x^4 + \dfrac{1}{24}x^8$

5 a $2x - \dfrac{4}{3}x^3 + \dfrac{4}{15}x^5 - \ldots$ b $1 - \dfrac{25x^2}{2} + \dfrac{625x^4}{24}$ c $1 + 8x + 32x^2 + \dfrac{256}{3}x^3 + \ldots$

   d $x^2 - \dfrac{1}{2}x^4 + \dfrac{1}{3}x^6 - \dfrac{1}{4}x^8 + \ldots$ e $-2x - 2x^2 - \dfrac{8}{3}x^3 - \ldots$

6 a x^2 b $1 + 4x + \dfrac{15}{2}x^2 + 9x^3 + \dfrac{63}{8}x^4$

   c $2 - 8x^2 + \dfrac{9}{4}x^4$ d $e(1 - \dfrac{1}{2}x^2 + \dfrac{1}{6}x^4)$ e $\ln(2 - \dfrac{1}{4}x^2 - \dfrac{1}{96}x^4)$

7 $1 + 2x^2 + 2x^4 + \dfrac{4}{3}x^6 + \ldots$

Exercise 2

1 $|x| < 3$ 2 1   3 3

4 $1 - \dfrac{1}{2!}x^6 + \dfrac{1}{4!}x^{12} + \ldots + (-1)^n \dfrac{x^{6n}}{(2n)!} + \ldots$ Valid for all values of x 5 Student's own response

Exercise 3

1 Value does not exist   2 $\dfrac{\pi}{2a}$ 3 Value does not exist   4 Value does not exist

5 $\dfrac{1}{25}$ 6 $\dfrac{1}{49}$ 7 $\ln\dfrac{40}{27}$ 8 $\ln\dfrac{12}{5}$

Review exercise

1 $-\dfrac{1}{2}$ 2 $\dfrac{1}{2\sqrt{2}}$ 3 a $7x + \dfrac{343}{3}x^3 + \dfrac{33614}{15}x^5$ b $-2x - 2x^2 + \dfrac{8}{3}x^3 - 4x^4$

4 $3x - \dfrac{9}{2}x^2 + \dfrac{17}{2}x^3 - \dfrac{75}{4}x^4$ limit is -3 5 $\sec x = 1 + \dfrac{1}{2}x^2 + \dfrac{5}{24}x^4$ $\tan x = x + \dfrac{1}{3}x^3 + \dfrac{2}{15}x^5$ limit is 6

1 a $1+2x-2x^2$ **b i** $\dfrac{1}{2}+\dfrac{1}{16}x+\dfrac{3}{256}x^2$ **ii** $-4 < x\,4$ **c** $\dfrac{1}{2}+\dfrac{17}{16}x-\dfrac{221}{256}x^2$

2 a $1+3x+4.5x^2$ **b** $3x^2$

3 a $\dfrac{\cos x}{1+\sin x}$ **b** Student's own response   **c** $-e^{-y}(\dfrac{dy}{dx})^2-(e^{-y})^2$ **d** $x-\dfrac{1}{2}x^2+\dfrac{1}{6}x^3-\dfrac{1}{12}x^4$

4 a Student's own response   **b** -1

5 a $\dfrac{x^3}{3}\ln x-\dfrac{x^3}{9}+c$ **b** Integrand is not defined at $x=0$ **c** $2\dfrac{e^3}{9}$

6 a i $1+x-\dfrac{1}{2}x^2-\dfrac{1}{6}x^3$ **ii** $3x-\dfrac{9}{2}x^2+9x^3$ **b i** $\dfrac{dy}{dx}=\sec^2 xe^{\tan x}$ **ii** 3

     iii Student's own response   **c** $\dfrac{1}{3}$

7 a The interval of integration is infinite   **b** $-\dfrac{1}{9}e^{-3x}-\dfrac{x}{3}e^{-3x}+c$ **c** $\dfrac{4}{9}e^{-3}$

20 De Moivre's Theorem

Exercise 1

1 $\cos 14\theta+i\sin 14\theta$ **2 a** $\cos 6\theta+i\sin 6\theta$ **b** $\cos 8\theta+i\sin 8\theta$ **c** -1 **d** $-i$

    e $\cos 8\theta-i\sin 8\theta$ **f** -1 **g** 1 **h** $-i$

3 a $\cos 10\theta+i\sin 10\theta$ **b** $\cos\theta-i\sin\theta$ **c** -1 **d** -8

4 a 16 **b** -1728 **c** -4 **d** 2^{12}

5 a $\cos 5\theta-i\sin 5\theta$ **b** $\cos 4\theta+i\sin 4\theta$ **c** $-\cos 6\theta-i\sin 6\theta$ **d** -1

6 $n=9$

Exercise 2

1 a i $\pm\sqrt{2}(1+/-i)$ **ii** $2e^{\frac{i\pi}{4}},\ 2e^{-\frac{i\pi}{4}},\ 2e^{\frac{3i\pi}{4}},\ 2e^{-\frac{3i\pi}{4}}$

   b i $2^{\frac{2}{3}}(1+i),\ 2^{\frac{7}{6}}\left[\cos\dfrac{11}{12}\pi+i\sin\dfrac{11}{12}\pi\right],\ 2^{\frac{7}{6}}\left[\cos\dfrac{5}{12}\pi-i\sin\dfrac{5}{12}\pi\right]$ **ii** $2^{\frac{7}{6}}e^{\frac{i\pi}{4}},\ 2^{\frac{7}{6}}e^{\frac{11i\pi}{12}},\ 2^{\frac{7}{6}}e^{-\frac{11i\pi}{12}}$

   c i $\dfrac{3}{2}(\sqrt{3}+i),\ \dfrac{3}{2}(-\sqrt{3}+i),-3i$ **ii** $3e^{\frac{i\pi}{6}},\ 3e^{\frac{5i\pi}{6}},\ 3e^{-\frac{i\pi}{2}}$

   d i $\pm 2\sqrt{2}(1+i)$ **ii** $4e^{\frac{i\pi}{4}},4e^{-\frac{3i\pi}{4}}$

   e i $-2,2\left[\cos\dfrac{\pi}{5}\pm i\sin\dfrac{\pi}{5}\right],2\left[\cos\dfrac{3\pi}{5}\pm i\sin\dfrac{3\pi}{5}\right]$ **ii** $2e^{\frac{i\pi}{5}},2e^{\frac{3i\pi}{5}},2e^{i\pi},2e^{-\frac{i\pi}{5}},2e^{-\frac{3i\pi}{5}}$

2 i $e^{\pm\frac{i\pi}{3}},e^{\pm\frac{2i\pi}{3}},e^{i0}$ or $1,-1,\pm\left(\dfrac{1}{2}\pm\dfrac{\sqrt{3}}{2}i\right)$

3 a $2-2i,-2-2i$ **b** $3,\pm i\sqrt{3}$ **c** $-\dfrac{1}{2}$ **d** $1+3i,\dfrac{1}{3}-i$

4 $e^{i0},e^{\frac{2i\pi}{7}},e^{\frac{4i\pi}{7}},e^{\frac{6i\pi}{7}},e^{-\frac{2i\pi}{7}},e^{-\frac{4i\pi}{7}},e^{-\frac{6i\pi}{7}}$ **5** $2e^{\frac{i\pi}{10}},2e^{\frac{i\pi}{2}},2e^{\frac{9i\pi}{10}},2e^{-\frac{3i\pi}{10}},2e^{-\frac{7i\pi}{10}}$

6 Student's own response   **7** Student's own response

8 a $\dfrac{1}{41}(4\cos 5x + 5\sin 5x)e^{4x} + c$ **b** $\dfrac{1}{58}(3\sin 7x - 7\cos 7x)e^{3x} + c$

   c $-\dfrac{1}{20}(2\sin 4x + 4\cos 4x)e^{-2x} + c$ **d** $\dfrac{1}{25}(3\sin 3x - 4\cos 3x)e^{-4x} + c$

Exercise 3

1 a $2i\sin 2\theta$ **b** $2\cos 4\theta$ **c** $2\cos 5\theta$

2 a $\dfrac{1}{2}\left(z^6 + \dfrac{1}{z^6}\right)$ **b** $\dfrac{-i}{2}\left(z^5 - \dfrac{1}{z^5}\right)$ **c** $\dfrac{1}{2}\left(z^4 + \dfrac{1}{z^4}\right)$

   d $\dfrac{i}{8}\left(z^3 - 3z + \dfrac{3}{z} - \dfrac{1}{z^3}\right)$ **e** $\dfrac{1}{16}\left(z^3 + \dfrac{1}{z^3}\right)$

3 a $32\cos^6\theta - 48\cos^4\theta + 18\cos^2\theta - 1$ **b** $8\cos^4\theta - 8\cos^2\theta + 1$ **c** $8\cos^3\theta - 4\cos\theta$

4 a $3\sin\theta - 4\sin^3\theta$ **b** $16\sin^5\theta - 20\sin^3\theta + 5\sin\theta$

   c $-64\sin^6\theta + 80\sin^4\theta - 24\sin^2\theta + 1$ **d** $16\sin^4\theta - 12\sin^2\theta + 1$

5 a $\dfrac{3}{4}\sin\theta - \dfrac{1}{4}\sin 3\theta$ **b** $\dfrac{1}{4}\cos 3\theta + \dfrac{3}{4}\cos\theta$ **c** $\dfrac{1}{16}\cos 5\theta + \dfrac{5}{16}\cos 3\theta + \dfrac{5}{8}\cos\theta$ **d** $\dfrac{1}{16}\sin 5\theta - \dfrac{5}{16}\sin 3\theta + \dfrac{5}{8}\sin\theta$

6 Student's own response   **7** Student's own response   **8** Student's own response

Review exercise

1 $\dfrac{5}{\sqrt{2}}(-1+i),\ \dfrac{5}{\sqrt{2}}(1-i)$ **2** $-3 - \dfrac{1}{2}i,\ -\dfrac{3}{2} + \dfrac{1}{4}i,\ -\dfrac{21}{10} + \dfrac{7}{10}i,\ -\dfrac{3}{2} - \dfrac{1}{2}i$

3 $\dfrac{1}{4}\left(-z^{10} + 2 - \dfrac{1}{z^{10}}\right)$ **4** $\dfrac{1}{32}\cos 6\theta + \dfrac{3}{16}\cos 4\theta + \dfrac{15}{32}\cos 2\theta + \dfrac{5}{16}$

Practice examination questions

1 a i $2e^{\frac{i\pi}{3}}, \sqrt{2}\,e^{-\frac{i\pi}{4}}$ **ii** $2^{\frac{21}{2}}e^{\frac{17\pi i}{12}}$ **b** $8\sqrt{2}\,e^{\frac{17\pi i}{36}}, 8\sqrt{2}\,e^{\frac{-7\pi i}{36}}, 8\sqrt{2}\,e^{\frac{-31\pi i}{36}}$

2 $\dfrac{\pi}{10}$ **3 a i** $\sin 5\theta = 5\cos^4\theta,\ \sin\theta - 10\cos^2\theta,\ \sin^3\theta + \sin^5\theta$ **ii** Student's own response

b $\tan\dfrac{k\pi}{5},\ k = 2, 3, 4$ **c** Student's own response

4 a $8e^{\frac{2i\pi}{3}}$ **b i** $2e^{\frac{-4\pi i}{9}}, 2e^{\frac{2\pi i}{9}}, 2e^{\frac{8\pi i}{9}}$ **ii** $3\sqrt{3}$ **c** Student's own response

5 a i Student's own response   **ii** w^2, w^3, w^4, w^5, w^6

6 i Student's own response   **ii** $\sin 3\theta = 3\cos^2,\ \sin\theta - \sin^3\theta$

b i Student's own response   **ii** $\tan\dfrac{5}{12}\pi \quad \tan\dfrac{9}{12}\pi$

21 Polar Coordinates

Exercise 1

1 a
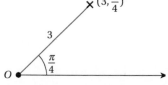

b
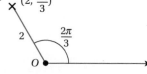

c

d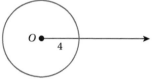

O

2

$\times \left(2, \dfrac{3\pi}{2}\right)$

e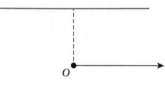

O

$\dfrac{\pi}{4}$

4

$\times \left(4, -\dfrac{\pi}{4}\right)$

2 a $x^2+y^2=16$ **b** $x=3$ **c** $y=7$ **d** $x^2+y^2=ax+a\sqrt{x^2+y^2}$ **e** $y^2=4-4x$

3 a $r=3$ **b** $r^2\sin 2\theta=32$ **c** $\dfrac{r^2\cos^2\theta}{9}+\dfrac{r^2\sin^2\theta}{16}=1$ **d** $r=6\cos\theta$ **e** $r^2=\cos 2\theta$

Exercise 2

1 a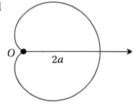

O • 4

b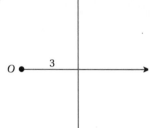

O • 3

c

O

d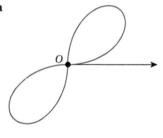

O 2a

e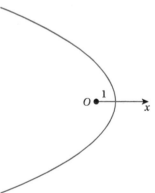

O • 1 → x

2 a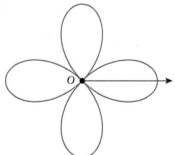

O

b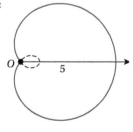

O

c

O • 5

d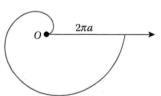

O • 2πa

e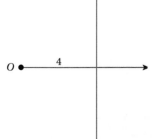

O • 4

3

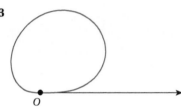

4 a Student's own response   **b** $32\sqrt{3}$

Exercise 3

1 $\dfrac{7\pi^3 a^2}{48}$

2 a $\dfrac{\pi a^2}{8}$ **b** $\dfrac{\pi a^2}{8}$ **c** $\dfrac{\pi a^2}{16}$

3 $\dfrac{\pi a^2}{4}$

4 $\dfrac{17\pi}{2}$

5 a $r^2 = \sin^4\theta$ **b i**

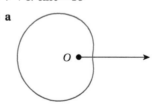

ii $\dfrac{3\pi}{8}$

6 $\left(1, \dfrac{\pi}{6}\right), \left(1, \dfrac{5\pi}{6}\right), \dfrac{7\pi}{3} - 4\sqrt{3}$ **7** $\dfrac{3\pi}{4}$

Review exercise

1 $x^2 + y^2 + ax = a\sqrt{x^2 + y^2}$ **2** $r^2 + 8r\sin\theta = 16$

3 a Student's own response **b** $r^2 = (1 - r\cos\theta)^2$ **4 a**   **b** $\dfrac{153\pi}{2}$

Practice examination questions

1 $\dfrac{13}{3}$ **2** $y^2 = \dfrac{16 + 24x - 7x^2}{16}$ **3** $r = 12\sin\theta - 16\cos\theta$ **4** $\dfrac{\pi}{2} - \ln 2$

5 a 11π **b i** $4\sqrt{3}$ **ii** $\dfrac{8\pi}{3} + 4\sqrt{3}$

6 a $y^2 = 4 - 4x$ **b** 2

7 a $x^2 + y^2 = 2y$ **b i** $\sqrt{3}, \dfrac{\pi}{3}$ **ii** Since $\sqrt{2} > 1$, A is further away from the pole than B.   **iii** $\dfrac{1}{2}\pi - \dfrac{3\sqrt{3}}{4}$

8 a i $x^2 + y^2 = 2(x - y)$ **ii** centre $(1, -1)$ radius $\sqrt{(2)}$

  b i 16.5π **ii** Student's own response   **iii** 14.5π

22 The Calculus of Inverse Trigonometrical Functions

Exercise 1

1 a $\dfrac{\pi}{6}$ **b** $-\dfrac{\pi}{6}$ **c** $\dfrac{5\pi}{6}$ **d** $\dfrac{\pi}{4}$ **e** $\dfrac{\pi}{4}$ **f** $18.4°$

2 a 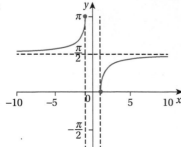   **b** 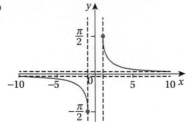   **c**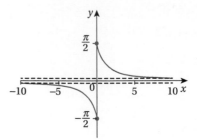

3 $\dfrac{\pi}{10}$ **4** Student's own response

Exercise 2

1 a $\dfrac{5}{\sqrt{1-25x^2}}$ **b** $\dfrac{3}{1+9x^2}$ **c** $\dfrac{\sqrt{2}}{\sqrt{1-2x^2}}$ **d** $\dfrac{2x}{\sqrt{1-x^4}}$ **e** $\dfrac{1-x^2}{1+3x^2+x^4}$

   f $\dfrac{1620(\tan^{-1}5x)^3}{1+25x^2}$ **g** $\dfrac{6(\sin^{-1}2x)^2}{\sqrt{1-4x^2}}$

2 a $\sin^{-1}\left(\dfrac{x}{2}\right)+c$ **b** $\sin^{-1}\left(\dfrac{x}{3}\right)+c$ **c** $\dfrac{1}{2}\sin^{-1}\left(\dfrac{2x}{5}\right)+c$ **d** $\dfrac{1}{3}\tan^{-1}\left(\dfrac{x}{3}\right)+c$

   e $\dfrac{1}{4}\tan^{-1}\left(\dfrac{x}{4}\right)+c$ **f** $\dfrac{1}{15}\tan^{-1}\left(\dfrac{5x}{3}\right)+c$

3 a $\dfrac{\pi}{2}$ **b** $\dfrac{\pi}{8}$ **c** $\dfrac{\pi}{2}$ **d** $\dfrac{\pi}{6\sqrt{3}}$ **e** $\dfrac{\pi}{5}$

4 a 0.0505 **b** 0.0444 **c** 0.615 **d** 0.0741 **e** 0.841 **f** 0.0207

5 $\dfrac{1}{4}\sin^{-1}\left(\dfrac{14}{25}\right)$

Review exercise

1 a $\dfrac{12}{16+9x^2}$ **b** $\dfrac{6(\sin^{-1}2x)^2}{\sqrt{1-4x^2}}$ **2 a** $\dfrac{1}{3}\sin^{-1}\left(\dfrac{3x}{4}\right)+c$ **b** $\dfrac{1}{20}\tan^{-1}\left(\dfrac{4x}{5}\right)+c$

3 $\dfrac{1}{12}\tan^{-1}\left(\dfrac{8}{3}\right)$ **4** $\dfrac{1}{20}\tan^{-1}4$ **5** $\dfrac{1}{\sqrt{2}}\sin^{-1}\left(\dfrac{\sqrt{2}(x-2)}{5}\right)+c$ **6** $\dfrac{1}{2\sqrt{24}}\tan^{-1}\left(\dfrac{(2x+1)}{\sqrt{24}}\right)+c$

Practice examination practice

1 a $9-2(x-1)^2$ **b** $\dfrac{\pi\sqrt{2}}{8}$ **2** $\dfrac{\pi}{3}$

3 a $-\dfrac{x}{\sqrt{1-x^2}}$ **b** $a=\dfrac{1}{6}$ $b=-\dfrac{1}{2}$

4 a $\dfrac{x}{1+x^2}+\tan^{-1}x$ **b** Student's own response

5 a Student's own response   **b** Student's own response

23 Arc Length and Area of Surface of Revolution

Exercise 1

1 $\dfrac{1}{27}(31\sqrt{31}-8)$ **2** $\dfrac{14\sqrt{14}-11\sqrt{11}}{9\sqrt{2}}$ **3** $ap\sqrt{1+p^2}+a\ln(p+\sqrt{1+p^2})$ **4** $4a$

Exercise 2

1 $\dfrac{12\pi}{5}(2+782\sqrt{17})$ **2** $\dfrac{8\pi}{3}(5\sqrt{5}-1)$ **3** $\dfrac{12\pi a^2}{5}$ **4** $\dfrac{5\pi}{6}\left(61^{\frac{3}{2}}-41^{\frac{3}{2}}\right)$

Review exercise

1 $\dfrac{13}{4}$ **2** $5+\ln\dfrac{7}{3}$ **3** $A=\dfrac{1}{3}\pi r^2$; area of whole cone: $A=\dfrac{1}{3}\pi r^2 h+\pi r^2$

Practice examination questions

1 Student's own response   **2 a** Student's own response   **b** $p=2$

3 Student's own response   **4 a** Student's own response   **b** $k=\dfrac{256\pi}{15}$

24 Hyperbolic Functions

Exercise 1

1 a i $\dfrac{1}{2}(e^2+e^{-2})$ **ii** 3.76   **b i** $\dfrac{1}{2}(e^3-e^{-3})$ **ii** 10.0   **c i** $\dfrac{e^8-1}{e^8+1}$ **ii** 0.999

d i $\dfrac{2}{(e^2+e^{-2})}$ **ii** 0.266   **e i** $\dfrac{2}{e^4-e^{-4}}$ **ii** 0.0366   **f i** $\dfrac{e^6+e^{-6}}{e^6-e^{-6}}$ **ii** 1.00

2 Student's own response

3 a Student's own response   **b** Student's own response   **c** Student's own response
4 a 0.540   **b** 0.693   **c** 0.457

5 $2,\dfrac{3}{2}$ and $\ln 2, \ln\dfrac{3}{2}$ **6** $p=7$

Exercise 2

1 a $2\sin 2x$ **b** $5\cosh 5x$ **c** $3\operatorname{sech}^2 3x$ **d** $45\sinh 3x\cosh^4 3c\ 64s\ h^3 8x\cosh 8x$

2 a $\dfrac{1}{3}\cosh 3x+c$ **b** $\dfrac{1}{4}\sinh 4x+c$ **c** $3\cosh\dfrac{x}{3}+c$ **d** $\dfrac{1}{4}\ln(\cosh 4x)+c$

3 a $-\operatorname{cosec} h^2 x$ **b** $\tanh x(-\operatorname{sech} x)$ **c** $10\operatorname{cosec} h 10x$

Exercise 3

1 a $\dfrac{5}{\sqrt{25x^2+1}}$ **b** $\sinh^{-1}\sqrt{2}$ **c** $\dfrac{3}{\sqrt{3x-4}\sqrt{3x+4}}$ **d** $\dfrac{2x}{\sqrt{x^4+1}}$ **e** $\dfrac{1}{(x\sqrt{(1-x^2)})}$ **f** $\dfrac{1}{1-x^2}$

2 a $\ln(\sqrt{x^2-4}+x)+c$ **b** $\dfrac{1}{2}\ln(\sqrt{4x^2-25}+2x)+c$ **c** $\sinh^{-1}\left(\dfrac{x}{3}\right)+c$ **d** $\dfrac{1}{4}\sinh^{-1}\left(\dfrac{4x}{5}\right)+c$

3 a $\ln(1+\sqrt{2})$ **b** $\ln(2+\sqrt{3})$ **c** $\left(\sqrt{\dfrac{3}{3}}\right)\ln(2+\sqrt{3})$

4 a $\dfrac{1}{5}\left(\ln\left(10+4\sqrt{6}\right)-\ln\left(5+\sqrt{21}\right)\right)$ **b** $\dfrac{1}{3}\left(\ln\left(3+\sqrt{10}\right)+\ln\left(\dfrac{2}{3+\sqrt{12}}\right)\right)$ **c** $\ln\left(1+\sqrt{\dfrac{2}{3}}\right)$

d $\dfrac{1}{2}\ln\left(\dfrac{1}{5}\left(4+\sqrt{21}\right)\right)$ **e** $\ln\left(1+\sqrt{\dfrac{2}{3}}\right)$ **f** $\left(\dfrac{1}{4}\right)\left[\ln\left(2\sqrt{71}+13\right)-\left(\dfrac{1}{2}\right)\ln\left(115\right)-\ln\left(2\,\dfrac{\sqrt{7}}{\sqrt{23}}+\left(\sqrt{\dfrac{5}{23}}\right)\right)\right]$

Review exercise

1 0.19

2 $\ln\left(1-\dfrac{2\sqrt{2}}{3}\right)$

3 a $112\cosh^3 7x\sinh 7x$ **b** $36\sinh^2 6x\cosh 6x$

4 a $\dfrac{\pi}{40}-\dfrac{1}{10}\tan^{-1}\dfrac{2}{5}$

b $\left(\dfrac{1}{2}\right)\left[\ln\left(\dfrac{5}{\sqrt{2}}+\sqrt{\dfrac{23}{2}}\right)-\ln\left(\sqrt{2}+1\right)\right]$

5 $\dfrac{1}{2}\sinh^{-1}\dfrac{2(1+x)}{7}$

Practice examination questions

1 a $\dfrac{e^x}{4+e^{2x}}$ **b** Student's own response   **2 a** Student's own response   **b** $3\ln2;\ -2\ln2$

3 a

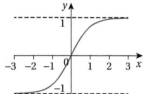

b Student's own response   **c i** Student's own response   **ii** $\dfrac{1}{2}\ln2$

4 a i Student's own response   **ii** Student's own response   **iii** Student's own response

b i Student's own response   **ii** $\dfrac{\pi}{6}$

25 Differential Equations of First and Second Order

Exercise 1

1 a x^2 **b** $\sqrt{x^2+1}$ **c** $\dfrac{1}{x^3}$ **d** $\sec x$ **e** $\sqrt{x^2-1}$ **f** 2^{3x}

2 $y=\dfrac{1}{3}x-\dfrac{1}{9}+ce^{-3x}$ **3** $y=-\dfrac{1}{3}e^{2x}+ce^{5x}$ **4** $y=\dfrac{x^2}{3}+\dfrac{c}{x}$ **5** $y=x^3+cx^2$

6 $y=5\left(x-1\right)^4\ln(x-1)+c\left(x-1\right)^4$ **7** $y=\dfrac{2}{5}e^{2x}\sin x-\dfrac{1}{5}e^{2x}\cos x+c$

Exercise 2

1 $y=Ae^{(3+\sqrt{17})x}+Be^{(3-\sqrt{17})x}$ **2** $y=Ae^{-x}+Be^{-2x}$ **3** $y=Ae^{2x}+Be^{-\frac{3}{2}x}$ **4** $y=Ae^{x}+Be^{-\frac{7}{3}x}$

5 $x=Ae^{8t}+Be^{-t}$ **6** $x=Ae^{7t}+Be^{4t}$ **7** $y=(A+Bx)\,e^{-2x}$ **8** $y=(A+Bx)\,e^{3x}$

9 $y=e^{-\frac{1}{2}x}\left[A\cos\dfrac{\sqrt{3}}{2}x+B\sin\dfrac{\sqrt{3}}{2}x\right]$ **10** $y=e^{-2x}\left(A\cos 2x+B\sin 2x\right)$

11 $x=Ae^{(3+\sqrt{2})t}+Be^{(3-\sqrt{2})t}$ **12** $x=e^{-2t}\left(A\cos 2\sqrt{3}\,t+B\sin 2\sqrt{3}\,t\right)$

Exercise 3

1 $y=Ae^{-8x}+Be^{x}-2x-\dfrac{7}{4}$ **2** $y=Ae^{-x}+Be^{-3x}-4e^{-2x}$ **3** $y=Ae^{-x}+Be^{\frac{5}{2}x}-2x^2+\dfrac{12}{5}x-\dfrac{81}{25}$

4 $y=Ae^{-x}+Be^{\frac{1}{3}x}-\dfrac{76}{1469}\sin 5x-\dfrac{10}{1469}\cos 5x$ **5** $x=Ae^{5t}+Be^{-t}-\dfrac{3}{8}e^{3t}$

6 $x=Ae^{5t}+Be^{3t}+\dfrac{55}{377}\cos 2t-\dfrac{80}{377}\sin 2t$ **7** $y=Ae^{-4x}+Be^{-x}+\dfrac{2}{3}xe^{-x}$

8 $y=Ae^{x}\sin\left(\sqrt{2}x\right)+Be^{x}\cos\left(\sqrt{2}x\right)+2e^{4x}$ **9** $y=Ae^{-3x}\sin x+Be^{-3x}\cos x+\dfrac{3}{2}e^{-4x}$

Review exercise

1 $ye^{7x}=\dfrac{1}{4}\left(1-e^{4x}\right)$ **2** $y=\dfrac{5}{2}x^2\,e^{3x}$ **3** $x=(1+t+2t^2)e^{t}$ **4** $x=\cos 4t-\dfrac{3\pi}{64}\sin 4t+\dfrac{3}{8}t\sin 4t$

1 a $p=\dfrac{1}{2}, q=\dfrac{5}{2}$

b $y=Ae^{-5x}+\dfrac{1}{2}\sin x+\dfrac{5}{2}\cos x$

2 a Student's own response

b $y=x^2\ln x+cx^2$

3 a $a=5, b=-2, c=-4$

b $y=3e^{-4x}+5-2\sin 2x-4\cos 2x$

4 $y=\dfrac{5}{2}e^{-x}-\dfrac{1}{2}xe^{-x}$

5 a $a=-\dfrac{2}{3}, b=-1, c=2$

b $y=Ae^{-3x}+Be^{x}-\dfrac{2}{3}-x+2\,xe^{-3x}$

c $y=\dfrac{5}{3}e^{-3x}-\dfrac{2}{3}-x+2\,xe^{-3x}$

26 Vectors and Three-Dimensional Coordinate Geometry

Exercise 1

1 a i $1:2:-2$ **ii** $\dfrac{1}{3}, \dfrac{2}{3}, -\dfrac{2}{3}$ **b i** $3:-4:-5$ **ii** $\dfrac{3}{5\sqrt{2}}, -\dfrac{4}{5\sqrt{2}}, -\dfrac{1}{\sqrt{2}}$

c i $3:2:-5$ **ii** $\dfrac{3}{\sqrt{38}}, \dfrac{2}{\sqrt{38}}, -\dfrac{5}{\sqrt{38}}$ **d i** $1:-2:-3$ **ii** $\dfrac{1}{\sqrt{14}}, -\dfrac{2}{\sqrt{14}}, -\dfrac{3}{\sqrt{14}}$

2 a 28 **b** 29 **c** 137 **d** 15 **e** $18a-36$ **f** $-23k-48$

Exercise 2

1 a $-5\mathbf{i}+7\mathbf{j}+11\mathbf{k}$ **b** $31\mathbf{i}+22\mathbf{j}+\mathbf{k}$ **c** $22\mathbf{i}+14\mathbf{j}-16\mathbf{k}$ **d** $-32\mathbf{i}+23\mathbf{j}-10\mathbf{k}$

2 $\sqrt{14}$ **3** 21 **4 a** $\begin{pmatrix} -2 \\ 16 \\ 4 \end{pmatrix}$ **b** $\dfrac{4\sqrt{69}}{69}$

Exercise 3

1 a i $\mathbf{r}=\begin{pmatrix} 2 \\ 4 \\ -7 \end{pmatrix}+t\begin{pmatrix} 2 \\ -5 \\ 1 \end{pmatrix}$ **ii** $\left(\mathbf{r}-\begin{pmatrix} 2 \\ 4 \\ -7 \end{pmatrix}\right)\times\begin{pmatrix} 2 \\ -5 \\ 1 \end{pmatrix}=0$

b i $\mathbf{r}=\begin{pmatrix} 2 \\ -5 \\ 4 \end{pmatrix}+t\begin{pmatrix} 2 \\ 3 \\ -7 \end{pmatrix}$ **ii** $\left(\mathbf{r}-\begin{pmatrix} 2 \\ -5 \\ 4 \end{pmatrix}\right)\times\begin{pmatrix} 2 \\ 3 \\ -7 \end{pmatrix}=0$

2 a $\mathbf{r}.\begin{pmatrix} 3 \\ -5 \\ 4 \end{pmatrix}=-13$ **b** $\mathbf{r}.\begin{pmatrix} 9 \\ 7 \\ -2 \end{pmatrix}=47$ **c** $\mathbf{r}.\begin{pmatrix} 28 \\ -17 \\ 18 \end{pmatrix}=41$

3 a $3x+y+7z=4$ **b** $2x+4y+3z=8$ **c** $-x+5y+3z+7=0$

4 a $68.5°$ **b** $43.2°$ **c** $28.0°$ **d** $48.5°$

5 $29.1°$ **6** $0°$ **7** $\mathbf{r}.\begin{pmatrix} \dfrac{3}{5\sqrt{2}} \\ \dfrac{4}{5\sqrt{2}} \\ \dfrac{-1}{5\sqrt{2}} \end{pmatrix}=2\sqrt{2}\,; 2\sqrt{2}$

8 $\dfrac{x}{7}=\dfrac{y+\dfrac{27}{7}}{10}=\dfrac{z+\dfrac{32}{7}}{9}\,; \dfrac{7}{\sqrt{230}}, \dfrac{10}{\sqrt{230}}, \dfrac{9}{\sqrt{230}}$

9 a $\begin{pmatrix} 6 \\ -5 \\ 7 \end{pmatrix}$ **b i** $\dfrac{6}{\sqrt{110}}, \dfrac{-5}{\sqrt{110}}, \dfrac{7}{\sqrt{110}}$ **ii** They measure respectively the cosines of the angles that the line makes with the x, y and z-axes

c $\left(\mathbf{r}-\begin{pmatrix} 1 \\ 3 \\ 5 \end{pmatrix}\right)\times\begin{pmatrix} 6 \\ -5 \\ 7 \end{pmatrix}=0$

10 a Student's own responses

b $\dfrac{3}{\sqrt{14}}, -\dfrac{1}{\sqrt{14}}, -\dfrac{2}{\sqrt{14}}$, angle $= 74.5°$

c $2\mathbf{i} - 5\mathbf{j} + 3\mathbf{k}$

d $77.5°$

e $(19, -3, -5)$

f $\dfrac{5\sqrt{38}}{38}$

Exercise 4

1 No   **2** -73 **3** 177   **4** 21   **5** 8   **6** 8

Review exercise

1 $\dfrac{9}{5\sqrt{2}}$

2 $\dfrac{1}{\sqrt{41}}, \dfrac{2}{\sqrt{41}}, -\dfrac{6}{\sqrt{41}}$

3 r. $\begin{pmatrix} 6 \\ 3 \\ -5 \end{pmatrix} = 14$

4 $40.1°$

5 r. $\begin{pmatrix} \dfrac{3}{\sqrt{26}} \\ \dfrac{-4}{\sqrt{26}} \\ \dfrac{1}{\sqrt{26}} \end{pmatrix} = \dfrac{8}{\sqrt{26}}$; $\dfrac{8}{\sqrt{26}}$

6 27

Practice examination questions

1 a $\dfrac{4}{9}, \dfrac{7}{9}, -\dfrac{4}{9}$ or $\dfrac{3}{7}, -\dfrac{2}{7}, \dfrac{6}{7}$

b The direction cosines are the cosines of the angles between the line and the coordinate axes.

2 a $4\mathbf{i} + 12\mathbf{j} - 3\mathbf{k}$

b i $\dfrac{4}{13}, \dfrac{12}{13}, -\dfrac{3}{13}$

ii The cosines of the angles between the line and the coordinate axes.

c $\mathbf{a} = \mathbf{i} - 2\mathbf{j} + \mathbf{k}$

3 Student's own response

4 a $\pm \begin{pmatrix} 1 \\ -5 \\ 4 \end{pmatrix}$

b $\begin{pmatrix} 3t+5 \\ 3t-16 \\ 3t-2 \end{pmatrix}$

c i 77

ii $\overrightarrow{OA}, \overrightarrow{OB}, \overrightarrow{OC}$ never coplanar

5 a i $\begin{pmatrix} -2 \\ 2 \\ -1 \end{pmatrix}$

ii $18 - a$

b i 1.5   **ii** $a = 18$

6 a $\begin{pmatrix} 16 \\ -16 \\ 0 \end{pmatrix}$

b $x - y = -1$

c 64

7 r. $= \begin{pmatrix} 0 \\ -8 \\ -11 \end{pmatrix} + \lambda \begin{pmatrix} 1 \\ 5 \\ 7 \end{pmatrix}$

8 a Student's own response

b Student's own response

c $(1, 1, -3)$

27 Solution of Linear Equations

Exercise 1

1 $x = 1, y = 3, z = -2$

2 $x = \dfrac{13}{5}, y = \dfrac{22}{5}, z = -0$

3 $x = 2, y = -3, z = 1$

Exercise 2

1 a -52 **b** 24

2 a Student's own answer   **b** $q = 25$

3 a Student's own answer   **b i** 1   **ii** 0

Review exercise

1 $x = \dfrac{2}{5}, y = -\dfrac{47}{5}, z = 9$

2 $a = \dfrac{-11 \pm 3\sqrt{41}}{4}$

3 $k = \pm 4; b = \dfrac{5}{9}$

Practice examination questions

1 a $x = 6, y = 1\frac{1}{2}, z = -2\frac{1}{2}$ **b i** $a = 1$ **ii** $b = -10$

2 a $a = 2, -5$ **b** $b = 4$

3 a $a = -3$ **b** $b \neq 5$ **c** linearly dependent since determinant is zero

4 a Student's own response **b i** one solution **b ii** infinitely many solutions **iii** no solutions

c i The single point of intersection of three planes. **ii** Three planes meet in a line (or form a sheaf).

iii Three planes form a prism (or have three parallel lines of intersection; or have no common intersection).

28 Matrix Algebra

Exercise 1

1 $PQ = \begin{pmatrix} 14 & 0 \\ 8 & 5 \end{pmatrix}$, $QP = \begin{pmatrix} 2 & -2 \\ 18 & 17 \end{pmatrix}$. Matrix multiplication is not commutative. **2** $\begin{pmatrix} -5 & 4 \\ 4 & -3 \end{pmatrix}$

3 $\begin{pmatrix} 4 & -7 \\ -1 & 2 \end{pmatrix}$ **4** $\begin{pmatrix} -20 & 4 & -9 \\ -5 & 1 & -2 \\ 11 & -2 & 5 \end{pmatrix}$ **5** $\begin{pmatrix} 0 & -1 & 2 \\ 1 & -1 & -3 \\ -2 & 3 & 5 \end{pmatrix}$ **6** $-\frac{1}{141} \begin{pmatrix} -21 & -12 & 18 \\ -17 & -3 & -19 \\ 5 & -24 & -11 \end{pmatrix}$

Exercise 2

1 $\begin{pmatrix} 1 & 0 & 0 \\ 0 & 1 & 0 \\ 0 & 0 & -1 \end{pmatrix}$ **2** Rotation about the y-axis of $\cos^{-1}\frac{12}{13}$ **3** $\begin{pmatrix} 2 & 0 & 0 \\ 0 & 2 & 0 \\ 0 & 0 & 2 \end{pmatrix}$; $\begin{pmatrix} -1 & 0 & 0 \\ 0 & 1 & 0 \\ 0 & 0 & 1 \end{pmatrix}$; $\begin{pmatrix} -2 & 0 & 0 \\ 0 & 2 & 0 \\ 0 & 0 & 2 \end{pmatrix}$

4 a $\begin{pmatrix} -1 & 0 & 0 \\ 0 & 1 & 0 \\ 0 & 0 & 1 \end{pmatrix}$; $\begin{pmatrix} 0 & 1 & 0 \\ 1 & 0 & 0 \\ 0 & 0 & 1 \end{pmatrix}$ **b** $\begin{pmatrix} 0 & 1 & 0 \\ -1 & 0 & 0 \\ 0 & 0 & 1 \end{pmatrix}$; **c** rotation of $\frac{3\pi}{2}$ about z-axis.

5 a $\begin{pmatrix} 5 & 0 & 0 \\ 0 & 5 & 0 \\ 0 & 0 & 5 \end{pmatrix}$ **b** $\begin{pmatrix} 0 & 1 & 0 \\ 1 & 0 & 0 \\ 0 & 0 & 1 \end{pmatrix}$

Exercise 3

1 $\mathbf{M}^3 - 6\mathbf{M}^2 + 37 = 0$ **2** $\mathbf{M}^3 - 3\mathbf{M}^2 + 12\mathbf{M} - 10 = 0$

3 All points on the line $z = 0$, $x = y$ are invariant.

4 a ± 7 **b** Student's own response **c** $\begin{pmatrix} 1 \\ -1 \end{pmatrix}$

6 Eigenvalues are 3, 9 and −3. Corresponding eigenvectors are $\begin{pmatrix} 1 \\ -2 \\ -2 \end{pmatrix}$, $\begin{pmatrix} 2 \\ 2 \\ -1 \end{pmatrix}$, $\begin{pmatrix} 2 \\ -1 \\ 2 \end{pmatrix}$ respectively.

Review exercise

1 $\frac{1}{16} \begin{pmatrix} 6 & -1 \\ -2 & 3 \end{pmatrix}$ **2** $-\frac{1}{33} \begin{pmatrix} -21 & 6 & -3 \\ -5 & 3 & -7 \\ 9 & -12 & 6 \end{pmatrix}$ **3** $1, \frac{1 \pm \sqrt{29}}{2}$

4 $\lambda = 3; \frac{3 \pm \sqrt{13}}{2}$ **5** $\begin{pmatrix} 0 & 0 & 1 \\ 0 & 1 & 0 \\ 1 & 0 & 0 \end{pmatrix}$

Eigenvector associated with $\lambda = 3$ is $\begin{pmatrix} 1 \\ -3 \\ 2 \end{pmatrix}$

There are no invariant points.

Practice examination questions

1 a x-axis
b $127°$

2 a i $\begin{pmatrix} 0 & 1 & 0 \\ 1 & 0 & 0 \\ 0 & 0 & 1 \end{pmatrix}$
ii $\begin{pmatrix} 0 & -1 & 0 \\ 1 & 0 & 0 \\ 0 & 0 & 1 \end{pmatrix}$
b i $\begin{pmatrix} -1 & 0 & 0 \\ 0 & 1 & 0 \\ 0 & 0 & 1 \end{pmatrix}$
ii reflection in $x = 0$ (or y-z plane)

3 a i $m = \dfrac{49}{9}$
ii $\dfrac{9}{49}\begin{pmatrix} -\dfrac{2}{3} & 2 & 1 \\ 1 & -\dfrac{2}{3} & 2 \\ 2 & 1 & -\dfrac{2}{3} \end{pmatrix}$

b i $\det A = k^3 - 6k + 9$
ii Student's own response
iii $k = 4$

4 a i $\lambda = 1, \begin{pmatrix} 3 \\ 4 \end{pmatrix}$
ii $m = \dfrac{4}{3}$ (since $\lambda = 1$)
iii Student's own response

5 $y = 0$ and $y = x$; $y = 0$ is a line of invariant points since $\lambda = 1$

29 Moment Generating Functions

Exercise 1

1 i $M_X(t) = \displaystyle\sum_{x=1}^{2a} \frac{1}{2a} e^{tx}$
ii $\dfrac{2a+1}{2}$

2 i Student's own response
ii $\dfrac{91}{21}$

3 Student's own response

4 2

5 $M_X(t) = \displaystyle\sum_{x=1}^{6} \frac{1}{6} e^{tx}$; $M_Y(t) = e^t \displaystyle\sum_{x=1}^{6} \frac{1}{6} e^{2tx}$

6 a Student's own responses
b $\dfrac{a}{2}, \dfrac{a^2}{12}$

7 a a_1
b $2a_2, 2a_2 - a_1^2$

Exercise 2

1 $M_X(t) = \dfrac{2}{2-t}$

2 $M_S(t) = q + pe^t$; $p, p(1-p)$

3 Student's own response

4 $M_X(t) = \dfrac{1}{1-300t}$

5 Student's own response

6 $e^{28(e^t - 1)}$; **28**

7 $\dfrac{3}{7e^{-t} - 4}; \dfrac{7}{3}$

Exercise 3

1 $e^{4\lambda(e^t - 1)}$

2 Student's own response

3 $\left(\dfrac{k}{k-t}\right)^3; \dfrac{3}{k}$

4 a $\{(1-p) + pe^t\}$
b $\{(1-p) + pe^t\}^n$ Binomial (n, p)

5 a $\{(1-p) + pe^t\}^n$
b $\{(1-p) + pe^t\}^{n_1 + n_2}$
c Binomial $(n_1 + n_1, p)$

6 i $M_X(t) = \dfrac{e^{-2t}}{4}\left(\dfrac{1-e^{8t}}{1-e^{2t}}\right)$
ii $M_Y(t) = \dfrac{e^{-4t}}{16}\left(\dfrac{1-e^{8t}}{1-e^{2t}}\right)^2$
7 Student's own response

1 i Student's own response **ii** $\dfrac{8}{3}$ **2 a** Student's own response **b** $e^{4t}\left(\dfrac{e^{2bt}-e^{2at}}{2(b-a)t}\right)$

3 a 3 **b** $M_X(t)=\dfrac{3}{3-t}; \dfrac{1}{3}$ **4** $e^{(\mu_X+\mu_Y)t+\frac{1}{2}(\sigma_X^2+\sigma_Y^2)t^2}; \mu_X+\mu_Y, \sigma_X^2+\sigma_Y^2$

Practice examination questions

1 a Student's own response **b** $M_X(t)=\dfrac{2}{2-t}$ **c** Student's own response

2 a $M_X(t)=\dfrac{4}{4-t}$ **b** $M_X(t)=e^{3t}\left(\dfrac{1}{1-t}\right)$ **c** 4, 1

3 a Student's own response **b** 10 **c i** $\exp\{10n(e^t-1)\}$ **ii** Poisson, mean $10n$.

4 a Student's own response **b** $X=\mu+\sigma Z$ **c** $e^{\mu t+\frac{1}{2}\sigma^2t^2}$

30 Estimators

Exercise 1

1 a statistic **b** statistic **c** random variable that is not a statistic **d** constant

2 a $0, \dfrac{\sigma^2}{n}$ **b** $\mu, \dfrac{\sigma^2}{6}$ **c** $\mu, 0.38\sigma^2$ **3** Student's own response

Exercise 2

1 T_1 and T_2 are both unbiased, T_3 is not unbiased. **2** $a+b=2$ **3** 13.87, 1.72

4 Student's own response **5 a** Student's own response **b** Student's own response

6 3.79, 0.31 **7 a** 10% **b** 1%

8 $\dfrac{\sigma^2}{n_X+n_Y}$ **9** Student's own response **10** $a_1=a_2=\dfrac{1}{2}$

11 a $\mu, \dfrac{\sigma^2}{24}; \mu, \dfrac{\sigma^2}{25}$ **b** $\dfrac{24}{25}$; T_2 is the most efficient estimator since $RE<1$ or T_2 has smaller variance.

Review exercise

1 a A statistic is a function of the random variables that make up the sample and which does not depend upon any unknown parameter.

b $\bar{X} \sim N(\mu, \dfrac{\sigma^2}{n})$; unbiased because $E[\bar{X}]=\mu$ and consistent because $\dfrac{\sigma^2}{n}\to 0$ as $n\to\infty$

2 a Student's own response **b** Student's own response

3 a Their mean values are both equal to p and the variance of T_1 is less than that of T_2. **b** Student's own response

4 a The expected value of T is t. **b** Student's own response

5 $\displaystyle\sum_{i=1}^{n}a_i^2\to 0$ as $n\to\infty$ **6 a** T_1 **b** 11

Practice examination questions

1 a π^2-8 **b i** $\pi, \dfrac{(\pi^2-8)}{n}$ **b ii** $E(\bar{X})=\pi$ unbiased, $\text{Var}(\bar{X})\to 0$ as $n\to\infty$ consistent

c i 0.565 or 0.566; prefer $\bar{X}$ since $RE(M\,\text{wrt}\,\bar{X})<1$ **ii** (A) $\pi>3.1$ (B) 3.20, 3.12

(C) $\bar{X}$ is the more efficient estimator, implying that for the majority of samples it will be closer than M to π. However, for this particular sample m is closer to π than $\bar{x}$.

2 a $E[T]=\theta$ **b i** 9 **ii** $\dfrac{\sigma^2}{81n}(64+9b)$ **iii** $RE(T_2\,\text{wrt}\,T_1)=\dfrac{81\times(16+81b)}{64+9b}>1$; T_2 more efficient than T_1.

3 a Unbiased estimator with the smaller variance would usually yield an estimate closer to the parameter.

b i $\dfrac{\sigma_1^2}{n_1}+\dfrac{\sigma_2^2}{n_2}$ **ii** Student's own response **iii** 100, 180.

4 a i Student's own response ‎ ‎ ‎ **ii** $\dfrac{\gamma^2}{8}$

‎ **b i** $\dfrac{4}{3}$ ‎ ‎ ‎ ‎ ‎ ‎ ‎ **ii** $\dfrac{8}{3}$ ‎ ‎ ‎ ‎ **iii** $\dfrac{3}{2}R$ because relative efficiency is greater than 1.

5 a Student's own response ‎ ‎ ‎ **b** Student's own response ‎ ‎ ‎ ‎ **c** $\dfrac{n_1}{n_1+n_2}$

‎ **d i** $\dfrac{n_1\bar{X}_1+n_2\bar{X}_2}{n_1+n_2}$ ‎ ‎ ‎ **ii** A weighted average of means ‎ ‎ ‎ **iii** $2\sigma^2\left(\dfrac{1}{n_1}+\dfrac{1}{n_2}\right)>0$; therefore minimum V.

31 Estimation

Exercise 1

1 a $(8.07, 10.93)$ ‎ ‎ ‎ **b** $(-1.93, 2.60)$ ‎ ‎ ‎ ‎ ‎ ‎ **2 a** $(7.37, 11.19)$ ‎ ‎ ‎ ‎ ‎ **b** $(-1.82, 1.98)$

3 a $(26.32, 28.28)$ ‎ **b** $(8.30, 9.30)$ ‎ ‎ ‎ ‎ **4 a** $(26.87, 28.93)$ ‎ ‎ **b** $(11.91, 12.69)$ ‎ ‎ **c** $(11.33, 14.47)$

5 a 0.95 ‎ ‎ ‎ ‎ ‎ ‎ **b** 0.99 ‎ ‎ ‎ ‎ ‎ ‎ ‎ **c** 0.025 ‎ ‎ ‎ ‎ ‎ ‎ **d** 0.1 ‎ ‎ ‎ ‎ ‎ ‎ ‎ **e** 0.09
6 a 2.131 ‎ ‎ ‎ ‎ ‎ **b** 2.896 ‎ ‎ ‎ ‎ ‎ ‎ **c** 1.306 ‎ ‎ ‎ ‎ ‎ ‎ **d** 10 ‎ ‎ ‎ ‎ ‎ ‎ ‎ **e** 1.328
7 a $(8.92, 13.64)$ ‎ ‎ **b** $(-4.90, 4.07)$ ‎ ‎ **c** $(27.41, 31.19)$

8 $(18.55, 25.07)$ ‎ ‎ ‎ **9** $(4.16, 4.84)$ ‎ ‎ ‎ ‎ **10** $(330.75, 332.85)$ ‎ ‎ ‎ **11** $(12.29, 13.25)$ ‎ ‎ ‎ ‎ **12** 46.15, 241.2

13 a $(97.16, 100.50)$ Sample is large enough to assume CLT operates and to provide a good estimate of population variance.
‎ **b** $(97.06, 100.60)$

14 $(4.04, 5.46)$

Exercise 2

1 a 17 ‎ ‎ ‎ ‎ ‎ ‎ ‎ ‎ ‎ **b** 8 ‎ ‎ ‎ ‎ ‎ ‎ ‎ ‎ ‎ **2 a** 19 ‎ ‎ ‎ ‎ ‎ ‎ ‎ **b** 12
3 a 23 ‎ ‎ ‎ ‎ ‎ ‎ ‎ ‎ ‎ **b** Not very safe because of the relatively small sample size.
4 a 8, 30 ‎ ‎ ‎ ‎ ‎ ‎ ‎ **b** No. The population is normally distributed with known variance. ‎ ‎ ‎ ‎ ‎ **5** 62

Exercise 3

1 i $(-0.35, 4.18)$ Reject H_0 ‎ ‎ ‎ ‎ **ii** $(3.92, 8.68)$ No reason to reject H_0 ‎ ‎ ‎ **iii** $(7.87, 10.13)$ No reason to reject H_0
2 a i $(21.83, 26.77)$ ‎ ‎ ‎ **ii** No reason to reject H_0 ‎ ‎ **b i** $(8.38, 9.22)$ ‎ ‎ ‎ ‎ ‎ ‎ ‎ **ii** Reject H_0
‎ **c i** $(11.15, 12.85)$ ‎ ‎ ‎ **ii** Reject H_0
3 a $(21.82, 22.98)$ Reject H_0 ‎ ‎ ‎ ‎ ‎ ‎ ‎ **b** $(13.15, 14.85)$ No reason to reject H_0
4 a $(8.29, 13.91)$ No reason to reject H_0 ‎ ‎ ‎ ‎ **b** $(57.09, 61.11)$ Reject H_0
‎ In both cases, the underlying normal distribution, sample standard deviation and small sample size suggest a t-distribution.
5 a $(78.45, 89.55)$; no reason to believe shop's performance has changed at 5% (although there is at 10%).
‎ **b** Conclusion about queuing times being unchanged is weak because the hypothesised mean is near the limit of the
‎ ‎ confidence interval. More research necessary.
6 a $(498.70, 514.90)$; no reason to believe that there is a change in the mean weight.
‎ **b** Assumed an underlying normal population.

Review exercise

1 $(3.85, 4.35)$ ‎ ‎ ‎ ‎ ‎ ‎ ‎ ‎ ‎ **2** $(2.77, 3.07)$ ‎ ‎ ‎ ‎ ‎ ‎ ‎ ‎ ‎ ‎ **3** 8
4 a $(1.74, 2.09)$; reject H_0 ‎ ‎ ‎ **b** Large sample and use of sample standard deviation suggests approximate normal distribution.

Practice examination questions

1 100 ‎ ‎ ‎ ‎ ‎ ‎ ‎ ‎ ‎ **2 a** $(377, 385)$ ‎ ‎ ‎ ‎ ‎ ‎ ‎ ‎ **b** 3
3 a $(881, 1015)$ ‎ ‎ **b i** 995 ‎ ‎ ‎ ‎ ‎ ‎ **ii** Because of the overlap by the confidence intervals, no definite conclusion is possible.
4 a $(51.86, 54.26)$ ‎ ‎ ‎ ‎ ‎ ‎ ‎ ‎ ‎ **b** Sample mean is lower than last year's mean so claim may be true. 53.41 lies within

32 Further Hypothesis Testing 1 – Means and Variances

Exercise 1

1 a $12.70<\bar{x}<17.30$ ‎ ‎ ‎ ‎ **b** $\bar{x}>9.03$
2 Reject hypothesis that mean is 20; 0.085; sample is large enough for the Central Limit Theorem to apply
3 Reject hypothesis that mean is 32; 0.554
4 i $\bar{x}<28.15$ ‎ ‎ ‎ ‎ ‎ ‎ ‎ **ii** 0.08 ‎ ‎ ‎ ‎ ‎ ‎ ‎ ‎ **iii** 92% ‎ ‎ ‎ ‎ ‎ ‎ ‎ ‎ **5** 96%

Exercise 2

1 a $z = -4.76$ Reject H_0 **b** $z = -3.60$ Reject H_0 **c** $z = 4.97$ reject H_0.
2 $z = -2.390$; reject H_0.
3 a $z = 0.77$; no reason to reject H_0 **b** $z = 3.20$; reject H_0 **c** $z = -2.10$; reject H_0.
4 $z = -3.04$; reject H_0
5 a $z = -3.56$; reject H_0 **b** $z = 1.74$; no reason to reject H_0 **c** $z = -1.95$; reject H_0
6 $z = -0.955$; no reason to reject equality of means.
   Not possible to say if test is exact or approximate, since if they are normal populations, then the test is exact but if not, then it is a good approximation because of the large sample size.
7 $z = -1.04$; no reason to reject H_0. Samples sufficiently large for the Central Limit Theorem to apply.
8 a $t = -2.76$; reject H_0 **b** $t = 1.674$; no reason to reject H_0 **c** $t = 1.84$; no reason to reject H_0
9 $t = -5.30$; reject equality of means   **10** $t = -1.197$; no reason to reject equality of means

Exercise 3

1 a $t = -2.12$; reject H_0 **b** $t = -0.910$; no reason to reject H_0 **c** $t = 1.406$; no reason to reject H_0
2 a $t = 2.57$; reject H_0 **b** $t = -0.8132$; no reason to reject H_0
3 a $t = 1.456$; no reason to reject H_0 **b** $t = -1.997$; reject H_0 **c** $t = -1.306$; no reason to reject H_0
4 $t = 2.010$; no reason to reject H_0. Heart rates are normally distributed.
5 a Differences in moisture content normally distributed.
   b $t = 2.23$; no reason to reject the hypothesis that mean moisture content is the same.

Exercise 4

1 a 0.01	**b** 0.975	**c** 0.05	**d** 0.925	**e** 0.025	**f** 0.9
2 a 29.141	**b** 6.408	**c** 10.982	**d** 3.940	**e** 26	**f** 1.735
3 a 21.20	**b** 0.097	**c** 0.198	**d** 0.430		
4 a 3.482	**b** 4.424	**c** 8.649	**d** 0.205	**e** 0.360	**f** 2.124

5 a $\chi^2 = 13.21$; reject H_0 **b** $\chi^2 = 24.40$; no reason to reject H_0
   c $\chi^2 = 11.84$; no reason to reject H_0 **d** $\chi^2 = 4.27$; no reason to reject H_0
6 $\chi^2 = 17.63$; reject H_0
7 $F_{20,15} = 1.87$; no reason to reject equality of variances. Independent, normally distributed populations.
8 $F_{50,50} = 0.396$; reject H_0 **9** $\chi^2 = 4.02$; no reason to reject H_0
10 $\chi^2 = 7.81$; no reason to reject suggestion. Assumed underlying normal distribution.

Review exercise

1 $\bar{x} > 16.88$
2 a $z = -4$; reject H_0 **b** 0.206   **c** 79.4%
3 $z = -2.27$; reject H_0 **4** $t = -3.76$; reject H_0 **5** $\chi^2 = 53.63$; no reason to reject H_0

Practice examination questions

1 $z = 1.85$; no evidence that $\mu_B \neq \mu_G$
2 a i $z = 1.82$; no evidence there is a difference.   **ii** Large sample so Central Limit Theorem applies.
   b i Student's own response   **ii** 0.14
3 a Independent random samples. Normal populations with common variance.
   b $t = 3.17$; Sufficient evidence to indicate that means are different at 2% level of significance.
4 $t = 2.71$; insufficient evidence to accept coach's belief.   **5** $\chi^2 = 2.22$; evidence to suggest that $\sigma \neq 0.7$ at 1%.

33 Further Hypothesis Testing 2 – Goodness of Fit Tests and Tests of No Association

Exercise 1

1 $\chi^2 = 12.35$; reject the hypothesis that the dice is fair.
2 Geometric, $p = 0.5$. $\chi^2 = 16.72$; reject the hypothesis that the coin is fair.
3 a 45.86, 55.11, 26.39   **b** $\chi^2 = 12.53$; Reject the null hypothesis.
4 a Random nature of email arrivals   **b** 3.267   **c** 13.29, 10.86, 7.09
   d Pool cells because of small frequency; $\chi^2 = 0.19$ (2 dp), df = 4; critical value 9.488. Evidence supports belief of a Poisson distributed number of emails per day.

Exercise 2

1 $\chi^2 = 5.50$ (2 dp); critical value 4.605. At 10%, evidence suggests that the operators are not equally effective.

2 a $\chi^2 = 7.34$ (2 dp); critical value 7.815. No reason to reject hypothesis of no association.

  b Observed χ^2 close to critical value, therefore more research advisable. Fewer than expected females plan to specialise in surgery.

3 a 12.6, 7.47, 14.4 (top to bottom left to right)

  b $\chi^2 = 3.08$ (2 dp); critical value 5.991. Evidence suggests that the two inspectors are consistent.

  c Type II error made.

4 a $\chi^2 = 0.069$; critical value 2.706. Evidence suggests no difference.

  b The evidence for the conclusion is very strong.

5 a 5.24, 17.81, 5.76 (top to bottom left to right)

  b $\chi^2 = 2.68$ (2 dp); critical value 5.991. Evidence supports the teacher's belief that there is no significant difference between classes.

  c Class B fail results are unexpectedly high, distinction results unexpectedly low but not sufficiently different to reject the null hypothesis.

Review exercise

1 Binomial, $n = 3$, $p = 0.3$. 2 degrees of freedom, $\chi^2 = 2.50$; no reason to reject hypothesis that $p = 0.3$.

2 $\chi^2 = 0.062$; no reason to reject the hypothesis that a Poisson distribution fits the data.

3 a 18.69, 6.51, 26.7, 9.3 (top: left to right, bottom: left to right)

  b $\chi^2 = 59.17$; reject null hypothesis and accept that there is an association between car colour and time of accident.

Practice examination questions

1 a Student's own response   **b** 0.0533, 0.0821

  c $\chi^2 = 0.970$; no reason to reject H_0. Exponential distribution may be suitable.

2 a i 0.0272, 0.1792, 0.4752, 0.8192, 1   **ii** 0.2960, 0.3440, 0.1808

  b $\chi^2 = 4.139$; no reason to reject H_0. Evidence to suggest claim is correct.

3 a 1.25   **b** 23.73, 39.55, 26.37, 8.79, 1.46, 0.1

  c $\chi^2 = 2.947$; no reason to reject null hypothesis. B(5, p) is a suitable model.

  d It gives some support. If, for example, the probabilities were different for a seed in the front row, say, then this would not be discernible from figures for 100 rows.

4 $\chi^2 = 3.3654$; no reason to reject H_0. No evidence at the 5% level of significance to support the claim that the drug is effective against sickness.

5 a H_0: no association between type of school and performance of 16 year olds in their GCSEs.   **b** 12.0 (1 dp)

  c No significant evidence to suggest an association between type of school and GCSE performance of 16 year olds.

  d More than expected gained at least/more than 5 GCSEs OR
  Fewer than expected gained at least/more than 1 GCSE but less than 5 GCSEs OR
  Fewer than expected gained no GCSEs.

  e Reject H_0 at 10% level of significance. Evidence to suggest an association between type of school and GCSE performance.

6 a (left to right, top row to bottom row)

   57, 44, 27, 17
   39, 43, 19, 4
   96, 87, 46, 21

  b $\chi^2 = 6.59$; reject H_0. There is significant evidence of an association between country and gender.

  c More females than expected from Scotland OR
  Fewer females than expected from Northern Ireland OR
  About the right number of females from England and/or Wales.

Index